Fractal Geometry and Analysis

NATO ASI Series

Advanced Science Institutes Series

A Series presenting the results of activities sponsored by the NATO Science Committee, which aims at the dissemination of advanced scientific and technological knowledge, with a view to strengthening links between scientific communities.

The Series is published by an international board of publishers in conjunction with the NATO Scientific Affairs Division

A Life Sciences **B Physics**	Plenum Publishing Corporation London and New York
C Mathematical and Physical Sciences **D Behavioural and Social Sciences** **E Applied Sciences**	Kluwer Academic Publishers Dordrecht, Boston and London
F Computer and Systems Sciences **G Ecological Sciences** **H Cell Biology** **I Global Environmental Change**	Springer-Verlag Berlin, Heidelberg, New York, London, Paris and Tokyo

NATO-PCO-DATA BASE

The electronic index to the NATO ASI Series provides full bibliographical references (with keywords and/or abstracts) to more than 30000 contributions from international scientists published in all sections of the NATO ASI Series.
Access to the NATO-PCO-DATA BASE is possible in two ways:

– via online FILE 128 (NATO-PCO-DATA BASE) hosted by ESRIN, Via Galileo Galilei, I-00044 Frascati, Italy.

– via CD-ROM "NATO-PCO-DATA BASE" with user-friendly retrieval software in English, French and German (© WTV GmbH and DATAWARE Technologies Inc. 1989).

The CD-ROM can be ordered through any member of the Board of Publishers or through NATO-PCO, Overijse, Belgium.

Series C: Mathematical and Physical Sciences - Vol. 346

Fractal Geometry and Analysis

edited by

Jacques Bélair

and

Serge Dubuc
Département de mathématiques et de statistique,
Université de Montréal,
Montréal, Québec, Canada

Proceedings of the NATO Advanced Study Institute and
Séminaire de mathématiques supérieures on
Fractal Geometry and Analysis
Montréal, Canada
July 3–21, 1989

Springer-Science+Business Media, B.V.

Proceedings of the NATO Advanced Study Institute and
Séminaire de mathématiques supérieures on
Fractal Geometry and Analysis
Montréal, Canada
July 3–21, 1989

ISBN 978-94-015-7933-9 ISBN 978-94-015-7931-5 (eBook)
DOI 10.1007/978-94-015-7931-5

Printed on acid-free paper

Originally published by Kluwer Academic Publishers in 1991
Softcover reprint of the hardcover 1st edition 1991

Table of Contents

Preface

This ASI – which was also the 28th session of the Séminaire de mathématiques supérieures of the Université de Montréal – was devoted to Fractal Geometry and Analysis. The present volume is the fruit of the work of this Advanced Study Institute. We were fortunate to have with us Prof. Benoît Mandelbrot – the creator of numerous concepts in Fractal Geometry – who gave a series of lectures on multifractals, iteration of analytic functions, and various kinds of fractal stochastic processes.

Different foundational contributions for Fractal Geometry like measure theory, dynamical systems, iteration theory, branching processes are recognized. The geometry of fractal sets and the analytical tools used to investigate them provide a unifying theme of this book. The main topics that are covered are then as follows.

Dimension Theory. Many definitions of fractional dimension have been proposed, all of which coincide on "regular" objects, but often take different values for a given fractal set. There is ample discussion on piecewise estimates yielding actual values for the most common dimensions (Hausdorff, box-counting and packing dimensions). The dimension theory is mainly discussed by Mendès-France, Bedford, Falconer, Tricot and Hata.

Construction of fractal sets. Scale invariance is a fundamental property of fractal sets. It is often reflected in a functional equation satisfied by the coordinates of the points in the sets. Both deterministic and random recurrent construction methods are presented, some extending the classical notions of random walk and interpolation function. In another instance, dynamical processes are involved in the study of generalized Cantor sets with nonuniform scaling laws. All these models are described by Dekking, Bedford, Hata and Vrscay.

Iterated Function Systems. The most novel scheme, which is also the most promising one for applications in computer graphics and image reconstruction, namely "Iterated Function Systems", is dealt with at length in a thoroughly transparent fashion by Vrscay. The attractors of iterated function systems are also discussed by Falconer, Hata and Dubuc.

Dynamics in the complex plane. Although geometrical in origin, fractal objects may embody intrinsically dynamical aspects. Iteration of complex mappings is discussed by Blanchard and major tools from the classical theory of one complex variable are shown to be of substantial value.

Multifractals. The so-called "$f(\alpha)$–spectrum" is presented in its natural setting, namely as a probability distribution rather than an ill-conceived "infinity of dimensions". In this volume, it is primarily the paper by Kahane that deals with the subject of multifractals.

Applications of fractal geometry to mathematical analysis. There are many connections between fractal geometry, and mathematical analysis. Many challenging problems in the field of analysis arise from fractal geometry, and there are many situations in analysis where fractal geometry provides a new insight. This situation is illustrated by Kahane (random coverings) and Dubuc (interpolation theory).

We feel that it is very important that mathematicians, especially those working in mathematical analysis, contribute to fractal geometry. As the number and diversity of scientists using fractal geometry increases (e.g., physicists, chemists, people in computer science and computer graphics), more and more valuable applications can be expected if there is a significant input by mathematicians to the field of fractal geometry. Some recent investigations in fractal geometry are presented in this book. We hope that these will be helpful to other branches of science.

We wish to express our sincere thanks to all lecturers and participants for having helped to make this ASI a success. Special thanks go to Aubert Daigneault, the Director of the ASI, and Ghislaine David, its secretary, for their excellent organizational work.

The ASI was funded in large part by NATO, with additional support from the Natural Sciences and Engineering Research Council of Canada, the Ministère de l'Éducation du Québec, and the Université de Montréal. We would like to thank all these organizations for their support. For their efforts on behalf of this ASI we are especially grateful to the Scientific Affairs Division of NATO, particularly to Dr. Luis V. da Cunha, the Director of the ASI Programme.

Jacques Bélair
Serge Dubuc
Scientific Directors

Participants

Simonetta ABENDA
Dipartimento di Fisica
Univ. degli Studi di Bologna
via Irnerio 46
40126 Bologna
Italy

Alaaddin AL-DHAHIR
Faculty of Applied Mathematics
University of Twente
P.O. Box 217
7500 AE Enschede
The Netherlands

Emin ANARIM
Dept. of Electrical &
Electronic Engineering
Bogaziçi University
80815 Bebek-Istanbul
Turkey

Tim BEDFORD
Mathematics and Informatics
Delft Univ. of Technology
Julianalaan 132
2628 BL Delft
The Netherlands

Jacques BÉLAIR
Département de mathématiques
et de statistique
Université de Montréal
C.P. 6128, Succ. A
Montréal, Québec, H3C 3J7
Canada

Fathi BEN NASR
Département de mathématiques
Faculté des Sciences
Université de Tunisie
5000 Monastir
Tunisia

Paul BLANCHARD
Department of Mathematics
Boston University
111 Cummington Street
Boston, MA 02215
U.S.A.

José BRANDÃO TIAGO
Fac. de Ciências Sociais e Humanas
Universidade Nova de Lisboa
Avenida de Berna, 24
1000 Lisboa
Portugal

Carlos A. CABRELLI
Dept. of Electrical Engineering
& Computer Science
University of California
Davis, CA 95616
U.S.A.

Robin CÔTÉ
Département de physique
Université Laval
Cité Universitaire
Québec, Québec, G1K 7P4
Canada

Anthony DAVIS
Department of Physics
McGill University
Ernest Rutherford Physics Bldg.
3600 University St.
Montréal, Québec, H3A 2T8
Canada

Michel DEKKING
Mathematics and Informatics
Delft Univ. of Technology
Julianalann 132
2628 BL Delft
The Netherlands

Gilles DESLAURIERS
Département de mathématiques appliquées
École Polytechnique
Case Postale 6079, Succ. A
Montréal, Québec, H3C 3A7
Canada

Serge DUBUC
Département de mathématiques et de statistique
Université de Montréal
C.P. 6128, Succ. A
Montréal, Québec, H3C 3J7
Canada

Jean-Paul DUFOUR
Institut de Mathématiques
Univ. des Sciences et Techniques de Languedoc
Pl. Eugène Bataillon
34060 Montpellier Cédex
France

Abdelkader ELQORTOBI
Département de mathématiques
Faculté des Sciences
Université Mohamed 1er
Oujda
Morocco

Manuel L. ESQUIVEL
Departamento de Matematica
Univ. Nova de Lisboa
Quinta da Torre
2825 Monte da Caparica
Portugal

Kenneth J. FALCONER
School of Mathematics
University of Bristol
Bristol BS8 1TW
U.K.

Fan ai HUA
Mathématique, Bâtiment 425
Université de Paris-Sud
Centre d'Orsay
91405 Orsay Cédex
France

Norman FICKEL
Mathematisches Institut
Universität Erlangen-Nürnberg
Bismarckstr. 1 1/2
W-8520 Erlangen
Germany

Jacqueline FLECKINGER
Mathématiques
Université Toulouse I
Place Anatole France
31042 Toulouse
France

Tony FOWLER
Department of Geology
University of Ottawa
Ottawa, Ontario, K1N 6N5
Canada

Simon FRASER
Chemical Physics Theory Group
Department of Chemistry
University of Toronto
Toronto, Ontario, M5S 1A1
Canada

Erland GADDE
Department of Mathematics
University of Umeå
S-901 87 Umeå
Sweden

Marcello GALEOTTI
Ist. di Matematica Applicata
Università di Firenze
Via Montebello, 7
50123 Firenze
Italy

Lucio GERONAZZO
Ist. di Matematica Applicata
Università di Firenze
Via Montebello, 7
50123 Firenze
Italy

William J. GILBERT
Department of Pure Mathematics
University of Waterloo
Waterloo, Ontario, N2L 3G1
Canada

O.L. GULDER
Div. de génie mécanique, M-9
Conseil National de Recherches
Ottawa, Ontario, K1A 0R6
Canada

Masayoshi HATA
Department of Mathematics
Faculty of Science
Kyoto University
Kyoto 606
Japan

John HOLBROOK
Dept. of Math. & Stat.
University of Guelph
Guelph, Ontario, N1G 2W1
Canada

Lawrence HUSCH
Department of Mathematics
University of Tennessee
121 Ayres Hall
Knoxville, TN 37996-1300
U.S.A.

Andrew J. IRWIN
81 Bayview Ridge
Willowdale, Ontario, M2L 1E3
Canada

Jean-Pierre KAHANE
Mathématiques
Bâtiment 425
Université de Paris-Sud
Centre d'Orsay
91405 Orsay Cédex
France

Michel LAPIDUS
Department of Mathematics
The University of Georgia
Athens, GA 30602
U.S.A.

Marko LEMIEUX
Dep. de génie électrique
École Polytechnique
Case Postale 6079, Succ. A
Montréal, Québec, H3C 3A7
Canada

Yi LI
Department of Mathematics
University of Alberta
Edmonton, Alberta, T6G 2G1
Canada

Paul C. LIU
Great Lakes Environmental
Research Laboratory, NOAA
2205 Commonwealth Blvd.
Ann Arbor, MI 48105
U.S.A.

Li-Shi LUO
School of Physics
Georgia Inst. of Technology
Atlanta, GA 30332-0430
U.S.A.

Mohammad A. MALIK
Department of Mathematics
Concordia University
Loyola Campus
7141 Sherbrooke Street West
Montréal, Québec, H4B 1R6
Canada

Benoît B. MANDELBROT
IBM Thomas Watson Research Center
P.O. Box 218
Yorktown Heights NY 10598
U.S.A.

Robert McCANN
1-368 Brock Street
Kingston, Ontario, K7L 1T1
Canada

Michel MENDÈS-FRANCE
Mathématiques et informatique
Université de Bordeaux I
351, cours de la Libération
33405 Talence Cédex
France

Ursula M. MOLTER
Department of Mathematics
University of California
Davis, CA 95616
U.S.A.

James MULDOWNEY
Department of Mathematics
University of Alberta
Edmonton, Alberta, T6G 2G1
Canada

Jiri MULLER
Institute for Energy Technology
Box 40
2007 Kjeller
Norway

Alexander NABUTOVSKY
Dept. of Theor. Mathematics
The Weizmann Inst. of Science
P.O. Box 26
Rehovot 76100
Israel

Fahima NEKKA
Département de mathématiques
et de statistique
Université de Montréal
C.P. 6128, Succ. A
Montréal, Québec, H3C 3J7
Canada

Van-Thanh-Van NGUYEN
Dept. of Civil Engineering
& Applied Mechanics
McGill University
817 Sherbrooke Street West
Montréal, Québec, H3A 2K6
Canada

K.-Georg NOLTE
Department of Mathematics
University of Ottawa
Ottawa, Ontario, K1N 6N5
Canada

Alec NORTON
Department of Mathematics
Boston University
111 Cummington Street
Boston, MA 02215
U.S.A.

Sema ONURLU
Materials Dept., Electron
Microscopy Lab.
Marmara Res. Inst. (TÜBITAK)
P.O. Box 21
41470 Gebze-Kocaeli
Turkey

Andrew OSBALDESTIN
Dept. of Mathematical Sciences
Loughborough Univ. of Technology
Loughborough, Leics. LE11 3TU
U.K.

Krzysztof OSTASZEWSKI
Department of Mathematics
University of Louisville
Louisville, KY 40292
U.S.A.

Mauro PICCIONI
Dipartimento di Matematica
Università di Roma
00173 Roma
Italy

Marcelo B. RIBEIRO
School of Mathematical Sci.
Queen Mary College
Mile End Road
London, E1 4NS
U.K.

Ricardo RIGON
Dpto. di Scienze Ambientali
Univ. degli Studi di Venezia
Calle Larga S. Marta
Dorsoduro 2137
30123 Venezia
Italy

Karsten RODENACKER
Institut für Strahlenschutz
GSF München
Ingolstadter Landstr. 1
W-8042 Neuherberg
Germany

Ramajayam SAHADEVAN
Ramanujan Institute for
Advanced Study in Mathematics
University of Madras
Madras 600 005
India

Jorge SALAZAR-SERRANO
Mathématiques - Bâtiment 425
Université de Paris-Sud
Centre d'Orsay
91405 Orsay Cédex
France

Michael SHALMON
INRS - Télécommunications
Université du Québec
3, Place du Commerce
Ile-des-Soeurs
Verdun, Québec, H3E 1H6
Canada

Frédérique SIMONDON
Laboratoire de Mathématiques
Fac. des Sciences & Techniques
Université de Franche-Comté
Route de Gray
25030 Besançon Cédex
France

Peter SINGER
Mathematisches Institut
Universität Erlangen-Nürnberg
Bismarckstr. 1 1/2
W-8520 Erlangen
Germany

Cathy SMILEY
Department of Mathematics
Auburn University
228 Parker Hall
Auburn, AL 36849
U.S.A.

Mark SMILEY
Department of Mathematics
Auburn University
228 Parker Hall
Auburn, AL 36849
U.S.A.

Donald SPEAR
Department of Mathematics
State Univ. of New York
Stony Brook, NY 11794
U.S.A.

Michael STIASSNIE
Dept. of Civil Engineering
Technion
Haifa 32000
Israel

Terry TREMAINE
Department of Mathematics
Queen's University
Kingston, Ontario, K7L 3N6
Canada

Claude TRICOT
Département de mathématiques
appliquées
École Polytechnique
Case Postale 6079, Succ. A
Montréal, Québec, H3C 3A7
Canada

Walter TRZCIENSKI
Département de géologie
Université de Montréal
Case Postale 6128, Succ. A
Montréal, Québec, H3C 3J7
Canada

Rémi VAILLANCOURT
Département de mathématiques
Université d'Ottawa
Ottawa, Ontario, K1N 6N5
Canada

Edward R. VRSCAY
Dept. of Applied Mathematics
University of Waterloo
Waterloo, Ontario, N2L 3G1
Canada

Jürgen WEITKÄMPER
FB Mathematik
Universität Marburg
Hans-Meerwein-Str.
W-3550 Marburg
Germany

Zhi-Xiong WEN
Mathématiques, Bâtiment 425
Université de Paris-Sud
Centre d'Orsay
91405 Orsay Cédex
France

Zhi-Ying WEN
Dept. of Mathematics
Wuhan University
Wuhan
People's Republic of China

John B. WILKER
Department of Mathematics
Scarborough College
University of Toronto
West Hill, Ontario, M1C 1A4
Canada

Peter WINGREN
Department of Mathematics
University of Umeå
S-90 187 Umeå
Sweden

Taiping YE
Dept. of Mathematics, U-9
University of Connecticut
Storrs, CT 06269-3009
U.S.A.

Peter ZAJDLER
Mathematisches Institut
Universität Erlangen-Nürnberg
Bismarckstr. 1 1/2
W-8520 Erlangen
Germany

Mohammed ZAOUI
Département de mathématiques
et de statistique
Université de Montréal
C.P. 6128, Succ. A
Montréal, Québec, H3C 3J7
Canada

Contributors

Tim BEDFORD
Faculty of Technical Mathematics
and Informatics
Delft University of Technology
P.O. Box 356
NL-2600 AJ Delft
The Netherlands

Paul BLANCHARD
Department of Mathematics
Boston University
111 Cummington Street
Boston, MA 02215
U.S.A.

Amy CHIU
Department of Mathematics
Boston University
111 Cummington Street
Boston, MA 02215
U.S.A.

Michel DEKKING
Faculty of Technical Mathematics
and Informatics
Delft University of Technology
P.O. Box 356
NL-2600 AJ Delft
The Netherlands

Serge DUBUC
Département de mathématiques
et de statistique
Université de Montréal
C.P. 6128, Succ. A
Montréal, Québec, H3C 3J7
Canada

Kenneth J. FALCONER
School of Mathematics
University of Bristol
Bristol BS8 1TW
U.K.

Masayoshi HATA
Department of Mathematics
Faculty of Science
Kyoto University
Kyoto 606
Japan

Jean-Pierre KAHANE
Mathématiques
Bâtiment 425
Université de Paris-Sud
Centre d'Orsay
91405 Orsay Cédex
France

Michel MENDÈS-FRANCE
Mathématiques et informatique
Université de Bordeaux I
351, cours de la Libération
33405 Talence Cédex
France

Claude TRICOT
Département de mathématiques
appliquées
École Polytechnique
Case Postale 6079, Succ. A
Montréal, Québec, H3C 3A7
Canada

Edward R. VRSCAY
Dept. of Applied Mathematics
University of Waterloo
Waterloo, Ontario, N2L 3G1
Canada

Applications of dynamical systems theory to fractals - a study of cookie-cutter Cantor sets

Tim BEDFORD
Department of Mathematics & Informatics
Delft University of Technology
P.O. Box 356
NL-2600 AJ Delft
The Netherlands

Abstract

Cookie-cutter Cantor sets in the line are studied as simple examples of fractals which are invariant sets of dynamical systems. The topics covered are: the characterization of a cookie-cutter via a dynamical system or an iterated function system (i.f.s.); introduction of measure theoretic and topological entropy and comparison with the concept of dimension; Bowen's formula for Hausdorff dimension; a flow canonically associated with a cookie-cutter Cantor set; the order two density of Hausdorff measure (a characterization of lacunarity); the multifractal spectrum for cookie-cutters; and, Sullivan's classification theorem for cookie-cutters.

Introduction

Simple examples of fractal sets are Cantor sets in the line which are invariant under certain dynamical systems. We make a detailed study of a class of systems which are frequently called cookie-cutter maps. This enables us to develop applications of some ideas from dynamical systems theory to fractal geometry. The same ideas can be extended to a much wider class of fractal sets but this would involve a more technical approach without giving further insight into the fractal geometry of the underlying sets.

The topics we shall cover are

- the characterization of a cookie-cutter via a dynamical system or an iterated function system (i.f.s.).
- introduction of measure theoretic and topological entropy and comparison with the concept of dimension.
- Bowen's formula for Hausdorff dimension.

J. Bélair and S. Dubuc (eds.), Fractal Geometry and Analysis, 1–44.

- a flow canonically associated with a cookie-cutter Cantor set; the order two density of Hausdorff measure (a characterization of lacunarity).
- the multifractal spectrum for cookie-cutters.
- Sullivan's classification theorem for cookie-cutters.

Much of the material presented here has either not appeared in print before, or has only appeared in a very different form.

1 Cookie-cutters as dynamical systems and i.f.s.'s

The aim of this first section is to define cookie-cutters and to introduce and explain the assumptions we shall make. The prototype for a cookie-cutter Cantor set is the standard middle-third set

$$C = \{\sum_{n=1}^{\infty} x_n 3^{-n} : x_n = 0, 2\}.$$

Definition 1.1 Write $I = [0,1]$, and take $0 < x_0 < x_1 < 1$. A *cookie-cutter map* is a mapping

$$S\colon\ [0,x_0] \cup [x_1,1] \to I$$

with the properties that

(a) $S|_{[0,x_0]}$ and $S|_{[x_1,1]}$ are $1-1$ maps onto I, and

(b) S is $C^{1+\gamma}$ differentiable, i.e. differentiable with a Hölder continuous derivative DS satisfying $|DS(x) - DS(y)| < c|x-y|^{\gamma}$ for some $c > 0$, and $|DS(x)| > 1$ for all $x \in [0,x_0] \cup [x_1,1]$.

The *cookie-cutter set* associated to S is the set

$$C = \{x \in [0,x_0] \cup [x_1,1] \colon\ S^n(x) \in [0,x_0] \cup [x_1,1]\ \forall n \geq 0\}.$$

Example If we take $x_0 = \frac{1}{3}, x_1 = \frac{2}{3}$ and $S(x) = 3x \bmod 1$, then C is the middle-third Cantor set.

The dynamics of a cookie-cutter map are very easy to describe in terms of a symbolic system called a full shift on two symbols.

Definition 1.2 Let $\Sigma = \{0,1\}^{\mathbb{Z}^+} = \{\underline{x} = (x_0, x_1, \ldots) \colon\ x_i = 0, 1\}$. The *full shift on two symbols* is the map

$$\sigma\colon\ \Sigma \to \Sigma$$

given by

$$(x_0, x_1, x_2, \ldots) \mapsto (x_1, x_2, \ldots).$$

With the metric $d(\underline{x}, \underline{y}) = e^{-m}$ where $m = \min\{k: x_k \neq y_k\}$ the map σ is a continuous map of a perfect compact, completely disconnected metric space. The dynamics of σ are chaotic (for example σ has a dense set of periodic points and points with dense orbits) but are still easy to understand because the action of σ on the symbol sequences is so simple [Wa3].

We now show the relationship of S with σ, and show that C is a self-similar set [Hu] or, equivalently, the attractor of an iterated function system [Ba]. This means that there are two contraction maps $\varphi_0, \varphi_1: I \to I$ such that

$$\varphi_0(C) \cup \varphi_1(C) = C.$$

Theorem 1.1 *Let S be a cookie-cutter map and C the corresponding cookie-cutter set. Then C is a self-similar Cantor set. Furthermore there is a homeomorphism $\pi: \Sigma \to C$ such that the following diagram commutes*

$$\begin{array}{ccc} \Sigma & \xrightarrow{\sigma} & \Sigma \\ \pi\downarrow & & \downarrow\pi \\ C & \xrightarrow{S} & C \end{array}$$

i.e. π is a topological conjugacy between σ and $S|_C$.

Proof. We show first how to define π^{-1}. For each point x of C we can define an associated symbol sequence via

$$x_n = \begin{cases} 0 & \text{if } S^n(x) \in I_0 \\ 1 & \text{if } S^n(x) \in I_1. \end{cases}$$

Since $S(I_0), S(I_1) \supset I_0, I_1$ one sees by induction that all sequences of 0's and 1's arise in this way. Two distinct points $x, y \in C$ have different associated symbol sequences because the expansive nature of S implies that for some n, $S^n(x)$ and $S^n(y)$ are in different intervals I_0, I_1. This shows that the map π exists and is a bijection

$$(x_0, x_1, \ldots) \overset{\pi}{\mapsto} x.$$

To check that π is a homeomorphism define

$$I_{x_0 \ldots x_n} = \{x \in I: x \in I_{x_0}, S(x) \in I_{x_1}, \ldots, S^n(x) \in I_{x_n}\}$$

and

$$c_{n+1}(\underline{x}) = \{\underline{y} \in \Sigma: x_0 = y_0, \ldots, x_n = y_n\}$$

so that

$$\pi c_{n+1}(\underline{x}) = I_{x_0 \ldots x_n} \cap C.$$

From the definition of the metric on Σ it is clear that π is continuous at $(x_0, x_1, \ldots)$ if and only if $|I_{x_0 \ldots x_n}| \to 0$ as $n \to \infty$. By induction one sees that the intervals $\{I_{x_0 \ldots x_n}\}$ for fixed n are disjoint and that S^{n+1} maps $I_{x_0 \ldots x_n}$ homeomorphically onto I. Since $|DS| > 1$ there exists $\lambda > 1$ with $|DS(x)| > \lambda > 1$ for all $x \in C$. This implies that $|I_{x_0 \ldots x_n}| < \lambda^{-(n+1)}$ and hence that π is continuous. The inverse π^{-1} is continuous since disjointness of the intervals $\{I_{x_0 \ldots x_n}\}$ implies that a small enough ball around any $x \in C$ intersects at most one $I_{x_0 \ldots x_n}$ so that π^{-1} of the ball is contained in $c_{n+1}(\underline{x})$ and is small. This shows that π is a homeomorphism and hence that C is, like Σ, a compact perfect and completely disconnected set, i.e. a Cantor set. See Figure 1.1 for the structure of the "Cantor intervals" $I_{x_0 \ldots x_n}$.

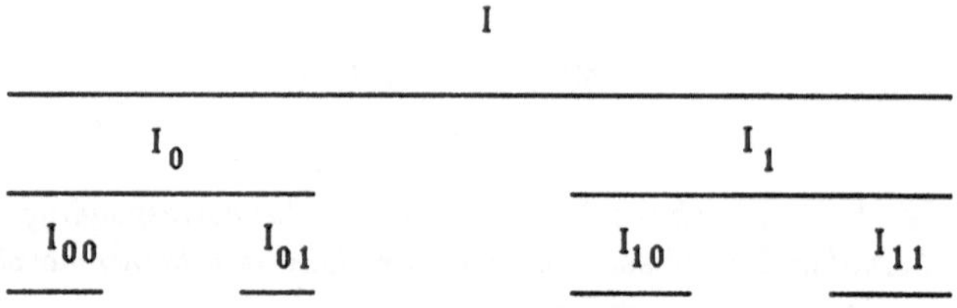

Figure 1.1: The structure of the Cantor intervals.

It is clear from the definition of π^{-1} that σ and $S|_C$ are conjugated by π.

Finally we must check that C is self-similar. From the symbolic description of S it is clear that

$$S(I_0 \cap C) = S(I_1 \cap C) = C.$$

Since $S|I_0$ and $S|I_1$ are 1-1 we can define $\varphi_0 : I \to I_0$ and $\varphi_1 : I \to I_1$ as the inverses of $S|_{I_0}$ and $S|_{I_1}$. The maps φ_0 and φ_1 are contractions since S is everywhere expanding. Furthermore $\varphi_0(C) = C \cap I_0$ and $\varphi_1(C) = C \cap I_1$, so that

$$C = \varphi_0(C) \cup \varphi_1(C).$$

This shows that C is a self-similar Cantor set. □

A consequence of the above theorem is that any two cookie-cutter systems $S|_C$ and $\tilde{S}|_{\tilde{C}}$ are topologically conjugate (because they are both topologically conjugate to $\sigma\colon \Sigma \to \Sigma$). Topological conjugacy does not tell us much about the fractal geometry of a cookie-cutter (by working slightly harder one can show that the conjugacy between two cookie-cutters is always Hölder continuous but this does not give information about dimension and other fractal properties). In order to have a stronger classification of the fractal geometry of cookie-cutters we need to consider other types of conjugacy– Lipschitz or differentiable conjugacies for example. Asking that the conjugating map be differentiable means that one is making very tight requirements on the local structure of the Cantor sets in question. We shall come back to the problem of this type of classification in the last part of these notes.

The most basic question one can ask about the fractal geometry of a Cantor set in the line is whether or not it has zero Lebesgue measure. It is possible to consruct cookie-cutters with positive Lebesgue measure if one allows the map to be C^1 rather than $C^{1+\gamma}$ (see [Bo3]). On the other hand any $C^{1+\gamma}$ cookie-cutter has zero Lebesgue measure. This fact is a consequence of a property of DS called *bounded variation* which we shall first show for a general Hölder continuous function.

Definition 1.3 A map $f\colon X \to \mathbf{R}$ (we shall take $X = I_0 \cup I_1$ or C) is *Hölder continuous of order* γ at $x \in X$ if for some $c > 0$ and all y in a neighbourhood of X

$$|f(x) - f(y)| < c|x-y|^{\gamma}.$$

Note that since we are taking X to be compact (X is either $I_0 \cap I_1$ or C), such a function is uniformly Hölder continuous of order γ. We shall often use the notation.

$$S_n f(x) = \sum_{i=0}^{n-1} f(S^i(x)).$$

Proposition 1.2 *(Bounded variation) There is a constant $M > 0$ such that for any $n > 0$, $x_0, \ldots, x_{n-1}$ and $x, y \in I_{x_0 \ldots x_{n-1}}$,*

$$|S_n f(x) - S_n f(y)| < M.$$

Proof. By uniform continuity of DS there is $\lambda > 1$ such that

$$|DS(z)| > \lambda > 1 \quad \text{for all } z \in I_0 \cup I_1.$$

This implies that $|D\varphi_i(z)| < \lambda^{-1} < 1 \quad (i = 0,1)$, and that

$$|I_{x_0 \ldots x_{n-1}}| = |\varphi_{x_0} \cdots \varphi_{x_{n-1}}(I)| < \lambda^{-n}.$$

In particular if $x, y \in I_{x_0 \ldots x_{n-1}}$ then $|x - y| < \lambda^{-n}$. But also $S(x), S(y) \in I_{x_1 \ldots x_{n-1}}$ so that $|S(x) - S(y)| < \lambda^{-(n-1)}$ and for general $i, 0 \le i < n, S^i(x), S^i(y) \in I_{x_i \ldots x_{n-1}}$ so that $|S^i(x) - S^i(y)| < \lambda^{-(n-i)}$. Hence

$$\begin{aligned}
|S_n f(x) - S_n f(y)| &= |\sum_{i=0}^{n-1} f(S^i x) - \sum_{i=0}^{n-1} f(S^i y)| \\
&\le \sum_{i=0}^{n-1} |f(S^i x) - f(S^i y)| \\
&\le \sum_{i=0}^{n-1} c|S^i(x) - S^i(y)|^{\gamma} \\
&\le c\sum_{i=0}^{n-1} \lambda^{-(n-i)\gamma} \le c\sum_{i=0}^{\infty} \lambda^{-i\gamma} := M. \quad \square
\end{aligned}$$

Bounded variation can be used to get good control over the sizes of Cantor intervals.

Corollary 1.3 *There is a constant $\zeta > 0$ such that for any $n > 0$, $x_0 \dots x_{n-1}$ and $z \in I_{x_0 \dots x_{n-1}}$,*

$$\zeta < |I_{x_0 \dots x_{n-1}}||DS^n(z)| < \zeta^{-1}.$$

Proof. Since S^n maps $I_{x_0 \dots x_{n-1}}$ diffeomorphically onto I, the mean value theorem implies that

$$\inf_{z \in I_{x_0 \dots x_{n-1}}} |DS^n(z)|.|I_{x_0 \dots x_{n-1}}| \leq |I| \leq \sup_{z \in I_{x_0 \dots x_{n-1}}} |DS^n(z)||I_{x_0 \dots x_{n-1}}|.$$

Now let x and x' be points of $I_{x_0 \dots x_{n-1}}$ on which $|DS^n|$ takes its infimum and supremum respectively. Define $f(x) = -\log|DS(x)|$. Since $|DS(x)|$ is a Hölder continuous function on a compact set and is bounded away from 0, $f(x)$ is also Hölder continuous. By bounded variation

$$|S_n f(x) - S_n f(x')| < M,$$

for some M independent of n, x and x'. But

$$\begin{aligned} S_n f(x) = \sum_{i=0}^{n-1} f(S^i x) &= \sum_{i=0}^{n-1} -\log|DS(S^i x)| \\ &= -\log(\prod_{i=0}^{n-1} |DS(S^i x)|) \\ &= -\log(|DS^n(x)|). \end{aligned}$$

This implies that

$$e^{-M} < \frac{|DS^n(x)|}{|DS^n(x')|} < e^M$$

and gives

$$\begin{aligned} & |DS^n(x)||I_{x_0 \dots x_{n-1}}| \\ > \; & e^{-M}|DS^n(x')||I_{x_0 \dots x_{n-1}}| \\ \geq \; & e^{-M} \end{aligned}$$

and similarly

$$|DS^n(x')||I_{x_0 \dots x_{n-1}}| < e^M.$$

Setting $\zeta = e^{-M}$ gives that for any $z \in I_{x_0 \dots x_{n-1}}$,

$$\zeta < |DS^n(z)||I_{x_0 \dots x_{n-1}}| < \zeta^{-1}$$

as claimed. □

We can also control the size of gaps in the Cantor intervals.

Corollary 1.4 *Let $G_{x_0\dots x_{n-1}}$ be the gap in the interval $I_{x_0\dots x_{n-1}}$, i.e.*

$$I_{x_0\dots x_{n-1}} = I_{x_0\dots x_{n-1}0} \cup G_{x_0\dots x_{n-1}} \cup I_{x_0\dots x_{n-1}1}.$$

There is a constant $\zeta_1 > 0$ such that for any $n > 0$ and $x_0 \dots x_{n-1}$,

$$|G_{x_0\dots x_{n-1}}| \geq \zeta_1 |I_{x_0\dots x_{n-1}}|.$$

Proof. Let G be the gap in I, i.e. $I = I_0 \cup G \cup I_1$. We know that S^n maps $G_{x_0\dots x_{n-1}}$ 1-1 onto G , and $I_{x_0\dots x_{n-1}}$ 1-1 onto I. By the argument of Corollary 1.3 there is a $\zeta' > 0$ such that

$$\zeta' < |DS^n(x)||G_{x_n\dots x_{n-1}}| < \zeta'^{-1}$$

for some $x \in G_{x_0\dots x_{n-1}} \subset I_{x_0\dots x_{n-1}}$. Hence

$$\frac{|G_{x_0\dots x_{n-1}}|}{|I_{x_0\dots x_{n-1}}|} \geq \frac{\zeta'|DS^n(x)|^{-1}}{\zeta|DS^n(x)|^{-1}} = \frac{\zeta'}{\zeta} := \zeta_1. \quad \square$$

We can now show that the Lebesgue measure of C, $\lambda(C)$, is zero.

Proposition 1.5 *For a $C^{1+\gamma}$ cookie-cutter map S, the cookie-cutter set C has $\lambda(C) = 0$.*

Proof. Clearly for any $n > 0$

$$C \subset \bigcup_{x_0\dots x_{n-1}} I_{x_0\dots x_{n-1}}$$

so that we can work with the Cantor intervals $I_{x_0\dots x_{n-1}}$. Now $I_{x_0\dots x_{n-1}}$ is a disjoint union,

$$I_{x_0\dots x_{n-1}} = I_{x_0\dots x_{n-1}0} \cup G_{x_0\dots x_{n-1}0} \cup I_{x_0\dots x_{n-1}1}$$

so that

$$\begin{aligned} \lambda(I_{x_0\dots x_{n-1}}) &= \lambda(I_{x_0\dots x_{n-1}0}) + \lambda(G_{x_0\dots x_{n-1}}) + \lambda(I_{x_0\dots x_{n-1}1}) \\ &\geq \lambda(I_{x_0\dots x_{n-1}0}) + \zeta_1\lambda(I_{x_0\dots x_{n-1}}) + \lambda(I_{x_0\dots x_{n-1}}) \end{aligned}$$

by Corollary 1.4. Hence

$$(1-\zeta_1)\lambda(I_{x_0\dots x_{n-1}}) \geq \lambda(I_{x_0\dots x_{n-1}0}) + \lambda(I_{x_0\dots x_{n-1}1})$$

which implies that

$$(1-\zeta_1)^n\lambda(I) \geq \lambda(\bigcup_{x_0\dots x_{n-1}} I_{x_0\dots x_{n-1}}) \geq \lambda(C)$$

for any $n > 0$. Letting $n \to \infty$ gives $\lambda(C) = 0$. $\square$

2 Entropy and Gibbs measures

The aim of this part of the course is to see how certain invariants coming from dynamical systems and ergodic theory are closely linked to the idea of fractal dimension (by which I mean either Hausdorff dimension or box dimension). In particular we shall use topological and measure theoretic entropies (see [Wa3] for general information about entropy in ergodic theory).

Recall that Hausdorff dimension is a geometric/combinatoric quantity defined using a metric. Lower bounds on the Hausdorff dimension of sets are usually calculated in terms of a probability measure via Frostman's lemma. One can state a sophisticated version of this in the following way. For a Borel probability measure ν on R^n and a Borel set $E \subset R^n$ define the Hausdorff dimension of ν on E to be

$$\dim_H(\nu, E) = \inf_{x \in E} \liminf_{\epsilon \to 0} \frac{\log \nu B(x, \epsilon)}{-\log \epsilon}.$$

Then one can estimate the Hausdorff dimension of E (see [Tr]) by

$$\dim_H E = \sup\{\dim_H(\nu, E) \colon \nu(E) = 1\}.$$

We shall see that something very similar to this relation holds for the concepts of topological and measure theoretic entropies. Very loosely stated, entropy can be interpreted as a dimension in which the usual notion of distance is replaced by a notion of distance that depends on the underlying dynamical system.

First we'll try to motivate the definition of topological entropy. Suppose $T \colon X \to X$ is a continuous mapping of a compact metric space. We want a quantity which measures how complicated the mapping T is. Let $\alpha = \{A_1, \ldots, A_n\}$ be a finite open cover of X (A_i are open sets and $\cup A_i = X$). Suppose that due to poor eyesight we can only distinguish parts of X which lie in seperate pieces of α. This means that given the orbit $x, T(x)$, we can distinguish it from the orbit $y, T(y)$, if either x and y are in seperate pieces of α or if $T(x)$ and $T(y)$ are in seperate pieces. Hence the number of distinct "orbits" we have found after two iterations is approximately

$$\operatorname{card} \alpha \vee T^{-1}\alpha$$

where the symbol $\vee$ means the *join* of two partitions, $\alpha \vee \beta = \{A \cap B \colon A \in \alpha, B \in \beta\}$ (Note that Tx, Ty are in seperate pieces of α if x, y are in seperate pieces of $T^{-1}\alpha$). The number of distinct "orbits" we can distinguish up to n iterations is thus approximately

$$\operatorname{card} \alpha \vee T^{-1}\alpha \vee \ldots \vee T^{-(n-1)}\alpha.$$

For "chaotic" maps this number grows exponentially fast as $n \to \infty$ and the exponential growth rate is the entropy of the map. One formal definition is as follows (for other equivalent definitions see [Wa3]).

Definition 2.1 With α, T, X as above let $N(\alpha)$ be the minimal possible cardinality of a subcover of α. The *topological entropy of T with respect to α* is

$$h(T,\alpha) = \lim_{n\to\infty} \frac{1}{n} \log N(\alpha \vee T^{-1}\alpha \vee \ldots \vee T^{-(n-1)}\alpha)$$

and the *topological entropy of T* is

$$h(T) = \sup_{\alpha} h(T,\alpha).$$

The dependence of $h(T,\alpha)$ is not so serious - for any "good" cover α (which means that the collection of open sets

$$\bigcup_{n\geq 1} \alpha \vee T^{-1}\alpha \vee \ldots \vee T^{-(n-1)}\alpha$$

generates the topology of X), we have $h(T) = h(T,\alpha)$. Topological entropy was introduced because it is an invariant of topological conjugacy: if the following diagram commutes

$$\begin{array}{ccc} X & \xrightarrow{T} & X \\ \phi\downarrow & & \downarrow\phi \\ Y & \xrightarrow{U} & Y \end{array}$$

where ϕ is a homeomorphism, then T and U are said to be topologically conjugate. One can easily check from the definition of topological entropy that in such a case $h(T) = h(U)$. In particular if $h(T) \neq h(U)$ then T and U cannot be topologically conjugate. It seems that topological entropy is a rather good invariant, but on the other hand it is far from being a complete invariant.

Notice the similarity between the definition of $h(T,\alpha)$ and the definition of box dimension. The sequence of covers

$$\alpha \vee T^{-1}\alpha \vee \ldots \vee T^{-(n-1)}\alpha$$

gives us a finer and finer mesh over the set. Counting the number of elements needed to cover the set X is precisely the kind of thing one does when calculating box dimension. If one uses the idea that elements of $\alpha \vee T^{-1}\alpha \vee \ldots \vee T^{-(n-1)}\alpha$ are of size e^{-n} then

$$h(T,\alpha) = \lim_{n\to\infty} \frac{\log N(\alpha \vee \ldots \vee T^{-(n-1)}\alpha)}{-\log(e^{-n})}$$

which is almost exactly the same as the definition of box dimension. The two concepts are of course different because in general the elements of $\alpha \vee \ldots \vee T^{-(n-1)}\alpha$ will not be of *geometric* size e^{-n}.

There is also a concept of measure-theoretic entropy. Take X as above and let ν be a Borel probability measure (so $\nu(X) = 1$). We assume that ν is a T- invariant measure, i.e. $\nu(T^{-1}B) = \nu(B)$ for all Borel sets B. Take a finite partition of X into Borel sets $\alpha = \{A_1, \ldots A_m\}$.

Definition 2.2 The entropy of α with respect to ν is

$$H_\nu(\alpha) = -\sum_{A\in\alpha} \nu(A)\log\nu(A).$$

The entropy of T with respect to ν and α is

$$h_\nu(T,\alpha) = \lim_{n\to\infty}\frac{1}{n}H_\nu(\alpha\vee T^{-1}\alpha\vee\ldots\vee T^{-(n-1)}\alpha),$$

and the entropy of T with respect to ν is

$$h_\nu(T) = \sup_\alpha h_\nu(T,\alpha).$$

The idea behind $H_\nu(\alpha)$ is that $H_\nu(\alpha)$ is higher if ν is evenly distributed with respect to α. Then $h_\nu(T,\alpha)$ measures how much ν is evenly distributed with respect to the refined sequence $\alpha\vee T^{-1}\alpha\vee\ldots\vee T^{-(n-1)}\alpha$. As with topological entropy it is possible to show that for "good" partitions $\alpha, h_\nu(T) = h_\nu(T,\alpha)$ ("good" now means that

$$\bigcup \alpha\vee T^{-1}\alpha\vee\ldots\vee T^{-(n-1)}\alpha$$

generates the underlying σ-algebra). With topological entropy we were able to see a close connection between its definition and the definition of box dimension. There is a similar kind of analogy one can make if one modifies the definition of box dimension to make it a suitable quantity for studying a measure rather than the topological support of the measure. This was suggested by Ledrappier (see [Yo]) who defined the following type of dimension:

Definition 2.3 The *L-box dimension* of ν is

$$\dim_L(\nu) = \lim_{\epsilon\to 0}\lim_{\delta\to 0}\frac{\log N(\delta;\epsilon)}{-\log\delta}$$

where $N(\delta;\epsilon)$ is the minimal number of δ-balls needed to cover a set of ν-measure larger than $1-\epsilon$ (when the limits exist).

(In many cases one does not need to take the limit $\epsilon\swarrow 0$ as $\lim_{\delta\to 0}\log N(\delta;\epsilon)/-\log\delta$ is independent of ϵ). The connection between this type of modified box dimension for measures and measure theoretic entropy is given by Shannon's Ergodic Theorem of Information which says that for large enough n, $1-\epsilon$ of the mass of ν lies on approximately $e^{nh_\nu(T,\alpha)}$ of the elements of $\alpha\vee T^{-1}\alpha\vee\ldots\vee T^{-(n-1)}\alpha$.

The analogy that we wish to point out is that Hausdorff dimension, $\dim_H$, is very similar to topological entropy h, and the dimension of a measure $\dim\nu$ is very similar to the entropy of a measure. At the beginning of this section we mentioned a version of Frostman's Lemma which says that $\dim_H E = \sup_{\nu(E)=1}\dim_H(\nu,E)$. In view of the connections we are making with entropy we should look for a similar result relating topological and measure theoretic entropies. This indeed exists and is due to Goodwyn, Dinaburg and Goodman (see [Wa3] Theorem 8.6).

Theorem 2.1 *Let $T: X \to X$ be a continuous map of a compact metric space, then*

$$h(T) = \sup h_\nu(T)$$

where the supremum is taken over all T-invariant Borel probability measures. □

For certain dynamical systems (including cookie-cutters) there is a unique invariant measure ν with $h_\nu(T) = h(T)$. This is called the measure of maximal entropy.

The topological entropy of a cookie-cutter is easy to calculate. An open cover which generates the usual topology on the C is $\alpha = \{I_0 \cap C, I_1 \cap C\}$. Then

$$\begin{aligned} \alpha \vee S^{-1}\alpha &= \{I_{00} \cap C, I_{11} \cap C, I_{01} \cap C, I_{10} \cap C\} \\ &\vdots \\ \alpha \vee \ldots \vee S^{-(n-1)}\alpha &= \{I_{x_0 \ldots x_{n-1}} \cap C : \ x_i = 0, 1\} \end{aligned}$$

so that $N(\alpha \vee \ldots \vee S^{-(n-1)}\alpha) = 2^n$ which implies that

$$\begin{aligned} h(S) = h(S, \alpha) &= \lim_{n \to \infty} \frac{1}{n} \log N(\alpha \vee \ldots \vee S^{-(n-1)}\alpha) \\ &= \log 2. \end{aligned}$$

The measure of maximal entropy for a cookie-cutter is also very easy to describe. It is the Borel measure ν supported on C such that

$$\nu(I_{x_0 \ldots x_{n-1}}) = \frac{1}{2^n} \quad \text{for all} \quad x_0 \ldots x_{n-1}.$$

Note that what we have just said applies for *all* cookie-cutters. We expect that in general different cookie-cutters will have different dimensions, so that there is not a direct equality between dimension and entropy. However we will see in part 4 that there is a dynamical system canonically associated with a cookie-cutter whose entropy equals the dimension of the cookie-cutter. In order to calculate the dimension of a cookie-cutter one needs to have a probability measure distributed nicely on the set to use in conjunction with Frostman's lemma. Any probability measure will give (via Frostman's lemma) a lower bound on the dimension of the set but we need a measure which will give us the true dimension in the estimate. We have seen that the measure of maximal entropy is probably not the "right" measure to use in general. On the other hand since the Cantor set has been generated dynamically one could expect that some measure from the class of invariant measures (i.e. invariant under S) would be the right one to use. In general this class of measures is too big to describe properly. There is a rich subclass of invariant Borel probability measures, called Gibbs measures, which generalizes the class of Markov measures. It turns out that the class of Gibbs measures is wide enough to provide the measures we need to calculate the dimension of cookie-cutters.

We shall now describe a way of generating Gibbs measures. This is a process which is very closely related to the "mass distribution" process that Falconer mentions in his notes in this volume, but the way I am presenting it is based on Keane's development of g-measures [Ke] (see also [Wa2]). A general reference for the construction of Gibbs measures is the book of Bowen [Bo2]. Although we shall discuss the construction of Gibbs measures in the symbol space Σ it will be clear that the construction is very closely related to Eltons construction of stationary measures for a Markov process associated with an iterated function system with place dependent probabilities [BDEG]. In the context of a cookie-cutter system Elton's approach and the Gibbs measures approach are essentially the same but they both generalize in different ways.

Suppose we have a Borel probability measure ν on Σ and we choose a random sequence $\underline{x} = x_0x_1x_2\dots$ according to the distribution ν. One can ask the question: What is the probability that x_0 is 0 given the rest of the sequence $x_1x_2\dots$,

$$P(x_0 = 0|x_1x_2\dots) \quad ?$$

This is a function of $x_1x_2\dots$, and it is clear that

$$P(x_0 = 0|x_1x_2\dots) + P(x_0 = 1|x_1x_2\dots) = 1.$$

Now let us reverse the problem and try to construct the measure ν given the conditional probabilities

$$P(x_0 = 0|x_1x_2\dots) \quad \text{and} \quad P(x_0 = 1|x_1x_2\dots).$$

For technical reasons we write these two functions as a single function $p(\underline{x})$ where

$$p(0x_1x_2\dots) = P(x_0 = 0|x_1x_2\dots)$$

and

$$p(1x_1x_2\dots) = P(x_0 = 1|x_1x_2\dots).$$

Consider the following sequence of random variables in Σ. Choose an initial point $\underline{x}^0$ in Σ according to any distribution ρ, and then set

$$\underline{x}^1 = \begin{cases} 0\underline{x}^0 & \text{with probability} \quad p(0\underline{x}^0) \\ 1\underline{x}^0 & \text{with probability} \quad p(1\underline{x}^0). \end{cases}$$

Inductively we define

$$\underline{x}^n = \begin{cases} 0\underline{x}^{n-1} & \text{with probability} \quad p(0\underline{x}^{n-1}) \\ 1\underline{x}^{n-1} & \text{with probability} \quad p(1\underline{x}^{n-1}). \end{cases}$$

It is intuitively plausible that the influence of ρ on the distribution of the random variable $\underline{x}^n$ decreases as $n \to \infty$. It turns out that if p is Hölder continuous (a weaker condition also works) then the distribution of $\underline{x}^n$ converges to a measure ν depending only on p. Furthermore the measure is Markov-like in the following sense. If we write $c_n(\underline{x}) = \{\underline{y} \in$

$\Sigma | x_0 = y_0, \ldots, x_{n-1} = y_{n-1}\}$ then the measure of $c_n(\underline{x})$ is given approximately by p: there exists $\zeta_0 > 0$ such that

$$\zeta_0 < \frac{\nu(c_n(\underline{x}))}{p(\underline{x}).p(\sigma\underline{x})\ldots p(\sigma^{n-1}\underline{x})} < \zeta_0^{-1} \tag{2.0}$$

for all $\underline{x} \in \Sigma$ and $n \geq 0$. (The fact that $p(\underline{y})p(\sigma\underline{y})\ldots p(\sigma^{n-1}\underline{y})$ is approximately the same as $p(\underline{x}).p(\sigma\underline{x})\ldots p(\sigma^{n-1}\underline{x})$ for any $\underline{y} \in c_n(\underline{x})$ is a consequence of bounded variation). The proof that ν exists works roughly like this. Remember that a measure is determined by how it integrates continuous functions (the Riesz representation theorem). Let $C(\Sigma) = \{\psi\colon \Sigma \to R | \psi \text{ is continuous}\}$. One can define a operator on $C(\Sigma)$ by

$$(L\psi)(\underline{x}) = \psi(0\underline{x})p(0\underline{x}) + \psi(1\underline{x})p(1\underline{x}).$$

This is a positive operator, i.e. if $\psi > 0$ then $L\psi > 0$, and it has the constant function $1(\underline{x}) \equiv 1$ as an eigenfunction with 1 as associated eigenvalue because p came from conditional probabilities. A version of the Perron-Frobenius theorem can be used to show that for any continuous function ψ the sequence $L^n\psi$ converges to an eigenfunction, i.e. to a constant. The measure ν is then defined by setting

$$\nu(\psi) = \int \psi d\nu := \lim_{n\to\infty} L^n\psi.$$

The operator L is related to the sequence of random variables we discussed above because the dual operator to L on the space of probability measures is exactly the Markov operator for the Markov chain $\underline{x}^n$.

One can start with an arbitrary Hölder continuous function $p\colon \Sigma \to \mathbf{R}$ satisfying

$$p > 0$$

and

$$p(0x_1x_2\ldots) + p(1x_1x_2\ldots) = 1, \tag{2.1}$$

and obtain a measure in this way. These measures were originally constructed by Keane [Ke] where they are called g-measures. The class of such measures is very rich. Unfortunately it is difficult in the case of a cookie-cutter to see what a natural choice of p should be. The only possible function we can use as information on how to distribute a measure on the Cantor set C is $|DS|$ which (as we saw in part 1) determines how big the intervals $I_{x_0\ldots x_{n-1}}$ are. The function $1/(|DS| \circ \pi)$ takes values in the interval (0,1) but need not satisfy relation (2.1). Fortunately there is still a way of constructing measures for a general function. Consider a Hölder continuous function $f\colon\ \Sigma \to R$. The function e^f is positive and thus one can consider a positive linear operator $L_f\colon\ C(\Sigma) \to C(\Sigma)$ given by

$$L_f\psi(\underline{x}) = \psi(0\underline{x})e^{f(0\underline{x})} + \psi(1\underline{x})e^{f(1\underline{x})} \tag{2.2}$$

(note the e^f now plays the role of p). The Ruelle-Perron-Frobenius theorem for L_f says that there exists a positive function $h \in C(\Sigma)$ and a real number $\lambda > 0$ such that

$$L_f h = \lambda h. \tag{2.3}$$

There is also an eigenmeasure for the induced map on probability measures

$$\frac{L_f^*\mu}{L_f^*\mu(1)} = \mu. \tag{2.4}$$

(The denominator on the left is just a normalization constant to ensure that the measure is a probability). This eigenmeasure μ will turn up later with a very natural geometric interpretation. For now, however, it is not the measure we were looking for because it is not (in general) σ-invariant. There is a simple trick which does give us a σ-invariant measure by changing the function f slightly. If we replace f by

$$\tilde{f}(\underline{x}) = f(\underline{x}) - \log\lambda + \log h(\underline{x}) - \log h(\sigma\underline{x}) \tag{2.5}$$

then one can easily check by using equations (2.2) and (2.3) that $e^{\tilde{f}}$ satisfies the requirements (2.1) of a conditional probability function. By the argument above we obtain a measure ν which we'll denote ν_f from now on. Now, the transformation (2.5) is a lot more natural than it looks at first sight because when one considers sums of f along orbits, $S_nf(\underline{x}) = \sum_{i=0}^{n-1} f(\sigma\underline{x})$, one has the following relationship between f and $\tilde{f}$,

$$S_n\tilde{f}(\underline{x}) = S_nf(\underline{x}) - n\log\lambda + \log h(\underline{x}) - \log h(\sigma^n\underline{x}).$$

Since h is bounded we know that $S_n\tilde{f}(\underline{x}) \approx S_nf(\underline{x}) - n\log\lambda$ for all n and $\underline{x}$. Hence one can get an expression like (2.0) for f: there exists $\zeta > 0$ such that

$$\zeta < \frac{\nu_f(c_n(\underline{x}))}{\exp\{-nP(f) + S_nf(\underline{x})\}} < \zeta^{-1} \tag{2.6}$$

for all $x \in \Sigma$ and $n > 0$, where we have written $P(f) = \log\lambda$. The property expressed in (2.6) is the *Gibbs property* and ν_f is called a *Gibbs measure.* The relationship between ν_f and $\mu_f = \mu$ (the eigenmeasure for L_f^*) is that $\nu_f \ll \mu_f$ and $\frac{d\mu_f}{d\nu_f} = h$. We'll see a geometric interpretation for this later on. The quantity $P(f)$ is called the *pressure* of f. It satisfies a variational principle which relates it to entropy,

$$P(f) = \sup\{h_\nu(\sigma) + \int f d\nu \colon \nu \quad \text{is } \sigma\text{-invariant}\}. \tag{2.7}$$

In particular pressure generalizes topological entropy because $h(\sigma) = P(0)$ by Theorem 2.1. Furthermore, when f is Hölder continuous there is exactly one measure taking the sup in (2.7) - the Gibbs measure ν_f. We can summarize the most important parts of the above discussion in the following theorem (see [Bo]).

Theorem 2.2 *Given a Hölder continuous map $f\colon \Sigma \to R$ there is a constant $P = P(f)$ and a unique invariant probability measure ν_f such that for some $\zeta > 0$*

$$\zeta < \frac{\nu_f(c_n(\underline{x}))}{\exp(-nP(f) + S_nf(\underline{x}))} < \zeta^{-1}$$

for any $\underline{x} \in \Sigma, n \geq 0$. Furthermore

$$P(f) = h_{\nu_f}(\sigma) + \int f d\nu_f = \sup\{h_\nu(\sigma) + \int f d\nu\}. \quad \square$$

The above description of Gibbs measures and pressure seems very abstract, but there is a very natural geometric interpretation for it all in the context of cookie-cutters. We'll see in the next part that Hausdorff dimension can be computed exactly in terms of the pressure of $\log|DS|$, the logarithmic derivate of the cookie-cutter map.

3 Bowen's formula for Hausdorff dimension

The method for calculating Hausdorff dimension that is presented here was discovered by Bowen [Bo4] and applied in the calculation of the Hausdorff dimension of quasi-circles - certain sets which appear in the study of Fuchsian groups. The argument is very easy to present with the help of Theorem 2.2 and Corollary 1.3.

We define $f: \Sigma \to \mathbf{R}$ by $f(\underline{x}) = -\log|DS(\pi\underline{x})|$ (recall $\pi: \Sigma \to C$ associates a point on C with its corresponding code in Σ). Since π and $|DS|$ are Hölder continuous maps of compact spaces, f is also Hölder continuous.

We consider the one parameter family of Hölder continuous maps $d\,f (d \in \mathbf{R})$. One can fairly easily show that the pressure functional $P(.)$ is convex, so that in particular the map

$$d \mapsto P(df)$$

is convex (see Walters [Wa3] Theorem 9.7 (v)). This map is very important for our development, so we'll make some further comments. From the variational principle and the fact that $|DS| > 1$ one has that $P(df) \to \pm\infty$ as $d \to \mp\infty$. Hence the graph of $d \mapsto P(df)$ is roughly as shown in Figure 3.1 In particular there is a unique value of d for which $P(df) = 0$.

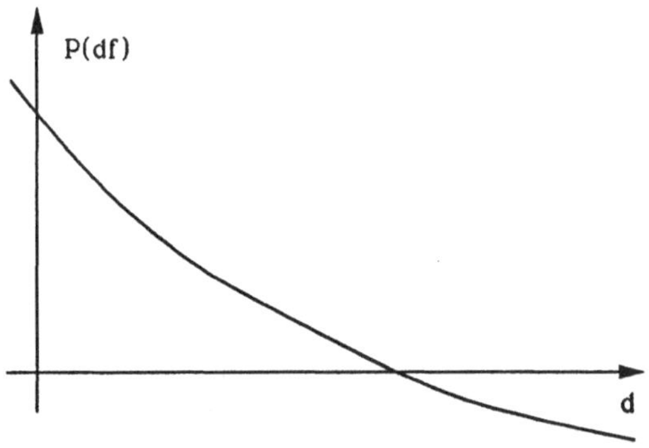

Figure 3.1: The graph of the pressure function $d \to P(df)$.

Theorem 3.1 *The Hausdorff dimension of C is the unique $d \in \mathbf{R}$ with $P(df) = 0$. Furthermore the d-dimensional Hausdorff measure of C is positive and finite and absolutely continuous with $\pi_* \nu_{df}$.*

Proof. Denote ν_{df} by ν. Since d is chosen with $P(df) = 0$ we must have, for $x = \pi(\underline{x}) \in C$,

$$\begin{aligned} \pi_*\nu(I_{x_0\ldots x_{n-1}}) &= \nu(c_n(\underline{x})) \\ &\leq \zeta^{-1} \exp(S_n df(\underline{x})) \text{ by Theorem 2.2} \\ &= \zeta^{-1} \prod_{i=0}^{n-1} |DS(S^i x)|^{-d}. \end{aligned}$$

Similarly one has

$$\pi_*\nu(I_{x_0\ldots x_{n-1}}) \geq \zeta \prod_{i=0}^{n-1} |DS(S^i x)|^{-d}.$$

To show that d is the Hausdorff dimension and to show also the claimed absolute continuity it is enough to show that there is a $\zeta_1 > 0$ such that for any $x \in C$ and $r > 0$

$$\zeta_1 < \frac{\pi_*\nu(B(x,r))}{r^d} < \zeta_1^{-1}.$$

To prove the left inequality, choose n such that

$$|I_{x_0\ldots x_n}| \leq r < |I_{x_0\ldots x_{n-1}}|$$

(where $x = \pi\underline{x}, \underline{x} = (x_0, x_1 \ldots)$). Then certainly $I_{x_0\ldots x_n} \subset B(x,r)$ so that

$$\begin{aligned} \pi_*\nu(B(x,r)) &\geq \pi_*\nu(I_{x_0\ldots x_n}) \\ &\geq \zeta \prod_{i=0}^{n} |DS(S^i x)|^{-d} \\ &\geq \zeta(\inf_{y \in C} |DS(y)|^{-d}) \prod_{i=0}^{n-1} |DS(S^i x)|^{-d} \\ &\geq \zeta^2(\inf |DS(y)|^{-d}) |I_{x_0\ldots x_{n-1}}|^d \\ &\geq \zeta^2(\inf |DS(y)|^{-d}) r^d. \end{aligned}$$

The right inequality can be proven in almost exactly the same way, first observing that at most two distinct Cantor intervals of length greater than r can intersect $B(x,r)$. □

Note that not only is Hausdorff measure on C absolutely continuous with $\pi_*\nu_{df}$, but also with $\pi_*\mu_{df}$. In fact we shall see in the last section that $\pi_*\mu_{df}$ is, up to a constant multiple, equal to the Hausdorff measure.

The formula $P(df) = 0$ is called Bowen's formula for Hausdorff dimension. One can show that the Hausdorff and box dimensions are equal for cookie-cutters, although this does not happen when one generalizes Bowen's formula, as we shall see later. We also remark that if one changes the cookie-cutter map S continuously in the $C^{1+\gamma}$ topology then the Hausdorff dimension of C changes continuously.

A special case of Theorem 3.1 is the case of a two scale linear Cantor set generated by the two contractions $\varphi_0(x) = r_0 x, \varphi_1(x) = r_1 x + (1 - r_1), (r_0 + r_1 < 1)$. Taking φ_0 and φ_1 as the inverse branches of a cookie-cutter S (see Figure 3.2) one has

$$DS(x) = \begin{cases} r_0^{-1} & \text{if} \quad x \in I_0 \\ r_1^{-1} & \text{if} \quad x \in I_1. \end{cases}$$

So

$$df(x) = -d \log DS(x) = \begin{cases} d \log r_0 & x \in I_0 \\ d \log r_1 & x \in I_1. \end{cases}$$

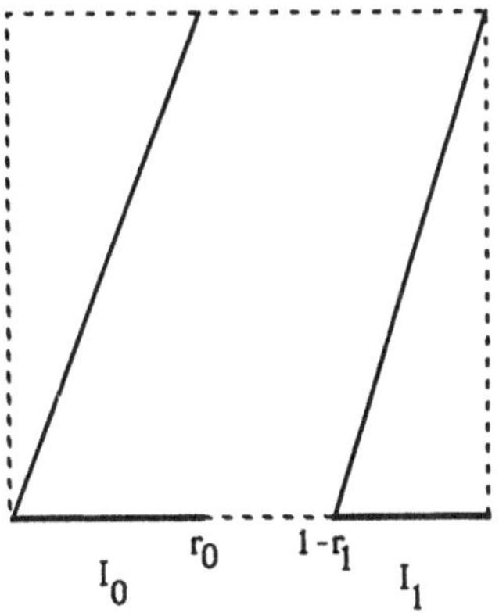

Figure 3.2: A two-scale linear cookie-cutter.

If one carries out the "mass distribution" scheme to calculate the measure ν_{df} one sees that ν_{df} is the (p_0, p_1)-Bernoulli probability where $p_0 = r_0^d$ and $p_1 = r_1^d$, so that

$$\pi_* \nu_{df}(I_{x_0 \dots x_{n-1}}) = p_{x_0} \cdots p_{x_{n-1}}.$$

The condition that $p_0 + p_1 = 1$ can be rewritten as $r_0^d + r_1^d = 1$, which is the well-known formula for the Hausdorff dimension of this type of Cantor set.

Bowen's method for calculating Hausdorff dimension has applications beyond cookie-cutters and quasi-circles. The limitations of the method, however, become obvious rather quickly. We now discuss some of the generalizations of Bowen's ideas.

(a) *Generalized cookie-cutters in* $\mathbf{R}$. Let $\varphi_0, \ldots, \varphi_{k-1} : \mathbf{R} \to \mathbf{R}$ be $C^{1+\gamma}$ contractions with associated self-similar set (i.f.s. attractor)

$$C = \varphi_0(C) \cup \varphi_1(C) \cup \ldots \cup \varphi_{k-1}(C).$$

The *open set condition* for the system requires the existence of an open set $U \subset \mathbf{R}$ such that

$$U \supset \bigcup_{i=0}^{k-1} \varphi_i(U), \text{ and}$$

$$\varphi_i(U) \cap \varphi_j(U) = \emptyset \text{ for } i \neq j.$$

If the open set condition holds then Bowen's methods can be applied to calculate Hausdorff dimension.

(b) *Recurrent i.f.s. in* $\mathbf{R}$. Let A be a $m \times m$ matrix of $0's$ and $1's$ such that $A^n > 0$ for some $n \geq 1$. We use this as a transition matrix for an iterated function system. Suppose that for each pair $0 \leq i, j \leq m-1$ we have a $C^{1+\gamma}$ contraction map $\varphi_{ij} \colon \mathbf{R} \to \mathbf{R}$. Then there is a unique collection of compact non-empty sets

$$\underline{C} = (C_0, \ldots, C_{m-1}), \qquad C_i \subset \mathbf{R}$$

such that

$$C_i = \bigcup_{a_{ij}=1} \varphi_{ij}(C_j).$$

This is a generalization of strict self-similarity to a type of Markovian self-similarity. The *open set condition* for such a system requires the existence of open sets $U_i \subset \mathbf{R}$ with

$$U_i \supset \bigcup_{a_{ij}=1} \varphi_{ij}(U_j)$$

and

$$\varphi_{ij}(U_j) \cap \varphi_{ik}(U_k) = \emptyset \quad \text{if} \quad i \neq k.$$

Bowen's argument can be used in this situation to calculate $\dim_H C_i$ ($\dim_H C_i = \dim_H C_j$ for each i, j here).

(c) *Hyperbolic Julia sets* The simplest example of such systems is the family of quadratic maps of the Riemann sphere

$$z \overset{f_c}{\mapsto} z^2 + c$$

parameterized by $c \in \mathbf{C}$ (c must have small modulus to ensure that f_c is hyperbolic on the Julia set). See the notes of Blanchard for more general information about Julia sets. When $c = 0$ the Julia set J_c for f_c is the unit circle $|z| = 1$. As c moves continuously away from 0, the Julia set J_c stays a topological circle but becomes fractal. Ruelle [Ru2] used Bowen's method to calculate the Hausdorff dimension of J_c and also showed that for small $|c|$, $\dim_H J_c$ is an analytic function of c.

(d) *Higher dimensions.* In examples a,b and c we restricted ourselves to fractal sets in a one-dimensional space (either **R** or **C**). One can generalize examples (a) and (b) to $\mathbf{R}^n$ with the condition that the maps φ_i are *conformal.* This means that the derivative $D\varphi_i(x)$ should be a similitude at every x, i.e. that φ_i carries infinitesimal circles to infinitesimal circles. This is important because Bowen's method works essentially by using the dynamics of the maps to take efficient coverings to efficient coverings by smaller sets. If the maps φ_i are affine then they take circles to ellipses: coverings with ellipses are usually poor which is why Bowen's method does not work in a non-conformal situation. We'll say more about this problem soon.

(e) *Recurrent curves* The fractal recurrent curves of Dekking [De1], [De2] (see his lectures) can be put into the framework of example (b). In other words a recurrent set is the attractor of a recurrent i.f.s. (this result is the main theorem of [Be2]). The resolvability condition used by Dekking turns out to be equivalent to the open set condition and the transition matrix A is derived from his matrix M_σ.

We can summarize the above discussion as follows. Bowen's method works if three elements are present:
conformal contractions,
open set condition, and
subshift of finite type (Markov) structure.

It is worth making a few remarks about the case of self affine (non-conformal) contractions. Firstly, although Bowen's argument does not work directly, Falconer [Fa2] has shown that the concept of pressure can be generalized to give a generalized version of Bowen's formula which holds for "almost all" systems of contractions when the corresponding self-affine sets are disconnected. Unfortunately in general one cannot check for particular systems whether or not the formula holds.

There are a few cases in which more specific results are known. There is a class of self-affine sets ([Mc], [Be1]; see also [Be5]) for which one can calculate both Hausdorff and box dimension, and for which the dimensions differ. Consider the unit square is $\mathbf{R}^2$ and divide it up into r equal columns and s equal horizontal strips ($r > s > 1$). Now shade a number of the resulting rectangles (see Figure 3.3). For each such rectangle there is an affine map of the form

$$\begin{pmatrix} x \\ y \end{pmatrix} \mapsto \begin{pmatrix} r^{-1} & 0 \\ 0 & s^{-1} \end{pmatrix} \begin{pmatrix} x \\ y \end{pmatrix} + \begin{pmatrix} a \\ b \end{pmatrix}$$

which takes the unit square to a shaded rectangle. The compact set C which is self-affine under these maps has Hausdorff dimension given by

$$\dim_H C = \log_s\left(\sum_{i=0}^{s-1} k_i^{\log_r s}\right)$$

where k_i in the number of shaded rectangles is the ith row. The box dimension is given by

$$\dim_B(C) = \log_s k + \log_r(\frac{\sum_{i=0}^{s-1} k_i}{k})$$

where k is the number of non-zero k_i. Box dimension and Hausdorff dimension agree if and only if for some j, $k_i = j$ or 0 for all i.

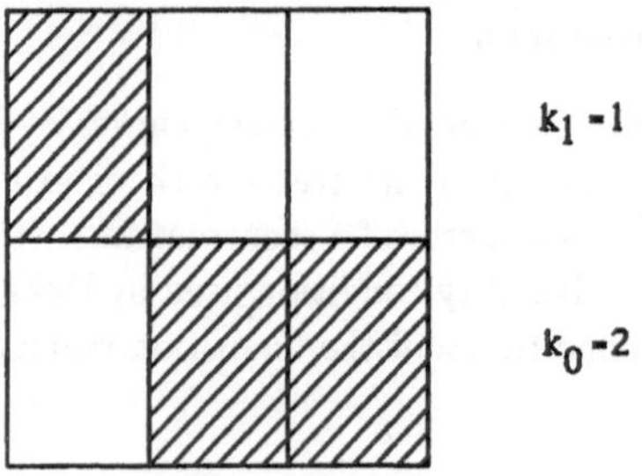

Figure 3.3: The generator for a self-affine set.

Przytycki and Urbanski [PU], and Bedford and Urbanski [BU] have considered a two parameter family of self affine subsets of $\mathbf{R}^2$ (see Figure 3.4) and have shown that there exist β and p values for which $\dim_H < \dim_B$, but that for a set of (β, p) with positive Lebesgue measure, $\dim_H = \dim_B$.

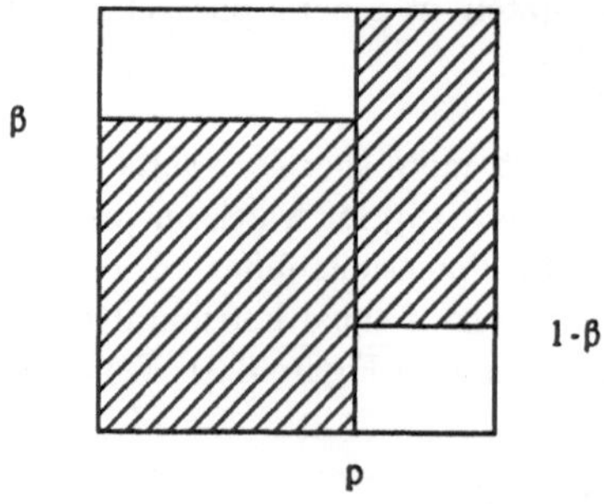

Figure 3.4: A two parameter family of self-affine sets.

Bedford [Be3] and Bedford and Urbanski [BU] have also shown that for certain kinds of self affine sets in $\mathbf{R}^2$ one can obtain a Bowen formula for box dimension. One defines two functions f_W, f_H which measure rates of contraction of the maps φ_i in different directions. A formula of the form $P(df_W + f_H) = 0$ can then be proven.

4 A flow dynamical system associated to a cookie-cutter, and density of the Hausdorff measure

In part 2 we saw that the definition of topological entropy is very similar to that of box dimension. On the other hand we know that every cookie-cutter map has entropy log 2, whilst the dimension of the cookie-cutter set can vary. It is therefore natural to ask whether there is another dynamical system associated with a cookie-cutter whose entropy equals the cookie-cutter dimension.

We first construct such a system and then explore some of the applications for the fractal geometry of the cookie-cutter.

We begin with the concept of a flow under a function (also called a suspension of a dynamical system).

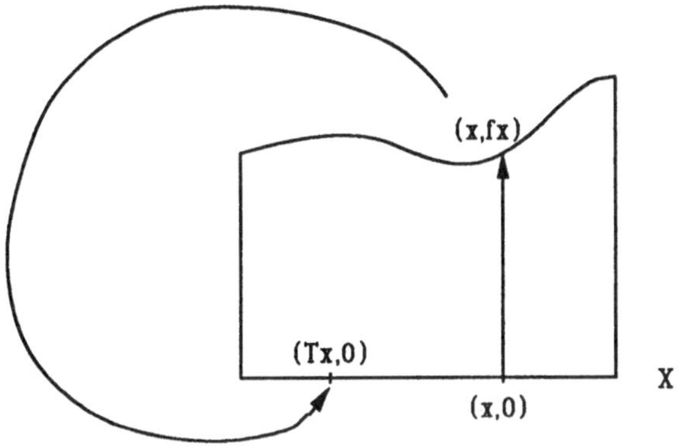

Figure 4.1: The flow under the function f.

Definition 4.1 Let $T: X \to X$ be a continuous map of a compact metric space and $f: X \to \mathbf{R}$ a continuous function. The *flow space* M is

$$M = \{(x,t): x \in X, t \in [0, f(x)]\}/\equiv$$

where $\equiv$ is the equivalence relation $(x, f(x)) \equiv (Tx, 0)$.

On M we define a flow Φ by

$$\Phi_s(x,t) = (x, t+s) \text{ for small } s \geq 0$$

where $(x, t+s)$ is taken to be $(Tx, t+s-f(x))$ if $t+s > f(x)$, and where Φ_s is defined for all $s \geq 0$ by the semigroup property that $\Phi_{s+t} = \Phi_s \circ \Phi_t$ (Φ is really a semiflow rather than a flow as it is not defined for inverse time). We call Φ the *flow under the function f* (see Figure 4.1). An invariant measure for the flow Φ is a measure that is invariant under each of the mappings Φ_s. An invariant measure m for Φ is always of the form

$$m(A \times B) = \frac{\mu(A)\lambda(B)}{\int f d\mu} A \subset X, B \subset \mathbf{R}, A \times B \subset M.$$

where μ is an invariant measure for T and λ is Lebesgue measure. The entropy of m with respect to the time s map Φ_s is related to μ by

$$h_m(\Phi_s) = s \frac{h_\mu(T)}{\int f d\mu}.$$

In particular the topological entropy of the time 1 map Φ_1 is

$$h(\Phi_1) = \sup_m h_m(\Phi_1) = \sup_\mu \frac{h_\mu(T)}{\int f d\mu}. \tag{4.1}$$

The entropy of the flow Φ is defined to be the entropy of the time 1 map.

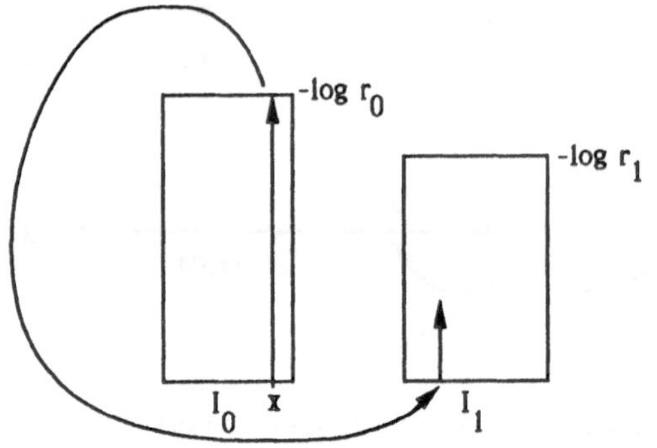

Figure 4.2: The semi-flow for a two-scale Cantor set.

We are going to apply this theory in the case of a cookie-cutter map $S\colon C \to C$ with the function $g\colon C \to \mathbf{R}$ given by

$$g(x) = \log |DS(x)|$$

($-g$ is equal to the function f we used in part 3). Consider the example of a two scale Cantor set C generated by maps $\varphi_0(x) = r_0 x, \varphi_1(x) = r_1 x + (1 - r_1)$. The space M is shown in Figure 4.2.

An example of an invariant measure for S is the (p,q)-Bernoulli measure ($q = 1 - p$) given by

$$\mu(I_{x_0 \dots x_{n-1}}) = p^i q^{n-i}$$

where i the number of 0's in the sequence $x_0 \dots x_{n-1}$. One can show fairly easily that $h_\mu(S) = -p\log p - q\log q$, and since

$$\int_C g d\mu = \int_{I_0} g d\mu + \int_{I_1} g d\mu = -p\log r_0 - q\log r_1$$

we see that

$$h_m(\Phi) = \frac{-p\log p - q\log q}{-p\log r_0 - q\log r_1}.$$

We now directly estimate the box dimension of C in this example, and shall see the same kind of formula arising. The argument we give will be hand waving but can be rigourized. Recall that

$$\dim_B(C) = \lim_{\varepsilon\to 0} \frac{\log N(\varepsilon)}{-\log\varepsilon}$$

where $N(\varepsilon)$ is the smallest number of ε-balls covering C. If we put $\varepsilon = e^{-t}$ then we have alternatively

$$\dim_B(C) = \lim_{t\to\infty} \frac{1}{t}\log N(e^{-t}).$$

Now the most natural way to cover C is via the Cantor intervals $I_{x_0 \dots x_{n-1}}$, so we should ask when such an interval has length approximately equal to e^{-t}. We shall write $a_i = -\log r_i$. We have

$$\begin{aligned} |I_{x_0 \dots x_{n-1}}| &= r_{x_0} r_{x_1} \dots r_{x_{n-1}} \\ &\approx e^{-t} \\ &\Leftrightarrow \sum_{i=0}^{n-1} a_i \approx t. \end{aligned}$$

So $N(e^{-t}) \approx \#\{x_0 \dots x_{n-1} \colon \sum_{i=0}^{n-1} a_i \approx t\}$ (the n can vary here!). (It is worth remarking that this number is also the number of periodic orbits for the flow that have period (i.e. length) equal to t). Now suppose that $a_0 > a_1$, then

$$\begin{aligned} \sum_{i=0}^{n-1} a_{x_i} \approx t \quad &\Leftrightarrow \quad n a_1 + \#\{0 \le i < n | x_i = 0\}(a_0 - a_1) \approx t, \\ &\Leftrightarrow \quad p := p(n) = \frac{\#\{0 \le i < n | x_i = 0\}}{n} \approx \frac{t/n - a_1}{a_0 - a_1}. \end{aligned}$$

This gives

$$\begin{aligned} N(e^{-t}) &\approx \sum_n \#\{x_0 \dots x_{n-1} \colon p(n) \approx \frac{t/n - a_1}{a_0 - a_1}\} \\ &\approx \sum_n \binom{n}{[np(n)]} p(n) = \frac{t/n - a_1}{a_0 - a_1}. \end{aligned}$$

Now Stirling's formula ([St]) implies that for $q = 1 - p$,

$$\frac{n!}{(np)!(nq)!} = o(n^{-1/2})(p^{-p}q^{-q})^n.$$

So using $n \approx \frac{t}{p(a_0-a_1)-a_1}$ and Stirling's formula we have

$$N(e^{-t}) \approx \sup_{p\in(0,1)} (p^{-p}q^{-q})^{t/(p(a_0-a_1)+a_1)}.$$

Hence

$$\begin{aligned}\dim_B(C) &= \lim_{t\to\infty} \frac{1}{t} \log N(e^{-t}) \\ &= \sup_{p\in(0,1)} \frac{-p\log p - q\log q}{-p\log r_0 - q\log r_1} \\ &= \sup\left\{\frac{h_\mu(S)}{\int g d\mu} : \mu \text{ a Bernoulli measure}\right\}.\end{aligned}$$

In fact the supremum over all invariant measures is given in this case by a Bernoulli measure so that we have

$$\dim_H C = \dim_B(C) = \sup_\mu \frac{h_\mu(S)}{\int g d\mu} = h(\Phi).$$

This fact generalizes to all cookie-cutters:

Theorem 4.1 *For a cookie-cutter $S\colon\ C \to C$ and the associated flow Φ one has*

$$\dim_H C = \dim_B C = h(\Phi) = \sup_{S\text{-invariant } \mu} \frac{h_\mu(S)}{\int g d\mu}. \quad \square \tag{4.2}$$

We have therefore found a dynamical system associated with a cookie-cutter C whose entropy is the same as the dimension of C. The expression for the dimension however is not the same as that given by Theorem 3.1. There we had $\dim_H C = d$ with $P(df) = 0$ or equivalently $P(-dg) = 0$ (since $g = -f$). The link between the two formulae is given by the variational principle for pressure,

$$P(df) = 0 = \sup_\mu \{h_\mu(S) - d \int g\, d\mu\}$$

which holds if and only if

$$d = \sup_\mu \frac{h_\mu(S)}{\int g d\mu}.$$

Hence the two expressions do give us the same number (luckily!). The difference between the flow Φ and the map S is that flowing along by Φ takes you through the Cantor set at a constant geometric rate. More precisely, if we define $\pi_M\colon\ M \to C$ by $\pi_M(x,t) = x$, then if $x, y \in C$ are very close together, then the sequences

$$\pi_M(\Phi_n(x)), \pi_M(\Phi_n(y))$$

diverge from each other at a rate that is independent of x and y, whilst the sequences

$$S^n(x), S^n(y)$$

diverge from each other at a rate which does depend on x and y. Another useful property of the flow Φ is that it enables us to make continuous estimates of certain quantities. This is especially useful for studying the density of Hausdorff measure.

Definition 4.2 The *d-dimensional lower and upper densities* of a Borel measure μ on $\mathbf{R}$ at $x \in \mathbf{R}$ are given by

$$\underline{D}(\mu, x) = \liminf_{\varepsilon \to 0} \frac{\mu(B(x,\varepsilon))}{(2\varepsilon)^d} \text{ and } \overline{D}(\mu, x) = \limsup_{\varepsilon \to 0} \frac{\mu(B(x,\varepsilon))}{(2\varepsilon)^d}.$$

If $\overline{D} = \underline{D}$ at x then we call the common value the *density* of μ at x.

For $d = 1$ and $\mu = \lambda$, Lebesgue measure, this is the Lebesgue density:

Theorem 4.2 (The Lebesgue Density Theorem) *If $E \subset \mathbf{R}$ is a Borel set with $\lambda(E) > 0$ then the Lebesgue density*

$$\lim_{\varepsilon \to 0} \frac{\lambda(B(x,\varepsilon) \cap \varepsilon)}{2\varepsilon}$$

exists λ a.e. and is equal to one for λ a.e. $x \in E$. □

If one takes a fractal set with μ being Hausdorff measure on the set, the density does not exist (see the discussion in Falconer's book [Fal] chapter 2). One can easily see this for the Hausdorff measure on a cookie-cutter Cantor set because the large gaps in the Cantor set cause large fluctuations in the ratio $\mu(B(x,\varepsilon))/(2\varepsilon)^d$ as $\varepsilon \to 0$. One could try applying an averaging method to

$$\frac{\mu(B(x,\varepsilon))}{(2\varepsilon)^d}$$

to smooth out these fluctuations. Applying an order-two averaging method gives the following definition.

Definition 4.3 The *(d-dimensional) order-two density* of μ at x is

$$D_2(x) = \lim_{T \to \infty} \frac{1}{T} \int_0^T \frac{\mu(B(x, e^{-t}))}{(2e^{-t})^d} dt$$

if the limit exists.

Order two density should be regarded as a measurement of the lacunarity of the measure μ. Using this definition Bedford and Fisher [BF] have the following generalization of the Lebesgue density theorem for cookie-cutters.

Theorem 4.3 *For a cookie-cutter C of dimension d let μ be d-dimensional Hausdorff measure restricted to C. Then the (d-dimensional) order-two density of μ exists μ a.e. and is constant a.s. (μ).* □

Unlike for the Lebesgue density theorem we don't know what the almost sure value is. The result is relatively easy to prove for the middle-third Cantor set. In this situation one clearly sees one of the main features of the proof, namely the conformal transformation property of Hausdorff measure. This property can be stated for linear transformations as follows:

If $r > 0$ and $E \subset \mathbf{R}^n$ is a Borel set then

$$HM^d(rE) = r^d\, HM^d(E).$$

(The analogue of this property for a general C^1 diffeomorphism $S : \mathbf{R} \to \mathbf{R}$ is that if $\mu = HM^d|_E$ then

$$HM^d(S(E)) = \int_E |DS(x)|^d\, d\mu(x). \tag{4.3}$$

With the help of this expression one can extend the proof we give below to the more general case of a non-linear cookie-cutter). We need to use one more concept–that of ergodicity. Above we said that a measure ν is invariant under a transformation T if for all measurable sets B,

$$\nu(T^{-1}B) = \nu(B).$$

Ergodicity says that only "trivial" sets are invariant.

Definition 4.4 An invariant Borel probability measure ν for a transformation T is *ergodic* if

$$T^{-1}B = B \text{ implies } \nu(B) = 0 \text{ or } 1, \text{ for any Borel set } B.$$

Ergodicity is a type of indecomposability condition on the measure: one can easily show that ergodic measures form the convex hull of the set of invariant measures. For flows (or semiflows) Φ we say that a measure is ergodic if it is ergodic for all the discrete time maps Φ_s.

Ergodicity is important because of the Birkhoff ergodic theorem ([Wa3], Theorem 1.14) which says that if $T\colon X \to X$ is a continuous map of a compact metric space, $f\colon X \to \mathbf{R}$ a continuous map, and ν an ergodic measure, then

$$\frac{1}{n}\sum_{i=0}^{n-1} f(T^i x) \to \int_X f\, d\nu \text{ as } n \to \infty$$

for ν a.a. $x \in X$. There is a similar statement for flows (or semi flows): if $\Phi_s\colon X \to X$ is a continuous flow with f and ν as above then

$$\frac{1}{T}\int_0^T f(\Phi_s(x))\, ds \to \int_X f d\nu$$

for ν a.a. $x \in X$.

Almost every measure that we have mentioned so far is ergodic. For example the Gibbs measures discussed in part 2 are ergodic under the shift. If ν is ergodic under $S: C \to C$ then the measure on M that is locally the product $\frac{\nu \times \lambda}{\int g d\mu}$ (introduced after Definition 4.1 above) is ergodic with respect to the semi-flow Φ.

After this brief excursion into ergodic theory let's return to the task of proving Theorem 4.3 for the middle-third set C. Consider the function

$$f:\ C \times (0, \infty) \to \mathbf{R}$$

given by

$$f(x,t) = \frac{\mu(B(x, e^{-t-\log 3}))}{(2e^{-t-\log 3})^d}$$

where μ is the measure giving

$$\mu(I_{x_0 \ldots x_{n-1}}) = 2^{-(n-1)} \text{ for all } x_0 \ldots x_{n-1}.$$

Recall that μ is exactly the Hausdorff measure on C, and that to calculate the order-two density of μ we just have to consider

$$\lim_{T \to \infty} \frac{1}{T} \int_0^T f(x,t)\, dt$$

at "typical" values of x (a little thought and an easy lemma that f is a bounded function shows that the above limit will be the same as that required for $D_2(x)$, if the limit exists).

Now take $x \in C$ and $e^{-t} \leq \frac{1}{3}$. Since the Cantor set is invariant under the map $S(x) = 3x$ mod 1, the structure of the set around $S(x)$ is a scaled up version of what one sees around x,

$$S(B(x, e^{-t}) \cap C) = B(S(x), e^{-t+\log 3}) \cap C.$$

We know from the conformal transformation property of μ that

$$\begin{aligned} &\mu(B(S(x), e^{-t+\log 3})) \\ &= \mu(3 \cdot B(x, e^{-t})) \\ &= 3^d \mu(B(x, e^{-t})). \end{aligned}$$

Hence

$$\begin{aligned} f(x,t) &= \frac{\mu(B(x, e^{-t-\log 3}))}{(2e^{-t-\log 3})^d} \\ &= \frac{\mu(B(S(x), e^{-t}))}{(2e^{-t})^d} \\ &= f(S(x), t - \log 3). \end{aligned}$$

This means that we can regard f as a function on the flowspace M, for M is here

$$M = \{(x,t)\colon\ x \in C, t \in [0, \log 3]\}/\equiv$$

where $\equiv$ is the identification

$$(x, \log 3) \equiv (Sx, 0).$$

Hence if we define $g\colon\ M \to \mathbf{R}$ by

$$g(x,t) = f(x, t - \log 3) t \in [0, \log 3)$$

then g is a continuous function since the identification $\equiv$ is respected. Furthermore the Cesaro time average of f is now the Cesaro flow-time average for g,

$$\frac{1}{T}\int_0^T f(x,t)dt = \frac{1}{T}\int_0^T g(\Phi_t(x,0))dt.$$

The Birkhoff ergodic theorem tells us that the limit of the last expression as $T \to \infty$ exists and is equal to $\int_M g d\mu$ for μ a.a. $x \in C$. This proves Theorem 4.3 for the middle-third Cantor set.

5 Cookie-cutters as multifractals: the spectrum of Hölder exponents

In this section we shall discuss a theory which has become known as "singularity spectrum" or "multifractal" analysis. In doing this we'll see in particular how it applies to the case of a cookie-cutter, and how it throws more light on the formula (4.2)

$$\dim_H C = \sup_\nu \frac{h_\nu(S)}{\int g d\nu}$$

that we derived from the flow representation of a cookie-cutter. We shall consider the coding $\pi\colon\ \Sigma \to C$ more carefully to see how it helps us calculate the dimension of subsets of C.

We can relate the dimension of a subset $V \subset \Sigma$ with the dimension of $\pi V \subset C$ by use of the following proposition.

Proposition 5.1 *Suppose X and Y are metric spaces and $p\colon\ X \to Y$ is bi-Hölder continuous of order α, i.e. for some $c_1, c_2 > 0$,*

$$c_1|x-y|^\alpha < |p(x) - p(y)| < c_2|x-y|^\alpha \text{ for all } x, y \in X$$

and p is invertible. Then

$$\dim_H X = \alpha \dim_H Y. \ \square$$

In principle it is easier to calculate dimension in Σ than in C (at least in terms of entropy) because the dynamics of the shift have a very simple effect on distances. Recall that the metric on Σ was defined to be

$$d(\underline{x}, \underline{y}) = e^{-m} \text{ where } m = \inf\{k: \ x_k \neq y_k\}.$$

It is easy to see that the shift map just expands distances by e. Now notice that if one takes the partition of Σ into $\alpha = \{\{\underline{x}|x_0 = 0\}, \{\underline{x}|x_0 = 1\}\}$ then

$$\alpha_\vee \ldots \vee \sigma^{-(n-1)}\alpha = \{c_n(\underline{x})\}$$

and furthermore each of the sets $c_n(\underline{x})$ has diameter e^{-n}. Checking back to the definition of topological entropy (Definition 2.1), we see that $h(\sigma, \alpha)$ is exactly equal to the box dimension of Σ. Rufus Bowen noticed the connections between entropy and dimension and gave a definition of topological entropy for non-compact sets [Bo1] which generalizes the usual notion of topological entropy, and which looks very much like Hausdorff dimension. With this definition one can check that for any σ-invariant Borel subset $E \subset \Sigma$,

$$\dim_H E = h(\sigma|_E).$$

In other words topological entropy and Hausdorff dimension coincide exactly in the space Σ. (Note that equality here depends on the choice of metric for Σ. If one changes the metric on Σ so that the diameter of $c_n(\underline{x})$ is k^{-n} then one has $\log k \dim_H E = h(\sigma|_E)$.) Bowen used his definition to relate the concepts of topological and measure theoretic entropy in a beautiful way. He showed that if G_ν is the set of generic points for an ergodic invariant measure ν,

$$G_\nu = \{\underline{x} \in \Sigma | \frac{1}{n} S_n g(\underline{x}) \to \int g \, d\nu \forall g \in C(\Sigma, \mathbf{R})\}$$

(recall that $\nu(G_\nu) = 1$ by the Birkhoff ergodic theorem), then

$$h(\sigma|_{G_\nu}) = h_\nu(\sigma).$$

The idea that we use in this section is to partition Σ into subsets on which the projection map $\pi: \ \Sigma \to C$ has different Hölder exponents, calculate the dimensions of these sets in Σ in terms of the entropy of some measure, and then to translate these expressions to the dimensions of subsets of C via Proposition 5.1. We first need a dynamical interpretation of the Hölder exponents of π.

For every $\underline{x} \in \Sigma$ and $n \geq 0$ we know that

$$c_n(\underline{x}) \xrightarrow{\pi} I_{x_0 \ldots x_{n-1}}$$

in a 1-1 fashion. Hence a set of diameter e^{-n} gets mapped into a set of diameter $|I_{x_0 \ldots x_{n-1}}|$.

Definition 5.1 The *local Hölder exponent* at $\underline{x} \in \Sigma$ is

$$\alpha(\underline{x}) = \liminf_{n \to \infty} \frac{1}{n} \log |I_{x_0 \ldots x_{n-1}}|.$$

(The idea of the local Hölder exponent $\alpha(\underline{x})$ is that for $\underline{y}$ close to $\underline{x}$ we have

$$|\pi\underline{x} - \pi\underline{y}| \approx d(\underline{x},\underline{y})^{\alpha(\underline{x})}).$$

Now by Corollary 1.3 we know that $|I_{x_0\dots x_{n-1}}| \approx |DS^n(x)|$ for $x = \pi\underline{x}$. Hence

$$\begin{aligned}\alpha(\underline{x}) &= \liminf_{n\to\infty} \frac{\log|DS^n(x)|}{n} \\ &= \liminf_{n\to\infty} \frac{1}{n}\sum_{i-0}^{n-1} \log|DS(S^i x)| \\ &= \liminf_{n\to\infty} \frac{1}{n} S_n g(\underline{x})\end{aligned}$$

where $g(\underline{x}) = \log|DS(\pi\underline{x})|$. This kind of limit is just the kind of expression that the Birkhoff ergodic theorem tells us about: it says that if ν is a σ-invariant ergodic measure then

$$\frac{1}{n}S_n g(\underline{x}) \to \int g d\nu$$

for ν a.a. $\underline{x} \in \Sigma$. However we need information about *all* points $\underline{x} \in \Sigma$, not just about ν almost all. To say something about this we need some of the theory of Gibbs states that comes from [Ru1] (though it is presented here in a much simplified form). Our presentation is loosely based on a paper of Rand and Bohr [RB] and the papers of Bedford [Be3] [Be4]. Although we are taking $g(\underline{x}) = \log|DS(\pi\underline{x})|$ most of what follows is directly applicable to general real-valued Hölder continuous maps of Σ.

We let $G_q = \{\underline{x} \in \Sigma | \frac{1}{n}S_n g(\underline{x}) \to q\}$ and $F_q = \{\underline{x} \in \Sigma | q \text{ is a limit point of } \frac{1}{n}S_n g(\underline{x})\}$. A rough box dimension-like estimate of the size of these sets can be made as follows. Given an open neighbourhood Q of q let

$$N_n(Q) = \#\{c_n(\underline{x}) | \exists \underline{y} \in c_n(\underline{x}), \frac{1}{n}S_n g(\underline{y}) \in Q\}.$$

The sequence $\log N_n(Q)$ is not quite subadditive, but is near enough to being subadditive that one can show that the limit

$$\lim_{n\to\infty} \frac{1}{n} \log N_n(Q)$$

exists. We call this limit $\mathcal{S}(Q)$ and define

$$\mathcal{S}(q) = \lim_{Q\downarrow q} \mathcal{S}(Q).$$

It turns out that $\mathcal{S}(q)$ is a continuous concave and even analytic function of q. When $g(\underline{x}) = \log|DS(\pi\underline{x})|$ we can interpret $\mathcal{S}(q)$ geometrically by:

There are roughly $e^{n\mathcal{S}(q)}$ intervals of the form $I_{x_0\dots x_{n-1}}$ with length approximately e^{-nq}.

Before studying $\mathcal{S}(q)$ further we check what its domain should be.

Define
$$R = \{q \in \mathbf{R} | F_q \neq \emptyset\}$$
and
$$T = \{q \in \mathbf{R} | G_q \neq \emptyset.$$
With the use of the so-called "shadowing" property of the shift map [Bo2] one can check the following Lemma.

Lemma 5.2 *R is a closed interval.* □

It turns out that $\mathcal{S}$ is most naturally defined on int R and that $T \cap$ int $R =$ int R. We remark though that there is a degenerate case in which R can be a single point. An example in which this occurs is the middle-third set. Here every interval scales like 3^{-n} and so $R = \{\log 3\}$.

The simplest non-trivial example is the two scale Cantor set discussed in part 4. The exact condition needed to avoid the degenerate case is that
$$g \neq c + h - h o \sigma$$
where $c \in \mathbf{R}$ and $h: \Sigma \to \mathbf{R}$ is a continuous function - compare this to equation (2.5). From now on we assume that we are in the non-degenerate case (the other case is easy to analyze along similar lines).

The way we proceed is to associate with every point $q \in$ int R an ergodic invariant measure which somehow best characterizes the set G_q. One convenient way to find candidate measures is via Theorem 2.2. In particular for the Gibbs state ν_g for $g: \Sigma \to \mathbf{R}$ one has, by the ergodic theorem that
$$\frac{1}{n} S_n g(\underline{x}) \to \int g \, d\nu_g \text{ for } \nu_g \text{ a.a. } \underline{x} \in \Sigma.$$

This suggests that ν_g characterizes (in some sense - we shall make this precise later) the set G_q with $q = \int g \, d\nu_g$. Now consider the family of measures $\nu_p := \nu_{pg}$ for $p \in \mathbf{R}$. Each measure is ergodic, so
$$\frac{1}{n} S_n g(\underline{x}) \to \int g \, d\nu_p \text{ for } \nu_p \text{a.a. } \underline{x} \in \Sigma.$$
We define a map $\Psi: \mathbf{R} \to R$ by
$$\Psi(p) = \int g \, d\nu_p.$$

This gives us a way of associating measures ν_p with sets $G_{\Psi(p)}$. Two remarkable facts about this map Ψ follow from the work of Ruelle [Ru1]:

(a) If $P(p) := P(pg)$ is the pressure of pg then $P(p)$ is a strictly convex analytic function of p and $\frac{dP}{dp} = \Psi$.

(b) The function Ψ maps $\mathbf{R}$ onto int R in an invertible and bi-analytic manner.

We can now easily establish a connection between P and $\mathcal{S}$.

Theorem 5.3 *For $p \in \mathbf{R}$ let $q = \Psi(p)$, then*

$$P(p) = \mathcal{S}(q) + pq.$$

In particular $\mathcal{S}$: int $R \to \mathbf{R}$ is real analytic and strictly concave.

Proof. Take an open ball Q around q. We know that

$$\frac{1}{n} S_n g(\underline{x}) \to q \text{ for } \nu_p \text{ a.a. } \underline{x} \in \Sigma.$$

Hence if D_n is the union of those $c_n(\underline{x})$ for which

$$\frac{1}{n} S_n g(\underline{y}) \in Q \text{ for some } \underline{y} \in c_n(\underline{x}).$$

then $\nu_p(D_n) \to 1$ as $n \to \infty$, and by definition of N_n, D_n contains $N_n(Q)$ sets. Now by Theorem 2.2,

$$\zeta < \frac{\nu_p(c_n(\underline{x}))}{\exp(-nP(p) + pS_n g(\underline{x}))} < \zeta^{-1} \text{ for all } n, \underline{x}.$$

In particular for $c_n(\underline{x})$ in D_n we have

$$\zeta < \frac{\nu_p(c_n(\underline{x}))}{\exp(-nP(p) + pnq)} < \zeta^{-1} \text{ for some } q \in Q.$$

This implies that

$$\zeta . N_n(Q) \inf_{q \in Q} \exp(-nP(p) + npq) < \nu_p(D_n) < \zeta^{-1} N_n(Q) \sup_{q \in Q} \exp(-nP(p) + npq)$$

so that taking the limit $n \to \infty$,

$$\mathcal{S}(Q) = \lim_{n \to \infty} \frac{1}{n} \log N_n(Q) \in \left[\inf_{q \in Q} (P(p) - pq), \sup_{q \in Q} (P(p) - pq) \right].$$

Taking the limit $Q \downarrow q$ gives us now

$$\mathcal{S}(q) = P(p) - pq$$

as claimed. □

We can now show that $\mathcal{S}(q)$ is indeed the dimension of G_q.

Theorem 5.4 $\mathcal{S}(q) = \dim_H G_q = \dim_H F_q$.

Proof. We first show $\dim_H F_q \le \mathcal{S}(q)$. It is enough to show that for a neighbourhood Q of q, $\dim_H F_q \le \mathcal{S}(Q)$. This we show by a direct covering argument.

If $\underline{x} \in F_q$ then infinitely often one has

$$\frac{1}{n} S_n g(\underline{x}) \in Q.$$

We can therefore cover F_q by

$$\{c_n(\underline{x}) \colon \; n > N, \exists \underline{y} \in c_n(\underline{x}) \text{ with } \frac{1}{n} S_n g(\underline{y}) \in Q\}$$

for any $N > 0$. For any n this cover contains $N_n(Q)$ n-cylinders $c_n(\underline{x})$. The estimate of $(\mathcal{S}(Q) + \delta)$-dimensional Hausdorff measure that this gives is

$$\sum_{n>N} e^{-n(\mathcal{S}(Q)+\delta)} N_n(Q).$$

Since $\frac{1}{n} \log N_n(Q) \to \mathcal{S}(Q)$ the above quantity is bounded as $N \to \infty$. This shows that $\dim_H F_q \le \mathcal{S}(Q) + \delta$. Letting $\delta \to O$ and $Q \downarrow q$ now gives $\dim_H F_q \le \mathcal{S}(q)$.

Since $G_q \subset F_q$ is it immediate that $\dim_H G_q \le \dim_H F_q$.

Finally we show that $\mathcal{S}(q) \le \dim_H G_q$. Comparing Theorem 5.3 with the variational principle for pressure we have

$$P(pg) = \mathcal{S}(q) + \int pg d\nu_p \text{ and } P(pg) = \sup_{\nu}\{h_\nu(\sigma) + \int pg d\nu\}.$$

This shows that $\mathcal{S}(q) = h_{\nu_p}$. Now notice that the set of generic points for $\nu_p, G_{\nu_p} \subset G_q$, so we have

$$\begin{aligned} h_{\nu_p}(\sigma) &= h(\sigma|G_{\nu_p}) \text{ by [Bo1]} \\ &\le h(\sigma|G_q) \text{ since } G_{\nu_p} \subset G_q \\ &= \dim_H(G_q) \end{aligned}$$

since topological entropy and dimension coincide here. This gives $\mathcal{S}(q) \le \dim_H G_q$ and completes the proof. □

We can summarize the most important parts of the above discussion as follows:

(a) $\mathcal{S}(q)$ is the dimension of $\{\underline{x} \in \Sigma | \frac{1}{n} S_n g(\underline{x}) \to q\}$ (Theorem 5.4).

(b) The Gibbs measure ν_p for pg is the measure of maximal dimension on $\{\underline{x} \in \Sigma | \frac{1}{n} S_n g(\underline{x}) \to q\}$ (this is a re-interpretation of the variational principle).

(c) $\mathcal{S}(q)$ is a concave real analytic function defined on int $\{q|\frac{1}{n}S_n g(\underline{x})$ has q as a limit point$\}$.

The above statements are true for general Hölder continuous functions $g\colon \Sigma \to \mathbf{R}$. One can generalize the theory to more general functions $g\colon \Sigma \to \mathbf{R}^k$, and by adapting it slightly can derive the $f(\alpha)$ formalism (see Rand [Ra]).

We now see how they can be applied to the situation of $g(\underline{x}) = \log|DS(\underline{x})|$. Recall from the discussion at the beginning of this section that

$$G_q = \{\underline{x} \in \Sigma | \frac{1}{n} S_n g(\underline{x}) \to q\}$$

is exactly the set where $\pi\colon \Sigma \to C$ has Hölder exponent q. By Theorem 5.4 together with Proposition 5.1,

$$\dim_H \{x \in C | \frac{1}{n} S_n g(\pi^{-1}x) \to q\} = \frac{\mathcal{S}(q)}{q}$$

so that one immediately has

$$\dim_H C \geq \sup_{q \in \mathbf{R}} \frac{\mathcal{S}(q)}{q}.$$

Working only a little bit harder with the sets F_q in place of G_q one can show that

$$\dim_H C = \sup_{q \in \mathbf{R}} \frac{\mathcal{S}(q)}{q}$$

(see [Be3] or [Be4] for the idea). This expression is really the same as we had before because firstly $\mathcal{S}(q) = h_{\nu_p}$ and $q = \int g d\nu_p$ so that

$$\dim_H C = \sup_{p \in \mathbf{R}} \frac{h_{\nu_p}}{\int g \, d\nu_p}.$$

Secondly, for each q, ν_p is the unique invariant measure of maximal entropy amongst those measures with $\int g d\nu_p = q$. Hence

$$\dim_H C = \sup_{\nu} \frac{h_\nu}{\int g d\nu}$$

which is the expression we had derived in part 4.

The function $\mathcal{S}(q)$ contains a great deal of geometric information about the way in which the cookie-cutter Cantor set C is built up. It contains the Hausdorff dimension, and information about the frequency all the asymptotic length scales in the Cantor intervals (i.e. how many intervals $I_{x_0 \ldots x_{n-1}}$ there are of length approximately e^{-nq}). Furthermore $\mathcal{S}$ is a fractal invariant if two cookie-cutters C and C' are Lipschitz conjugate then they have the same $\mathcal{S}(q)$ function. A natural question is therefore:

Does $\mathcal{S}(q)$ determine the fractal geometry of C?

In other words if two cookie-cutters C and C' have the same $\mathcal{S}(q)$ functions, is C Lipschitz conjugate to C'? This is *not* true. In fact there are C^∞ cookie-cutters arbitrarily close to each other (in the C^∞ topology) which are not Lipschitz conjugate but which have the same $\mathcal{S}(q)$. This shows that $\mathcal{S}(q)$ is not such a good fractal invariant after all!

6 Sullivan's classification of cookie-cutters

In this part we shall present some recent results of Sullivan ([Su1], [Su2]) on the differentiable structure of cookie-cutter Cantor sets. These results give

(i) a classification of $C^{1+\gamma}$ cookie-cutters up to $C^{1+\gamma}$ conjugacy by the so-called *scale function* on a dual Cantor set ([Su1]),

(ii) a classification of $C^{1+\gamma}$ cookie-cutters up to Lipschitz conjugacy in terms of dimension and a Gibbs measure ([Su2]),

(iii) a classification of *non-linear* C^k cookie-cutters up to C^{k-1} conjugacy by a Gibbs measure ([Su2]).

The first result (i) was used by Sullivan to show that if two quadratic like maps of the interval have the same renormalization limit under the Feigenbaum renormalization operator, then their invariant Cantor sets are $C^{1+\gamma}$ conjugate.

Results (ii) and (iii) are remarkable because they show that for non-linear cookie-cutter Cantor sets the differential geometry of the sets is determined by their fractal (or Lipschitz) geometry. It is even more surprising that (iii) only holds for non-linear cookie-cutters and is actually false for linear (or piecewise linear) cookie-cutters.

We begin with a discussion of (i).

Definition 6.1 For a cookie-cutter C and a sequence $x_0 \ldots x_{n-1}$, the *ratio geometry* of $x_0 \ldots x_{n-1}$ is the triple

$$\left(\frac{|I_{x_0\ldots x_{n-1}0}|}{|I_{x_0\ldots x_{n-1}}|}, \frac{|I_{x_0\ldots x_{n-1}1}|}{|I_{x_0\ldots x_{n-1}}|}, 1 - \frac{|I_{x_0\ldots x_{n-1}0}| + |I_{x_0\ldots x_{n-1}1}|}{|I_{x_0\ldots x_{n-1}}|}\right).$$

By Corollaries 1.3 and 1.4 these numbers are uniformly bounded away from zero. Sullivan calls this property *bounded geometry*. It is clear that if one knows the ratio geometry for all $x_0 \ldots x_{n-1}$ then the Cantor set can be reconstructed exactly. It turns out however that to reconstruct the Cantor set up to $C^{1+\gamma}$ conjugacy (i.e. so that the reconstructed set is $C^{1+\gamma}$ conjugate to the original) one only needs asymptotic information about the ratio geometries.

To see why asymptotic (i.e. as $n \to \infty$) ratio geometries are invariants of differentiable conjugacy consider the effect of a C^1 diffeomorphism φ on the ratio geometries around a point $x \in C$. Very close to x, the map φ is almost equal to its derivative $D\varphi$, so that the lenghts of intervals $I_{x_0 \dots x_{n-1}}$ close to x are (approximately) multiplied by $|D\varphi(x)|$ when transformed by φ. This means that the ratio's of such lengths do not change (in the limit as $n \to \infty$).

Unfortunately, for a given $x \in C$, there is no well-defined asymptotic ratio geometry at x. To see this consider two contractions φ_0, φ_1 generating C as the attractor of an iterated function system (as discussed in part 1). The local structure of Cantor intervals around x at level n, $I_{x_0 \dots x_{n-1}}, I_{x_0 \dots x_{n-1}0}, I_{x_0 \dots x_{n-1}1}$ is the image of the local structure at level 0, I, I_0, I_1, under the composition $\varphi_{x_0} \cdots \varphi_{x_{n-1}}$ (see Figure 6.1). Note the order in which the maps are applied. $\varphi_{x_{n-1}}$ is applied first and so is the largest contributor to the non-linear distortion of the ratio geometry. On the other hand φ_{x_0} is applied last and has almost no effect on the ratio geometry as the interval $I_{x_1 \dots x_{n-1}}$ is already small so that φ_{x_0} acts almost linearly. Now, as $n \to \infty$, the map $\varphi_{x_{n-1}}$ keeps changing so that the ratio geometry does not converge as $n \to \infty$.

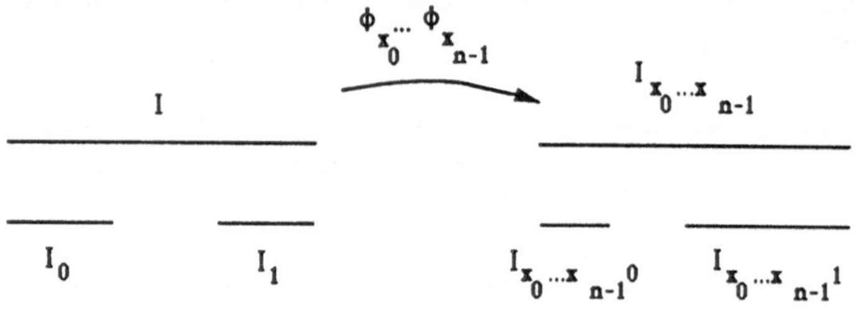

Figure 6.1: How the ratio geometry is transformed.

If we compose maps in the opposite direction then there is convergence, but now the sequence of ratio geometries that are converging nicely do not correspond to any point of the Cantor set. To indicate this fact, Sullivan defines a "dual Cantor set" or symbol space Σ^*, in which sequences are written to the right instead of to the left.

Definition 6.2 The *dual Cantor set* is

$$\Sigma^* = \{(\dots, x_n, \dots, x_1, x_0) \colon \; x_i = 0, 1\}.$$

The *scaling function* for a cookie-cutter Cantor set C is the map

$$\sigma_s \colon \; \Sigma^* \to \{(q_0, q_1, q_2) \colon \; q_i \geq 0, \sum_{i=0}^{2} q_i = 1\}$$

given by

$$\sigma_s(\dots x_{n-1} \dots x_0) =$$

$$= \lim_{n\to\infty} \left(\frac{|I_{x_{n-1}\dots x_0 0}|}{|I_{x_{n-1}\dots x_0}|}, \frac{|I_{x_{n-1}\dots x_0 1}|}{|I_{x_{n-1}\dots x_0}|}, 1 - \frac{|I_{x_{n-1}\dots x_0 0}| + |I_{x_{n-1}\dots x_0 1}|}{|I_{x_{n-1}\dots x_0}|} \right)$$

Sullivan [Sul] proves the following result.

Theorem 6.1 *The scaling function is a complete invariant of $C^{1+\gamma}$ conjugacy for cookie-cutters.* □

The other two results of Sullivan lie closer to the main themes of these notes. His results can be (incompletely) stated as follows.

Theorem 6.2 *Let ν be a Gibbs measure on Σ, and $d \in (0,1)$. Then there is a $C^{1+\gamma}$ cookie-cutter map S such that*

(i) *$\pi_*\nu$ is (up to a constant multiple) the d-dimensional Hausdorff measure on X, and in particular $d = \dim_H C$,*

(ii) *the pair d,ν determines the Lipschitz conjugacy class of C (i.e. if $\tilde{C}$ has dimension d and Hausdorff measure absolutely continuous to $\pi_*\nu$ then C and $\tilde{C}$ are Lipschitz conjugate),*

(iii) *if, furthermore, S is of class $C^k (k \geq 2)$ and $D^2S \neq 0$ at some point of C, then the C^{k-1} conjugacy class of C is determined by ν.* □

To explain the techniques used in the proof of this theorem (Sullivan doesn't give a proof himself, but sketches the argument), we have to go back to the discussion of Ruelle-Perron-Frobenius theory that we had in part 2, and in particular back to equations (2.3) and (2.4). We used the construction of a Gibbs measure for the function df where $f(\underline{x}) = -\log|DS(\pi\underline{x})|$ and $P(df) = 0$. Equations (2.3) and (2.4) then read

$$L_{df}h = h \tag{6.1}$$

and

$$L^*_{df}\mu = \mu \tag{6.2}$$

(recall that $\log\lambda = P(df)$).

It turns out that $\pi_*\mu$ is (up to a constant multiple) equal to the d-dimensional Hausdorff measure. One sees this as follows. Firstly the two measures are absolutely continuous with respect to each other (they are both equivalent to $\pi_*\nu$). Secondly they both satisfy the conformal transformation property (4.3) which then implies that the Radon-Nikodyn derivative of one measure with respect to the other is $\pi_*\nu$ almost everywhere S-invariant (one should formally use the Jacobian derivatives of the measures which will be defined soon). Finally, by the Ergodic Theorem, any S-invariant function is constant $\pi_*\nu$ almost everywhere.

In general the two measures μ and ν are different, but for each d and ν one can construct a "canonical" cookie-cutter in which $\mu = \nu$ (or equivalently $h \equiv 1$), in other words for which the d-dimensional Hausdorff measure is invariant under S. An example of this is the middle-third set. The general construction of the "canonical" cookie-cutter is quite straight forward (one just has to be careful in showing that the corresponding map S is differentiable of class $C^{1+\gamma}$, for some $\gamma > 0$). To construct the Cantor set one defines the Cantor intervals inductively by setting

$$\ell_{x_0 \dots x_{n-1}} = \nu(c_n(\underline{x}))^d \text{ for any } \underline{x} \in \Sigma, n > 0$$

and then defining

$$I = [0,1], I_0 = [0, \ell_0], I_1 = [1 - \ell_1, 1]$$

and in general $I_{x_0 \dots x_{n-1} 0}$ and $I_{x_0 \dots x_{n-1} 1}$ to be the intervals of length

$$\ell_{x_0 \dots x_{n-1} 0} \text{ and } \ell_{x_0 \dots x_{n-1} 1}$$

which are, respectively, the left- and right-most subintervals of $I_{x_0 \dots x_{n-1}}$ (see Figure 6.2).

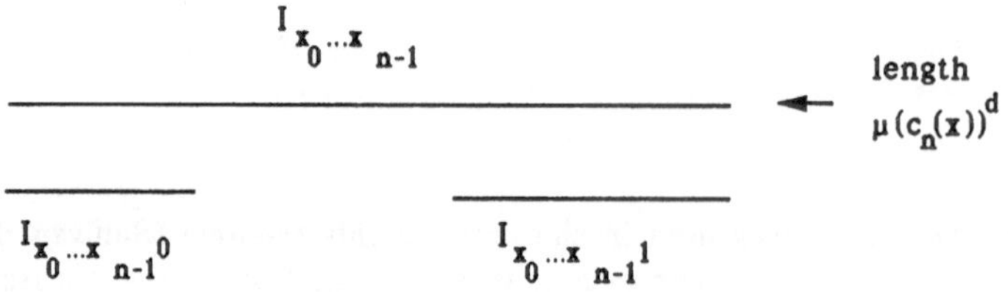

Figure 6.2: Constructing the canonical cookie-cutter Cantor set.

With this construction one can prove part (i) of Theorem 6.2.

Proving part (ii) of Theorem 6.2 is also relatively straight forward. Given two cookie-cutters $S_1\colon\ C_1 \to C_1$ and $S_2\colon\ C_2 \to C_2$ with the same ν and d we just have to show that the topological conjugacy $\phi\colon\ C_1 \to C_2$

$$\begin{array}{ccc} C_1 & \xrightarrow{S_1} & C_1 \\ \phi \downarrow & & \downarrow \phi \\ C_2 & \xrightarrow{S_2} & C_2 \end{array}$$

is a Lipschitz map. The following Lemma is easy to check.

Lemma 6.3 *ϕ is Lipschitz if and only if for some $\alpha > 0$,*

$$\alpha < \frac{|I^{(1)}_{x_0 \dots x_{n-1}}|}{|I^{(2)}_{x_0 \dots x_{n-1}}|} < \alpha^{-1}$$

for all $\underline{x} \in \Sigma$ and $n > 0$, where $I^{(i)}_{x_0 \dots x_{n-1}}$ is a Cantor interval of $C_i (i = 1, 2)$. □

Part (ii) of Theorem 6.2 now follows immediately out of Corollary 1.3 and Theorem 2.2 because they imply that for some $\zeta_1 > 0$,

$$\zeta_1 < \frac{|I^{(i)}_{x_0 \dots x_{n-1}}|}{\nu(c_n(\underline{x}))} < \zeta_1^{-1}$$

for any $\underline{x} \in \Sigma$, $n > 0$ and $i = 1, 2$.

The last part, (iii), of Theorem 6.2 is very surprising because it breaks two very general rules of dynamical systems theory:

(1) Differentiable conjugacy classes are usually hard to characterize.

(2) Many results are proven first for "linear" systems and then with various degrees of difficulty are extended to "non-linear" systems.

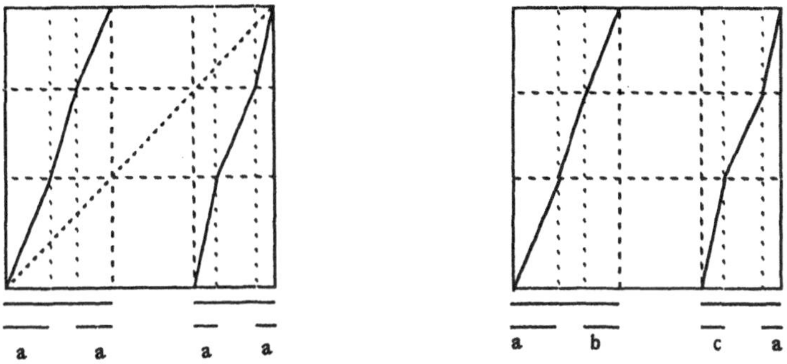

Figure 6.3:
Two cookie cutters that are Lipschitz but not differentiably conjugate. The slopes on second level Cantor intervals are shown and b and c are chosen with $bc = a^2$.

An example of two piecewise-linear cookie-cutters that are Lipschitz conjugate but not differentiably conjugate is shown in Figure 6.3. One can easily check that the conjugacy ϕ satisfies Lemma 6.3, (and so must be Lipschitz) but is not differentiable at the point with code 010101 ....

To see why part (iii) of Theorem 6.2 holds (we only sketch part of the proof here) consider the conjugacy equation

$$S_1\phi = \phi S_2. \tag{6.3}$$

Ideally one would like to show that ϕ is differentiable by expressing it as a composition of other differentiable maps. Unfortunately we cannot manipulate (6.3) to get ϕ on its own. In order to obtain another equation for ϕ we need a new tool.

Definition 6.3 A Borel measure m on C is *non-singular* with respect to S if for any Borel set A,

$$m(A) > 0 \text{ implies } m(S(A)) > 0.$$

All of the measures on C that we have considered are non-singular. We use the following concept which gives us a way to describe how much a measure is locally expanded by S.

Definition 6.4 The *Jacobian derivative* of a non-singular Borel measure m under S is the Radon-Nikodyn derivative $\frac{dS_*m}{dm}$.

When there is no confusion about the measure, we shall denote this Jacobian by J_S. Intuitively

$$J_S(x) = \lim_{A \downarrow x} \frac{m(SA)}{m(A)}$$

(see Figure 6.4).

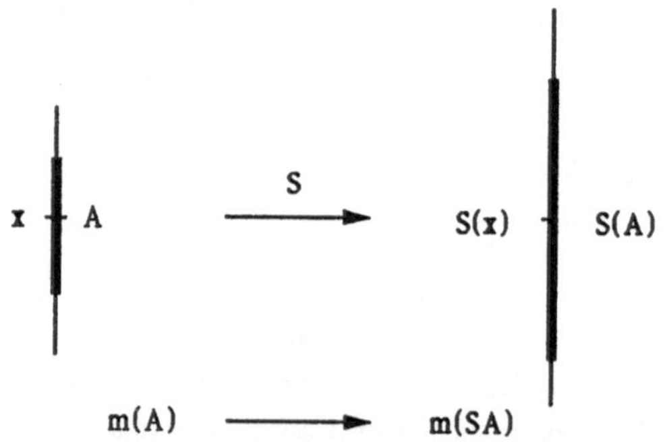

Figure 6.4: An invariant measure is expanded under forward iteration of S.

See [PW] for more information about Jacobian derivatives.

Jacobian derivatives work like ordinary derivatives but are only useful for studying the properties of non-invertible maps, for if m is an invariant measure for an invertible map T then the Jacobian derivative $J_T = \frac{dTm}{dm}$ is constant and equal to 1 almost surely.

If we differentiate the conjugacy relation $S_1\phi = \phi S_2$ in the sense of Jacobian derivative with respect to the measure $\pi_*\nu$ we get

$$J_{S_1} \circ \phi + J_\phi = J_\phi \circ S_2 + J_{S_2}$$

by the chain rule. Since ϕ is invertible, $J_\phi \equiv 1$ a.s. so that

$$J_{S_1} \circ \phi = J_{S_2}.$$

From this equation one has

$$\phi = J_{S_1}^{-1} \circ J_{S_2}$$

if J_{S_1} is invertible. If *furthermore* $J_{S_1}^{-1}$ and J_{S_2} are differentiable then ϕ is differentiable. This is precisely where the proof breaks down for piecewise linear cookie-cutters, for there J_{S_1} is constant on sets of the form $C \cap I_{x_0 \dots x_{n-1}}$, and hence nowhere invertible. Let us suppose now that S_1 is the canonical cookie-cutter for which $\nu = \mu$, i.e. for which Hausdorff measure is invariant. We know then precisely how $\pi_*\nu$ transforms under S because this is given by the conformal transformation property (4.3). Hence

$$J_{S_1}(x) = |DS_1(x)|^d.$$

Now the assumption that $D^2S_1 \neq 0$ somewhere on the Cantor set implies that J_{S_1} is invertible with differentiable inverse on some open subset of the Cantor set. One next has to show that J_{S_2} is also differentiable - this is nontrivial and requires consideration of the function h from (6.1) to show that it can be defined as a differentiable function on the Cantor set. The strategy of using the Jacobian derivative was used by Shub and Sullivan to study expanding maps of the circle [SS].

The last step in the proof is to show that ϕ is differentiable everywhere on C. We have seen that ϕ is differentiable on some open subset U of C. Now the equation $S_1\phi = \phi S_2$ implies that if ϕ is differentiable on U then ϕ is differentiable on $S_2(U)$ and so by induction on $S_2^n(U)$ for $n \geq 0$. By the expanding dynamics of S_2,

$$C \cap S_2^n(U) = C \text{ for some } n.$$

This then shows that ϕ is differentiable on the whole of C. Similar considerations enable one to show that ϕ has the higher derivations claimed in Theorem 6.2.

References

[Ba] M.F. Barnsley (1988), *Fractals Everywhere*, Academic Press, Boston.

[Be1] T. Bedford (1984), Ph.D Thesis, University of Warwick.

[Be2] T. Bedford (1986), Dimension and dynamics of fractal recurrent sets, *J. London Math. Soc. (2)* **33**, 89-100.

[Be3] T. Bedford (1988), Hausdorff dimension and box dimension in self-similar sets, *Proceedings of Topology and Measure V, Ernst-Moritz-Arndt Universität*, Greifswald.

[Be4] T. Bedford (1989), The box dimension of self-affine graphs and repellers, *Nonlinearity* **2**, 53-71.

[Be5] T. Bedford (1989), On Weierstrass-like functions and random recurrent sets, *Math. Proc. Camb. Philos. Soc.* **106**, 325-342.

[Bo1] R. Bowen (1973), Topological entropy for noncompact sets, *Trans. Amer. Math. Soc.* **184**, 125-136.

[Bo2] R. Bowen (1975), *Equilibrium States and the Ergodic Theory of Anosov Diffeomorphisms*, Lecture Notes in Mathematics 470, Springer, Berlin.

[Bo3] R. Bowen (1975), A horseshoe with positive measure, *Invent. Math.* **29**, 203-204.

[Bo4] R. Bowen (1979), Hausdorff dimension of quasi-circles, *Publications Mathématiques (I.H.E.S., Paris)* **50**, 11-25.

[BDEG] M.F. Barnsley, S. Demko, J. Elton & J. Geronimo (1988), Invariant measures for Markov processes arising from function iteration with place dependent probabilities, *Ann. Inst. H. Poincaré*.

[BR] T. Bohr and D.A. Rand (1986), The entropy function for characteristic exponents, *Physica* **25D**, 387-398.

[BU] T. Bedford and M. Urbański, The box and Hausdorff dimension of self-affine sets, *Ergodic Theory Dynamical Systems*, to appear.

[De1] F.M. Dekking (1982), Recurrent sets, *Advances in Math.* **44**, 78-104.

[De2] F.M. Dekking (1982), Recurrent sets: a fractal formalism, *Delft University of Technology Report of the Department of Mathematics* **82-32**.

[Fa1] K.J. Falconer (1985), *The Geometry of Fractal Sets*, Cambridge University Press,Cambridge.

[Fa2] K.J. Falconer (1988), A subadditive thermodynamic formalism for mixing repellers, *J. Phys.* **21A**, 737-742.

[Hu] J.E. Hutchinson (1981), Fractals and self-similarity, *Indiana Univ. Math. J.* **30**, 713-747.

[Ke] M. Keane (1972), Strongly mixing g-measures, *Invent.Math.* **16**, 309-324.

[Ma] B. Mandelbrot (1983), *The Fractal Geometry of Nature*, W.H. Freeman

[Mc] C. McMullen (1984), The Hausdorff dimension of general Sierpński carpets, *Nagoya Math J.* **96**, 1-9.

[PU] F. Przytycki and M. Urbański, On Hausdorff dimension of some fractal sets, *Studia Math.*, to appear.

[Ra] D.A. Rand (1989), The singularity spectrum (α) for cookie cutters, *Ergodic Theory Dynamical Systems* **9**, 527-541.

[RB] D. Rand and T. Bohr (1986), The entropy function for characteristic exponents, *Physica* **25D**, 387-398,

[Ru1] D. Ruelle (1978), Thermodynamic formalism: the mathematical structures of classical equilibrium statistical mechanics. *Encyclopedia of Mathematics and its Applications* 5, Addison-Wesley 1978.

[Ru2] D. Ruelle (1982), Repellers for real analytic maps, *Ergodic Theory Dynamical Systems* **2**, 99-107.

[St] J. Stirling, Methodus differentialis: sive tractus de summatione et interpolatione serierum infinitarum , London, 1730.

[Su1] D. Sullivan, Differentiable structures on fractal-like sets, determined by intrinsic scaling functions on dual Cantor sets, in Proceedings of the Herman Weyl Symposium, Duke University, *Proceedings of Symposia in Pure Mathematics* **48**.

[Su2] D. Sullivan (1986), Quasiconformal homeomorphisms in dynamics, topology, and geometry., *Proc. Int. Cong. Berkeley* **2**, 1216-1228.

[SS] M. Shub and D. Sullivan (1985), Expanding endomorphisms of the circle revisited, *Ergodic Theory Dynamical Systems* **5**, 285-289.

[Tr] C. Tricot (1982), Two definitions of fractional dimension, *Math. Proc. Cambridge Philos. Soc.* **91**, 57-74.

[Wa1] P. Walters (1973), Some results on the classification of measure preserving transformations, *Recent Advances in Topological Dynamics*, Lecture Notes in Mathematics 319, Springer, Berlin, 255-276.

[Wa2] P. Walters (1975), Ruelle's operator theorem and g-measures, *Trans. Amer. Math. Soc.* **214**, 375-387.

[Wa3] P. Walters (1982), *An Introduction to Ergodic Theory*, Graduate Texts in Mathematics 79, Springer, New York.

[Yo] L.S. Young (1982), Dimension, entropy and Lyapunov exponents, *Ergodic Theory Dynamical Systems* **2**, 109-124.

Complex dynamics: an informal discussion

Paul BLANCHARD
Amy CHIU
Department of Mathematics
Boston University
Boston, Massachusetts 02215, USA

Abstract

This paper is an informal introduction to the theory of complex-analytic dynamical systems. We survey the general theory, and then discuss two important classes of examples - quadratics and Newton's method. We also present the Douady-Hubbard theory of polynomial-like mappings.

Introduction

In the study of complex-analytic dynamical systems, simple functions spawn complicated systems. In turn, these systems are successfully analyzed using the powerful techniques of complex analysis. Due to this unusual combination of simplicity and complexity, many mathematicians and scientists are keenly interested in the subject. These notes are the result of discussions we have had with both groups.

There are a number of introductions to the subject for those with specialized backgrounds, and we do not intend to duplicate these efforts. Rather, we present an informal (and incomplete) introduction. Although we provide the basic definitions and fundamental results, we mainly focus on interesting and representative examples.

The notes are divided into four sections. The first contains basic definitions and elementary examples. In Section 2, we present a general view and suggest how complex analysis

Lecture notes based on lectures given by Paul Blanchard at the NATO ASI on Fractals at the Université de Montréal in the summer of 1989 and at the Regional Institute in Dynamical Systems at Boston University in the summer of 1990.

J. Bélair and S. Dubuc (eds.), Fractal Geometry and Analysis, 45–98.

provides handy tools that assist in the analysis. In Section 3, we describe a basic dichotomy within the family of quadratic polynomials. This division is the basis for the definition of the Mandelbrot set, a catalogue for the dynamics of quadratic polynomials. We also indicate how Douady and Hubbard use Riemann maps to describe the structure of many quadratic Julia sets as well as the Mandelbrot set. Finally, in Section 4, we describe the results of a computer experiment involving Newton's method. Not only are the results quite surprising, but they are a beautiful illustration of the Douady-Hubbard theory of polynomial-like maps.

We have tried to be reasonably precise, but we do not hesitate to skip details if they do not conveniently fit into our exposition. On the other hand, we have provided a large number of references so that the interested reader can pursue a more detailed study of the subject.

We would like to thank both the organizers of the NATO Conference on Fractals at the Université of Montréal in the summer of 1989 and those who organized the Regional Institute in Dynamics at Boston University in the summer of 1990 for their hospitality.

Section 1: Basic definitions and examples

We study dynamical systems that are generated by iterating a complex-valued function $f(z)$ of one complex variable. The function can be a polynomial such as $p(z) = z^2 - 1$, a rational function such as $R(z) = (2z^3+1)/(3z^2)$ (Newton's method for solving the equation $z^3 = 1$), or an entire transcendental function such as $\sin z$. Given a starting point z_0, we form its (forward) *orbit*, the sequence $\{z_n\}$, via repeated application of f

$$z_{n+1} = f(z_n),$$

and we describe the long-term behavior of this sequence.*

Let

$$\overline{\mathbf{C}} = \mathbf{C} \cup \{\infty\}$$

be the Riemann sphere. More precisely, we view the Riemann sphere as the two-sphere in $\mathbf{R}^3$

$$S^2 = \{(x_1, x_2, x_3) \mid x_1^2 + x_2^2 + x_3^2 = 1\}$$

equipped with an atlas of charts containing the chart $\phi_1 : \mathbf{C} \to S^2 - \{(0,0,1)\}$

$$\phi_1(z) = \phi_1(x_1 + ix_2) = \left(\frac{2x_1}{x_1^2 + x_2^2 + 1}, \frac{2x_2}{x_1^2 + x_2^2 + 1}, \frac{x_1^2 + x_2^2 - 1}{x_1^2 + x_2^2 + 1} \right)$$

*Due to the essential singularity at infinity, the theory for entire transcendental functions is somewhat different. In fact, a few of these differences are quite surprising. In these notes, we focus on the case where $f(z)$ is rational, but we recommend studying the transcendental case as well (see [DD] and [EL2]).

and a similar chart $\phi_2 : \mathbf{C} \to S^2 - \{(0,0,-1)\}$. These charts, determined by "stereographic projection", define $\overline{\mathbf{C}}$ as a Riemann surface. Throughout these notes, we refer to z when we really mean $\phi_1(z)$ and to ∞ when we mean $(0,0,1)$.

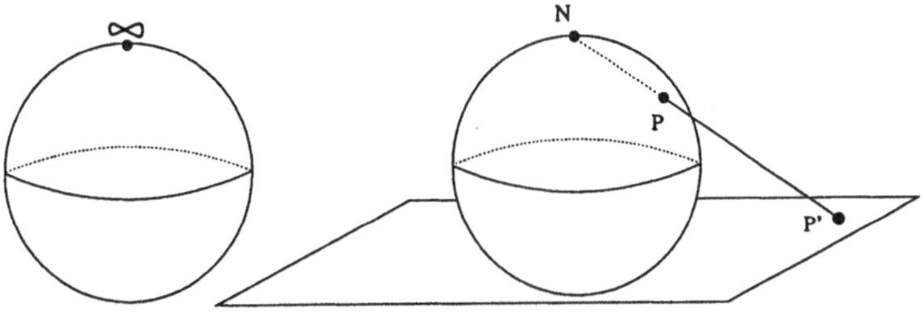

Figure 1.1: The Riemann Sphere $\overline{\mathbf{C}}$ and stereographic projection.

A complex-analytic map from $\overline{\mathbf{C}}$ to $\overline{\mathbf{C}}$ is a rational function

$$R(z) = \frac{p(z)}{q(z)}$$

where $p(z)$ and $q(z)$ are polynomials without common factors, and its degree is defined to be

$$\deg(R) = \max\{\deg p,\ \deg q\}.$$

Roughly speaking, the degree indicates how many times the map wraps the sphere around itself, and it equals the number (counted with multiplicity) of inverse images of any given point.

The dynamical behavior of Möbius transformations, rational functions whose degree is unity, is both well understood and more elementary than that of any other degree (see [A1]). In these notes, we focus on those rational maps R with $\deg R \geq 2$.

There are two ways to begin an exposition of the theory, and following Fatou [F1], we base our approach on Montel's theory [Mon] of normal families (see also [A1]).

Definition A family $\mathcal{F}$ of meromorphic functions defined on a domain $U \subset \overline{\mathbf{C}}$

$$\mathcal{F} = \{f_i\colon U \to \overline{\mathbf{C}} \mid f_i \text{ is meromorphic}\}$$

is a *normal family* if every sequence $\{f_n\} \subset \mathcal{F}$ has a subsequence $\{f_{n_i}\}$ that converges uniformly on every compact subset of U.

When a sequence of analytic functions converges uniformly on compact sets, it converges to an analytic function. Frequently, the limiting functions are constant functions, and it is important to remember that the function which is constantly equal to ∞ is an allowable limit.

1.2 Remark It is both useful and informative to reinterpret the above definition in terms of equicontinuous families. Recall that a family of functions $\mathcal{F}$ from a metric space X to a metric space Y is equicontinuous if, for each $\epsilon > 0$, there exists a $\delta > 0$ such that $d(x_1, x_2) < \delta$ implies that $d(f(x_1), f(x_2)) < \epsilon$ for *all* $f \in \mathcal{F}$. For our purposes, the spherical metric on $\overline{\mathbf{C}}$ is the most appropriate one. That is, consider the metric induced on S^2 by the Euclidean metric on $\mathbf{R}^3$, and interpret it as a metric on $\overline{\mathbf{C}}$. Then, as points approach ∞ in any direction, the distances between them approach zero. When one uses the spherical metric on $\overline{\mathbf{C}}$, then $\mathcal{F}$ is normal if and only if $\mathcal{F}$ is equicontinuous on all compact subsets of U.

When we use the theory of normality, we consider the family $\mathcal{F}$ of iterates

$$\{R^n \mid n = 1, 2, 3, \dots\}$$

of the rational map. In this case, equicontinuity implies that nearby starting points do not diverge under iteration (measured using the spherical metric).

Fatou [F1] and Julia [J] started their investigations with the realization that the family of iterates $\{R^n\}$ could not possibly be a normal family unless the domain of R was restricted to a subset of $\overline{\mathbf{C}}$. In fact, if one considers the question of normality for the family of iterates

$$\{R^n : U \to \overline{\mathbf{C}}\},$$

the choice of U plays a crucial role.

Definitions A point $z \in \overline{\mathbf{C}}$ belongs to the *Fatou set* F (or the *Domain of Normality** if there exists a neighborhood U of z such that the family of iterates

$$\mathcal{F} = \{R^n : U \to \overline{\mathbf{C}} \mid n = 1, 2, \dots\}$$

is a normal family on U. The *Julia set* J is the complement of F, i.e.,

$$J = \overline{\mathbf{C}} - F.$$

Throughout these notes, J or J_R represents the Julia set of a rational map R, and F or F_R represents the domain of normality. One should note that both J and F are forward and backward invariant by the map R. Also, using the equivalence of normality with equicontinuity (as discussed in 1.2 above), one should note that the Fatou set is exactly the set on which the dynamics is stable. Therefore, the Julia set is the locus of chaos for R.

In the following example, we illustrate an elementary consideration that determines the membership of a given point z_0 in either F or J.

---

*In [B1], this set was termed the "Fatou set" in light of the fact that Fatou extensively analyzed it in his papers [F1]. Other authors use the terms "domain of normality", "domain of equicontinuity", and "stable set". Although we still prefer to recognize Fatou's contributions in this way, we also use the more descriptive terminology.

1.3 Example Consider the dynamical system generated by $p(z) = 2z + z^2$. The origin is a fixed point. Also, $p'(0) = 2$. Therefore,

$$(p^n)'(0) = 2^n,$$

and the iterates p^n cannot converge to an analytic function in any neighborhood of 0 since $2^n \to \infty$ as $n \to \infty$. Therefore, $0 \in J_p$. On the other hand, if $|z| > 2$, then $|p(z)| > |z|$. It follows that the family $\{p^n\}$ is normal on the disk $D = \{z \mid |z| > 2\}$, and therefore, $D \subset F$. In particular, ∞ is a fixed point in F. □

This example suggests that an analysis of fixed points and periodic points is useful.

Definition Let $R(z_0) = z_0$ and $R'(z_0) = \lambda$. If $|\lambda| > 1$, z_0 is a *repelling* fixed point. If $|\lambda| < 1$, z_0 is an *attracting* fixed point.

1.4 Proposition *Repelling fixed points belong to the Julia set, and attracting fixed points belong to the domain of normality.* □

Using the chain rule and the fact that $J_{R^n} = J_R$, one can define attracting and repelling periodic points (points such that $R^n(z_0) = z_0$) and obtain an analogous result.

Since the Fatou set is open by definition, the Julia set is closed. Therefore, it follows that

$$\overline{\{\text{repelling periodic points}\}} \subset J.$$

1.5 First Fundamental Theorem (Fatou and Julia) The Julia set is the closure of the set of repelling periodic points.

Remark Actually, Julia started his memoire [J] by focusing on the closure of the repelling periodic points (what we call the Julia set), and he showed that its complement is the union of the domains on which $\mathcal{F}$ is normal. Thus, he proved this result even though he does not begin his study using normal families.

We will not present all of the details involved in the proof of this theorem, but in the next section, we discuss many of the essential ingredients. A complete proof can be found in [B1].

In the remainder of this section, we describe six more examples that we use repeatedly throughout these notes.

1.6 Example Perhaps the most important example in the entire subject is the simple quadratic function

$$T(z) = z^2.$$

Then, $T(\infty) = \infty$ and $T(0) = 0$. Since $T'(0) = 0$ and $T'(\infty) = 0$ (one must compute $T'(\infty)$ in the local coordinate system in a neighborhood of infinity), T has two so-called *superattracting* fixed points. Under T, the magnitude of a point is squared. Points inside the unit circle iterate to 0, and points outside iterate to ∞. The chaotic dynamics is confined to the unit circle S^1, which is the Julia set J. If we parameterize points on J by their angles θ (that is, we write $z = e^{2\pi i\theta}$ where $\theta \in \mathbf{R}$), the map $T|S^1$ is given by the equation $\theta \mapsto 2\theta$ (mod 1). Therefore, we can conveniently describe the dynamics of a point using θ. In particular, if $\theta \in \mathbf{Q}$, then the point is eventually periodic. Also, since $|T'(z)| = 2$ for any $z \in S^1$, all such periodic points are repelling. □

1.7 Exercise Let $\theta = p/q \in \mathbf{Q}$. What is the dynamics of θ under the map $\Delta(\theta) = 2\theta$? In other words, what is the smallest nonnegative integer n such that $\Delta^n(\theta)$ is a periodic point? What is the period of this orbit?

Remark The circle is almost the only Julia set that is not a fractal.

1.8 Example Now consider the quadratic

$$p(z) = z^2 + \epsilon$$

where ϵ is nonzero but small ($|\epsilon| < 1/4$ suffices.)

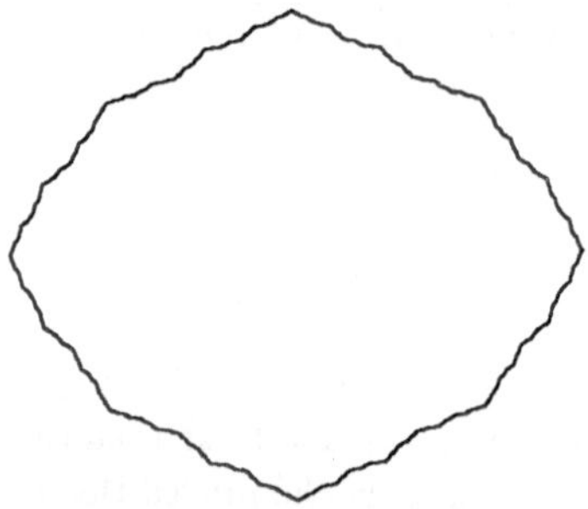

Figure 1.9: The Julia set of $z^2 + \epsilon$.

From a topological point of view, the dynamics on the Julia set in this example is essentially the same as that of $z \mapsto z^2$. Outside the Julia set, points iterate to ∞, and therefore, the exterior of the Julia set is one component of the Fatou set. However, the fixed point z_0 contained in the finite domain bounded by the Julia set is attracting but not superattracting ($|p'(z_0)| \neq 0$). By the inverse function theorem, there exists a neighborhood of z_0 on which the map is injective. Note that this is a property that does not hold on any neighborhood of 0 for the map $z \mapsto z^2$. Hence, in a neighborhood of the finite attracting fixed points, Examples 1.6 and 1.8 are topologically distinct. But this is the only topological difference between these two systems. □

1.10 Example Consider the quadratic

$$z \mapsto z^2 + \tfrac{3}{10}.$$

This is a complex version of the cookie-cutter map discussed by Bedford elsewhere in these proceedings. In Section 3, we prove that the Julia set of this example is a Cantor set in the complex plane (see Figure 1.11).

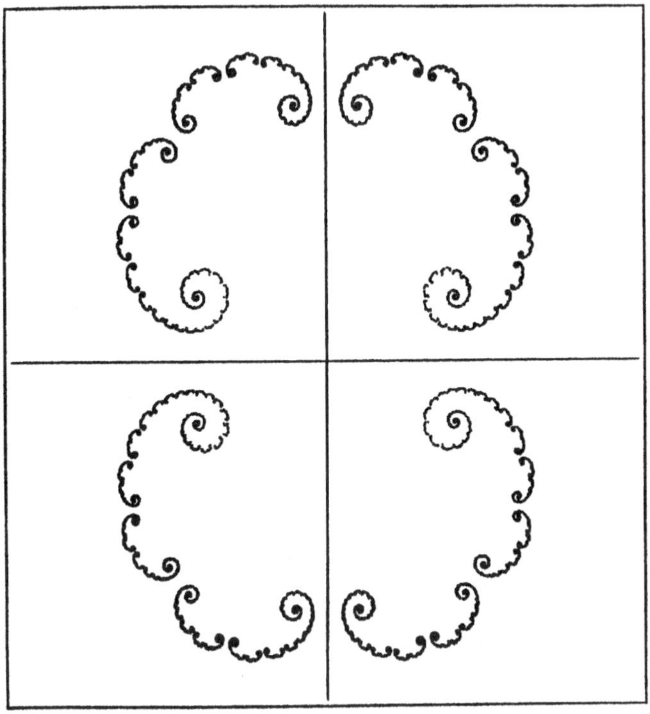

Figure 1.11: The Julia set of $z \mapsto z^2 + \frac{3}{10}$.

The Fatou set F consists of exactly one component, and all points in F iterate to ∞. However, as we can easily see from Figure 1.11, this component is not simply-connected. □

1.12 Example Recall that Newton's method involves iterating

$$N(z) = z - \frac{p(z)}{p'(z)}$$

in order to numerically solve the equation $p(z) = 0$. If we apply the method to the polynomial equation $p(z) = z^3 - 1 = 0$, we obtain the rational map

$$N(z) = z - \frac{z^3 - 1}{3z^2} = \frac{2z^3 + 1}{3z^2}.$$

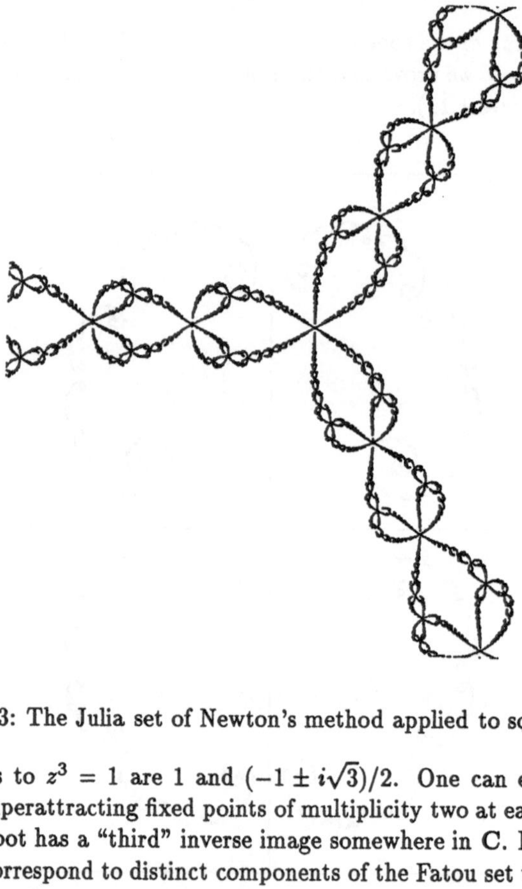

Figure 1.13: The Julia set of Newton's method applied to solving $z^3 = 1$

The three solutions to $z^3 = 1$ are 1 and $(-1 \pm i\sqrt{3})/2$. One can easily verify that this rational map has superattracting fixed points of multiplicity two at each of the roots. Since $\deg(N) = 3$, each root has a "third" inverse image somewhere in $\mathbf{C}$. In this example, these "prefixed" points correspond to distinct components of the Fatou set which map injectively onto the components that contain the roots.

Note that, unlike a polynomial function, the inverse image of ∞ in this example contains more than one point. That is, $N^{-1}(\infty) = \{0, \infty\}$. In fact, as Figure 1.13 suggests, the preorbit of ∞ (all points that eventually map to ∞ under iteration) is dense in the Julia set. □

1.14 Example Consider the quadratic

$$p(z) = z^2 - 1.$$

In Section 3, we will see that all quadratics have the dynamics of $z \to z^2$ in a neighborhood of infinity.

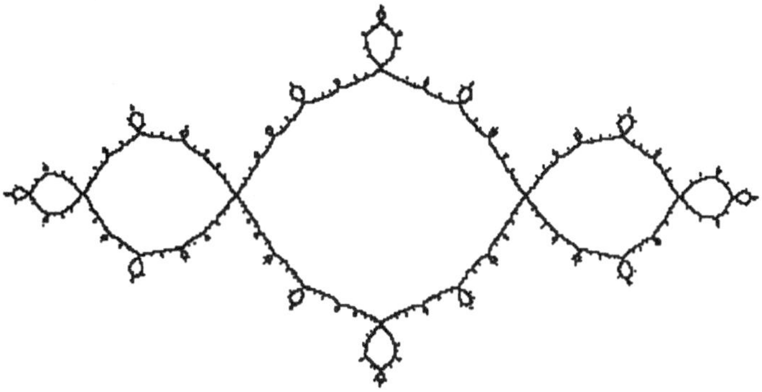

Figure 1.15: The Julia set for $z^2 - 1$.

Note that $0 \mapsto -1 \mapsto 0$. Also, since $p'(0) = 0$, $(p^2)'(0) = 0$ for z in the orbit $\{0, -1\}$. Thus, this map possesses a superattracting orbit of period two. The domain of normality consists of two primary components, each containing one element of the orbit $\{0, 1\}$, as well as countably many components that eventually iterate onto one of these two primary components (see Figure 1.15). The dynamics in the infinite component of F is essentially identical to that of the corresponding components in Examples 1.6 and 1.8. □

At this point, you may be wondering if both J and F are always non-empty, as is the case in all of the above examples.

1.16 Theorem *The Julia set of a rational map is nonempty.*

Proof. Suppose $J = \emptyset$. Then $F = \overline{\mathbf{C}}$, and there exists a subsequence $\{R^{n_i}\}$ that converges to an analytic function S uniformly on $\overline{\mathbf{C}}$. Therefore, since S is analytic on $\overline{\mathbf{C}}$, it is a rational function. On the other hand,

$$\deg(R^{n_i}) = [\deg(R)]^{n_i} \to \infty,$$

which contradicts the fact that $\deg(S)$ is finite. □

There are, however, some completely chaotic rational maps, maps whose Fatou set is empty.

1.17 Example Lattès's Example [L]. Consider the rational map

$$R(z) = \frac{(z^2+1)^2}{4z(z^2-1)}. \qquad (1.18)$$

The Julia set of this map is the entire Riemann sphere. Although we omit a few formidable details, the following description of R suggests why this is the case.

Given $w_1, w_2 \in \mathbf{C}$ such that $w_2/w_1 \notin \mathbf{R}$, let L be the lattice in $\mathbf{C}$ given by

$$\{n_1 w_1 + n_2 w_2 \mid n_i \in \mathbf{N}\}.$$

Note that L is invariant under integer multiplication. Hence, any integer multiplication induces an endomorphism of the complex torus $T^2 \cong \mathbf{C}/L$. In this example, we employ the multiplication $M(z) = 2z$ and obtain a map $M : T^2 \to T^2$ whose degree is 4. Also, the Weierstrass $\wp$-function

$$\wp(z) = \frac{1}{z^2} + \sum_{\substack{w \in L \\ w \neq 0}} \left(\frac{1}{(z-w)^2} - \frac{1}{w^2} \right)$$

induces an analytic function (actually a two-fold branched cover) from T^2 to $\overline{\mathbf{C}}$ (see [A1] and [Coh]). Since $M: \mathbf{C} \to \mathbf{C}$ preserves the inverse images of $\wp: T^2 \to \overline{\mathbf{C}}$, we obtain a rational map R that satisfies the following commutative diagram:

$$\begin{array}{ccc} T^2 & \xrightarrow{M} & T^2 \\ \downarrow{\wp} & & \downarrow{\wp} \\ \overline{\mathbf{C}} & \xrightarrow{R} & \overline{\mathbf{C}} \end{array}$$

We claim that R possesses a dense set (in $\overline{\mathbf{C}}$) of repelling, eventually periodic points. In fact, we can describe this set as follows. Fix a positive integer n and let A_n be the lattice in $\mathbf{C}$ defined by

$$A_n = \left\{ \left(\frac{a_1}{b_1}\right) w_1 + \left(\frac{a_2}{b_2}\right) w_2 \;\middle|\; a_i, b_i \in \mathbf{N} \text{ and } b_i \leq n,\ i = 1, 2 \right\}.$$

Also, let B_n denote the projection of A_n onto the torus T^2. In other words, B_n is the finite set A_n/L. It follows that $M(B_n) \subset B_n$, and hence, every point of B_n is eventually periodic. Moreover, for $z \in B_n$, $\wp(z) \in J_R$ since $M'(z) = 2$ for all $z \in T^2$. By considering all positive integers n, we have exhibited a dense set of repelling, eventually periodic points, and therefore, $J = \overline{\mathbf{C}}$. Formula (1.18) depends on the choice of lattice L and the choice of multiplication M. See Lattès [L] or Herman [H1] for more details about how to derive such formulae.

Section 2: Fundamental results

In this section, we discuss the implications of two basic results from complex analysis, the Schwarz Lemma and Montel's Theorem. We illustrate their utility in complex dynamics by discussing certain local and global aspects of the Fatou-Julia theory. We also discuss what we call the Second Fundamental Theorem, the classification of the dynamics on the Fatou set.

A. The Schwarz lemma and neutral periodic points

We start with the Schwarz Lemma, a consequence of the maximum principle for analytic functions.

2.1 The Lemma of Schwarz *Let* $\mathbf{D}$ *represent the open unit disk* $\{z \mid |z| < 1\}$ *and let* $f: \mathbf{D} \to \mathbf{D}$ *be a complex-analytic function such that* $f(0) = 0$. *Then*

(a) $|f'(0)| \leq 1$, *and*

(b) $|f(z)| \leq |z|$ *for all* $z \in \mathbf{D}$.

Moreover, if either $|f'(0)| = 1$ *or* $|f(z)| = |z|$ *for* $z \neq 0$, *then*

$$f(z) = \lambda z$$

for some λ *with* $|\lambda| = 1$. □

Interpreted dynamically, this lemma establishes that, if a complex analytic function maps $\mathbf{D}$ into itself and fixes the origin, then the map is either (1) a rigid rotation about the origin or (2) every forward orbit limits on the attracting fixed point at the origin.

We know that attracting periodic points are in the Fatou set and that repelling periodic points are in the Julia set. Now we consider *neutral* fixed points, those whose derivatives have magnitude 1. More precisely, assume that $R(z_0) = z_0$ and $R'(z_0) = \lambda$ where $|\lambda| = 1$. We use the Schwarz Lemma to investigate the dynamics of R in a neighborhood of z_0, and to do so, we need the notion of a local conjugacy.

Definition Let U be an open subset of $\overline{\mathbf{C}}$ and $R: U \to U$ be a rational function with a fixed point at z_0. Suppose there is an analytic homeomorphism ϕ that maps U to some neighborhood V of 0 with $\phi(z_0) = 0$. Moreover, suppose there is a map $f: V \to V$ such that $R = \phi^{-1} \circ f \circ \phi$. Then we say that R is *locally, analytically conjugate* to f.

2.2 Proposition *Let* R *be a rational function with a neutral fixed point at* z_0. *Then* $z_0 \in F$ *if and only if* R *is locally conjugate to its derivative* $z \mapsto \lambda z$ *on some neighborhood of* z_0.

Proof. If there is a local conjugacy between R and $z \mapsto \lambda z$, then considering the definition of F, we have $z_0 \in F$.

Conversely, suppose that $z_0 \in F$, and let U be the component of F that contains z_0. For simplicity, we discuss the proof only in the case where U is simply-connected. Using the Riemann mapping theorem, we have $\phi: U \to \mathbf{D}$ such that $\phi(z_0) = 0$. Consider the map

$$f = \phi \circ R \circ \phi^{-1}.$$

We see that $f(0) = 0$ and $|f'(0)| = 1$. Now, apply the Schwarz Lemma to f to conclude that

$$f(z) = \lambda z.$$

□

2.3 Remark If U is not simply-connected, then one can prove Proposition 2.2 using the same reasoning applied to $\tilde{U}$, the universal cover of U.

2.4 Corollary *Suppose $\lambda = e^{2\pi i\theta}$ where $\theta \in \mathbf{Q}$. Then $z_0 \in J$.*

Proof. Suppose $z_0 \notin J$. Write $\theta = \frac{p}{q}$ where p and q are relatively prime integers, and consider R^q. By Proposition 2.2, we have

$$\begin{array}{ccc} U & \xrightarrow{R^q} & U \\ \downarrow{\scriptstyle\phi} & & \downarrow{\scriptstyle\phi} \\ U & \xrightarrow{Id} & V \end{array}$$

and thus $R^q = \phi^{-1} \circ Id \circ \phi = Id$ on U. Since R is analytic, it follows that $R^q = Id$ on $\overline{\mathbf{C}}$, which contradicts our standing hypothesis that R is a rational map whose degree is greater than one. □

2.5 Remark Neutral fixed points z_0 such that $R'(z_0) = e^{2\pi i\theta}$ where $\theta \in \mathbf{Q}$ are called ***rationally*** neutral fixed points. The Julia sets in Example 2.6 – 2.8 contain rationally neutral fixed points at 0.

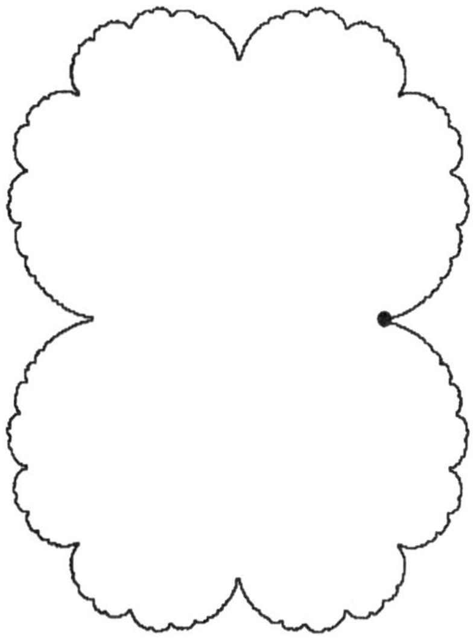

2.6. Example. The Julia set of $z \mapsto z + z^2$. The dot identifies the fixed point at 0.

2.7. Example. The Julia set of $z \mapsto -z + z^2$. The dot identifies the fixed point at 0.

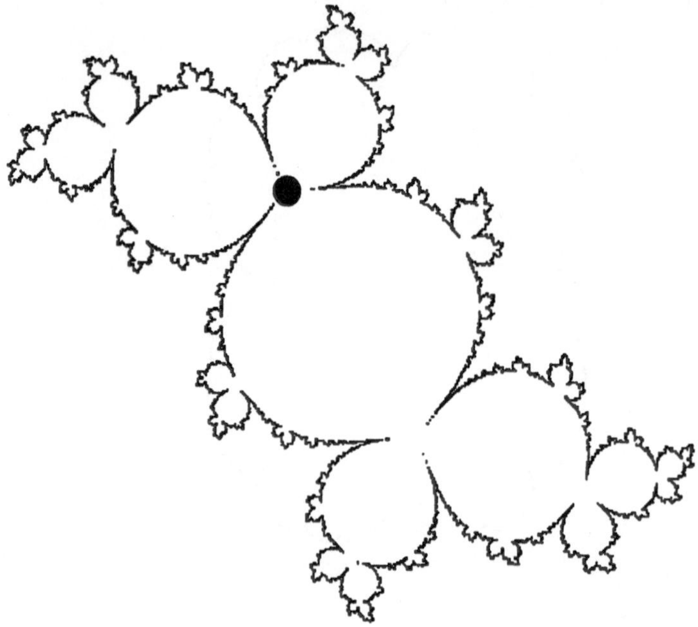

2.8. Example. The Julia set of $z \mapsto \lambda z + z^2$ where $\lambda = e^{2\pi i/3}$. The dot identifies the fixed point at 0.

When $\theta \in \mathbf{Q}$ as in Corollary 2.4, there are flowers in the Fatou set.

2.9 The Flower Theorem *Suppose z_0 is a rationally neutral fixed point and $\theta = p/q$ as above. Then there exist analytic curves that bound petals which are pairwise tangent at z_0. The union of these petals is forward invariant, and any orbit in a petal limits on z_0. Moreover, the number of petals is a multiple of q.* □

Until 1942, no one knew if neutral fixed points were ever in the Fatou set. Then Siegel [Si] established a number-theoretic condition on the derivative which guarantees that a rational map is locally conjugate to its derivative in a neigborhood of a neutral fixed point.

2.11 Theorem (Siegel) *Let R be a rational function with $R(z_0) = z_0$ and $R'(z_0) = e^{2\pi i\theta}$ where $\theta \in \mathbf{R}$. If there exist $a, b \in \mathbf{R}^+$ such that*

$$\left|\theta - \frac{p}{q}\right| > \frac{a}{q^b} \tag{2.12}$$

holds for all $p \in \mathbf{N}$ and $q \in \mathbf{N}^+$, then $R(z)$ is locally conjugate to its derivative in a neighborhood of z_0. □

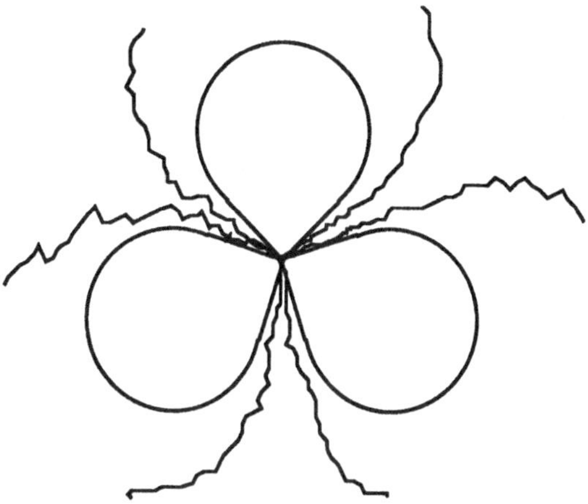

Figure 2.10:
A sketch of the petals specified in the Flower Theorem. In this case, $q = 3$. The regions enclosed by the petals are subsets of the Fatou set, and the Julia set is represented by the jagged curves that approach the fixed point. We note that the petals are not uniquely defined. As we see later in this section, the Julia set is perfect, and therefore, it must accumulate on the fixed point. It does so by making its way between the petals. The dynamics illustrated in this figure are realized in Example 2.8.

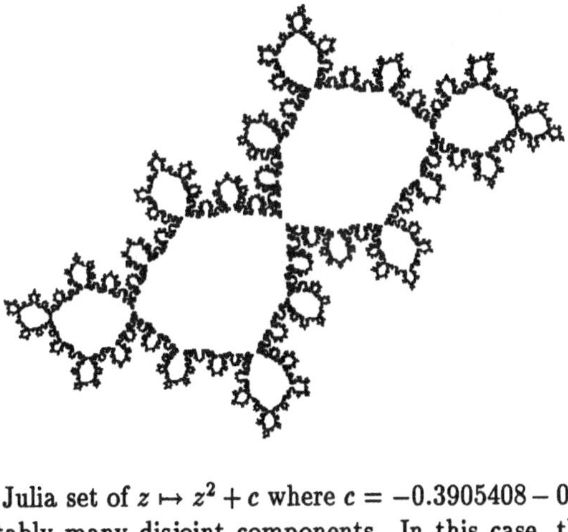

2.13. Example. The Julia set of $z \mapsto z^2 + c$ where $c = -0.3905408 - 0.5867881i$. The Fatou set consists of countably many disjoint components. In this case, the largest component D on the left contains a neutral fixed point, and the map restricted to D is analytically conjugate to the derivative map at the fixed point. All other finite components of F are preimages of D under some iterate of the map.

2.14 Remarks:

(a) The set of irrationals θ that satisfy (2.12) is a set of full measure in the unit interval.

(b) It is possible to find a dense set (actually a generic set) of irrationals θ for which the map

$$z \mapsto e^{2\pi i\theta} z + z^2$$

is not conjugate to

$$z \mapsto e^{2\pi i\theta} z.$$

These examples are due to Cremer [Cr] (see also [B1]).

(c) Recently, Yoccoz [Y1] has completely resolved the local conjugacy question for neutral fixed points. He showed that a number theory condition, more general than (2.12), on the derivative determines the existence of a local conjugacy between the map and its derivative.

(d) Using the chain rule applied to a periodic orbit, we can define neutral periodic orbits just as we define attracting or repelling periodic orbits. All of the results that we have discussed regarding neutral fixed points also apply to neutral periodic orbits.

B. Montel's theorem and omitted points

We now change our point of view and use a theorem of Montel to obtain several global results.

2.15 Montel's Theorem *Let $\mathcal{F} = \{f \mid f : U \to \overline{\mathbf{C}}\}$ be a family of analytic functions. If there exist three points $\alpha, \beta, \gamma \in \overline{\mathbf{C}}$ such that*

$$\left[\bigcup_{f\in\mathcal{F}} f(U)\right] \subset \overline{\mathbf{C}} - \{\alpha, \beta, \gamma\},$$

then $\mathcal{F}$ is a normal family. □

This theorem gives us a geometric condition that implies normality, and here we use it to find the so-called exceptional points of a rational map (if they exist).

Definition Let U be any nonempty, open subset of $\overline{\mathbf{C}}$. We define the set of *omitted points* of the rational map $R|U$ to be the set

$$E_U = \overline{\mathbf{C}} - \bigcup_{n\geq 0} R^n(U).$$

The set of omitted points for the "germ" of R at z is the union

$$E_z = \bigcup_{U\in\Omega} E_U$$

where Ω is the set of all open neighborhoods of z.

Montel's theorem has the following consequence.

2.16 Corollary *Let $z \in J$ and U be an open neighborhood of z.*

(a) The set E_U contains at most two points.

(b) The set E_z contains at most two points.

Proof. Statement (a) follows immediately from Montel's Theorem, and Statement (b) is also straightforward given the observation that, if U and V are two neighborhoods of z, then $E_{U \cap V}$ contains both E_U and E_V. □

2.17 Example Consider the map $z \mapsto z^2$ whose Julia set is the unit circle. Let U be any neighborhood, not containing 0 and ∞, of any point $z \in J$. Under iteration, the images of U spread over the entire complex plane with the exception of 0 and ∞.

2.18 Example If R is a polynomial, then $\infty \in E_z$ for all $z \in J$.

In order to understand omitted sets in general, we need the concept of a global conjugacy.

Definition Let R_1 and R_2 be two rational maps. Suppose that there exists a Möbius transformation M (an analytic homeomorphism of $\overline{\mathbf{C}}$ to $\overline{\mathbf{C}}$) such that

$$R_1 = M^{-1} \circ R_2 \circ M.$$

Then we say that R_1 and R_2 are *(globally, analytically) conjugate* by the conjugacy M.

2.19 Remark Just as local conjugacies indicate that two dynamical systems are locally equivalent, global conjugacies establish the equivalence of two systems on the entire Riemann sphere. For instance, if R_1 and R_2 are globally conjugate by the conjugacy M as in the above definition, then $M(J_{R_1}) = J_{R_2}$.

2.20 Proposition *The omitted set E_z is independent of the choice of $z \in J$.* □

2.21 Remarks:

(a) To prove Proposition 2.20, one first observes that the set E_z is R^{-1} invariant. Then, (1) if E_z contains exactly one point, we conjugate R to a rational map S whose corresponding omitted set is $\{\infty\}$; or (2) if E_z contains two points, we conjugate R to a rational map S whose corresponding omitted set is $\{0, \infty\}$. In case (1), S is a polynomial, and in case (2), S is conjugate to the map $z \mapsto z^{\pm d}$ where $d = \deg(R)$. See [B1] for more details.

(b) Given Proposition 2.20, we drop the z from E_z, and we call this omitted set the set of *exceptional points* E (or E_R) of R.

2.22 Corollary *If* $Int(J) \neq \phi$, *then* $J = \overline{\mathbf{C}}$.

Proof. Let $U \subset \text{Int}(J)$ be an open set. Then,

$$\left[\bigcup_{n\geq 0} R^n(U)\right] \cup E = \overline{\mathbf{C}}.$$

However, since $U \subset J$ and J is invariant under R, we have $[\cup_{n\geq 0} R^n(U)] \subset J$. Since E contains at most two points and since J is closed, $J = \overline{\mathbf{C}}$. □

In Section 1, we stated the First Fundamental Theorem. Its proof is based on two results that, in turn, are proved using Montel's Theorem.

2.23 Proposition *The Julia set J is a perfect set. That is, J is a nonempty, closed set, and every point in J is an accumulation point of J.* □

2.24 Proposition *The Julia set is contained in the closure of the set of periodic points of R.*

Complete proofs of Propositions 2.23 and 2.24 are contained in [B1].

C. Bounds for the number of nonrepelling periodic orbits

At this point, only one additional ingredient is needed to complete the proof of the First Fundamental Theorem. We need to know that there are only finitely many attracting and finitely many neutral periodic orbits.

To get a bound on the number of attracting periodic points, we study the dynamics of the critical points of the rational map.

Definition A point z_0 is a *critical point* if $R'(z_0) = 0$. Its value $w_0 = R(z_0)$ is called a *critical value*.

The image under R of any sufficiently small circle around a critical point z_0 always wraps more than once around the critical value w_0.

One of the remarkable aspects of complex-analytic dynamics is the "fact" that the dynamics of the critical orbits often determines the dynamics of all orbits. In the remainder of this section as well as in Sections 3 and 4, we hope to provide some insight into this bold assertion.

Definition Let z_0 be an attracting fixed point. Its *basin of attraction* is defined to be

$$W^s(z_0) = \{z \mid R^n(z) \to z_0 \text{ where } n \to \infty\},$$

and its *immediate basin of attraction* $A(z_0)$ is the component of $W^s(z_0)$ that contains z_0.

2.25 Theorem *The immediate basin $A(z_0)$ of a fixed point contains at least one critical point.*

Proof. Again we only discuss the proof in the representative case where $A(z_0)$ is simply-connected.

Let $\phi: A(z_0) \to \mathbf{D}$ be a Riemann map with $\phi(z_0) = 0$ and let $f = \phi \circ R \circ \phi^{-1}$.

$$\begin{array}{ccc} A(z_0) & \xrightarrow{R} & A(z_0) \\ \downarrow \phi & & \downarrow \phi \\ \mathbf{D} & \xrightarrow{f} & \mathbf{D} \end{array}$$

If we assume that $R|A(z_0)$ does not possess a critical point, then neither does $f: \mathbf{D} \to \mathbf{D}$. Thus, f is a conformal automorphism of the disk. There are 3 types of such automorphisms:

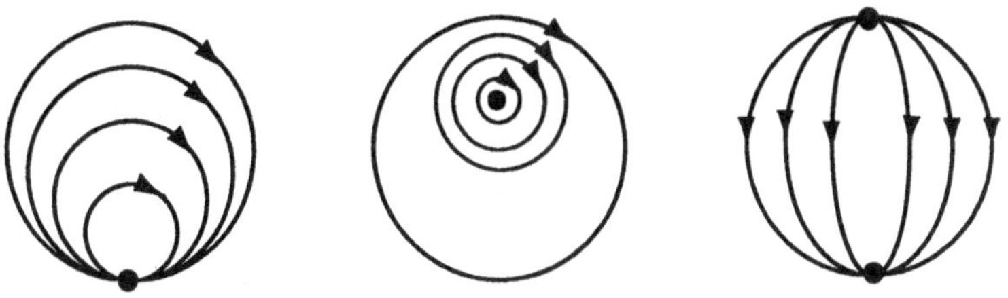

Figure 2.26:
The three types of conformal automorphisms of the disk: parabolic, elliptic, and hyperbolic.

The parabolic and hyperbolic possibilities are eliminated by the fact that $f(0) = 0$. The elliptic case is ruled out by the fact that $|R'(z_0)| < 1$ and, hence, $|f'(0)| < 1$. □

We combine Theorem 2.25 with the following elementary observation regarding the number of critical points.

2.27 Proposition *A rational map of degree d has at most $2d - 2$ critical points.* □

We can define basins and immediate basins for attracting periodic orbits and similarly prove that each such basin contains at least one critical point. Thus, we obtain a bound for the number of attracting periodic orbits.

2.28 Corollary *A rational map of degree d has at most $2d-2$ attracting periodic orbits.* □

Theorem 2.25 is also useful as a computational device for locating attracting periodic orbits. Any such orbit must attract a critical orbit; therefore, to find the attracting periodic orbits we simply follow the forward orbits of the critical points.

To complete the proof of the First Fundamental Theorem, we also need a bound on the number of neutral orbits. Such a bound was known to Fatou and Julia, and it has recently been sharpened by Douady-Hubbard [D1] and by Shishikura [Sh1]. For our purposes, the following statement will suffice.

2.29 Proposition *For a rational map of degree d,*

$$\#\{\text{attracting orbits}\} + \left(\tfrac{1}{2}\right) \#\{\text{neutral orbits}\} \leq 2(d-1).$$

□

D. The Second Fundamental Theorem

In 1982, Sullivan [S1,S2] provided the final ingredients for a classification theorem for the dynamics of a rational map on its Fatou set. Using the theory of quasi-conformality (especially the measurable Riemann mapping theorem of Morrey, Ahlfors, and Bers — see [A3]), he developed a technique for constructing analytic maps using what is essentially topological data. We will discuss a similar construction in detail in Section 4 — the Douady-Hubbard theory of polynomial-like maps. For now, we conclude this section with a brief statement of the classification, a theorem we consider to be the Second Fundamental Theorem.

Definition Let D be a component of the Fatou set. The domain D is *periodic* if there exists n such that $R^n(D) = D$. The component D is *eventually periodic* if there is k such that $R^k(D)$ is periodic.

2.30 The No Wandering Domains Theorem (Sullivan [S1]) *All components of the Fatou set are eventually periodic. Moreover, there are only finitely many periodic components.* □

Combining this theorem with previously known results (Fatou [F1], Julia [J], Siegel [Si], Herman [H3]), Sullivan [S2] described all of the possibilities for the dynamics of a rational map R on a periodic component of its Fatou set (see also [MSS]).

2.31 Theorem *Let D be a component of the Fatou set of R such that $R(D) = D$. Then the dynamics of $R|D$ is one of the following five types.*

(1) Superattracting: The domain D is the immediate basin of attraction $A(z_0)$ for some fixed point z_0 in D. The point z_0 is also a critical point. Thus, near z_0, D is foliated by simple closed curves, and this foliation is preserved by R. In other words, each simple closed curve in the foliation is wrapped by R in a many-to-one fashion to another curve in the foliation which is closer to z_0. The basin of infinity for any polynomial is an example of a typical superattracting domain.

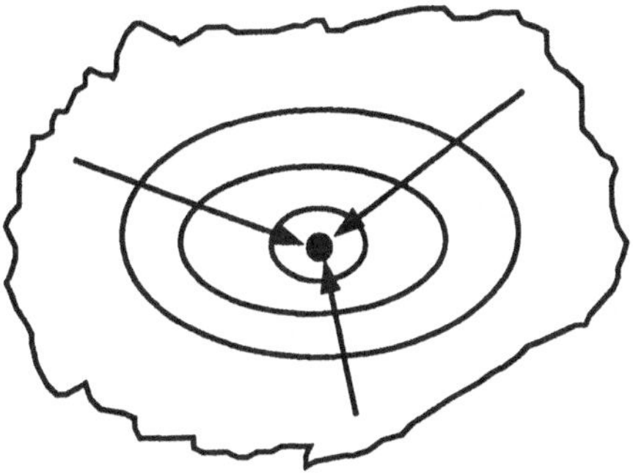

Figure 2.32:
A superattracting domain: near z_0, the map is locally, analytically conjugate to the map $z \mapsto z^k$ for $k > 1$.

(2) Attracting: The domain D is again an immediate basin of a fixed point z_0, but in this case, the map R is locally injective near z_0. Even though Theorem 2.17 guarantees that a critical point is contained in this domain, the fixed point is not a critical point. See Figure 1.9.

(3) Parabolic: The petals in the Flower Theorem are subsets of distinct components of the basin of attraction for a rationally neutral fixed point. A parabolic domain D is one of these components. The neutral fixed point is on the frontier of D and, hence, in J. See Examples 2.6 – 2.8. This type of domain also must contain a critical point.

The final two cases involve rotation rather than attraction, and these domains are often referred to as rotation domains.

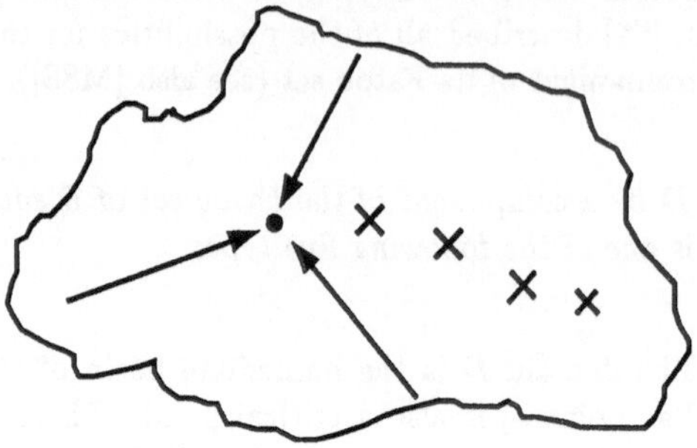

Figure 2.33:
An attracting domain: the crosses indicate the forward orbit of a critical point that converges to the fixed point (as do all other orbits in the domain).

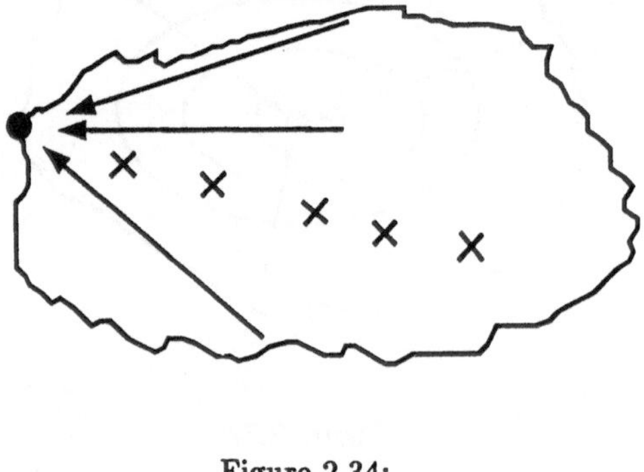

Figure 2.34:
A parabolic domain: the crosses indicate the forward orbit of a critical point that converges to the rationally neutral fixed point on the boundary of D.

(4) Siegel disk: The domain D is a component of F that contains a neutral fixed point z_0, and the map $R|D$ is analytically conjugate to its derivative map at z_0. In this case, all orbits "rotate" about z_0 on invariant, simple closed curves. Every orbit except $\{z_0\}$ forms a dense subset of its invariant curve. The frontier of D is contained in the closure of the forward orbits of the critical points of R. See Example 2.13.

(5) Herman ring: Of the five types, this is the only one that does not correspond to a fixed point. In this case, the domain D is conformally equivalent to an annulus, and every orbit in D lies on an invariant, simple closed curve that winds once around the annulus. The

orbit of every point on the curve is dense in that curve, and the frontier of D is contained in the closure of the forward orbits of the critical points. Using the maximum principle, one can show that a polynomial cannot have a Herman ring in its Fatou set.

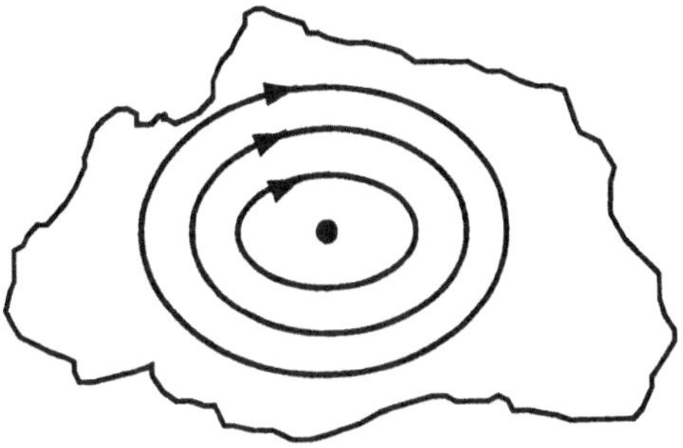

Figure 2.35:
A Siegel disk: the domain is conformally equivalent to a disk, and the rational function is analytically conjugate to an irrational rotation of this disk. See Example 2.13.

Figure 2.36:
A Herman ring: it is conformally equivalent to an annulus and is foliated by invariant curves on which the rational map is analytically conjugate to an irrational rotation.

Remark Of course, Sullivan's No Wandering Domains Theorem (Theorem 2.30) and Theorem 2.31 (as applied to periodic domains as well as fixed domains) combine to yield a complete classification of the dynamics of any rational map on its Fatou set. We call this classification the Second Fundamental Theorem.

Corollary 2.37 is a curious consequence of the Second Fundamental Theorem.

2.37 Corollary *Suppose that all critical points of R are preperiodic (eventually periodic but not periodic). Then $J = \overline{\mathbf{C}}$.*

Proof. One can show that $F = \emptyset$ because cases (2) – (5) of Theorem 2.31 require the existence of critical points with infinite forward orbits. □

2.38 Example Consider the map

$$z \mapsto \left(\frac{z-2}{z}\right)^2.$$

The numbers 0 and 2 are the only critical points, and

$$2 \mapsto 0 \mapsto \infty \mapsto 1.$$

Compare this example to Example 1.17. □

Section 3: The dynamics of quadratic polynomials

When one studies a particular dynamical system, it is often extremely useful to be able to consider that system as one among a family of similar systems. Within the subject of complex dynamics, the most successful use of this strategy has been the analysis of the quadratic family, both via computer graphics by Mandelbrot [M1] * and via the techniques of complex analysis by Douady and Hubbard [DH2]. In this section, we discuss a number of their results to illustrate this interplay between parameter space and individual dynamical systems.

First, we briefly apply the ideas from Section 2 to a general polynomial p with $\deg p \geq 2$.

3.1 Remark Observe that ∞ is a superattracting fixed point and $p^{-1}(\infty) = \{\infty\}$. As defined in the previous section, the basin of attraction for ∞ is the set

$$W^s(\infty) = \{z \in \overline{\mathbf{C}} \mid p^n(z) \to \infty \text{ as } n \to \infty\}.$$

This basin is somewhat special in that it is connected. In other words, the basin has one component, and consequently, $A(\infty) = W^s(\infty)$. As indicated in the following definition, these observations yield another dynamical decomposition of $\overline{\mathbf{C}}$, one which is slightly different from the decomposition $F \cup J$.

*For those readers interested in computer graphics, we recommend the book by Peitgen and Richter [PR] and the article by Peitgen [P].

3.2 Definition The *filled* Julia set K of p is the complement of the basin of infinity

$$K = \overline{\mathbf{C}} - W^s(\infty).$$

Many of the color pictures of quadratic Julia sets (see [PR] and [P]) are produced using a coloring scheme or a shading scheme that distinguishes the basin of infinity. In these schemes, the filled Julia set is usually colored solid black. We use such a scheme in Figure 4.16, where the black region indicates the filled Julia set for $z \mapsto z^2 - 1$.

3.3 Remark Using Montel's Theorem, we observe that

$$\partial W^s(\infty) = \partial K = J.$$

Since ∞ is superattracting for a polynomial p with $\deg p \geq 2$, we are able to use the following theorem to specify the dynamics of p in a neighborhood of ∞.

3.4 Theorem *Let $R(z_0) = z_0$ and $R'(z_0) = R''(z_0) = \ldots = R^{(k)}(z_0) = 0$ but $R^{(k+1)}(z_0) \neq 0$. Then, $R(z)$ is locally conjugate to $z \mapsto z^{k+1}$ in some neighborhood of z_0.* □

3.5 Corollary *A polynomial of degree d is locally conjugate to $z \mapsto z^d$ on some neighborhood of ∞.* □

3.6 Remark Before focusing on the quadratic case exclusively, we mention the implications of Theorem 2.25 for polynomials p of degree d. Theorem 2.25 associates at least one critical point to each attracting periodic orbit, and therefore, an arbitrary rational map has at most $2d-2$ attracting periodic orbits. However, when the map is a polynomial p, this bound can be sharpened. The point at infinity is a critical point of multiplicity $d-1$, and therefore, p has at most $d-1$ finite critical points. From Theorem 2.25, we conclude that p can have at most $d-1$ finite, attracting periodic orbits.

Given these general remarks, we turn our attention to the quadratic case for the remainder of this section. The family of all quadratic polynomials $z \mapsto Az^2 + Bz + C$ is a family with three parameters. However, we eliminate uninteresting duplication (using global conjugacies as defined in Section 2) to establish that quadratics are, from a dynamical point of view, really a one-parameter family.

3.7 Example Let $p_1(z) = z + z^2$ and $p_2(z) = z^2 + \frac{1}{4}$. The maps p_1 and p_2 are conjugate by a rigid translation by $\frac{1}{2}$. Namely, if $t(z) = z + \frac{1}{2}$, the diagram

$$\begin{array}{ccc} z & \xrightarrow{p_1} & z + z^2 \\ \downarrow t & & \downarrow t \\ z + \frac{1}{2} & \xrightarrow{p_2} & (z + \frac{1}{2})^2 + \frac{1}{4} \end{array}$$

commutes. Therefore, the Julia set of p_2 is simply a translation of the Julia set of p_1 by $\frac{1}{2}$. □

3.8 Remark Since ∞ is a distinquished point for any polynomial, one usually does not conjugate a polynomial by a Möbius transformation that moves ∞. In other words, one usually conjugates by affine transformations (maps of the form $M(z) = az + b$) so that $M(\infty) = \infty$.

3.9 Exercise For any arbitrary quadratic $q(z) = Az^2 + Bz + C$, there exists an affine map $M(z) = az + b$ such that

$$M \circ q \circ M^{-1}(z) = z^2 + c$$

for some $c \in \mathbf{C}$. Note that the resulting map $z \mapsto z^2 + c$ has its finite critical point located at 0.

Therefore, we follow the standard practice of focusing on the family of quadratics of the form $p_c(z) = z^2 + c$, and Exercise 3.9 justifies our assertion that this one-parameter family is completely representative.

As Douady and Hubbard observed, the first step in a detailed analysis of this family is a careful understanding of the "natural" domain of definition of the conjugacy, denoted ϕ_c, of p_c to $z \mapsto z^2$ defined on some neighborhood of infinity.

3.10 Exercise Let $c = -2$. For this c-value, we can explicitly exhibit ϕ_c^{-1}, which we denote by ψ_c. In fact,

$$\psi_c(z) = z + \frac{1}{z}.$$

Check that $p_c(\psi_c(z)) = \psi_c(z^2)$. Note also that ψ_c is invertible on $D = \overline{\mathbf{C}} - \overline{\mathbf{D}}$ and that $\psi_c(D)$ is the entire Riemann sphere with the exception of the interval $[-2, 2]$ on the real line. In this case, we say that the natural domain of definition of ϕ_c is $W^s(\infty)$.

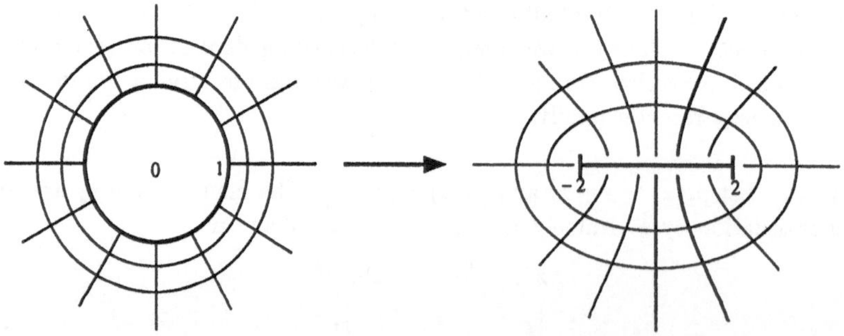

Figure 3.11: The conjugacy ψ on $W^s(\infty)$ between $z \mapsto z^2$ and $z \mapsto z^2 - 2$.

In general, we can express ϕ_c as an infinite product expansion. Given a point z_0 near ∞ (more precisely, given $|z_0| \geq R_c = 1 + |c|$), we define

$$\phi_c(z_0) = z_0 \prod_{k=0}^{\infty} \left(1 + \frac{c}{z_k^2}\right)^{\frac{1}{2^{k+1}}} \tag{3.12}$$

where z_k denotes the k-th iterate of z_0. In (3.12), we are using the branch of 2^{k+1}-th root such that, when $z_0 = \infty$,

$$\left(1 + \frac{c}{z_k^2}\right)^{\frac{1}{2^{k+1}}} = 1.$$

3.13 Exercise Verify that equation (3.12) actually yields a conjugacy. In other words, verify that it converges for $|z_0| \geq R_c$ and that $\phi_c(z_1) = [\phi_c(z_0)]^2$.

From our point of view, there are two distinct types of quadratic polynomials, those with connected Julia sets (like $z \mapsto z^2 - 1$) and those whose Julia sets are totally-disconnected (like $z \mapsto z^2 + 1$). In fact, one can distinquish these types using the orbit of the finite critical point.

3.14 Theorem *Let J_c represent the Julia set of p_c and let $\{c_1, c_2, c_3, \ldots\}$ be the forward orbit of the critical point*

$$0 \mapsto c = c_1 \mapsto c^2 + c = c_2 \mapsto (c^2 + c)^2 + c = c_3 \mapsto \ldots$$

Then J_c is disconnected if and only if $c_k \to \infty$ as $k \to \infty$. If so, J_c is a Cantor set. In particular, J_c is totally-disconnected.

Proof (sketch). There are two common ways to make pictures of the Julia sets of polynomials: (1) take a point in the Julia set and plot its backward orbit; or (2) iterate every point on a lattice in the complex plane and color that point according to the number of iterations necessary for its orbit to escape a certain bound. When the second method is used, we are really approximating level curves for a harmonic function $h_c(z)$, which is closely related to the conjugacy $\phi_c(z)$ mentioned earlier.

Let $h_c: \overline{\mathbf{C}} \to [0, \infty]$ be defined by

$$h_c(z) = \lim_{k \to \infty} \tfrac{1}{2^k} \log_+ |p_c^k(z)|, \tag{3.15}$$

where $\log_+ : \mathbf{R}^+ \to [0, \infty]$ equals $\max\{\log z, 0\}$. Note that $h_c(z) > 0$ if and only if $z \in W^s(\infty)$, and consequently, $h_c(z) = 0$ for $z \in K$. From the definition, it also follows that

$$h_c(p_c(z)) = 2h_c(z). \tag{3.16}$$

Whenever z is in the domain of ϕ_c, we have

$$h_c(z) = \log|\phi_c(z)|. \tag{3.17}$$

From (3.17), we see that the level sets of h_c are equal to the images under $\phi_c^{-1} = \psi_c$ of level sets of the absolute value function. In other words, for r large enough so that $h^{-1}(r)$ belongs to the domain of the conjugacy ϕ_c, we have

$$h_c^{-1}(r) = \phi_c^{-1}(\{w \mid |w| = e^r\}).$$

We can use h_c to indicate why J_c is disconnected when $c_k \to \infty$. For r near ∞, the level curves $h_c^{-1}(r)$ are simple closed curves that are very close to circles centered at ∞. Using (3.16) and the mapping properties of analytic maps (away from critical points), one can verify that the level curve containing the critical value c is also a simple closed curve passing through c. Since the critical point 0 maps to its critical value c locally as a two-to-one branched covering map, the level curve of h_c containing 0 is a "pinched circle," also sometimes referred to as a "figure 8."

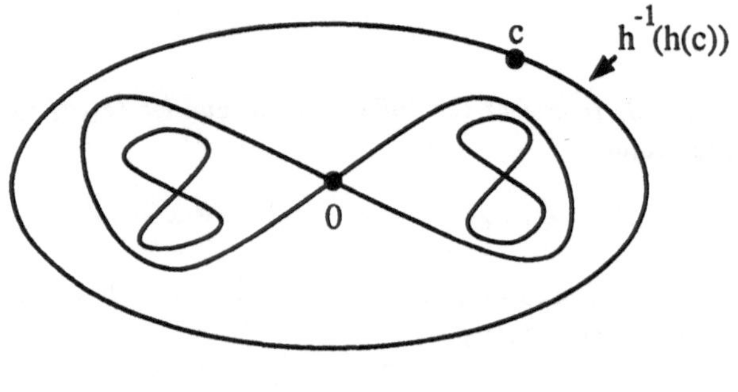

Figure 3.18: The level curves of h_c for $c \notin M$.

Using (3.16) and the fact that a quadratic polynomial is a two-fold branched cover of $\overline{\mathbb{C}}$ where branching only occurs at the critical points, we observe that the level sets

$$h_c^{-1}\left(\frac{h_c(0)}{2^k}\right)$$

consist of 2^k pinched circles.

If we visualize h_c as a height function, we get the following interpretation of $W^s(\infty)$ (see Fig. 3.19 on the next page). The "ends" of $W^s(\infty)$ are the points of the Julia set.

On the other hand, if $\{c_k\}$ is bounded, then all nonzero levels of h_c will be simple closed curves that surround K_c and exhaust $W^s(\infty)$. It follows that K_c is connected and that $J_c = \partial K_c$ is connected. □

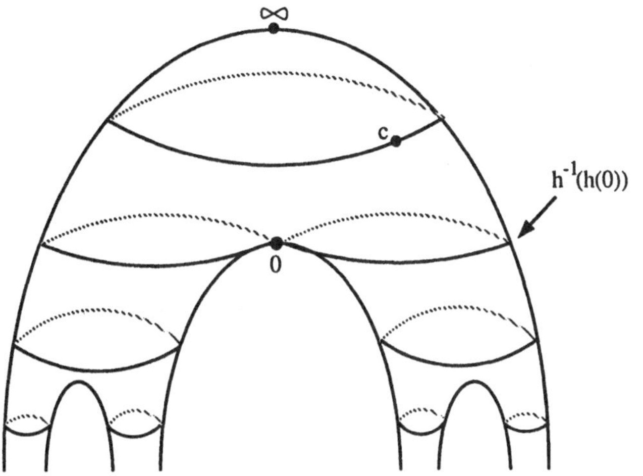

Figure 3.19: The level curves of h viewed as if h were a height function.

Theorem 3.14 establishes a basic dichotomy for quadratics. The Julia set of a quadratic is either connected or a Cantor set. This distinction leads to the definition of the Mandelbrot set, which serves as the encyclopedia for the dynamics of quadratics.

Definition The *Mandelbrot set* M is the subset of **C** consisting of those c-values for which J_c is connected. It can also be characterized as

$$\begin{aligned} M &= \{c \mid \text{the orbit } p_c^k(c) \text{ is bounded}\} \\ &= \{c \mid h_c(0) = 0\}. \end{aligned}$$

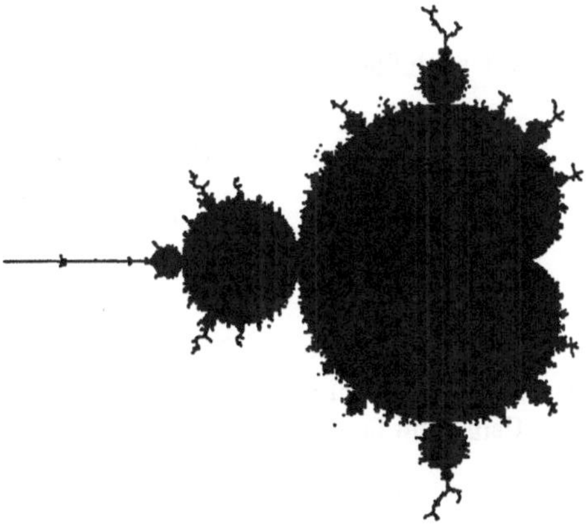

Figure 3.20: The Mandelbrot set.

For the remainder of this section, we briefly mention some of the recent work of Douady and Hubbard pertaining both to J_c when $c \in M$ and to the structure of M itself. As mentioned above, their work begins with a careful analysis of ϕ_c. There are two cases:

(3.21) The number $c \in M$. Then the conjugacy can be extended to a map $\phi_c: W^s(\infty) \to \overline{\mathbf{C}} - \overline{\mathbf{D}}$. See Exercise 3.10.

(3.22) The number $c \notin M$. As was suggested in the "proof" of Theorem 3.14, the natural domain of definition of ϕ_c does not exhaust $W^s(\infty)$. In fact, the level set of h_c through 0 bounds this domain. To be more precise, let

$$K'_c = \{z \mid h_c(z) \leq h_c(0)\}.$$

In other words, K'_c consists of the pinched circle $h_c^{-1}(h_c(0))$ and its finite interior (see Figure 3.18). Then, we have

$$\phi_c: \overline{\mathbf{C}} - K'_c \to \overline{\mathbf{C}} - \overline{\mathbf{D}_r}$$

where $\log r = h(0)$, i.e., ϕ_c has an analytic extension to $\overline{\mathbf{C}} - K'_c$.

Ironically, Douady and Hubbard use (3.22) to show that M is connected. In fact, they use the conjugacies ϕ_c to construct the Riemann map of $\overline{\mathbf{C}} - M$.

3.23 Theorem (Douady-Hubbard [DH1]) *The Mandelbrot set M is connected, and the Riemann map Φ of $\overline{\mathbf{C}} - M$ can be defined by*

$$\Phi(c) = \phi_c(c). \tag{3.24}$$

Proof (sketch). The main point is to show that the map

$$\Phi: \overline{\mathbf{C}} - M \to \overline{\mathbf{C}} - \overline{\mathbf{D}}$$

with $\Phi(\infty) = \infty$ and $\Phi'(\infty) = 1$ as defined by (3.24) is a Riemann map.

First, note that $c \notin K'_c$, so $\phi_c(c)$ exists. Using formula (3.12) for ϕ_c, we see that

$$\frac{\Phi(c)}{c} = \prod_{k=0}^{\infty} \left(1 + \frac{c}{c_k^2}\right)^{\frac{1}{2^{k+1}}}.$$

This infinite product converges uniformly for $|c| \geq 4$, and since all factors tend to 1 as $|c| \to \infty$, Φ can be extended to a holomorphic map defined on $\overline{\mathbf{C}} - M$, i.e., Φ has a removable singularity at ∞. Using the fact that Φ is a proper map, $\deg(\Phi) = 1$ because

$$\Phi^{-1}(\infty) = \{\infty\}$$

with multiplicity 1. □

Remark Intuitively, we regard ϕ_c^{-1} as a map that induces a dynamically significant "polar coordinate" system on $\overline{\mathbf{C}} - K_c'$ for all $c \notin M$. Since $c \in K_c'$, c has coordinates $\phi_c(c)$. Douady and Hubbard's proof establishes the fact that these radial and angular coordinates for c uniquely determine the map $z \mapsto z^2 + c$.

At this point, we have discussed two types of polar coordinate systems arising in the study of quadratics. In (3.21) where $c \in M$, ϕ_c yields polar coordinates on $W^s(\infty)$. We say that these coordinates are in the dynamic plane or phase plane, and of course, they depend on the value of c. The coordinates arising from (3.24) are fundamentally different. They are coordinates in the parameter plane, and they serve to describe the dynamics of all quadratics. Douady and Hubbard continue their analysis by studying the question of extending these coordinate systems in both cases. In other words, in the dynamic plane, they consider extending the coordinates for $W^s(\infty)$ to the Julia set J_c. In parameter space, they consider extending the coordinates for $\overline{\mathbf{C}} - M$ to ∂M.

Remark Due to a classical theorem of Carathéodory [C2], the question of extending these Riemann maps can be rephrased in topological terms: "Is J_c or is M locally-connected?" More precisely, let U be a simply-connected domain that is conformally equivalent to the unit disk $\mathbf{D}$ by the Riemann map $f : U \to \mathbf{D}$. Then Carathéodory's theorem states that, if ∂U is locally-connected, the inverse map f^{-1} extends to a continuous map $g: \overline{\mathbf{D}} \to (U \cup \partial U)$. However, g is *not* necessarily injective on $S^1 = \partial\overline{\mathbf{D}}$. (See Milnor [Mi2] for a more detailed discussion of this result.)

Definition Given a Riemann map $f: U \to \mathbf{D}$ for a simply-connected domain $U \subset \mathbf{C}$ with $f(z_0) = 0$ and $f'(z_0) \in \mathbf{R}^+$, the *radial arc* of angle θ is the curve $f^{-1}(re^{2\pi i\theta})$ where $0 \leq r < 1$.

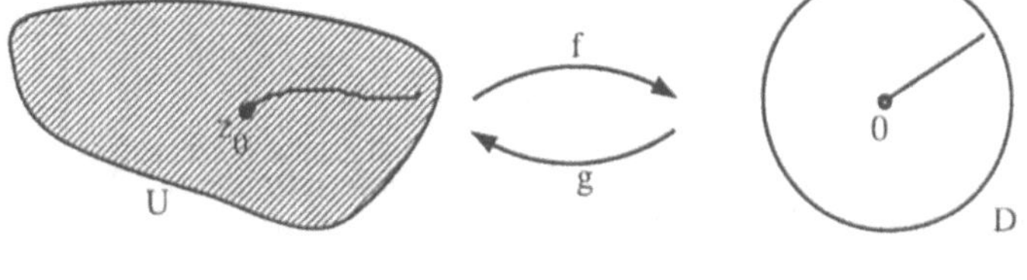

Figure 3.25: A radial arc in the simply-connected domain U.

3.26 Example In Example 3.10, we exhibited the semi-conjugacy ϕ_c^{-1} when $c = -2$. Therefore, for the map $z \mapsto z^2 - 2$, we can explicitly see that the radial arcs limit to unique points in J_c, the interval $[-2, 2]$ on the real axis. (See Figure 3.11.) Note that, with the exception of the two endpoints -2 and 2, each point in J_c is the limit of exactly two radial arcs. Only one radial arc limits on each of the two endpoints. □

We conclude this section with a brief summary of Douady and Hubbard's work [DH2] on radial limits. First, we state their results for Julia sets.

3.27 Theorem (Douady-Hubbard) *Let $p_c(z) = z^2 + c$. If either*
(1) c is attracted to a periodic attracting orbit;
(2) c is attracted to a rationally neutral orbit; or
(3) c is preperiodic, i.e., eventually periodic but not periodic;
then K_c and J_c are locally-connected. □

Remark Theorem 3.27 implies that there exists a continuous map

$$g: \overline{\mathbf{C}} - \mathbf{D} \to W^s(\infty) \cup J$$

that equals ϕ_c^{-1} on $\overline{\mathbf{C}} - \overline{\mathbf{D}}$. The map g is an example of a semiconjugacy between the map $z \mapsto z^2$ on S^1 and the map p_c on J_c. In other words, the diagram

$$\begin{array}{ccc} S^1 & \xrightarrow{z \mapsto z^2} & S^1 \\ \downarrow g & & \downarrow g \\ J_c & \xrightarrow{p_c} & J_c \end{array}$$

commutes, but the map g is not necessarily injective. Therefore, g is not a conjugacy as defined in Section 2.

Remark There are values of c in ∂M for which it is known that K_c is not locally-connected. See [D1] and [S4] for details.

Douady and Hubbard also have a result for radial arcs in the complement of M.

3.28 Theorem (Douady-Hubbard) *Let θ be a rational angle (one that is a rational multiple of 2π) and let $\theta(r)$ with $1 < r \leq \infty$ be the corresponding arc for the map*

$$\Phi^{-1}: \overline{\mathbf{C}} - \overline{\mathbf{D}} \to \overline{\mathbf{C}} - M.$$

Then, as $r \to 1$, the radial arc $\theta(r)$ limits on a unique number c_θ. Moreover, c_θ satisfies either case (2) or case (3) of Theorem 3.26, and the dynamics of $z \mapsto z^2 + c_\theta$ can be determined from θ. □

More detailed versions of the following figure can be found in the 1982 CRAS note by Douady and Hubbard [DH1] and the excellent survey of quadratics by Branner [Br1].

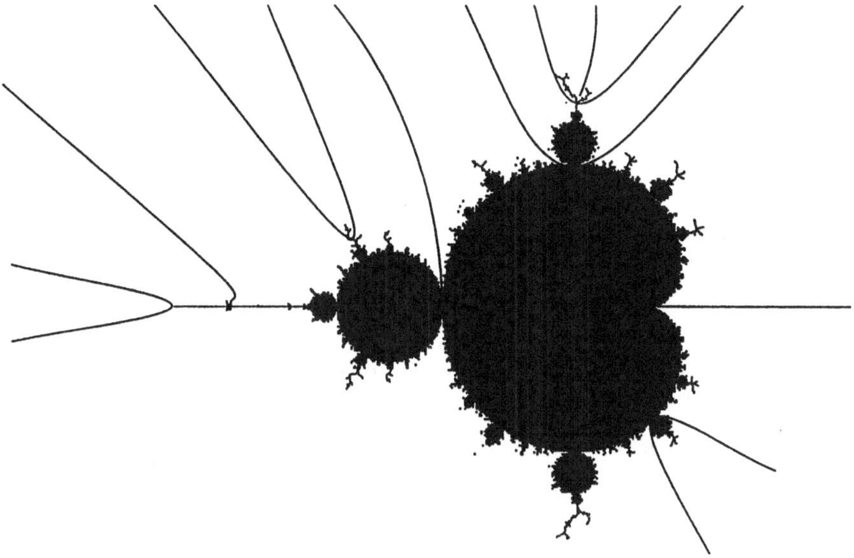

Figure 3.29: A few radial arcs in $\overline{\mathbf{C}} - M$ that limit on M.

Section 4: Newton's method and the Douady-Hubbard theory of polynomial-like maps

In this section, we study the dynamics of a different type of rational function. We consider those rational functions that are obtained from Newton's method as applied to a polynomial equation. Such maps are interesting for two reasons:

(1) they form a natural family of non-polynomial examples; and

(2) their dynamical properties are related to their utility as numerical algorithms.

After reviewing some basic facts, we describe a one-parameter family of third degree rational functions derived from Newton's method applied to cubic equations in one variable. Then, in order to explain the results of a related computer experiment, we present the remarkable theory of polynomial-like mappings, due to Douady and Hubbard.

Using Newton's method to find the roots of a polynomial equation

$$p(z) = a_d z^d + a_{d-1} z^{d-1} + \ldots + a_0 = 0$$

is identical to computing individual orbits of the dynamical system generated by the Newton map $N(z)$ or $N_p(z)$

$$N(z) = z - \frac{p(z)}{p'(z)} \tag{4.1}$$

(recall Example 1.12). First, we summarize a few elementary facts regarding the dynamics of N_p.

4.2 Remarks:

(a) From (4.1), we see that the roots of $p(z)$ correspond to the finite fixed points of $N(z)$.

(b) The point at infinity is a fixed point, and since $N'(\infty) = d/(d-1)$, it is repelling. Therefore, if Newton's method produces a point near ∞, its successive iterates tend towards the origin.

(c) The derivative of N is

$$N'(z) = \frac{p(z)p''(z)}{[p'(z)]^2},$$

and therefore, the simple roots of $p(z)$ are superattracting fixed points of $N(z)$. This is a desirable property for a numerical root finding algorithm because, in a neighborhood of its superattracting fixed points, the algorithm is locally conjugate to $z \mapsto z^k$ for some $k > 1$. Thus, local convergence is very rapid. In fact, the number of decimal places of accuracy at least doubles with each iteration.

(d) Multiple roots are attracting fixed points, but they are not superattracting. Thus, the resulting linear rate of attraction does not yield an effective numerical algorithm. More precisely, for a multiple root of multiplicity m, the derivative of Newton's method is $(m-1)/m$, which implies slow convergence for roots of high multiplicity.

(e) For a generic polynomial of degree d, the Newton's map is a rational map of degree d. However, when the polynomial has multiple roots, $\deg(N) < d$. In this section, we usually consider polynomials that do not have multiple roots.

(f) In Sections 2 and 3, we have seen that dynamical properties of a complex-analytic function are often determined by the dynamics of its critical points. Therefore, it is important to note that the critical points of $N(z)$ are the simple roots as well as the inflection points of $p(z)$.

(g) Likewise, since ∞ is a (slowly) repelling fixed point for Newton's method, we note that the poles of $N(z)$ are the critical points of $p(z)$. Consequently, orbits that avoid the critical points of $p(z)$ have the best chance of converging rapidly to a root.

(h) Unfortunately, given (f) and (g), we must keep two sets of critical points in mind. Note that it is the critical points of $N(z)$ (a subset of the roots and inflection points of $p(z)$) that predict the overall dynamics of $N(z)$.

Given (4.2g), one wonders how the critical points of $p(z)$ are related to its roots.

4.3 Theorem (Lucas, 1874) *The critical points of $p(z)$ are contained in the convex hull of the roots of $p(z)$.* □

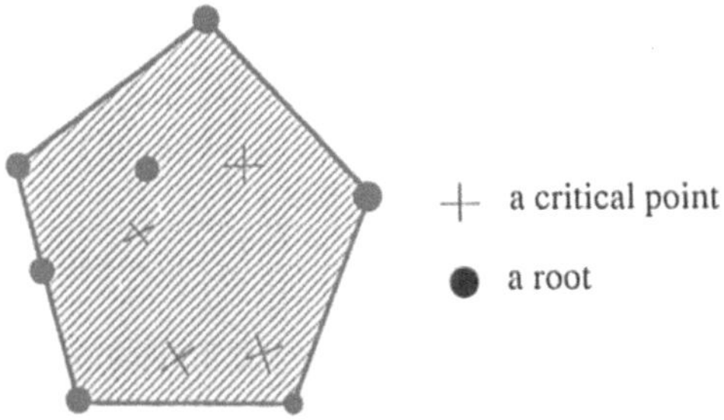

Figure 4.4:
The critical points of $p(z)$ (i.e., the poles of $N(z)$) are contained in the convex hull of the roots of $p(z)$.

It is also useful to consider the manner in which Newton's method for a given polynomial is related to Newton's method for a "rescaled" polynomial.

4.5 Remark Let $T(z) = \alpha z + \beta$ where $\alpha \neq 0$ and let $q(z) = p(T(z))$. Then

$$T \circ N_q \circ T^{-1} = N_p.$$

In other words, we can transform the roots by an affine map without qualitatively changing the dynamics of the corresponding Newton's function.

Given these general observations, we are now ready to discuss global convergence properties. Due to its simplicity, we start with the quadratic case. The result we describe has been known since at least 1870 and is explicitly discussed in the papers of Schröder [Sch] and Cayley [Ca1–4].

4.6 Theorem *Let $p(z)$ be a quadratic with distinct roots. Then Newton's method $N_p(z)$ is globally, analytically conjugate to $z \mapsto z^2$.*

Proof. We can establish this result without doing any calculation. Denote the roots of the quadratic by α and β and consider the Möbius transformation

$$h(z) = \frac{z - \beta}{z - \alpha}.$$

Note that $h(\infty) = 1$, $h(\beta) = 0$, and $h(\alpha) = \infty$. Then, $h \circ N_p \circ h^{-1}$ is a rational map of degree 2 that has superattracting fixed points at 0 and ∞, and it fixes 1. It must be $z \mapsto z^2$. □

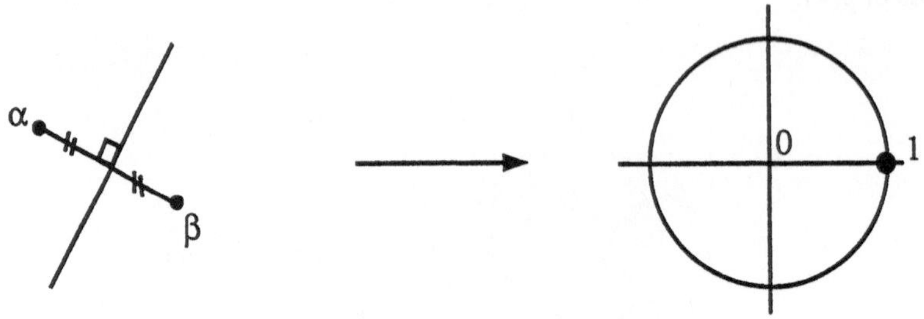

Figure 4.7: The conjugacy h in Theorem 4.6.

Note that the Julia set for $z \mapsto z^2$ corresponds under the conjugacy h to the line that is the perpendicular bisector of the line segment from α to β. Along this line, N_p has the "angle doubling" dynamics of the map $z \mapsto z^2$ restricted to the unit circle.

As we mentioned above, the critical points for Newton's method are the roots and the inflection points of the polynomial. Since the roots are always fixed, the only "free" critical points are the inflection points. In the quadratic case, there are no inflection points, and Newton's method is always conjugate to the rational map $z \mapsto z^2$.

The analysis of Newton's method becomes dramatically more complicated as soon as we increase the degree of the polynomial equation to 3. To see why, we describe work done in the early 80's by Curry, Garnett, and Sullivan [CGS] and by Douady and Hubbard [DH4].

4.8. A Computer Experiment:

To study Newton's method for cubics using computer graphics, we use (4.5) to eliminate duplication. Given three distinct roots in $\mathbf{C}$, there is an affine map that transforms this "triangular" configuration to one that is entirely located in the half plane

$$\{z \mid \mathrm{Im}(z) \geq 0\}$$

with its longest side equal to the interval $[0, 1]$. Therefore, we need only consider polynomials of the form

$$p_\rho(z) = z(z-1)(z-\rho), \tag{4.9}$$

where $\mathrm{Im}(\rho) \geq 0$, $|\rho| \leq 1$, and $|\rho - 1| \leq 1$. In other words, we use a one-parameter family of polynomials whose roots are 0, 1, and ρ, where ρ is the parameter.

Moreover, by identifying the triangles in Figure 4.10 that are congruent via a conformal affine map, we see that this parameter space is homeomorphic to the two-sphere S^2 with one puncture.

Given this representative family of cubics, we now explore the dynamics of the associated

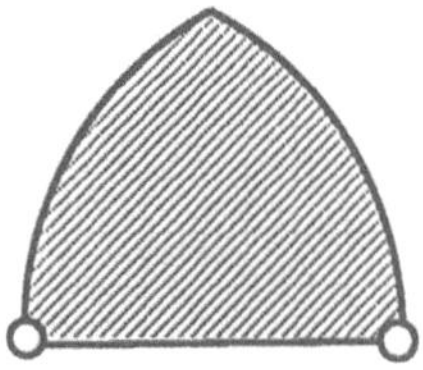

Figure 4.10:
This figure is a sketch of the region in the ρ-plane that corresponds to a completely representative collection of cubic polynomials in the form of (4.9). Given an arbitrary cubic q, there is a number ρ in this region such that N_q and N_{p_ρ} are globally conjugate using (4.5).

Newton functions. In general, given a polynomial $p(z)$, the Julia set of N_p does not play an important role for two reasons. In all known examples, it has measure zero, and it usually repells nearby points. Therefore, numerical errors force most orbits to move away from the Julia set.

Consequently, we focus on the structure of the Fatou set F. Computer evidence suggests that the basins of the roots are relatively large components of F. However, F may contain basins of other attracting periodic orbits. If such orbits exist, they do not correspond to roots of $p(z)$, and every starting point in such a basin leads to an unsuccessful application of Newton's method. The following experiment ([CGS] and [DH4]) is designed to locate such orbits.

Recall that all periodic attracting orbits attract at least one critical point (Theorem 2.25). Thus, we use critical points to locate attracting periodic orbits. We follow the orbit of the "free" critical point of N_p, the inflection point of p. In Figures 4.11– 4.13, we shade the parameter value ρ according to the number of iterates it takes for the orbit of the associated inflection point to converge to one of the three roots. If it does not converge, we plot a black point at ρ.

Figures 4.11 – 4.13 describe portions of parameter space. We use these figures to determine interesting values of the parameter ρ. Figures 4.12 and 4.13 illustrate the structure of the Fatou set for a value of ρ chosen from the black region in Figure 4.13.

The results of this experiment are remarkable. We are studying the dynamics of maps that bear little relationship with quadratics, and we are considering a parameter space that presumably has nothing to do with the quadratic family. Why do we obtain images of the Mandelbrot set? What is their significance?

Douady and Hubbard [DH4] answered both questions when they developed their theory of polynomial-like mappings.

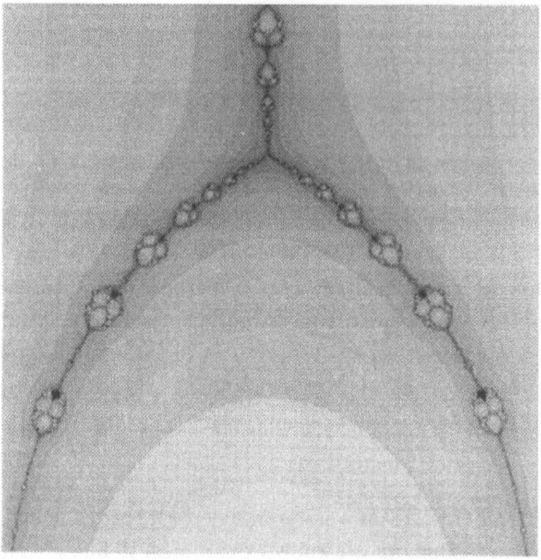

Figure 4.11: This figure contains the unit square $\{z \mid 0 \leq \mathrm{Re}(z) \leq 1,\ 0 \leq \mathrm{Im}(z) \leq 1\}$ in the ρ-plane. Thus it contains the region sketched in Figure 4.10, and it is shaded using the scheme described earlier.

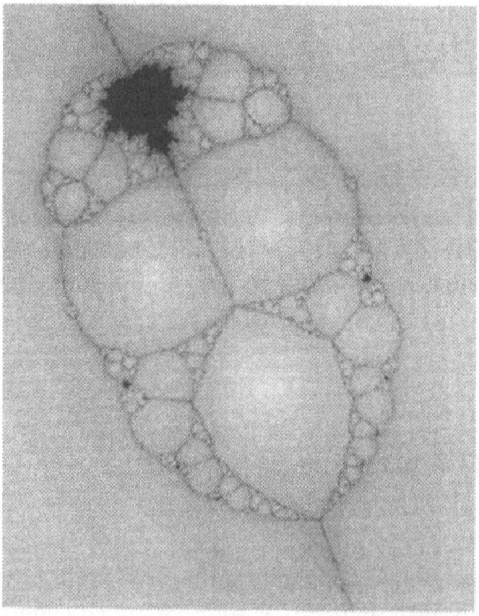

Figure 4.12: This figure is an enlargement of a small rectangle from Figure 4.11. Precise coordinates are presented in an appendix to this section. Recall that the color black corresponds to parameters for which the orbit of the inflection point does not converge to one of the three roots.

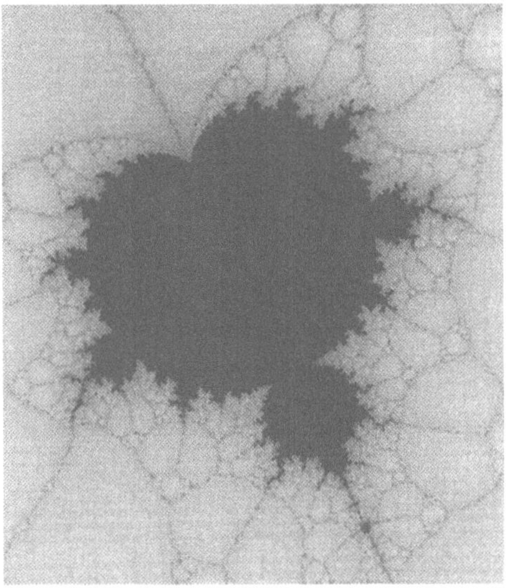

Figure 4.13: This figure is an enlargement of a rectangle from Figure 4.12. It illustrates the largest of black regions in Figure 4.13.

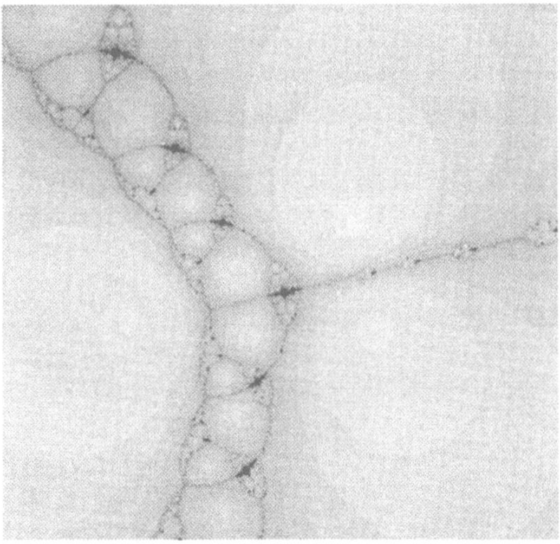

Figure 4.14: This picture illustrates the structure of the Fatou set for $\rho = 0.909419 + 0.416106i$. Recall that the roots are located at 0, 1 and ρ and that we shade a point corresponding to the number of iterates necessary for its orbit to converge to one of the roots (up to a reasonable accuracy). If the orbit does not converge after a prescribed number of iterates, a black dot is plotted at the initial point.

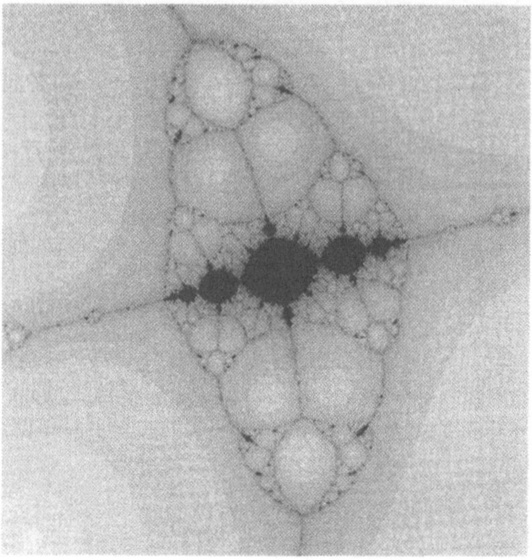

Figure 4.15: This figure is an enlargement of a rectangle containing one of the black regions in Figure 4.14. The large black region K that resembles the filled Julia set of $z \mapsto z^2 - 1$ is invariant under the second iterate of N. The union $K \cup N(K)$ contains a superattracting periodic orbit whose period is 4. All of the black regions in Figure 4.14 are preimages of K under some iterate of N.

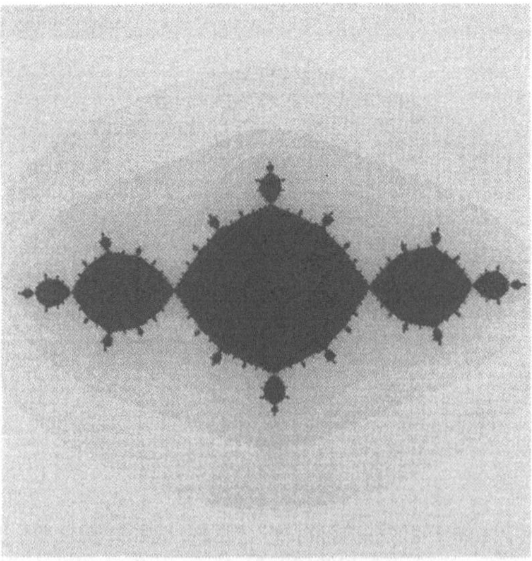

Figure 4.16: The filled Julia set of $z \mapsto z^2 - 1$ (Example 1.14). We include this figure here to emphasize the similarity between Figure 4.15 and this filled Julia set. Recall Example 1.14.

Definition Suppose that U' and U are simply-connected domains and that U' is a relatively compact subset of U. A map

$$f: U' \to U$$

that is analytic and proper is called a *polynomial-like* map.

Remarks

(a) A map f is *proper* if the inverse image of a compact set is compact.

(b) A polynomial-like map has a finite degree which can be determined by counting inverse images with multiplicity.

(c) The definition of a polynomial-like map depends both on the analytic map and the choice of U'. For example, it is possible to find cubics that are, of course, cubic-like on disks of large radius around the origin but that are quadratic-like on smaller domains. Also, transcendental functions can be polynomial-like if the domain U' is chosen appropriately. See Examples 4.17 and 4.19.

4.17 Example Given a cubic polynomial $p(z)$, we define a harmonic function $h_p : \overline{\mathbf{C}} \to [0, \infty]$ in much the same way as we did in (3.15). That is,

$$h_p(z) = \lim_{k \to \infty} \tfrac{1}{3^k} \log_+ |p^k(z)|.$$

In general, p has two distinct critical points c_1 and c_2. For any cubic p that has $c_1 \in W^s(\infty)$ and $c_2 \notin W^s(\infty)$, we obtain a a polynomial-like map q of degree 2 by restricting p. Let $v = h(c_1)$; then $h(p(c_1)) = 3v$. We set

$$U = \overline{\mathbf{C}} - h^{-1}[3v, \infty]$$

and choose U' to be the component of $p^{-1}(U)$ that contains c_2 (see Figure 4.18). Then $p : U' \to U$ is a polynomial-like map of degree 2. □

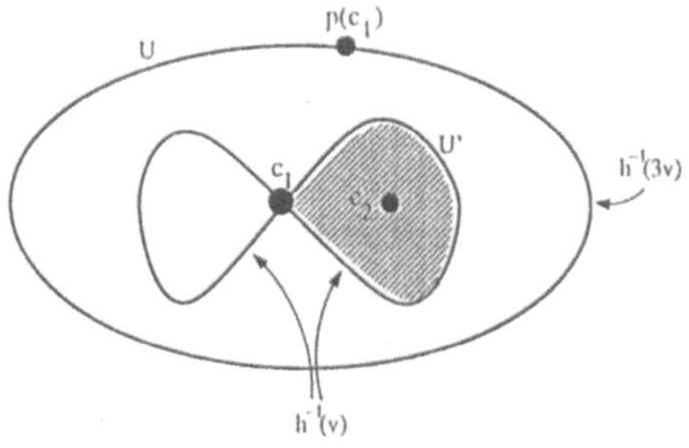

Figure 4.18: The levels sets of level $3v$ and v for h along with the domains U' and U.

4.19 Example Let $f(z) = (\cos z) - 2$ and

$$U' = \{z \mid |\mathrm{Re}(z)| < 2,\ |\mathrm{Im}(z)| < 3\}.$$

The map $f|U'$ is polynomial-like of degree 2 on U' even though f is transcendental on **C**. □

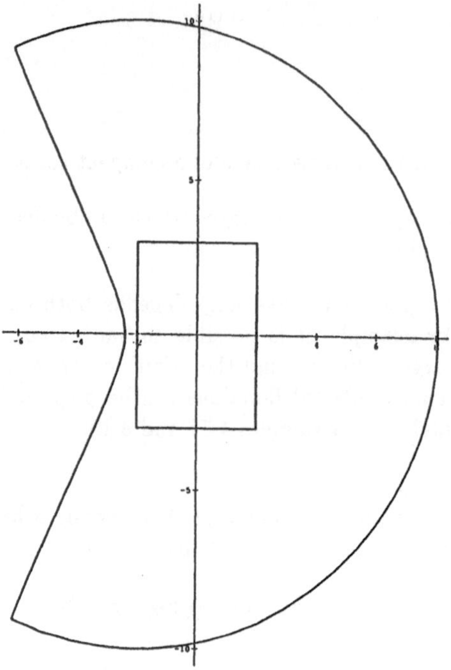

Figure 4.20:
The rectangle U' and its image U under $f(z) = (\cos z) - 2$. On U', the transcendental function f is polynomial-like of degree 2.

In the last section, we defined the filled Julia set of a polynomial as the set of points whose orbits do not converge to infinity. A similar concept exists for polynomial-like maps.

Definition If $f : U' \to U$ is a polynomial-like map, then the filled-in Julia set K_f for f is

$$K_f = \{z \in U \mid f^n(z) \in U' \text{ for all } n \in \mathbf{N}^+\}.$$

4.21 Theorem (Douady-Hubbard) *Associated to each polynomial-like map f is a polynomial q such that the dynamics of f on a neighborhood of K_f is topologically conjugate to the dynamics of q on a neighborhood of K_q.* □

This theorem explains why the black region in Figure 4.15 resembles the black region in Figure 4.16. The map N^2 is polynomial-like of degree 2 on a region U' contained in the rectangle shown in Figure 4.15. The corresponding quadratic given by Theorem 4.21 is $z \mapsto z^2 - 1$.

Although Theorem 4.21 explains why we see quadratic Julia sets in Newton's method, it does not explain why there is a region in Figure 4.13 that resembles the Mandelbrot set. However, the Douady-Hubbard paper also contains results that apply to parameter space.

Before we state the relevant result from [DH4], we recall a special case of Exercise 3.9, which illustrates the result in a familiar case.

4.22 Example Let f_λ be the quadratic polynomial

$$f_\lambda(z) = \lambda z + z^2.$$

Since any quadratic is conjugate to one of the form $z \mapsto z^2 + c$, we have c as a function of λ. In fact, this function is

$$c = \frac{\lambda}{2} - \frac{\lambda^2}{4},$$

which is a branched covering map. □

With Example 4.22 in mind, we turn to the general result. Let Λ be a simply-connected domain in $\mathbf{C}$ and $\{f_\lambda \mid \lambda \in \Lambda\}$ be a one-parameter family of degree two polynomial-like maps. We define

$$\Psi: \Lambda \to \mathbf{C}$$

where

$$q_\lambda(z) = z^2 + \Psi(\lambda)$$

is related to f_λ by Theorem 4.21. We also define M_Λ as $\Psi^{-1}(M)$.

4.23 Theorem (Douady-Hubbard) *If Ψ is not constant and M_Λ is compact, then*

$$\Psi: M_\Lambda \to M$$

is a branched covering map. □

Remark Theorem 4.23 explains why we see a Mandelbrot set in Figure 4.13. On a simply-connected domain contained in the rectangle illustrated in Figure 4.13, the family of second iterates of the Newton's method functions is a polynomial-like family of degree 2 (on the appropriate domains in $\overline{\mathbf{C}}$). Thus, Theorem 4.23 indicates that we will see branched covers of the Mandelbrot set in parameter space.

The Douady-Hubbard paper also indicates how to determine the degree of the covering map from Theorem 4.23. Let A be a closed subset homeomorphic to $\overline{\mathbf{D}}$ such that $A \subset \Lambda$ and $M_\Lambda \subset \operatorname{int}(A)$.

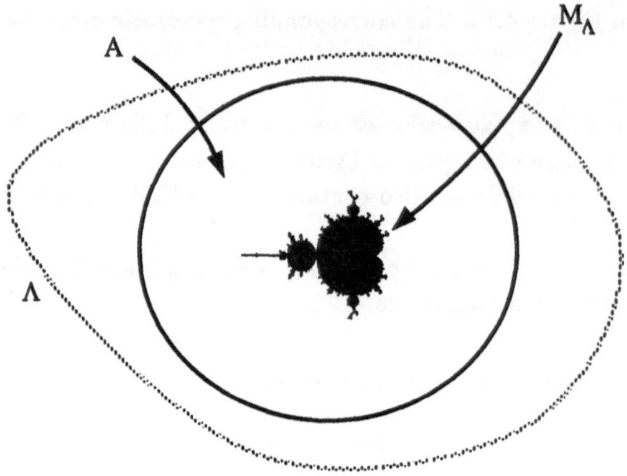

Figure 4.24: The closed, simply-connected region A that contains M_Λ.

4.25 Theorem *Let f_λ and Ψ represent the maps in Theorem 4.23 and let w_λ denote the critical point of f_λ. Then,* $\deg(\Psi\colon M_\Lambda \to M)$ *equals the winding number of $f_\lambda(w_\lambda) - w_\lambda$ about 0 as λ traverses once around the boundary of A.* □

Therefore, when the winding number is 1, M_Λ is homeomorphic to the Mandelbrot set.

Polynomials whose degree is greater than three usually have move than one inflection point. Thus, it is reasonable to expect examples of Newton's method with more than one periodic attracting orbit. Hurley [Hu] has shown that, for $d \geq 3$, there is a polynomial of degree d with $d-2$ distinct attracting periodic orbits.

4.26 Theorem (Hurley) *For each $d \geq 3$, there exists a polynomial $p(z)$ of degree d whose corresponding Newton's method function has $d-2$ distinct attracting periodic orbits of period greater than one.* □

Appendix

The following list provides detailed specifications for Figures 4.11 – 4.15.

Figure 4.11:
- Lower left corner: -0.1
- Upper right corner: $1.1 + 1.2i$

Figure 4.12:
- Lower left corner: $0.869039 + 0.323172i$
- Upper right corner: $0.971966 + 0.450939i$

Figure 4.13:
- Lower left corner: $0.894704 + 0.407193i$
- Upper right corner: $0.919835 + 0.435468i$

Figure 4.14:
- Lower left corner: $-0.790167 - 0.907198i$
- Upper right corner: $1.83213 + 1.55605i$

Figure 4.15:
- Lower left corner: $0.471402 - 0.023518i$
- Upper right corner: $0.807454 + 0.318906i$

References

[A1] L. Ahlfors, *Complex Analysis*, McGraw-Hill, 1979.

[A2] L. Ahlfors, *Conformal Invariants: Topics in Geometric Function Theory*, McGraw-Hill, 1973.

[A3] L. Ahlfors, *Lectures on Quasiconformal Mappings*, Van Nostrand, 1966.

[Ab1] W. Abikoff, *The Real Analytic Theory of Teichmüller Space*, Lecture Notes in Math. **820**, Springer-Verlag, 1980.

[Ab2] W. Abikoff, The uniformization theorem, *Amer. Math. Monthly* **88**(1981), 574–592.

[Ar] V. Arnold, Small denominators I: On the mappings of the circumference onto itself, *Amer. Math. Soc. Transl.*(2) **46**(1965), 213–284.

[At] P. Atela, *Bifurcations of Dynamic Rays in Complex Polynomials of Degree Two*, (preprint) University of Colorado, Boulder, Colorado, 1990.

[B1] P. Blanchard, Complex analytic dynamics on the Riemann sphere, *Bull. Amer. Math. Soc. (New Series)* **11**(1984), 85–141.

[B2] P. Blanchard, Disconnected Julia sets, *Chaotic Dynamics and Fractals*, ed. Barnsley and Demko, Academic Press, 1986, 181–201 .

[Bak1] I. Baker, The existence of fixpoints of entire functions, *Math. Z.* **73**(1960), 280–284.

[Bak2] I. Baker, Multiply connected domains of normality in iteration theory, *Math Z.* **81**(1963), 206–214.

[Bak3] I. Baker, Fixpoints of polynomials and rational functions, *J. London Math. Soc.* **39**(1964), 615–622.

[Bak4] I. Baker, Sets of non-normality in iteration theory, *J. London Math. Soc.* **40**(1965), 499–502.

[Bak5] I. Baker, The distribution of fixpoints of entire functions, *Proc. London Math. Soc.*(3) **16**(1966), 493–506.

[Bak6] I. Baker, Repulsive fixpoints of entire functions, *Math. Z.* **104**(1968), 252–256.

[Bak7] I. Baker, Limit functions and sets of non-normality in iteration theory, *Ann. Acad. Sci. Fenn. Ser. A* **467**(1970), 1–11.

[Bak8] I. Baker, Completely invariant domains of entire functions, *Mathematical Essays Dedicated to A.J. MacIntyre*, Ohio. Univ. Press, 1970.

[Bak9] I. Baker, The domains of normality of an entire function, *Ann. Acad. Sci. Fenn. Ser. A* **1**(1975), 277–283.

[Bak10] I. Baker, An entire function which has wandering domains, *J. Austral. Math. Soc. (A)* **22**(1976), 173–176.

[Bak11] I. Baker, The iteration of polynomials and transcendental entire functions, *J. Austral. Math. Soc. (A)* **30**(1981), 483–495.

[Barn] M. Barnsley, *Fractals Everywhere*, Academic Press, 1988.

[BD] B. Branner and A. Douady, Surgery on complex polynomials, *Holomorphic Dynamics*, Lecture Notes in Math. **1345**, Springer-Verlag, 1988, 11–72.

[BH1] B. Branner and J.H. Hubbard, The iteration of cubic polynomials, Part I: the global topology of parameter space, *Acta Math.* **160** (1988), 143–206.

[BH2] B. Branner and J.H. Hubbard, The iteration of cubic polynomials, Part II: patterns and parapatterns, *Acta Math.*, to appear.

[Bi] B. Bielefeld (ed.), *Conformal Dynamics Problem List*, preprint #1990/1, SUNY StonyBrook, Institute for Mathematical Sciences.

[BL] P. M. Bleher and M. Yu. Lyubich, *The Julia Sets and Complex Singularities in Hierarchical Ising Models*, preprint #1990/12, SUNY StonyBrook, Institute for Mathematical Sciences.

[BM] R. Brooks and P. Matelski, The dynamics of 2-generator subgroups of PSL(2,C), *Riemann Surfaces and Related Topics*, Proceedings 1978 Stony Brook Conference, ed. Kra & Maskit, *Ann. Math. Stud.* **97**, Princeton U. Press, 1981, 65–71.

[Br1] B. Branner, The Mandelbrot set, *Chaos and Fractals*, ed. Devaney and Keen, *Proc. Symp. Applied Math.* **39**, Amer. Math. Soc., 1989, 75-105.

[Br2] B. Branner, The parameter space for complex cubic polynomials, *Chaotic Dynamics and Fractals*, ed. Barnsley and Demko, Academic Press, 1986, 169–179.

[Bro] H. Brolin, Invariant sets under iteration of rational functions, *Arkiv für Matematik* **6**(1965), 103–144.

[Böt] L.E. Bötkher, The principal laws of convergence of iterates and their application to analysis (Russian) *Izv. Kazan Fiz.-Mat. Obshch.* **14**(1904), 155–234.

[Bry] A.D. Bryuno, Convergence of transformations of differential equations to normal forms, *Dokl. Akad. Nauk. USSR* **165**(1965), 987–989.

[Bu] R. Burckel, Iterating analytic self-maps of discs, *Amer. Math. Monthly* **88** (1981), 396–407.

[C1] C. Carathéodory, *Theory of Functions*, Vols. 1,2, Chelsea, New York, 1960.

[C2] C. Carathéodory, Über die Begrenzung einfach zusammenhängender Gebiete, *Math. Ann.* **73**(1913), 323–370. (Gesam. Math. Schr., v.4.)

[Ca1] A. Cayley, The Newton–Fourier Imaginary Problem, *Amer. J. Math* **2** (1879), 97.

[Ca2] A. Cayley, On the Newton–Fourier Imaginary Problem, *Proc. Cambridge Phil. Soc.* **3** (1880), 231–232.

[Ca3] A. Cayley, Application of the Newton–Fourier method to an imaginary root of an equation, *Quart. J. Pure Appl. Math.* **16** (1879), 179–185.

[Ca4] A. Cayley, Sur les racines d'une équation algébrique, *C. R. Acad. Sci. Paris* **110** (Janvier–juin 1890), 174–176, 215–218.

[Cam] C. Camacho, On the local structure of conformal mappings and holomorphic vector fields, *Astérisque* **59–60**(1978), 83–94.

[CE] P. Collet and J. Eckmann, *Iterated Maps of the Interval as Dynamical Systems*, Birkhäuser, 1980.

[CGS] J. Curry, L. Garnett and D. Sullivan, On the iteration of a rational function: Computer experiments with Newton's method, *Comm. Math. Phys.* **91**(1983), 267–277.

[CH] S. Chow and J. Hale, *Methods of Bifurcation Theory*, Springer-Verlag, 1982.

[Coh] H. Cohn, *Conformal Mapping on Riemann Surfaces*, McGraw-Hill, 1967.

[Col] P. Collet, Local C^∞ conjugacy on the Julia set for some holomorphic perturbations of $z \to z^2$, *J. Math. Pures Appl.*(9) **63**(1984), 391–406.

[Cr] H. Cremer, Zum Zentrumproblem, *Math. Ann.* **98**(1928), 151–163.

[D1] A. Douady, Systèmes dynamiques holomorphes, Séminaire Bourbaki 1982/1983, Exposé 599, *Astérisque* **105–106**(1983), Societé math. de France.

[D2] A. Douady, Disques de Siegel et anneaux de Herman, *Séminaire Bourbaki* **39** (1986–87), no. 677.

[D3] A. Douady, Julia sets and the Mandelbrot set, *The Beauty of Fractals*, by Peitgen and Richter, Springer-Verlag, 1986, 161–173.

[D4] A. Douady, Chirurgie sur les applications holomorphes, *Proceedings of the I.C.M.*, Berkeley (1986), Amer. Math. Soc.,724–738.

[D5] A. Douady, Algorithm for computing angles in the Mandelbrot set, *Chaotic Dynamics and Fractals*, ed. Barnsley and Demko, Academic Press, 1986, 155–168.

[DD] R. Devaney and M.B. Durkin, The exploding exponential and other chaotic bursts in complex dynamics, *Amer. Math. Monthly* (to appear).

[De1] R. Devaney, The structural instability of Exp(z), *Bull. Amer. Math. Soc. (New Series)* **11**(1984), 85–141.

[De2] R. Devaney, *An Introduction to Chaotic Dynamical Systems*, §3, Addison-Wesley, 1985, 1989.

[De3] R. Devaney, Exploding Julia sets, *Chaotic Dynamics and Fractals*, ed. Barnsley and Demko, Academic Press, 1986, 141–154.

[Den] A. Denjoy, Sur l'itération des fonctions analytiques, *C.R. Acad. Sci. Paris* **182**(1926), 255–257.

[DG] R. Devaney and L. Goldberg, Uniformization of attracting basins for exponential maps, *Duke Math. J.* **55**, No.**2** (1987), 253–266.

[DH1] A. Douady and J. Hubbard, Itération des polynômes quadratiques complexes, *C.R. Acad. Sci. Paris* **294**(1982), 123–126.

[DH2] A. Douady and J. Hubbard, *Étude dynamique des polynômes complexes*, Publications mathématiques d'Orsay, Université de Paris-Sud.

[DH3] A. Douady and J.H. Hubbard, A proof of Thurston's topological characterization of rational functions, preprint, Mittag-Leffler Institute, 1984.

[DH4] A. Douady and J.H. Hubbard, On the dynamics of polynomial-like mappings, *Ann. Sci. École Norm. Sup.* **18**(1985), 287–343.

[DK1] R. Devaney and L. Keen, Dynamics of meromorphic maps: Maps with polynomial Schwarzian derivative, *Ann. Sci. École Norm. Sup.* (4) **22** (1989), 55–79.

[DK2] R. Devaney and L. Keen, Dynamics of tangent, *Dynamical Systems* (College Park, MD, 1986-87), Lecture Notes in Math. **1342**, Springer-Verlag, 1988, 105–111.

[DKr] R. Devaney and M. Krych, *Dynamics of* $\exp(z)$, *Ergodic Theory Dynamical Systems* **4**(1984), 35–54.

[DM] P. Doyle and C. McMullen, *Solving the Quintic by Iteration*, (preprint) Princeton University, 1988.

[dL] R. de la Llave, A simple proof of a particular case of C. Siegel's center theorem, *J. Math. Phys.* **24**(1983), 2118–2121.

[E] J. Écalle, Théorie itérative: introduction á la théorie des invariants holomorphes, *J. Math. Pures Appl.* **54**(1975), 183–258.

[EL1] A. E. Eremenko and M. Yu. Lyubich, Dynamical Properties of Some Classes of Entire Functions, preprint #1990/4, SUNY StonyBrook, Institute for Mathematical Sciences.

[EL2] A. E. Eremenko and M. Yu. Lyubich, The dynamics of analytic transformations, *Leningrad Math. J.* **1** (1990), 563–634.

[F1] P. Fatou, Sur les équations fonctionelles, *Bull. Soc. Math. France* **47**(1919), 161–271; **48**(1920), 33–94 and 208–314.

[F2] P. Fatou, Sur l'itération des fonctions transcendantes entières, *Acta Math.* **47**(1926), 337–370.

[F3] P. Fatou, Sur les solutions uniformes de certaines équations fonctionnelles, *C.R. Acad. Sci. Paris* **143**(1906), 546–548.

[Fi] Y. Fisher, Exploring the Mandelbrot set, *The Science of Fractal Images*, ed. Peitgen and Saupe, Springer-Verlag, 1989, 287–296.

[FK] H. Farkas and I. Kra, *Riemann Surfaces*, Graduate Texts in Math. **71**, Springer-Verlag, 1980.

[G] L. R. Goldberg, *Fixed Points of Polynomial Maps, Part I: Rotation Subsets of the Circle*, preprint #1990/14, SUNY StonyBrook, Institute for Mathematical Sciences.

[GK] L. Goldberg and L. Keen, *The Mapping Class Group of a Generic Quadratic Rational Map and Automorphisms of the 2-Shift*, preprint.

[GM] L. R. Goldberg and J. Milnor, *Fixed Points of Polynomial Maps, Part II: Fixed Point Portraits*, preprint #1990/14, SUNY StonyBrook, Institute for Mathematical Sciences.

[Gu] J. Guckenheimer, Endomorphisms of the Riemann Sphere, *Proc. Sympos. Pure Math.* **14** (S.S.Chern and S.Smale, eds.), Amer. Math. Soc. (1970), 95–123.

[H1] M. Herman, Exemples de fractions rationnelles ayant une orbite dense sur la sphère de Riemann, *Bull. Soc. Math. France* **112** (1984), 93–142.

[H2] M. Herman, *Lecture notes on an easy proof of Siegel's theorem.* See also Appendice VII of [H1].

[H3] M. Herman, Sur la conjugation différentiable des difféomorphismes du cercle à des rotations, *Pub. I.H.E.S.* **49** (1979), 5–233.

[Hi] E. Hille, *Analytic Function Theory,* Vols. 1,2, Chelsea, New York, 1973.

[HP] F. von Haeseler and H.-O. Peitgen, Newton's method and complex dynamical systems, *Acta Appl. Math.* **13** (1988), 3–58.

[Hu] M. Hurley, Multiple attractors in Newton's method, *Ergodic Theory Dynamical Systems* **6** (1986), 561–569.

[J] G. Julia, Memoire sur l'itération des fonctions rationnelles, *J. Math. pures et app.* **8**(1918), 47–245. See also *Oeuvres de Gaston Julia* (Gauthier-Villars, Paris) **1**, 121–319.

[Ja1] M. Jakobson, Structure of polynomial mappings on a singular set, *Mat. Sb.* **80**(1968), 105–124; English transl. in *Math. USSR-Sb.* **6**(1968), 97–114.

[Ja2] M. Jakobson, On the problem of the classification of polynomial endomorphisms of the plane, *Mat. Sb.* **80**(1969), 365–387; English transl. in *Math. USSR-Sb.* **9**(1969), 345–364.

[K] L. Keen, Julia sets, *Chaos and Fractals, the Mathematics Behind the Computer Graphics*, ed. Devaney and Keen, *Proc. Symp. Appl. Math.* **39**, Amer. Math. Soc., 1989, 57–75.

[Ko] G. Koenigs, Recherches sur les intégrales de certaines équations fonctionnelles, *Ann. Sci. École Norm. Sup. (3)* **1**(1884) supplém., 1–41.

[L] S. Lattès, Sur l'itération des substitutions rationnelles et les fonctions de Poincaré, *C.R. Acad. Sci. Paris* **166**(1918), 26–28.

[Le] L. Leau, Étude sur les équations fonctionnelles à une ou plusieurs variables, *Ann. Fac. Sci. Toulouse* **11**(1897).

[Lei] T. Lei, *Cubic Newton's Method of Thurston's Type*, (preprint) Institut für dynamische Systeme, Universität Bremen, FRG.

[Lev] S. Levy, *Critically Finite Rational Maps*, Ph.D. Dissertation, Princeton University, 1985.

[LV] O. Lehto and K. Virtanen, *Quasiconformal Mappings in the Plane*, Springer-Verlag, 1973.

[Ly1] M. Lyubich, The dynamics of rational transforms: the topological picture, *Russian Math. Surveys* **41:4** (1986), 43–117.

[Ly2] M. Lyubich, Some typical properties of the dynamics of rational maps, *Russian Math. Surveys* **38**(1983), 154–155.

[Ly3] M. Lyubich, An analysis of the stability of the dynamics of rational functions, *Selecta Math. Sovietica* **9**(1990), 69–90. (Russian original published in 1984.)

[M1] B. Mandelbrot, *The Fractal Geometry of Nature*, Freeman, 1982.

[M2] B. Mandelbrot, Fractal aspects of the iteration of $z \mapsto \lambda z(1-z)$ for complex λ and z, *Ann. New York Academy of Sci.* **357**(1980), 249–259.

[Mat] J. Mather, Topological proofs of some purely topological consequences of Carathéodory's theory of prime ends, *Selected Studies*, ed. T. and G. Rassias, North-Holland, 1982, 225–255.

[Mc1] C. McMullen, Area and Hausdorff dimension of Julia sets of entire functions, *Trans. Amer. Math. Soc.* **300** (1987), 329–342.

[Mc2] C. McMullen, Families of rational maps and iterative root-finding algorithms, *Annals of Math.* **125** (1987), 467–493.

[Mc3] C. McMullen, Automorphisms of rational maps, *Holomorphic functions and moduli*, Vol. I (Berkeley, CA, 1986), Math. Sci. Res Inst. Publ. **10**, Springer-Verlag, 1988, 31–60.

[Mi1] J. Milnor, Self-similarity and hairiness in the Mandelbrot set, *Computers in Geometry and Topology*, ed. Tangora, Lect. Notes Pure Appl. Math. **114**, Dekker 1989, (cf. p.218 and §5), 211–257 .

[Mi2] J. Milnor, *Dynamics in One Complex Dimension: Introductory Lectures*, preprint #1990/5, SUNY StonyBrook, Institute for Mathematical Sciences.

[Mi3] J. Milnor, *Remarks on Iterated Cubic Maps*, preprint #1990/6, SUNY StonyBrook, Institute for Mathematical Sciences.

[Mis] M. Misiurewicz, On iterates of e^z, *Ergodic Theory Dynamical Systems* **1**(1981), 103–106.

[Moe] R. Moeckel, Rotations of the closures of some simply connected domains, *Complex Variables Theory Appl.* **4** (1985), no. 3, 285–294.

[Mon] P. Montel, *Leçons sur les familles normales*, Gauthier-Villars, Paris, 1927.

[MSS] R. Mañé, P. Sad and D. Sullivan, On the dynamics of rational maps, *Ann. Sci. École Norm. Sup.* (4) **16** (1983), 193–217.

[MT] J. Milnor and W. Thurston, Iterated maps of the interval, *Dynamical Systems* (Maryland 1986–87), ed. J.C. Alexander, Lect. Notes in Math. **1342**, Springer-Verlag, 1988, 465–563.

[My1] P. Myrberg, Inversion der Iteration für rationale Funktionen, *Ann. Acad. Sci. Fenn. Ser. A* **292**(1960).

[My2] P. Myrberg, Sur l'itération des polynômes réels quadratiques, *J. Math. Pures Appl. (9)* **41**(1962), 339–351.

[My3] P. Myrberg, Iteration der Polynome mit reellen Koeffizienten, *Ann. Acad. Sci. Fenn. Ser. A* **348**(1964).

[Na] M. Narasimhan, *Riemann Surfaces*, Tata Inst. Fund. Res. Math. Pamphlets No.1, 1963.

[O] M. Ohtsuka, *Dirichlet Problem, Extremal Length and Prime Ends*, van Nostrand, 1970.

[P] H.-O. Peitgen, Fantastic deterministic fractals, *The Science of Fractal Images*, ed. Peitgen and Saupe, Springer-Verlag, 1989, 169–218.

[Pf] G.A. Pfeifer, On the conformal mapping of curvilinear angles; the functional equation $\phi[f(x)] = a_x\phi(x)$, *Trans. Amer. Math. Soc.* **18**(1917), 185–198.

[PR] H.-O. Peitgen and P.H. Richter, *The Beauty of Fractals*, Springer-Verlag, 1986.

[R1] M. Rees, Ergodic rational maps with dense critical point forward orbit, *Ergodic Theory Dynamical Systems* **4**(1984), 311–322.

[R2] M. Rees, Positive measure sets of ergodic rational maps, *Ann. Sci. École Norm. Sup.* (4) **19**(1986), 383–407.

[R3] M. Rees, *Hyperbolic Rational Maps*, preprint.

[Ri] J. Ritt, On the iteration of rational functions, *Trans. Amer. Math. Soc.* **23**(1920), 348–356.

[Ro] P. Rosenbloom, L'itération des fonctions entières, *C.R. Acad. Sci. Paris* **9**(1948), 382–383.

[Ru] D. Ruelle, Repellers for real analytic maps, *Ergodic Theory Dynamical Systems* **2**(1982), 99–108.

[Rü] H. Rüssman, Kleine Nenner, II: Bemerkungen zur Newtonschen Methode, *Nachr. Akad. Wiss. Göttingen Math. Phys.* **K1**(1972), 1–20.

[S1] D. Sullivan, Quasiconformal mappings and dynamics I, *Annals of Mathematics* **122**(1985), 401–418.

[S2] D. Sullivan, *Quasiconformal mappings and dynamics III: Topological conjugacy classes of analytic endomorphisms*, preprint.

[S3] D. Sullivan, Seminar on conformal and hyperbolic geometry, *IHES Seminar notes*, March 1982.

[S4] D. Sullivan, Conformal dynamical systems, *Geometric Dynamics*, ed. Palis, Lecture Notes in Math. **1007**, Springer-Verlag, 1983, 725–752.

[Sch] E. Schröder, Ueber iterirte Functionen, *Math. Ann.* **3**(1871); see p.303.

[Sh1] M. Shishikura, On the quasiconformal surgery of rational functions, *Ann. Sci. École Norm. Sup.* **20**(1987), 1–29.

[Sh2] M. Shishikura, *Configuration of Herman rings and dynamical systems on trees*, (preprint) Department of Mathematics, Kyoto Univ., Kyoto, Japan.

[Sh3] M. Shishikura, Surgery of complex analytic dynamical systems, *Dynamical Systems and Non-linear Oscillations* (Japanese) (Kyoto, 1985), 93–105, World Sci. Adv. Ser. Dyn. Syst. **1**, World Publishing, Singapore, 1986.

[Sh4] M. Shishikura, *A new tree associated with Herman rings*, (preprint) Department of Mathematics, Kyoto Univ., Kyoto, Japan.

[ShT] M. Shishikura and T. Lei, *A family of cubic rational maps and matings of cubic polynomials*, (preprint) Max-Planck-Institut für Mathematik, Bonn.

[Si] C. Siegel, Iteration of analytic functions, *Annals of Math.* (2) **43**(1942), 607–612.

[SM] C. Siegel and J. Moser, *Lectures on Celestial Mechanics*, Springer-Verlag, 1971.

[Sm] S. Smale, Differentiable dynamical systems, *Bull. Amer. Math. Soc.* **73** (1967), 747–817.

[ST] D. Sullivan and W. Thurston, Extending holomorphic motions, *Acta Math.* **157** (1986), 243–257.

[Su] S. Sutherland, *Finding Roots of Complex Polynomials with Newton's Method*, (preprint) Institute for Mathematical Sciences, SUNY Stony Brook.

[T] W. Thurston, *On the dynamics of iterated rational maps*, preprint, Princeton University.

[Tö] H. Töpfer, Über die Iteration der ganzen transzendenten Funktionen, insbesondere von sin z und cos z, *Math. Ann.* **117** (1939), 65–84.

[UvN] S. Ulam and J. von Neumann, On combinations of stochastic and deterministic processes, *Bull. Amer. Math. Soc.* **53**(1947), 1120.

[Y1] J.-C. Yoccoz, Linéarisation des germes de difféomorphismes holomorphes de (**C**,0), *C.R. Acad. Sci. Paris* **306**(1988), 55–58.

[Y2] J.-C. Yoccoz, Conjugation différentiable des difféomorphismes du cercle dont le nombre de rotation vérifie une condition diophantienne, *Ann. Sci. École Norm. Sup.* (4) **17**(1984), 333–359.

[Z] E. Zehnder, A simple proof of a generalization of a theorem of C.L. Siegel, Lecture Notes in Math., **597**, Springer-Verlag, 1977, 855–866.

Substitutions, branching processes and fractal sets

F. Michel DEKKING
Faculty of Technical Mathematics and Informatics
Delft University of Technology
P.O. Box 356
NL-2600 AJ Delft
The Netherlands

Abstract

Substitutions and branching processes provide analytical methods to describe many deterministic, respectively random sets. Fractal curves generated by substitutions are introduced, followed by a discussion of the determination of their Hausdorff dimension. We give a necessary and sufficient variant of the "open set condition", and an application of this condition in the quasi-lattice case. Dynamical systems generated by substitutions are defined, and it is shown how this leads to an interpretation of many fractal curves as realizations of generalized random walks. We then define random substitutions, and mention their relation to branching processes. Finally we discuss a particular random fractal set known as Mandelbrot percolation.

Introduction

In this series of lectures I shall treat a large class of deterministic and random fractal sets from the viewpoint of substitutions. The fractal sets generated by substitution are called *recurrent sets* in [De 2]. This class coincides with the collection of attractors (also called invariant sets) of what usually is called *recurrent iterated function systems* (RIFS) of linear maps [BEH]. These have also been named *mixed invariant sets* [Ba1], or *graph self-similar sets* [Ed], or even *graph directed constructions* [MW2]. The fact that these definitions lead to the same class of sets has been shown first by Bedford [Be1], although not in complete generality. See [Ba1], [Ba2] for a full treatment.

The viewpoint of RIFS's leads to an attractive mathematical theory, but I believe that the viewpoint of substitutions leads to more insight in the structure of the sets involved

J. Bélair and S. Dubuc (eds.), Fractal Geometry and Analysis, 99–119.

especially that structure which is not tied up with the contraction aspect of the construction. An important example of this is the *open set condition* ([Hu], [Mo]), which is essential to determine the Hausdorff dimension. It seems probable that no algorithmic equivalent of this condition will be found and this is illustrated by the class of examples in section 3.

In sections 5 and 6 we show how many fractal curves may be interpreted as realizations of generalized random walks. Surprisingly, the open set condition is equivalent to transience of these random walks. The viewpoint of substitutions also leads to a natural generalization of random fractal sets by considering random substitutions. This has been proposed by Mandelbrot [Ma1], and analyzed further by Peyrière [Pe1, Pe2, Pe3]. In this way the theory of branching processes can be used as a tool in fractal analysis. The problem which remains is that branching process theory is in general purely analytical (cf. [AN]), and it may require some effort to handle the geometrical aspects of the fractal sets (this is the same type of problem as with the open set condition). In the final section we discuss the structure of a random fractal set known as the random Sierpinski carpet.

1 Curves and substitutions

We consider plane polygonal curves whose vertices lie on the lattice $\mathbf{Z}^2$. Our favorite examples are the paperfolding curves $K_1, K_2, \ldots$, obtained by unfolding a piece of paper folded n times at right angles, and projecting the result on the plane, cf. Figure 1 (and see [DMP]). There are only four possible directions for the segments of the curves. These will be coded by a, b, c and d, cf. Figure 2.

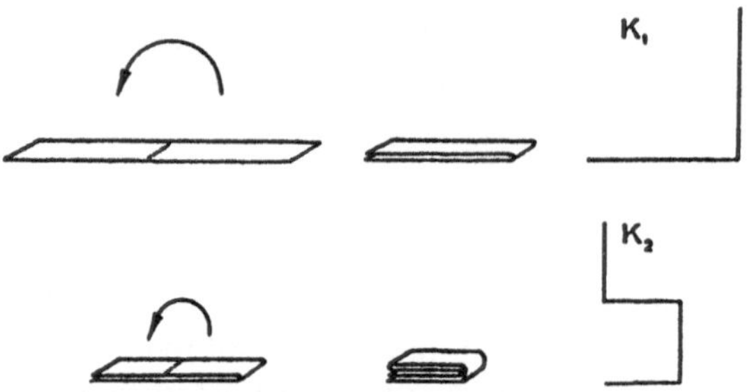

Figure 1: Paperfolding and paperfolding curve

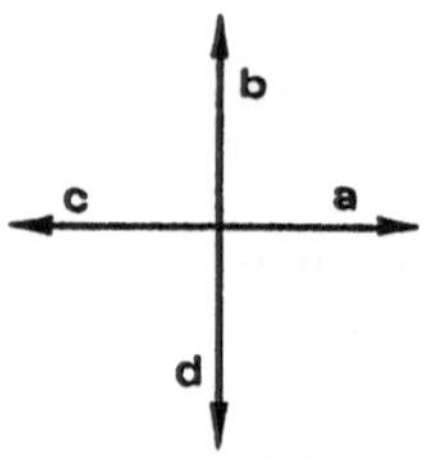

Figure 2: Coding of directions

In this way K_1 is coded by ab, K_2 by $abcb$, etc. More formally, we are considering a set of symbols $\{a, b, c, d\}$, and *words* $w = w_1 \ldots w_n$, the elements of S^n, for $n \geq 1$. Words $u = u_1 \ldots u_m$ and $v = v_1 \ldots v_n$ can be *concatenated* to $uv = u_1 \ldots u_m v_1 \ldots v_n \in S^{n+m}$. This gives $S^* = \cup_{n \geq 0} S^n$ a semigroup structure. A *substitution* σ is a map $\sigma\colon\ S^* \to S^*$, which respects the semigroup structure. This simply means that for all words $w_1 \ldots w_n$

$$\sigma(w_1 \ldots w_n) = \sigma(w_1) \ldots \sigma(w_n),$$

i.e., substitute for each symbol w_i its image $\sigma(w_i)$ for $i = 1, \ldots, n$ to obtain the image of $w_1 \ldots w_n$.

The substitution which is of interest for the paperfolding example is defined by

$$\sigma(a) = ab,\ \sigma(b) = cb,\ \sigma(c) = cd,\ \sigma(d) = ad. \tag{1.1}$$

The interest lies herein, that if K_n is coded by the word $w = w_1 \ldots w_{2^n}$, then K_{n+1} is coded by the word $\sigma(w)$. (This can be proved easily by induction.) As we would rather like to handle the curves than their codes, we define a map $f:\ S^* \to \mathbf{R}^d$, which is a *homomorphism*, i.e., has the property $f(vw) = f(v) + f(w)$ for all words v, w. We call f an *embedding* of S^* into $\mathbf{R}^d$ (here d is the dimension of the Euclidean space, usually $d = 2$).

For the paperfolding curves one would take

$$f(a) = (1,0), f(b) = (0,1), f(c) = (-1,0), f(d) = (0,-1). \tag{1.2}$$

If σ is as in (1.1), and $w = w_1 \ldots w_{2^n} = \sigma^n(a)$ (here σ^n denotes the nth iterate of σ), then $f(w_1 \ldots w_k)$ is the kth vertex of the paperfolding polygon K_n for any k with $1 \le k \le 2^n$. This still does not give us a curve. To achieve this, we define a map $K\colon\ S^* \to \mathcal{K}(\mathbf{R}^d)$ (the non-empty compact subsets of $\mathbf{R}^d$), which satisfies for all words v, w

$$K(vw) = K(v) \cup (K(w) + f(v)),$$

where for $A \subset \mathbf{R}^d, A + x = \{a + x\colon\ a \in A\}$.

A map K which will yield polygons is given by

$$K(s) = \{\alpha f(s)\colon\ 0 \le \alpha \le 1\}, \quad s \in S. \tag{1.3}$$

So $K(s)$ is the line segment joining the origin to the point $f(s)$. With this choice we can finally write

$$K_n = K(\sigma^n a), \qquad n = 1, 2, \ldots$$

In Figure 3 there is a picture of $K(\tau^5 a)$ (with corners rounded off), where $\tau\colon \{a,b,c,d\}^* \to \{a,b,c,d\}^*$ is the substitution given by

$$\tau(a) = abadab, \tau(b) = cb,$$

$$\tau(c) = cbcd, \tau(d) = adcd,$$

and f and K are as in (1.2) and (1.3). In this section we have seen how the substitution σ in (1.1) defines longer and longer words which code the paperfolding curves K_n. The first half of each K_{n+1} equals K_n, and this is reflected in σ by the fact that $\sigma(a)$ starts with a. Moregenerally, if for a substitution σ, and some symbol t we have $\sigma(t) = t \ldots$, then it makes sense to talk of the *infinite sequence* $\sigma^\infty(t) = \lim_{n \to \infty} \sigma^n(t)$ *generated from* t.

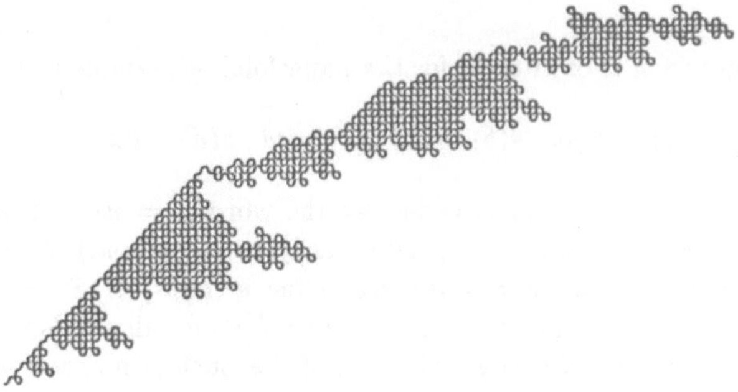

Figure 3: A curve $K(\tau^5 a)$.

2 Incorporating a scaling structure

The paperfolding curve $K_{n+1} = K(\sigma^{n+1}(a)) = K(\sigma^n(ab))$ consists of two curves $K_n = K(\sigma^n(a))$ and $K'_n = K(\sigma^n(b)) + f(\sigma^n(a))$. The curve K'_n is a rotation followed by a translation of the curve K_n. This "self-similarity" property gives the possibility of scaling the K_n and obtaining a limiting curve, the well known dragon curve.

It happens more generally that for a substitution σ on a set of symbols S, and an embedding $f\colon\ S^* \to \mathbf{R}^d$ there exists a linear map $L = L_\sigma = L_{\sigma,f}$ on $\mathbf{R}^d$ such that

$$f \circ \sigma = L \circ f. \tag{2.1}$$

We call L a *representation* of σ in $\mathbf{R}^d$. For the paperfolding substitution σ, with the embedding (1.2), L_σ has matrix representation $\begin{pmatrix} 1 & -1 \\ 1 & 1 \end{pmatrix}$, since

$$\begin{aligned} L_\sigma((1,0)) &= f(\sigma(a)) = f(a) + f(b) = (1,1), \text{ and} \\ L_\sigma((0,1)) &= f(\sigma(b)) = f(c) + f(b) = (-1,1). \end{aligned}$$

The existence of a representation of a substitution is closely connected to the structure of the *matrix* of a substitution, defined by

$$M_\sigma = (m_{st})_{s,t\in S},\ m_{st} = \mathrm{Card}\{k\colon\ \sigma(s)_k = t\} \tag{2.2}$$

where $\sigma(s)_k$ denotes the kth symbol of $\sigma(s)$. Any substitution has at least one representation: just take $f(s) = e_s$, the sth canonical basis vector in $\mathbf{R}^d$, where $d = \mathrm{Card}(S)$, and $L_\sigma = M_\sigma^T$, the transpose of the matrix of σ.

A situation which occurs frequently is that the substitution σ is *symmetric*, which means that $\sigma\pi = \pi\sigma$ for some non-trivial permutation π of the symbols of S. If $S = \{0,\ldots,m-1\}$

with $\pi(j) = j + 1$ modulo m, then a symmetric σ has a representation L_σ in $\mathbf{R}^2$ which is simply multiplication by $f(\sigma(0))$ (identifying $\mathbf{R}^2$ with $\mathbf{C}$), with the embedding $f\colon\ S^* \to \mathbf{C}$ given by $f(j) = \exp(2\pi j/m)$.

The following result from [De2] (with a small correction in the conditions) gives a rather general recipe to obtain fractal curves.

Theorem 2.1 *Let $\sigma\colon\ S^* \to S^*$ be a primitive substitution with representation L_σ, such that L_σ^{-1} is a contraction. Then there is for each $s \in S$ a unique compact set $A_\sigma = A_\sigma(s)$ such that $L_\sigma^{-n}K(\sigma^n s) \to A_\sigma$ as $n \to \infty$ in the Hausdorff metric, and A_σ is a Jordan curve.*

Here a substitution σ is *primitive* if its matrix M_σ is primitive, i.e. there exists an N such that $M_\sigma^N > 0$ (all components strictly positive). By a Jordan curve we mean a continuous image of [0,1]. We call A_σ a *recurrent curve*.

Theorem 2.1 has an extension to the case where we consider substitutions σ of the free group $G(S) \supset S^*$ generated by S ([De2]). Here one requires of course that not only $\sigma(vw) = \sigma(v)\sigma(w)$ for all words v, w but also $(\sigma(s^{-1})) = (\sigma(s))^{-1}$ for all symbols $s \in S$. The simplest example is $S = \{a, b\}$, with σ defined by $\sigma(a) = ab$, $\sigma(b) = a^{-1}b$, and the canonical embedding in $\mathbf{R}^2$ $(f(a) = (1,0), f(b) = (0,1))$. Here $A_\sigma(aba^{-1}b^{-1})$ is the boundary of the set obtained by joining four dragon curves head to tail.

Substitutions on free groups are useful to describe tilings ([De1]), Markov partitions ([Be2]), and in general boundaries of fractal sets. In this context we like to mention the substitution $\rho : \{1,2,3\}^* \to \{1,2,3\}^*$ defined by $\rho(1) = 12, \rho(2) = 13, \rho(3) = 1$.

Rauzy has shown ([Ra]) how this substitution describes the distribution of the rotation $\binom{\alpha}{\beta} \to \binom{\alpha}{\beta} + \binom{\zeta}{\zeta^2}$ on the torus, where ζ is the unique real root of $x + x^2 + x^3 = 1$ For any $n \geq 1$ the point $n\binom{\zeta}{\zeta^2}$ modulo $\binom{1}{1}$ is in the region Ω_i (see Figure 4) where $i = w_n$ if $w = \rho^\infty(1) =$ 121312112... is the infinite sequence generated from 1. Ito [It] has computed the Hausdorff dimension of the boundaries of the regions Ω_i (which is larger than 1), using the inverse (!) of the Rauzy substitution ρ, which is θ given by $\theta(1) = 3$, $\theta(2) = 3^{-1}1$, $\theta(3) = 3^{-1}2$.

By inverse we mean that $\rho\theta = \theta\rho = Id$ holds. In other words ρ and θ are automorphisms of the free group generated by $\{1,2,3\}$. Clearly $M_{\sigma^{-1}} = (M_\sigma)^{-1}$ if σ is an automorphism of $G(S)$, so a necessary condition for σ to be an automorphism is $\det(M_\sigma) = \pm 1$. Question: what are sufficient conditions?

See [Ar] for a further application of Rauzy's result to dynamical systems theory, and [Th] for work related to that of Rauzy and Ito.

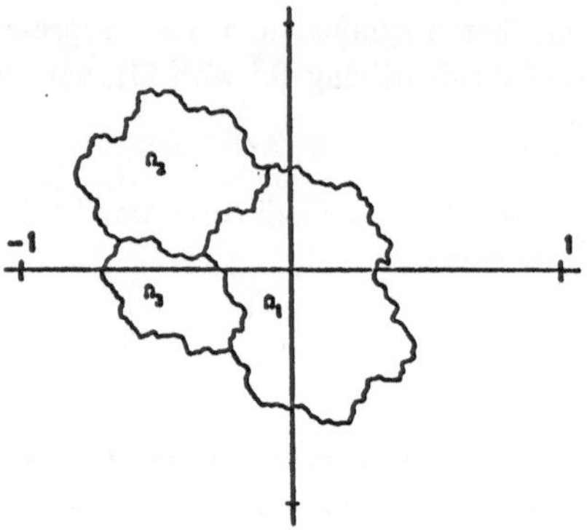

Figure 4: Regions $\Omega_1, \Omega_2, \Omega_3$

3 Hausdorff dimension

For the definition of Hausdorff dimension, and a survey of its main properties, see the lectures of Falconer and Tricot. Our next result gives a formula for the Hausdorff dimension for recurrent curves A_σ generated by substitutions under an extra condition. A situation which occurs in many examples is that σ has a *lattice representation*. This means that the embedding $f\colon\ S^* \to \mathbf{R}^d$ has the property that $f(s) \in \mathbf{Z}^d$ for all $s \in S$.

Theorem 3.1 ([De2], see also [Bel]) *Let $\sigma:\ S^* \to S^*$ be a primitive substitution with lattice representation L_σ, such that L_σ^{-1} is a contraction with contracting factor r, and let λ be the largest eigenvalue of the matrix M_σ of σ. Let A_σ be a recurrent curve generated by σ. Then*

$$\dim A_\sigma = \frac{\log \lambda}{-\log r} \iff \textit{no symbol duplicates} .$$

Here we say that a symbol $s \in S$ *duplicates* if there exists a word $w \in S^*$ such that sws is a σ-word and $f(sw) = 0 \in \mathbf{R}^d$. A σ-word is a word $v \in S^*$ which occurs in some $\sigma^n(s)$ for some $n \geq 1$, and $s \in S$. Our terminology comes from the fact that the segment from 0 to $f(s)$ is traversed twice by the curve $K(sws)$ if s duplicates (where $K(.)$ is the "polygon"-map of (1.3)).

Theorem 3.1 has an extension to the case of general embeddings with the duplication property replaced by a so called resolvability property as in [DMF]; this is proved in [Bel]. Because of the intractability of this resolvability condition (which is equivalent to Hutchinson's [Hu] open set condition), we shall restrict ourselves to the lattice case, making however an extension to what we shall call the quasi lattice case. Before doing so, we remark that there is no (known) algorithm to decide whether for a given σ and lattice embedding f no letter duplicates. There is a special case (which includes many interesting examples) where this is known. Let $S = \{a, b, c, d\}$, $f(a) = -f(c) = (1, 0)$ and $f(b) = -f(d) = (0, 1)$, and suppose σ: $S^* \to S^*$ leaves the subsemigroup T^* invariant (i.e., $\sigma(T^*) \subset T^*$), where T^* is generated by the words ab, ad, cb and cd. Suppose moreover that $f(\sigma(abcd)) = 0$. Then no symbol duplicates iff no symbol duplicates in both $\sigma(abcd)$ and $\sigma(adcb)$ (cf.[DMP]). This result applies for instance to the dragon curve (where $\lambda = 2$ and $r = 1/\sqrt{2}$).

Suppose a substitution σ has a representation L_σ (i.e. $f \circ \sigma = L_\sigma \circ f$ for a f: $S^* \to \mathbf{R}^d$) such that there exist $\delta > 0$ such that the embedding f satisfies either $f(v) = 0$ or $\| f(v) \| > \delta$ for any σ-word v. Then we say that σ has a *quasi lattice representation*. Theorem 3.1 also holds for substitutions with quasi lattice representations. The essential ingredient for the proof ([De4]) is that, as in the lattice case, a finite disk can only contain a bounded number of points of the unscaled curves. A well known example of a quasi lattice fractal set is the Penrose tiling (see e.g. [De2], and [Go] for a recent example of an aperiodic tiling). The following lemma, which is due to François Parreau, yields many more examples of quasi lattice fractal sets. For a symmetric substitution σ on $S = \{0, \dots, m-1\}$ we write $P(z) = \sum_{j=0}^{m-1} m_{0j} z^j$ for the generating function of the first row of M_σ (cf. (2.2)).

Lemma 3.2 *Let σ be a symmetric (with respect to the cyclic permutation) substitution on $\{0,1,2,3,4\}^*$, and let $\zeta = e^{2\pi i/5}$. If $|P(\zeta)| > 1$, and $|P(\zeta^2)| < 1$, then σ has a quasi lattice representation, with the embedding $f(j) = \zeta^j, j \in S$.*

Proof. The condition $|P(\zeta)| > 1$ is there to ensure that L_σ^{-1} is a contraction (L_σ is multiplication by $f(\sigma(0)) = P(\zeta)$). Let $\ell = P(1) = |\sigma(0)|$, where $|w|$ denotes the length of a word w. To check the quasi lattice condition, we first consider words u which occur at the *beginning* of some $\sigma^n(s)$. Write

$$|u| = \sum_{k=0}^{N} \epsilon_k \ell^k, \qquad 0 \le \epsilon_k < \ell.$$

Since u occurs at the beginning of $\sigma^n(s)$, there exist words $w^{(k)}$ with length ϵ_k such that

$$u = \sigma^N(w^{(N)})\sigma^{N-1}(w^{(N-1)}) \dots \sigma(w^{(1)})w^{(0)}.$$

Hence $f(u) = \sum_{k=0}^N f(\sigma^k(w^{(k)})) = \sum_{k=0}^N L_\sigma^k f(w^{(k)}) = \sum_{k=0}^N (P(\zeta))^k f(w^{(k)})$. Since $|f(w^{(k)})| \le \epsilon_k < \ell$ for all k, and $|P(\zeta^2)| < 1$, this yields that $|\tau f(u)| \le \ell/(1 - |P(\zeta^2)|) =: M$, where τ is the conjugation map which replaces ζ by ζ^2. Now let v be an arbitrary σ-word. Then there are u and u' such that $uv = u'$, and u' occurs at the beginning of some $\sigma^n(s)$. For any polynomial $Q(z)$ with integer coefficients, $|Q(\zeta)|^2 |Q(\zeta^2)|^2 = Q(\zeta)Q(\zeta^2)Q(\zeta^3)Q(\zeta^4)$ is an integer (since this expression is invariant for the four conjugation maps). In particular $|f(v)|^2 |\tau f(v)|^2$ is an integer. But then, since $|\tau f(v)| = |\tau f(u') - \tau f(u)| \le 2M, |f(v)|^2 \ge 1/2M$ if $f(v) \ne 0$. Hence σ has a quasi lattice representation.

As an example consider the symmetric substitution on $\{0, \dots, 4\}$ given by $\sigma(0) = 01$ (then $\sigma(1) = 12, \sigma(2) = 23$, etc.). Here $P(z) = 1 + z$, and

$$|P(\zeta)| = \frac{1+\sqrt{5}}{2} > 1, |P(\zeta^2)| = \frac{\sqrt{5}-1}{2} < 1.$$

Hence σ has a quasi lattice representation.

See Figure 5 for $K(\sigma^6 0)$. Also, as remarked in [DMF], no symbol duplicates (if $w = \sigma^\infty(0)$, let $s_k = f(w_0 \dots w_{k-1})$, then, since $f(w_{2k}w_{2k+1}) = f(\sigma(w_k)) = L_\sigma f(w_k) = (1+\zeta)f(w_k), s_{2k} = (1+\zeta)s_k$ and $s_{2k+1} = (1+\zeta)s_k + f(w_{2k}) = (1+\zeta)s_k + f(w_k)$. It then follows easily by induction that $s_k = s_\ell$ and $s_{k+1} = s_{\ell+1}$ imply $k = \ell$). The extended version of Theorem 3.1 then gives that

$$\dim A_\sigma = \frac{\log 2}{\log(\frac{1+\sqrt{5}}{2})}.$$

This result gives a partial answer to a question of Hata ([Ha], Example 1, p.409).

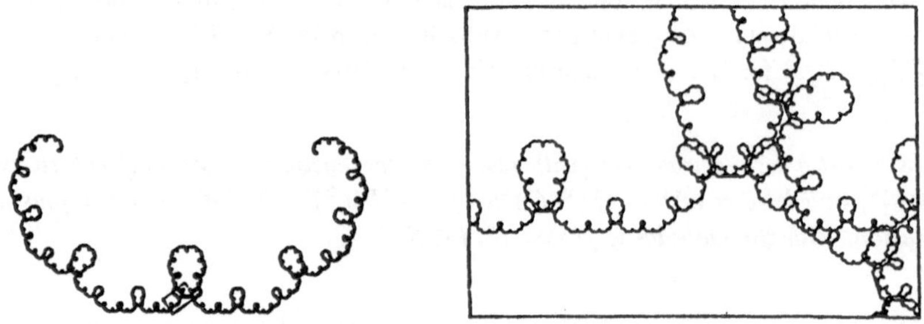

Figure 5: A quasi lattice fractal curve, and a detail

4 Binary bit parity

In this section we consider an elementary number theory problem with unexpected fractal aspects. It also serves as an introduction to the notion of generalized random walk, which will be treated in section 6.

The *binary bit parity* $s(n)$ of a natural number n equals the number of ones in its binary expansion modulo 2. For example,

$$s(0) = 0, s(1) = 1, s(2) = 1, s(3) = 0, s(6) = 0, s(9) = 0, s(12) = 0, s(15) = 0, s(18) = 0.$$

There must be something wrong here, as the multiplies of three have no particular reason to have an even numbers of ones in their expansion (we shall give a precise mathematical meaning to this statement in section 6). In the long run there should be on the average as many $s(3k) = 0$ as there are $s(3k) = 1$. To study this, we consider

$$T_n = \sum_{k=0}^{n-1} (-1)^{s(3k)}.$$

It has been shown for the first time by D.J. Newman [Ne] that the even's always stay ahead of the odds in the following strong sense. There exist positive real constants α, c_*, c^* such that for all $n \geq 1$

$$c_* n^\alpha \leq T_n \leq c^* n^\alpha. \tag{4.1}$$

Here $\alpha = \log 3/\log 4 < 1$, so indeed $\frac{1}{n}T_n \to 0$ as $n \to \infty$. The analysis of T_n is performed via the generating function

$$Z_n = \sum_{k=0}^{n-1} (-1)^{s(k)} \xi^k, \tag{4.2}$$

where $\xi = \exp(2\pi i/3)$. The behaviour of T_n can be deduced from that of Z_{3n} (cf.[De5]). Now consider the symmetric substitution σ on the symbol set $S = \{\pm 1, \pm\xi, \pm\xi^2\}$ defined by

$$\sigma(1) = 1, -\xi, -\xi^2, 1, \sigma(\xi^j) = \xi^j \sigma(1), \sigma(-s) = -\sigma(s),\ s \in S.$$

Let $\sigma^\infty(1) = w_0 w_1 w_2 \ldots$. Then one can see that $w_k = (-1)^{s(k)}\xi^k$, the kth term of the sum Z_n in (4.2). (This is an exercise, note first that $(s(k))$ itself is generated by the substitution θ on $\{0,1\}$ given by $\theta(0) = 01, \theta(1) = 10$, and hence also by θ^2.) Now using this description, it is easy to prove that the points $Z_n, n = 1, 2, \ldots$ traverse the unscaled von Koch curve (putting $Z_0 = 0$), cf. Figure 6. With a little more effort one shows that it is justified to interpret the exponent α in (4.1) as the inverse of the Hausdorff dimension of the von Koch curve.

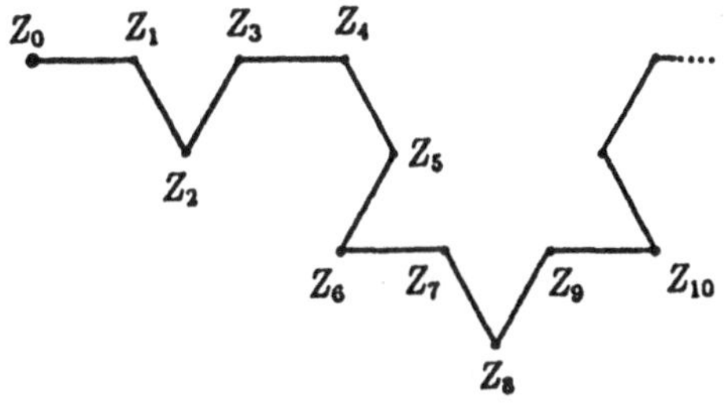

Figure 6: Binary bit parity and von Koch curve

A similar observation has been made by Coquet [Co], who computed the best possible c_* and c^* in (4.1). See [Du] for the behaviour of the binary bit parity in other arithmetical sequences.

5 Dynamical systems generated by substitutions

See the monograph by Queffelec [Qu] for the details and proofs of the results in this section. Let $X = S^{\mathbb{N}}$ be the space of one-sided infinite sequences over S; X is equipped with the map $T: X \to X$, called the *shift*, defined by $(Tx)_k = x_{k+1}$ for $k \geq 0$. Given a primitive substitution σ on S^*, let X_σ be the subset of X consisting of all sequences in which only σ-words occur. By primitivity of σ this set equals the closure of the orbit (under T) of the infinite sequence $\sigma^\infty(s)$, for any s such that $\sigma(s)$ starts with s.

Theorem 5.1 *Let σ be a primitive substitution. There is a unique T-invariant probability measure μ_σ on X_σ, and*

(a) *μ_σ is ergodic (by uniqueness).*

(b) *μ_σ is diffuse (i.e., has no atoms), if X_σ is not finite.*

(c) *for $w = w_1 \dots w_m \in S^*$, if $[w] := \{x \in X_\sigma:\ x_1 \dots x_m = w\}$ then*

$$\mu_\sigma([w]) = \lim_{n\to\infty} \frac{\text{Card}\{i:\ w \text{ occurs in } \sigma^n(s) \text{ at position } i\}}{|\sigma^n(s)|}$$

for all $s \in S$.

(d) *if λ is the largest eigenvalue of M_σ and $\pi M_\sigma = \lambda\pi$, where $\sum_{s\in S}\pi_s = 1$, then*

$$\mu_\sigma([s]) = \pi_s \quad \text{for } s \in S.$$

6 Generalized random walks

A sequence of random variables $(S_n)_{n=0}^\infty$ such that $S_0 = 0$, and $S_n = X_1 + \dots + X_n$, where the X_k are independent identically distributed random variables with values in $\mathbf{R}^d$ is called a *random walk*. If the $(X_n)_{n=1}^\infty$ only form a stationary and ergodic sequence, then $(S_n)_{n=0}^\infty$ is called a *generalized random walk* (GWR). Any shift dynamical system (X, T, μ), where $X = S^{\mathbf{N}}$ with μ T-invariant and ergodic, together with a function $f:\ S \to \mathbf{R}^d$ yields a GWR in $\mathbf{R}^d$ by putting for $n \geq 1$

$$X_n(x) = f(x_n) \quad \text{for } x = x_1x_2 \dots \in S^{\mathbf{N}}.$$

The classical example is *simple random walk* on $\mathbf{Z}$ given by $S = \{0,1\}, \mu = \Pi_{n=1}^\infty(p, 1-p)$ (product measure) for some $p \in (0,1)$, and $f(0) = 1, f(1) = -1$. For simplicity we consider GRW's on the d-dimensional lattice, given by a $f:\ S \to \mathbf{Z}^d$.

A GRW (S_n) is called *recurrent* if

$$\mu\{x \in X:\ \exists n > 0 \quad S_n(x) = 0\} = 1,$$

otherwise (S_n) is called is *transient*. These two types of behaviour are, as in the random walk case, very different, since

(S_n) recurrent iff $\mu\{x \in X: \; S_n(x) = 0 \quad \infty \text{ - often}\} = 1$ (Exercise)

(S_n) transient iff $\mu\{x \in X: \| S_n(x) \| \to \infty\} = 1$ ([Sc]).

To decide which type of behaviour occurs, one often uses the notion of the *range* R_n of a GWR, defined by

$$R_n(x) = \mathrm{Card}\{S_1(x), S_2(x), \ldots, S_n(x)\}, \qquad n = 1, 2, \ldots$$

See e.g. [De3] for the proofs of the following facts.

Theorem 6.1 *Let $D = \{x \in X: \; S_n(x) \neq 0, n = 1, 2, \ldots\}$ be the collection of paths which never return to the origin. Then $\frac{1}{n} ER_n := \frac{1}{n}\int R_n(x)d\mu(x) \to \mu(D)$, and $\frac{1}{n}R_n(x) \to \mu(D)$ for μ-almost x.*

Theorem 6.2 *A GRW (S_n) is recurrent iff $\mu(D) = 0$.*

Theorem 6.3 *If $d = 1$, and $E|X_1| = \int |f(x_1)| d\mu(x) < \infty$, then (S_n) recurrent iff $EX_1 = 0$.*

For example, simple random walk on **Z** is recurrent if and only if $p = \frac{1}{2}$.

We now return to the binary bit parity in the multiples of three, i.e., the behaviour of $T_n = \sum_{k=0}^{n-1}(-1)^{s(3k)}$. We shall show that T_n is a realization of a recurrent GRW (S_n) (this means that $T_n = S_n(x)$ for some $x \in X$, and some shift dynamical system (X, T, μ)). The substitution θ defined by $\theta(0) = 01, \theta(1) = 10$, which generates the sequence $s(k)$ ($s(k)$ equals the kth symbol of $\theta^\infty(0)$), induces a substitution σ on the set S of θ-words of length 3 as follows

$$\begin{aligned}\sigma([011]) = [011][010], \sigma([010]) = [011][001], \sigma([001]) = [010][110]\\ \sigma([110]) = [101][001], \sigma([101]) = [100][110], \sigma([100]) = [100][101].\end{aligned}$$

Hence if we define for all $i, j \in \{0, 1\}, i \neq j$

$$f([0ij]) = 1, f([1ij]) = -1 \tag{6.1}$$

then $(-1)^{s(3k)}$ equals $f(w_k)$, where $w = \sigma^\infty([011])$, and $T_n = \sum_{k=0}^{n-1}(-1)^{s(3k)}$ is the realization $T_n = S_n(w)$ of the GWR determined by $(X_\sigma, T, \mu_\sigma)$ and f in (6.1). The matrix M_σ has constant column sums, hence (1,1,1,1,1,1) is a left eigenvector of M_σ with the maximal eigenvalue 2 of M_σ. By Theorem 5.1.d), $\mu_\sigma([s]) = \frac{1}{6}$ for all $s \in S$, and hence

$$EX_1 = \int f(x_1) d\mu_\sigma(x) = \frac{1}{6}\sum_{s \in S} f(s) = 0. \tag{6.2}$$

By Theorem 6.3, (S_n) is recurrent. For another example, let $S = \{a, b, c, d\}$, and let σ be the paperfolding substitution of section 1. Then M_σ has again column sums constantly equal to 2, so for the paperfolding measure μ_σ given by Theorem 5.1, $\mu_\sigma([a]) = \mu_\sigma([b]) = \mu_\sigma([c]) = \mu_\sigma([d]) = \frac{1}{4}$. Hence if we consider what might be called the *paperfolding random walk*, given by $(X_\sigma, T, \mu_\sigma)$ and the same f as in (1.2) $(f(a) = -f(c) = (1,0), f(b) = -f(d) = (0,1))$, then the increments X_n have (as in (6.2)) mean zero. However, the paperfolding random walk is transient. To see this, consider $x = x_1x_2\ldots$ in the support of μ_σ (the smallest closed set with full μ_σ- measure).

Then $\mu_\sigma([x_1 \ldots x_k]) > 0$ for all $k \geq 1$ (otherwise the complement of the cylinder $[x_1 \ldots x_k]$ is a closed set of measure 1). But then, by Theorem 5.1 c), the word $x_1 \ldots x_k$ occurs in $\sigma^n(a) = y_1y_2\ldots y_{2^n}$ for n large enough. Geometrically this means that the "walk" $f(x_1), f(x_1x_2), \ldots, f(x_1x_2\ldots x_k)$ is a "subwalk" of $f(y_1), f(y_1y_2), \ldots, f(y_1y_2\ldots y_{2^n})$. But it is obvious that this "walk" does not visit any lattice point more than twice, and hence the paperfolding random walk is transient. (Exercise. This really is a μ_σ-almost sure statement; show that there exist $x \in X_\sigma$ such that $f(x_1 \ldots x_n) = 0$ for infinitely many n). See Figure 7 for a realization of (S_n). For the GRW's $(S_n) = (S_n^{\sigma,f})$ generated by a primitive substitution σ and an embedding f there is a criterion for transience.

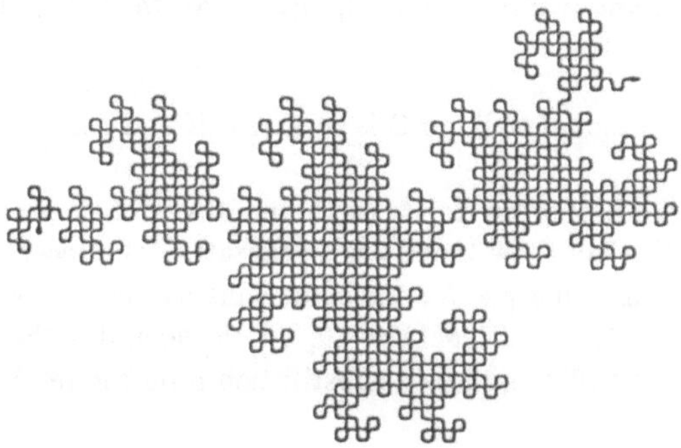

Figure 7: A realization of the paperfolding random walk

Theorem 6.4 *$(S_n^{\sigma,f})$ is transient iff no symbol duplicates.*

Proof (Outline, cf. p.464 of [De3]). If no symbol duplicates, then it follows (with an argument similar to the one above for the paperfolding random walk) that no point is visited more than Card(S) times, for μ_σ-almost all $x \in X_\sigma$, so $(S_n^{\sigma,f})$ is transient. On the other hand, if a symbol t duplicates, then virtualizing as in [De4] or [Be1], one can see that this implies $R_{|\sigma^n(s)|}/|\sigma^n(s)| \to 0$ exponentially fast as $n \to \infty$ for any $s \in S$. This implies that $\frac{1}{n}R_n(x) \to 0$ for all $x \in X_\sigma$. So by Theorem 6.1 and 6.2, $(S_n^{\sigma,f})$ is recurrent.

For an example where segments are traversed twice but in opposite directions consider

$S = \{a, b, c, d\}$, and the substitution on S defined by

$$\sigma(a) = abdadba, \sigma(b) = bcabacb,$$

$$\sigma(c) = cbdcdbc, \sigma(d) = dcadacd,$$

and let f be the embedding of (1.2). It is easily proved by induction that no symbol duplicates, hence $(S_n^{\sigma,f})$ is transient. See Figure 8 for a realization of this GRW. Exercise: Show that the probability of no return $\mu_\sigma(D)$ of this GRW equals 11/30 (Hint: construct a Markov chain whose state space consists of lattice points with configurations of edges to compute $7^{-n}R_{7^n}(\sigma^\infty(a))$).

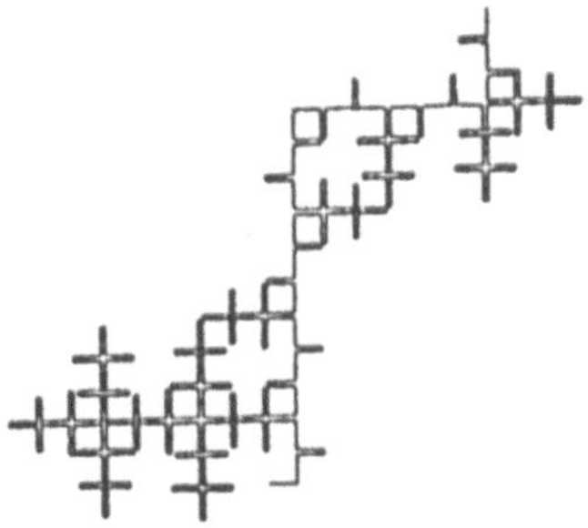

Figure 8: A realization of a transient GRW

Finally we remark that there are GRW's $(S_n^{\sigma,f})$ which are extremely recurrent in the sense that $\| S_n^{\sigma,f} \|$ is bounded for all realizations. For examples see [De3] and [MFS].

7 Random substitutions and branching processes

As Benoît Mandelbrot has stressed ([Ma2]), an effective modelling of (fractal) natural phenomena will often require an element of randomness (see also the title of his first essay on the subject: "Les objects fractals. Forme, chance et dimension"). Many people will have seen the stunning imitations of landscapes by fractional Brownian motion. For imitations of rivers less randomness and more control is required. As an example we consider Mandelbrot's river model ([Ma2], §24]). The formulation in terms of random substitutions is due to Peyrière ([Pe1, Pe3]). A river flowing from west to east is approximated by polygons R_n on a nested sequence (Δ_n) of trigonal lattices, cf. Figure 9.

On each level there are two types of triangles traversed by R_n: the A-type where the single level $n+1$ triangle is on the left, and the V-type where it is on the right of the traversed polygon, cf. Figure 10.

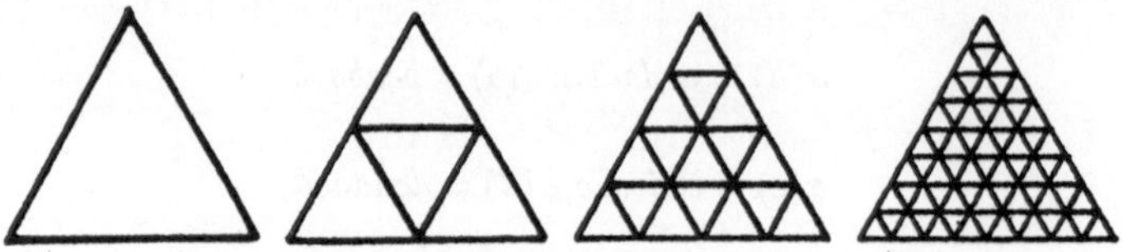

Figure 9: Trigonal lattices $\Delta_1, \Delta_2, \Delta_3$ and Δ_4

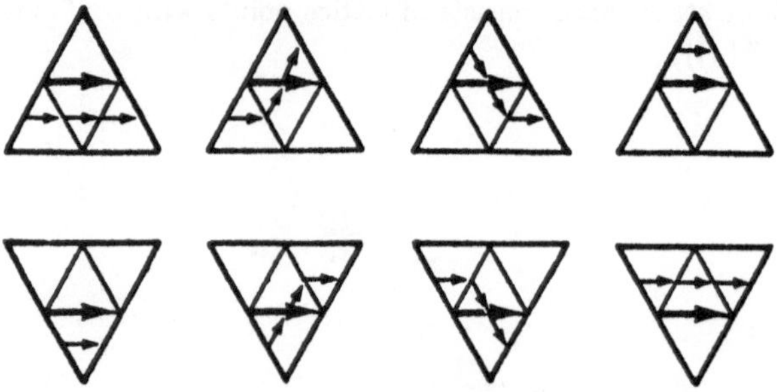

Figure 10: Triangles of A-type (top), and V-type(bottom), and crossings.

Let p with $0 < p < 1$ be a fixed parameter. At each level the left side L is chosen with probability p, the right side R with probability $1 - p$ independently for each side which is crossed by R_n, and R_{n+1} is obtained by connecting the midpoints of the chosen sides in the obvious way, cf. Figure 10. Clearly the R_n converge uniformly to a limiting curve R. See Figure 11 for two realizations of R_9. The random substitution which describes how to obtain R_{n+1} from R_n (with the obvious embedding) is given by

$$\begin{array}{ll} \sigma_{L,L}(A) = A & \sigma_{L,L}(V) = VAV \\ \sigma_{L,R}(A) = VAA & \sigma_{L,R}(V) = VVA \\ \sigma_{R,L}(A) = AAV & \sigma_{R,L}(V) = AVV \\ \sigma_{R,R}(A) = AVA & \sigma_{R,R}(V) = V. \end{array}$$

(Actually the substitution is symmetric, and there are 6 different A-type and V-type segments corresponding to the 6 directions in the trigonal lattice, here we have, for simplicity, identified all directions). Here the randomness resides in the pairs $(L, L), \ldots, (R, R)$. Writing σ for the random substitution we have for instance

$$P[\sigma = \sigma_{L,R}] = p(1 - p).$$

The expectation of the matrix of σ is given by

$$EM_\sigma = p^2 \begin{pmatrix} 1 & 0 \\ 1 & 2 \end{pmatrix} + 2p(1-p) \begin{pmatrix} 2 & 1 \\ 1 & 2 \end{pmatrix} + (1-p)^2 \begin{pmatrix} 2 & 1 \\ 0 & 1 \end{pmatrix}$$

$$= \begin{pmatrix} 2-p^2 & 1-p^2 \\ 1-(1-p)^2 & 2-(1-p)^2 \end{pmatrix}.$$

The largest eigenvalue of EM_σ is $\lambda_p = 2 + 2p(1-p)$. Peyrière shows ([Pe1], [Pe3]) that the random version of Theorem 3.1 holds, i.e.,

$$\dim(R) = \frac{\log \lambda_p}{\log 2} \quad \text{almost surely.}$$

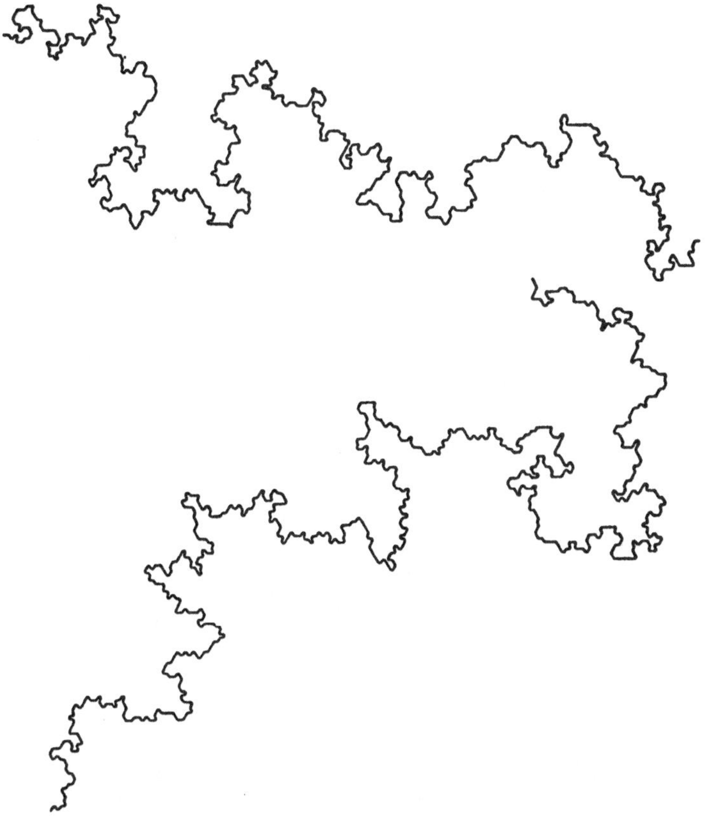

Figure 11: Two realizations of ninth order approximants to random river

This is an example of a much more general result proved in [Pe1], [Pe2] and [Pe3]. It is noteworthy that in these results dependence between neighbouring symbols (as in the river example) is allowed. In other work on random fractal sets ([Ha], [Fa], [Gr], [MW2]) independence is usually assumed.

If we forget the geometric structure, and just look at the number N_n of segments (each having length 2^{-n}) of the nth polygon R_n, then it is not hard to see that (N_n) is a branching process (for $p = \frac{1}{2}$), cf. [Ma1], [Pe1]. A *branching process* (Z_n) is a collection of random variables with $Z_0 = 1$, $Z_{n+1} = 0$ if $Z_n = 0$, and for $Z_n > 0$

$$Z_{n+1} = \sum_{k=1}^{Z_n} X_{k,n},$$

where the $X_{k,n}$ for $k = 1, \ldots, Z_n$ are independent random variables distributed as Z_1. Thus the (distributional) properties of the process are determined by the distribution of Z_1 which is called the *offspring distribution.* For the general case, the total number of occurrences of a symbol s in a word generated by a random substitution is described by a *multitype branching process*, where there is a offspring distribution $Z_1(s)$ for each symbol (type) s, a vector which gives the number of offspring of s of type t for each $t \in S$ (see [AN]).

8 Mandelbrot percolation and the random Sierpinski carpet

Here we shall discuss the geometry of a simple class of random fractal sets generated by random substitutions. I shall call these sets random Cantor sets, despite the fact that this is an abuse of language: the sets need not be totally disconnected. The terminology refers rather to the usual construction of the middle third Cantor set $C = \cap C_n$, obtained by leaving out intervals of C_n to obtain C_{n+1} for $n + 1, 2, \ldots$. The whole point about Mandelbrot percolation is that by merely changing the probability structure infinitesimally (keeping the substitution the same), the set may change from being totally disconnected to having large connected components with positive probability.

By way of introduction, let us generate the middle third Cantor set C with a substitution. Let $S = \{a, \overline{a}\}$, and let $\sigma\colon S^* \to S^*$ be defined by

$$\sigma(a) = a\overline{a}a, \quad \sigma(\overline{a}) = \overline{a}\overline{a}\overline{a}.$$

Let $f\colon S^* \to \mathbf{R}$ be the embedding given by $f(a) = f(\overline{a}) = 1$, and consider $K\colon S^* \to \mathcal{K}_0(\mathbf{R})$ (the collection of all compact subsets of the real line), defined by

$$K(a) = [0,1], \quad K(\overline{a}) = \emptyset.$$

Note that $f \circ \sigma = L \circ f$, where $L(x) = 3x$. It is rather obvious that

$$C = \lim_{n\to\infty} L^{-n} K(\sigma^n(a)).$$

For an example in the plane, let $S = \{a, b, c, d, \overline{a}, \overline{b}, \overline{c}, \overline{d}\}$ and let σ_1 be the symmetric ($\sigma\pi = \pi\sigma$, for π given by $\pi(a, b, c, d) = (b, c, d, a)$, $\pi(\overline{a}, \overline{b}, \overline{c}, \overline{d}) = (\overline{b}, \overline{c}, \overline{d}, \overline{a})$) substitution defined by

$$\sigma_1(a) = abad\overline{c}daba, \quad \sigma_1(\overline{a}) = \overline{a}\overline{a}\overline{a}.$$

Let f be the canonical embedding (as in (1.2)), and let $K(\overline{a}) = K(\overline{b}) = K(\overline{c}) = K(\overline{d}) = \emptyset$, $K(s)$ the square with diagonal $[0, f(s)]$ for $s = a, b, c, d$. See Figure 12 for $K(\sigma(a))$. The $L^{-n}K(\sigma^n(a))$ now converge to the Sierpinski carpet, where L is multiplication by 3 in the plane. Let $\sigma_0 : \ S^* \rightarrow S^*$ be the symmetric substitution (π as above) defined by $\sigma_0(a) = \overline{aaa}$, $\sigma_0(\overline{a}) = \overline{aaa}$. For a fixed $p \in (0,1)$, we consider the random substitution $\boldsymbol{\sigma}$ defined by

$$P[\boldsymbol{\sigma} = \sigma_0] = 1 - p, \ P[\boldsymbol{\sigma} = \sigma_1] = p.$$

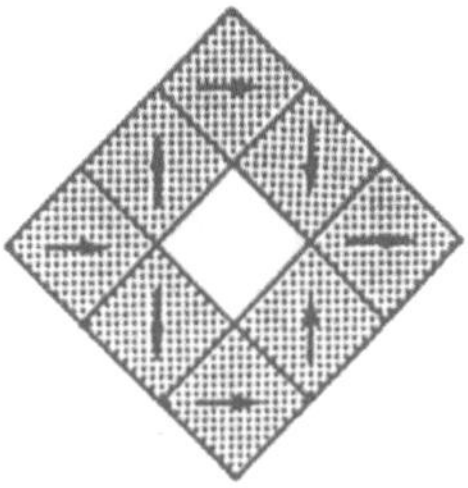

Figure 12: First order approximant to Sierpinski carpet

We now iterate $\boldsymbol{\sigma}$, replacing symbols S in $\boldsymbol{\sigma}^n(a)$ by words $\boldsymbol{\sigma}(s)$, independently of each other. Geometrically this means that n^{th} order triadic squares in the set $K(\boldsymbol{\sigma}^n(a))$ are removed with probability $1-p$ or kept, except for the middle $(n+1)^{th}$ order triadic square, with probability p, independently of the other squares. We call the limiting set $A = A_p = \lim_{n\rightarrow\infty} L^{-n}K(\boldsymbol{\sigma}^n(a))$ the *random Sierpinski carpet* with parameter p. This is a variation on *Mandelbrot percolation* ([CCD]). See Figure 13 for a realization of a $K(\boldsymbol{\sigma}^6 a)$. The set A is almost surely empty for $p \leq \frac{1}{8}$ (the corresponding branching process dies out with probability 1), and has Hausdorff dimension

$$\dim(A) = \frac{\log(8p)}{\log 3}$$

almost surely for $p > \frac{1}{8}$. This follows by applying the results of any of [Pe1], [Ha], [Fa], [MW2], [Gr]. The set $A = A_p$ passes through several phases of increasing denseness if the parameter p increases from 0 to 1. Without proofs we quote from [DM] (see also [DG]).

I. $A = \emptyset$ almost surely, for $0 \leq p \leq \frac{1}{8}$
II. $P[A = \emptyset] > 0$, but $\dim(\pi A) = \dim(A)$, for $\frac{1}{8} < p \leq 54^{-\frac{1}{4}}$
III. $\dim(\pi A) < \dim(A)$ a.s. given $A \neq \emptyset$, but $\lambda(\pi A) = 0$ a.s., for $54^{-\frac{1}{4}} < p \leq 18^{-\frac{1}{3}}$
IV. $0 < \lambda(\pi(A)) < 1$ a.s. given $A \neq \emptyset$,
V. $P[\lambda(\pi(A)) = 1] > 0$, but A does not percolate a.s.,
VI. A percolates with positive probability.

Here π is the projection on the x-axis, λ is Lebesgue measure on [0,1], and A *percolates* means that A contains a connected set which has a non-empty intersection with the left side

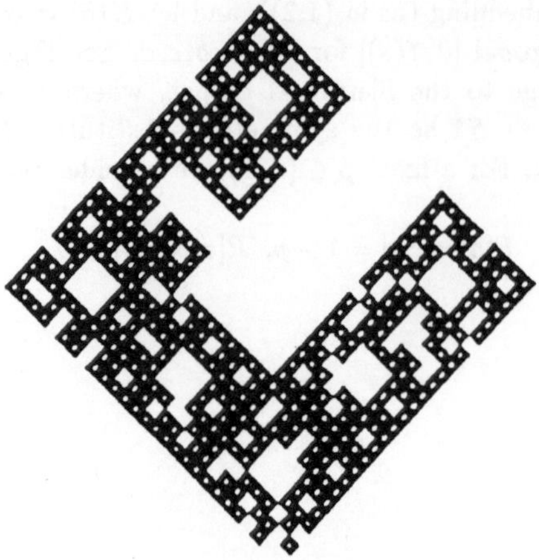

Figure 13:
A realization of the random Sierpinski carpet (third order approximant)

and the right side of $[0,1] \times [0,1]$. It is known that there are non-empty intervals of p-values so that IV, V and VI occur. E.g. A percolates if $p > (\frac{7}{8})(\frac{49}{48})^6$ ([DM]). The percolation phase was conjectured by Mandelbrot [Ma2] (for a similar model), and proven to exist in [CCD]. See [Me] for more results on the geometric structure of A in the percolation phase.

References

[AN] Athreya, K.B., Ney, P.E. *Branching Processes.* Berlin-Heidelberg-New York: Springer 1972.

[Ar] Arnoux, P. Un exemple de semi-conjugaison entre un échange d'intervalles et une translation sur le tore. *Bull. Soc. Math. France* **116** (1988), 489-500.

[Ba1] Bandt, Ch. Self-similar sets. I. Topological Markov chains and mixed self-similar sets. *Math. Nachr.* **142** (1989), 107-123.

[Ba2] Bandt, Ch. Self-similar sets III. Constructions with sofic systems. *Monatsh. Math.* **108** (1989), 89-102.

[BEH] Barnsley, F., Elton, J.H., Hardin, D. Recurrent iterated functions systems. *Constr. Approx.* **5** (1989), 3-39.

[Be1] Bedford, T. Dynamics and dimension for fractal recurrent sets. *J. London Math. Soc. (2)* **33** (1986), 89-100.

[Be2] Bedford, T. Generating special Markov partitions for hyperbolic toral automorphisms using fractals. *Ergodic Theory Dynamical Systems* **6** (1986), 325-333.

[Be3] Bedford, T. On Weierstrass-like functions and random recurrent sets. *Math. Proc. Cambridge Philos. Soc.* **106** (1989), 325-342.

[CCD] Chayes, J.T., Chayes, L. and Durrett, R. Connectivity properties of Mandelbrot's percolation process. *Probab. Theory Rel. Fields* **77** (1988), 307-324.

[Co] Coquet, J. A summation formula related to the binary digits. *Invent. Math.* **73** (1983), 107-115.

[De1] Dekking, F.M. Replicating superfigures and endormorphisms of free groups. *J. Combin. Theory Ser. A* **32** (1982), 315-320.

[De2] Dekking, F.M. Recurrent sets. *Adv. in Math.* **44** (1982), 78-104.

[De3] Dekking, F.M. On transience and recurrence of generalized random walks. *Z. Wahrsch. Verw. Geb.* **61** (1982), 459-465.

[De4] Dekking, F.M. Recurrent sets: a fractal formalism. Report 82-32. Technische Hogeschool Delft, Delft, 1982.

[De5] Dekking, F.M. On the distribution of digits in arithmetic sequences. *Sém. Théorie des Nombres Bordeaux* **82-83** (1983), 3201-3211.

[DG] Dekking, F.M., Grimmett, G.R. Superbranching processes and projections of random Cantor sets. *Probab. Theory Rel. Fields* **78** (1988), 335-355.

[DM] Dekking, F.M., Meester, R.W.J. On the structure of Mandelbrot's percolation process and other random Cantor sets. *J. Statist. Phys.* **58** (1990), 1109-1125.

[DMF] Dekking, F.M., Mendès France, M. Uniform distribution modulo one: a geometrical viewpoint. *J. Reine Angew. Math.* **329** (1981), 143-153.

[DMP] Dekking, F.M., Mendès France, M., v.d. Poorten, A. Folds ! *Math. Intelligencer* 4 (1982), 130-138, 173-181, 190-195.

[Du] Dumont, J. Discrépance des progressions arithmetiques dans la suite de Morse. *C.R. Acad. Sci. Paris Sér. I* **297** (1983), 145-148.

[Ed] Edgar, G.A. *Measure, topology and fractal geometry.* Berlin Heidelberg New York, Springer 1990.

[Fa] Falconer, K.J. Random fractals. *Math. Proc. Cambridge. Philos. Soc.* **100** (1986), 559-582.

[Go] Godrèche, C. The Sphinx: a limit-periodic tiling of the plane. *J. Phys. A: Math. Gen.* **22** (1989), L1163-L1166.

[Gr] Graf, S. Statistically self-similar fractals. *Probab. Theory Rel. Fields* **74** (1987), 357-392.

[Ha] Hata, M. On the structure of self-similar sets. Japan. *J. Appl. Math.* **2** (1985), 381-414.

[Hu] Hutchinson, J.E. Fractals and self-similarity. *Indiana Univ. Math. J.* **30** (1981), 713-747.

[Hw] Hawkes, J. Trees generated by a simple branching process. *J. London Math. Soc. (2)* **24** (1981), 373-384.

[It] Ito, S. Weyl automorphisms, substitutions and fractals (1989), Preprint.

[Ma1] Mandelbrot, B. Colliers aléatoires et une alternative aux promenades au hasard sans boucle: les cordonnets discrets et fractals. *C.R. Acad. Sci. Paris Sér. I* **286** (1978), 933-936.

[Ma2] Mandelbrot, B. *The Fractal Geometry of Nature.* New York: Freeman 1983.

[Me] Meester, R.W.J. Connected components in supercritical Mandelbrot percolation. (1990). Preprint.

[MFS] Mendès France, M., Shalitt, J. Wire bending. *J. Combin. Theory Ser. A* **50** (1989), 1-7.

[Mo] Moran, P.A.P. Additive functions of intervals and Hausdorff measures. *Math. Proc. Cambridge Philos. Soc.* **42** (1946), 15-23.

[MW1] Mauldin, R.D., Williams, S.C. Random recursive constructions: asymptotic geometric and topological properties. *Trans. Amer. Math. Soc.* **295** (1986), 325-346.

[MW2] Mauldin, R.D., Williams, S.C. Hausdorff dimensions in graph directed constructions. *Trans. Amer. Math. Soc.* **309** (1988), 811-829.

[Ne] Newman, D.J. On the number of digits in a multiple of three. *Proc. Amer. Math. Soc.* **21** (1969), 719-721.

[Pe1] Peyrière, J. Mandelbrot random beadsets and birth processes with interaction. I.B.M. research report RC-7417 (1978).

[Pe2] Peyrière, J. Substitutions aléatoires itérées. *Sém. Théorie des Nombres Bordeaux* **80-81** (1981), 1701-1709.

[Pe3] Peyrière, J. Processus de naissance avec interaction des voisins, évolution des graphes. *Ann. Inst. Fourier* **31** (1981), 187-218.

[Qu] Queffélec, M. *Substitution Dynamical Systems - Spectral Analysis.* Lecture. Notes in Math. 1294, Berlin, Springer, 1987.

[Ra] Rauzy, G. Nombres algébriques et substitutions. *Bull. Soc. Math. France* **110** (1982), 147-178.

[Sc] Schmidt, K. *Lectures on Cocycles of Ergodic Transformation Groups.* Lect. in Math. 1. Delhi-Bombay-Calcutta-Madras: Mac Millan 1977.

[Th] Thurston, W.P. *Groups, Tilings and Finite State Automata.* Lecture notes. Coll. Lect. Amer. Math. Soc. (1989), Boulder, Colorado.

Interpolation fractale

Serge DUBUC
Département de mathématiques et de statistique
Université de Montréal
C.P. 6128, Succ. A
Montréal, Qué., H3C 3J7
Canada

Résumé

Motivés par la courbe de von Koch et d'autres fonctions fractales, nous définissons un procédé d'interpolation itérative pour des données fournies sur un sous-groupe discret fermé de l'espace euclidien. Nous décrivons les principales propriétés algébriques de ce procédé. À l'aide de la transformée de Fourier, on détermine des conditions suffisantes pour la continuité de chaque fonction d'interpolation selon un procédé donné d'interpolation itérative. Un critère de continuité aussi général que possible est trouvé. On s'attarde ensuite au sujet de l'interpolation fractale sur la droite réelle. Dans ce cas, on détermine les procédés qui sont localement de carré sommable. Puis on fait l'étude d'une classe d'interpolations fractales qui seraient susceptibles de concurrencer les fonctions splines, c'est l'interpolation itérative symétrique de Lagrange.

Abstract

As extension of the construction of von Koch curves and of some fractal functions, an iterative interpolation process for data spread over a closed discrete subgroup of the Euclidean space is defined. The main algebraic properties of this process are described. By means of the Fourier transform, sufficient conditions for continuity of each interpolation function of a given iterative interpolation process are found. A criterion for continuity, as general as possible, is given. Fractal interpolation on the real line is the next topic. In this case, processes which are locally square integrable are characterized. Afterwards, a class of fractal interpolation, which could be as useful for numerical analysis as spline functions, is studied, viz. the class of Lagrange iterative symmetric interpolation processes.

J. Bélair and S. Dubuc (eds.), Fractal Geometry and Analysis, 121–220.

Introduction

J'ai achevé un monument plus durable que le bronze, plus haut que la royale architecture des Pyramides et que ne sauraient détruire ni la pluie rongeuse, ni l'Aquilon emporté, ni la chaîne innombrable des ans, ni la fuite des âges.

Horace, Ode XXX, Livre III

Les formules de Newton, les polynômes de Lagrange et les polynômes orthogonaux forment les piliers de la théorie de l'interpolation. De façon plus récente, se sont ajoutés de nouveaux outils, les fonctions splines et la méthode des éléments finis. Le développement du calcul numérique par les ordinateurs n'est pas étranger à cette évolution. Sous l'influence du calcul automatique et aussi de la géométrie fractale, nous proposons une nouvelle méthode d'interpolation: l'interpolation dite itérative ou fractale. Ce schéma permet de réaliser des interpolations qui sont tantôt dérivables, tantôt non dérivables selon le choix des paramètres. Dans la version unidimensionnelle, on peut engendrer des courbes non-dérivables comme la courbe de von Koch ou des courbes plus régulières, dérivables une ou deux fois ou même plus.

Le chapitre I tient un rôle à part. Nous y exposons la construction des attracteurs de systèmes de contractions à l'aide de cinq algorithmes d'approximation. Une autre approche pour construire certaines fractales et pour établir leurs propriétés provient d'équations fonctionnelles. C'est ainsi que les courbes de von Koch peuvent se voir comme attracteurs d'un système approprié de contractions et aussi comme solution de l'équation fonctionnelle de Bajraktarević. On remarque ici que plusieurs fonctions fractales se construisent par interpolations successives.

Le chapitre II expose le modèle retenu de l'interpolation itérative à une ou plusieurs variables. Ce procédé d'interpolation part de données fournies sur un sous-groupe discret G fermé de l'espace euclidien $\mathbf{R}^d$ et prolonge une fonction f définie sur G en une fonction g définie sur un sous-groupe plus grand G_∞ de $\mathbf{R}^d$. Après avoir exposé quelques exemples, nous décrirons les principales propriétés algébriques de ce procédé, la linéarité, l'invariance sous les translations et une invariance d'échelle. Nous déterminerons une condition suffisante pour que le sous-groupe G_∞ soit dense. Si jamais la fonction g est uniformément continue sur ce sous-groupe, le prolongement de g à $\mathbf{R}^d$ achève le procédé d'interpolation.

Le chapitre III introduit la fonction fondamentale $F(x)$ d'un procédé donné d'interpolation itérative. Après avoir déterminé sommairement le support de celle-ci, nous relevons comment la fonction fondamentale permet de résoudre tout problème d'interpolation dans le cadre du procédé d'interpolation itérative. Nous identifions les principales équations fonctionnelles dont la fonction fondamentale est la solution. Par la suite, nous entreprenons ce qui pourrait s'appeler l'analyse harmonique de la fonction fondamentale; en particulier, sous des hypothèses bénignes, nous introduisons une fonction $\Phi(\xi)$ qui se manifeste comme la transformée de Fourier de la fonction fondamentale dans un certain sens. Si $\Phi(\xi) \geq 0$, alors Φ est intégrable. Si Φ est intégrable, alors $F(x)$ est uniformément continue sur le sous-groupe G_∞ et se prolonge donc par continuité à l'espace entier. Dans le chapitre IV, nous déterminons des conditions nécessaires et suffisantes pour que la fonction fondamentale d'un procédé d'interpolation itérative soit continue. Quelques exemples illustreront la théorie.

Dans les chapitres V et VI, nous nous restreignons à des procédés d'interpolation itérative unidimensionnelle, l'interpolation ayant lieu sur la droite réelle. Nous découvrirons les conditions précises pour que la distribution associée à l'interpolante fondamentale soit celle d'une fonction de carré sommable. Une première condition est que la transformée de Fourier Φ soit de carré sommable. Une autre condition fera suite à l'examen des valeurs propres d'une matrice. Nous appliquerons ces résultats aux courbes de von Koch pour obtenir un critère simple pour qu'une telle courbe soit de carré sommable et nous aurons un développement de Fourier pour chacune de celles-ci.

Le chapitre VI définit le procédé d'interpolation itérative symétrique de Lagrange. Ce procédé dépend de deux paramètres entiers, une base et un nombre pair de noeuds mobiles. On y étudie la fonction fondamentale F. Nous démontrons qu'elle est une fonction continue définie positive. Nous déterminons presque précisément la classe de Lipschitz à laquelle appartiennent les dérivées successives de F. Nous effectuons le calcul de F'. Enfin, nous fournissons des bornes sur l'erreur dans ce procédé d'interpolation itérative. C'est ce procédé qui pourrait s'avérer le plus utile à l'analyse numérique, en approximation, à la compression des données d'une fonction ou à la théorie des ondelettes.

Dans le chapitre VII, nous comparons l'interpolation fractale à d'autres méthodes d'interpolation. Cette comparaison est entreprise sous deux volets, un volet théorique par un rappel de propriétés et un volet expérimental à l'examen d'une douzaine de fonctions plus ou moins dérivables. Dans tous les cas, l'interpolation fractale supporte bien la comparaison avec les autres procédés.

Quelques autres commentaires généraux. Afin d'inviter le lecteur à une participation active, nous avons ajouté quelques problèmes à chacun des chapitres. Nous avons aussi préparé plusieurs figures, celles-ci servent diverses fins. Une de ces fins est d'illustrer la théorie et d'atténuer la sécheresse de certains lieux de l'exposé. La présence de ces figures est aussi une invitation au lecteur à observer avant de se poser d'autres questions. De plus, nous sommes convaincus que l'analyse mathématique des images, digitales ou non, deviendra de plus en plus fréquente dans l'avenir. Avec l'avènement de l'informatique, la confection et l'analyse des images devient une discipline scientifique autonome.

Je termine en remerciant Gilles Deslauriers, qui a eu partie liée dans mes recherches des dernières années et qui a lu avec soin ce manuscrit.

Chapitre I
Courbes de von Koch et autres fractales

Sommaire. Nous reprenons la définition d'une classe générale de fractales, les attracteurs d'un système de contractions. Le principe des contractions de Banach justifie l'existence de ces attracteurs et permet une première construction de ceux-ci. Nous exposerons ensuite cinq algorithmes d'approximation d'attracteurs en insistant sur la majoration des erreurs commises. Nous illustrons deux de ces approximations par la fougère de Barnsley et la courbe de Peano. Une autre approche pour construire certaines fractales et pour établir leurs propriétés provient d'équations fonctionnelles. C'est ainsi que nous étudions l'équation fonctionnelle de Bajraktarević, nous déterminons des conditions suffisantes de continuité pour les solutions bornées de cette équation. Le chapitre se termine par l'examen de courbes de von Koch. Celles-ci peuvent se voir comme attracteurs d'un système approprié de contractions et aussi comme solution de l'équation fonctionnelle de Bajraktarević.

1 Attracteurs

On définit une grande famille de fractales de la façon suivante. (X, d) est un espace métrique complet. On dispose d'un système S de m contractions définies sur X : $f_1, f_2, \ldots, f_m$. Une portion A de l'espace X est dite *invariante relativement à S* si $A = f_1(A) \cup f_2(A) \cup \ldots \cup f_m(A)$.

On dit parfois d'une portion invariante qu'elle est un *attracteur*. Nous illustrons cette définition d'attracteur par un exemple: le napperon de Sierpiński (1916). Celui-ci est le seul attracteur compact (non vide) pour le système suivant de trois transformations du plan:

$$f_1(x,y) = (\frac{x}{2}, \frac{y}{2}) \qquad f_2(x,y) = (\frac{1}{4} + \frac{x}{2}, \frac{\sqrt{3}}{4} + \frac{y}{2}) \qquad f_3(x,y) = (\frac{1}{2} + \frac{x}{2}, \frac{y}{2})$$

Le napperon de Sierpiński est formé de trois répliques de lui-même, chacune des répliques étant deux fois plus petite que le napperon lui-même. Si $(0,0), (\frac{1}{2}, \frac{\sqrt{3}}{2})$ et $(1,0)$ est le triangle équilatéral des pointes du napperon, les transformations sont les trois homothéties de rapport $\frac{1}{2}$ admettant les sommets du triangle comme centre d'homothétie.

Des ensembles invariants correspondant à divers systèmes de contractions ont été étudiés par Williams, Mandelbrot, Hutchinson, Hata, Barnsley et bien d'autres personnes. Williams et Hutchinson ont montré que pour un ensemble donné S de contractions, il existe un et un seul sous-ensemble compact non vide A de X qui soit invariant sous S. Étant donné un ensemble compact invariant à déterminer, au moins trois suites distinctes d'approximations ont été proposées par Williams, Hutchinson et Barnsley-Demko respectivement. Nous décrirons ces façons d'approcher un tel ensemble invariant, nous discuterons aussi de deux autres approximations plus efficaces. Cependant avant d'en arriver là, nous rappelons quelques résultats relatifs aux contractions d'un espace métrique.

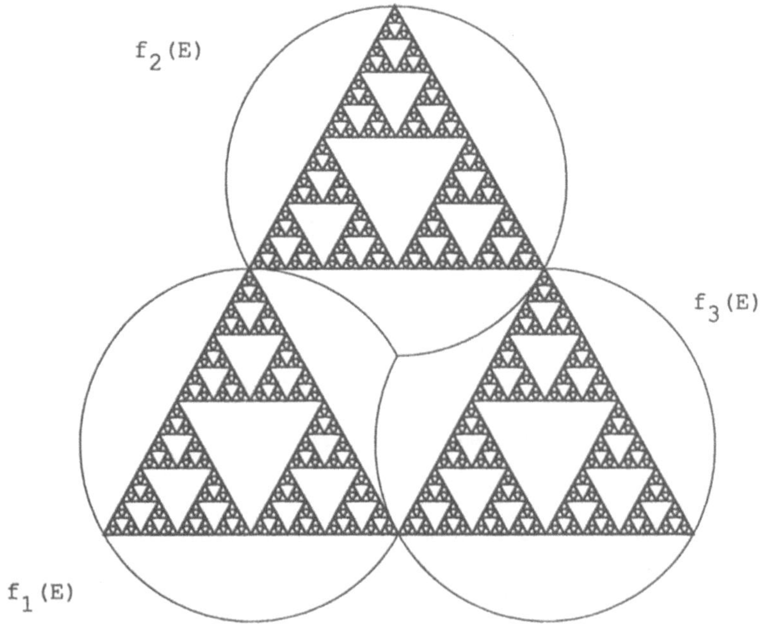

Figure 1:
Le napperon de Sierpiński interprété comme un attracteur d'un système de trois contractions.

2 Métrique de Hausdorff et contractions

Une transformation T d'un espace métrique (X, d) est *lipschitzienne* s'il existe un nombre λ tel que $d(Tx, Ty) \leq \lambda d(x, y)$ pour tout x et y de X. L'infimum de ces nombres λ est appelé la *constante de Lipschitz* de $T, L(T)$. Si $L(T)$ est inférieur à 1, on dit que T est une *contraction*. Selon l'usage, T^n est la $n$ème puissance fonctionnelle de T.

Théorème 1 (Principe de contraction, voir Ostrowski, p. 216 ou encore Kantorovich-Akilov, p. 165). *Si T est une contraction d'un espace métrique complet (X, d) avec L comme constante de Lipschitz, si x est un point donné de X, alors la suite $T^n x$ converge vers un point fixe ξ de T. De plus, pour $n = 0, 1, \ldots, d(T^n x, \xi) \leq d(Tx, x)L^n/(1 - L)$. Le point fixe de T est unique.*

Nous rappelons ce qu'est la métrique de Hausdorff telle que proposée par Hausdorff lui-même. Si A est une partie bornée d'un espace métrique (X,d), alors $d(x,A) = \inf\{d(x,a)\colon a \in A\}$; ce qui s'appelle la distance (inférieure) de x à A. On pose $\rho(A,B) = \sup\{d(b,A)\colon b \in B\}$. Par définition, la *distance de Hausdorff*, D(A,B), entre deux parties bornées A et B de l'espace métrique X est le plus grand des deux nombres $\rho(A,B)$ et $\rho(B,A)$. On considère que deux parties bornées sont équivalentes si $D(A,B) = 0$; $(2^X)_m$ est l'espace-quotient sous cette relation d'équivalence et c'est un espace métrique. Hahn a établi que $(2^X)_m$ est un espace métrique complet.

Supposons maintenant que $S = \{f_1, f_2, \ldots, f_m\}$ est un ensemble de m contractions de X. Ce système de contractions induit une transformation F définie sur $(2^X)_m$: si A est une partie bornée de X, alors on définit $F(A)$ comme $\bigcup_{i=1}^m f_i(A)$.

Lemme 2 (Hutchinson) *Si L est la plus grande des constantes de Lipschitz $L(f)$ lorsque f varie dans S, alors pour tout couple de parties bornées A et B,*

$$D(F(A),F(B)) \leq LD(A,B).$$

La constante de Lipschitz de F ne dépasse pas la plus grande des constantes de Lipschitz $\{L(f)\colon f \in S\}$. Si l'on se sert de la métrique de Hausdorff, alors F est une contraction de $(2^X)_m$. Un ensemble invariant pour S est par définition un ensemble A tel que $F(A) = A$; un ensemble invariant est un point fixe de la transformation F. Selon le principe des contractions de Banach, il existe un unique ensemble borné fermé invariant.

Théorème 3 (Hutchinson) *La famille S admet un et un seul attracteur fermé borné A. Si B est une partie bornée, alors A est la limite de $F^n(B)$ $(F^0(B) = B, F^{n+1}(B) = F(F^n(B)), n = 0,1,2,\ldots)$.*

Ce théorème est une simple conséquence du théorème 1 et du lemme 2.

3 Description de cinq algorithmes

Nous décrivons cinq algorithmes qui produisent des approximations d'attracteurs.

Algorithme A (approximation de Hutchinson) Pour toute partie B, la suite $B_n = F^n(B)$ est une suite de parties qui converge dans la métrique de Hausdorff vers un attracteur borné A. Cet algorithme est simple, particulièrement lorsque B est un singleton. Un bon choix pour B est de retenir comme singleton le point fixe d'une des contractions f de S. Dans ce cas, $F^n(\{x\})$ est une suite croissante de parties. Nous introduisons une suite de familles de transformations de l'espace métrique X. S^{01} est l'ensemble S de contractions.

Si $n \geq 2$, alors S^{on} est l'ensemble de toutes les compositions de n transformations tirées de $S, f_n \ldots \circ f_2 \circ f_1$ où f_k appartient à S. S^* est la réunion de tous les S^{on}:

$$S^* = \bigcup_{n \geq 1} S^{on}.$$

Théorème 4 *Si L_n est la plus grande des constantes de Lipschitz des transformations g de S^{on} et si A est un attracteur borné de B une partie bornée, alors*

$$D(A, F^n(B)) \leq L_n D(A, B).$$

Démonstration. Si A est un attracteur par rapport à S, alors A est aussi un attracteur par rapport à S^{on}. Si $T(B)$ est l'union des $g(B)$ pour g de S^{on}, T est une contraction. Selon le lemme 2, pour toute partie bornée B de X, $D(A, T(B)) = D(T(A), T(B)) \leq L_n D(A, B)$. Puisque $T(B) = F^n(B)$, le théorème est démontré. □

Algorithme B (construction de Barnsley-Demko) On se sert d'une suite de variables aléatoires indépendantes $\{\Phi_n\}_{n \geq 1}$ telles que les valeurs sont dans S et pour tout f de S, pour tout entier n, $\Pr(\Phi_n = f) > 0$. Un point x_0 est choisi dans X; on définit alors une suite de points aléatoires: $x_n = \Phi_n(x_{n-1})$. Presque sûrement, l'ensemble des points d'accumulation de la suite x_n coïncide avec l'unique attracteur compact A. Cette construction est très simple. Cependant deux difficultés se présentent dans cette construction. D'abord, comment choisir la distribution des variables aléatoires Φ_n? Quand les constantes de Lipschitz de chaque contraction sont identiques, il est naturel de se servir de variables aléatoires uniformes; mais si tel n'est pas le cas, il devient nécessaire d'expérimenter sur le choix de Φ_n. Et puis quel est le temps d'arrêt approprié qui permet d'approcher l'attracteur compact avec une précision donnée? Néanmoins, la construction de l'attracteur proposée par Barnsley-Demko est vraiment surprenante.

Algorithme C (Formule de Williams) Nous rappelons que S^* est l'ensemble de toutes les compositions finies de transformations de S. Si g est dans S^*, nous désignons par $\mathrm{Fix}(g)$ l'unique point fixe de g. Comme Hata l'a remarqué, Williams a démontré en 1971 que l'adhérence de $\bigcup_{g \in S^*} \mathrm{Fix}(g)$ est invariante sous S et qu'il n'existe pas d'autre ensemble fermé borné invariant. Pour intégrer ce résultat dans un algorithme, nous choisissons un nombre positif ε et nous introduisons une famille $S(\varepsilon)$ de contractions: g est dans $S(\varepsilon)$ s'il existe une suite finie de fonctions de S, $f_1, f_2, \ldots, f_n$ telles que premièrement, si $k < n$, alors la constante de Lipschitz de $f_1 \circ f_2 \circ \ldots \circ f_k$ est supérieure à ε, deuxièmement, g est la composition $f_1 \circ f_2 \ldots \circ f_n$ et la constante de Lipschitz de g est inférieure ou égale à ε. Alors l'ensemble $B = \{\mathrm{Fix}(g) : g \in S(\varepsilon)\}$, l'ensemble des points fixes des fonctions g de $S(\varepsilon)$, est une approximation de l'attracteur A.

Lemme 5 *Si A est un attracteur par rapport à S et si $\varepsilon > 0$, alors A est un attracteur par rapport à $S(\varepsilon)$.*

Démonstration. Soit ε un nombre positif donné. On peut trouver une valeur entière N telle que la constante de Lipschitz de tout produit fonctionnel de N applications de S ne dépasse pas ε. Si A est un attracteur par rapport à S, alors A est un attracteur pour $S^{\circ N}$. Toute transformation f de $S^{\circ N}$ admet une factorisation $g \circ h$, avec g dans $S(\varepsilon) \cap S^{\circ k}$ et h dans la famille $S^{\circ N-k}$ pour un certain entier k. Si g appartient à $S(\varepsilon) \cap S^{\circ k}$ alors A est la réunion des $h(A)$ pour les transformations h de $S^{\circ N-k}$, si bien que $g(A)$ est la réunion des $g(h(A))$ alors que h parcourt $S^{\circ N-k}$. La réunion de tous les $g(A)$ alors que g parcourt $S^*(\varepsilon)$ est A. □

Théorème 6 *Si* $B = \{\mathrm{Fix}(g): \ g \in S(\varepsilon)\}$ *et si* A *est un attracteur borné, alors*

$$D(A,B) \leq \varepsilon \ \mathrm{diam}(A).$$

($\mathrm{diam}(A)$ est le *diamètre* de A: $\sup\{d(x,y): \ x \in A \ et \ y \in A\}$).

Démonstration. Si A est un attracteur pour S, alors il est un attracteur pour $S(\varepsilon)$ (selon le lemme 5).

a) Estimation de $\rho(A,B)$. Si x est dans B, alors $g(x) = x$ pour au moins une application g de $S(\varepsilon)$. Mais $g(A)$ est contenu dans A, si bien que l'adhérence de A est stable sous g. D'où il s'ensuit que x appartient à l'adhérence de A: $d(x,A) = 0$. $\rho(A,B) = 0$.

b) Une borne pour $\rho(B,A)$. Si y est dans A, parce que A est un attracteur pour $S(\varepsilon)$, alors il existe une transformation g de $S(\varepsilon)$ et un point z de A tel que $y = g(z)$. Si x est le point fixe de $g, g(x) = x$. D'où

$$d(y,x) = d(g(z),g(x)) \leq \varepsilon \ d(z,x) \leq \varepsilon \ \mathrm{diam}(A).$$

$d(y,B) \leq \varepsilon \ \mathrm{diam}(A)$ et $\rho(B,A) \leq \varepsilon \ \mathrm{diam}(A)$. □

Une variante de l'algorithme contenu dans le dernier théorème est fournie par le prochain théorème.

Théorème 7 *Si* x *est le point fixe d'une des contractions de* S, *si* $B = \{g(x): \ g \in S(\varepsilon)\}$ *et si* A *est un attracteur borné, alors* $D(A,B) \leq \varepsilon D(\{x\},A)$.

Démonstration. Si A est un attracteur pour S, alors il est aussi un attracteur pour $S(\varepsilon)$ (selon le dernier lemme).

(a) Estimation $\rho(A,B)$. B est adhérent à A. $\rho(A,B) = 0$.

(b) Une borne $\rho(B,A)$. Si y est dans A, puisque A est un attracteur pour $S(\varepsilon)$, alors il existe un g de $S(\varepsilon)$ et un point z de A tel que $y = g(z)$. Si x est un point fixe, alors $d(y,g(x)) = d(g(z),g(x) \leq \varepsilon d(z,x) \leq \varepsilon D(\{x\},A)$. $d(y,B) \leq \varepsilon D(\{x\},A)$ et $\rho(B,A) \leq \varepsilon D(\{x\},A)$. □

Algorithme D (par factorisation) Dans cet algorithme, on se sert non seulement de contractions de $S(\varepsilon)$ et du point fixe d'une contraction f de S, mais aussi d'un opérateur d'arrondissement R dans l'espace métrique X. Comme attracteur approché, nous proposons la partie $B = \{h \circ R \circ g(x)\colon\ g \text{ et } h \in S(\varepsilon)\}$.

Théorème 8 *Si A est un attracteur borné, si x est le point fixe d'une contraction f de S, si $B = \{h \circ R \circ g(x)\colon\ g \text{ et } h \in S(\varepsilon)\}$ et si pour tout y de X, $d(Ry, y) \leq \delta$, alors*

$$D(A, B) \leq \varepsilon^2 D(\{x\}, A) + \varepsilon\delta.$$

Démonstration. Si $F_\varepsilon(B) = \bigcup_{g \in S(\varepsilon)} g(B)$, alors F_ε est une contraction de l'espace des parties bornées de X. La constante de Lipschitz de F_ε ne dépasse pas ε. L'attracteur A de S est le point fixe de F_ε. Nous posons $C = \{R \circ g(x)\colon\ g \in S(\varepsilon)\}$. D'où $D(A, B) = D(F_\varepsilon A, F_\varepsilon C) \leq \varepsilon D(A, C)$. Si $C' = F_\varepsilon(\{x\})$, alors $D(A, C) \leq D(A, C') + D(C', C)$. Mais $D(A, C') = D(F_\varepsilon A, F_\varepsilon\{x\}) \leq \varepsilon D(\{x\}, A)$ et $D(C', C) \leq \delta$. D'où $D(A, B) \leq \varepsilon^2 D(\{x\}, A) + \varepsilon\delta$. □

Algorithme E (un algorithme graphique) Soit δ un nombre positif. Nous choisissons ce que l'on peut appeler une maille $M(\delta)$. $M(\delta)$ est une partie de X qui remplit les deux conditions suivantes

(a) pour tout x de $X, d(x, M(\delta)) < \delta$;

(b) dans toute boule de X, on ne retrouve qu'un nombre fini de points de $M(\delta)$.

Par récurrence, nous définissons deux suites de parties de X, B_n et C_n pour $n = 0, 1, 2, \ldots$ jusqu'à ce qu'une partie vide C_{n+1} soit produite. B_0 est un singleton, une approximation, qui est un des points de $M(\delta)$, à moins de δ du point fixe d'une des contractions f de S. $C_0 = B_0$. Si B_n et C_n ont été calculées et si C_n n'est pas vide, B_{n+1} et C_{n+1} sont calculées à partir des règles suivantes.

(1) Un point x est choisi dans C_n.

(2) Une partie temporaire T est initialement vide. Une boucle sur S est faite. Pour toute contraction f de S pour laquelle la distance de $f(x)$ à $B_n \cup T$ n'est pas inférieure à δ, on choisit un point x' de $M(\delta)$ tel que $d(f(x), x') < \delta$; ce point x' est ajouté à la partie T.

(3) $B_{n+1} = B_n \cup T$ et $C_{n+1} = C_n \cup T - \{x\}$.

Nous étudions les propriétés de cet algorithme. Le nombre L reste encore la plus grande des constantes de Lipschitz $L(f)$ si f parcourt S.

Lemme 9 *Si A est un attracteur borné, alors pour tout $n, \rho(A, B_n) < \delta/(1 - L)$.*

Démonstration. Rappelons que $\rho(A,B) = \sup\{d(b,A) : b \in B\}$. Si x' appartient à $B_{n+1} - B_n$, il y a un x de B_n et une contraction f de S tels que $d(f(x),x') < \delta$. D'où

$$d(x',A) \leq d(x',f(x)) + d(f(x),A) < \delta + d(f(x),f(A)) \leq \delta + Ld(x,A).$$

Par récurrence sur n, on obtient l'inégalité $d(x',A) < \delta + L\rho(A,B_n) < \delta/(1-L)$. Et l'on établit l'inégalité $\rho(A,B_{n+1}) < \delta/(1-L)$. □

Théorème 10 *Si X est localement compact, alors pour tout nombre positif δ, il existe un entier positif N tel que C_N est vide et dans ce cas, la distance entre B_N et l'attracteur borné A est inférieure à $\delta/(1-L)$.*

Démonstration. La suite B_n est non décroissante, bornée et est contenue dans $M(\delta)$. La cardinalité de B_n est donc une suite bornée monotone. Il existe un entier N_0 pour lequel B_n est un ensemble constant lorsque $n > N_0$. La suite C_n est décroissante si $n > N_0$. Si bien qu'il existe un entier N tel que C_N est vide. Puisque C_N est vide, alors $D(F(B_N),B_N) < \delta$. Si $x \in B_N$ et si $f \in S$, alors $d(f(x),B_N) < \delta$. D'où $\rho(B_N,F(B_N)) = \sup\{d(b,B_N) : b \in F(B_N)\} < \delta$.

$$\begin{aligned} \rho(B_N,A) &\leq \rho(B_N,F(B_N)) + \rho(F(B_N),A) < \delta + \rho(F(B_N),F(A)) \\ \rho(B_N,A) &< \delta + L\rho(B_N,A) < \delta/(1-L). \end{aligned}$$

Vu le lemme 9, nous obtenons que $D(B_N,A) = \max(\rho(B_N,A),\rho(A,B_N))$ et $D(B_N,A) < \delta/(1-L)$. □

4 Illustration des algorithmes

Nous illustrons les deux derniers algorithmes, l'un par la fougère de Barnsley, l'autre par la courbe de Peano. La fougère de Barnsley (Barnsley-Ervin-Hardin-Lancaster), est l'attracteur d'un système de quatre transformations affines du plan, $\{f_k(x,y)\}_{1\leq k\leq 4}$, selon les formules suivantes: $(0,\ 0,16y),(0,849x + 0,037y,\ 1,6 - 0,037x + 0,849y)(0,197x - 0,257y,\ 1,6 + 0,226x + 0,223y),(-0,15x + 0,283y,\ 0,44 + 0,26x + 0,238y)$

La fougère de Barnsley fournie par la figure 2 a été tracée à partir de l'algorithme D. Une modification de cet algorithme permet d'effectuer un zoom sans augmentation indue du temps de calcul de l'image. Nous n'indiquerons pas les détails de cette modification, mais pour donner l'esprit de celle-ci, nous nous attardons à la création d'un zoom avec l'algorithme C.

Théorème 11 *Soit A un attracteur borné d'un système S de contractions d'un espace métrique, soit x le point fixe d'une des contractions, on pose $B = \{g(x) : g \in S(\varepsilon)\}$. Soit*

Figure 2: Fougère de Barnsley vue avec un zoom.

Ω une partie de l'espace métrique, on pose $L = \{g \in S(\varepsilon) : d(g(x), \Omega) \leq \varepsilon D(\{x\}, A)\}$ et $Z = \{h(g(x)) : g \in S(\varepsilon)$ et $h \in L\}$, alors tout point de $A \cap \Omega$ est approché par un des points de Z à $\varepsilon^2 D(\{x\}, A)$ près.

Démonstration. Remarquons d'abord que $D(\{x\}, A) = \sup\{d(a, x) : a \in A\}$. En effet, $D(\{x\}, A) = \max \rho(\{x\}, A), \rho(A, \{x\})$. Or $\rho(A, \{x\}) = d(x, A)$; mais parce que x appartient à A, alors $d(x, A) = 0$. D'autre part, $\rho(\{x\}, A) = \sup\{d(a, x) \colon a \in A\}$. D'où $D(\{x\}, A) = \sup\{d(a, x) \colon a \in A\}$.

Vérifions maintenant que $A \cap \Omega = \bigcup_{g \in L} g(A) \cap \Omega$. On sait déjà par le lemme 5 que A est un attracteur pour $S(\varepsilon)$. Il suffirait de montrer que $g(A)$ ne rencontre pas Ω si $g \notin L$. Si g n'appartient pas à la liste L, alors $d(g(x), \Omega) > \varepsilon D(\{x\}, A)$. Raisonnons par l'absurde en supposant pour un instant que $d(g(x), \Omega) > \varepsilon D(\{x\}, A)$ et $g(A) \cap \Omega$ n'est pas vide. Si $a \in A$ et $g(a) \in \Omega$, alors $\varepsilon D(\{x\}, A) < d(g(x), g(a)) < \varepsilon d(x, a)$; d'où $D(\{x\}, A) < d(x, a)$, ce qui contredit la remarque du début de la démonstration.

D'autre part, la distance entre $\bigcup_{g \in L} g(A)$ et $\bigcup_{g \in L} g(B)$ selon la métrique de Hausdorff est majorée par $\varepsilon D(A, B)$ et depuis le théorème 7, l'on sait que $D(A, B) \leq \varepsilon D(\{x\}, B)$. De tout cela, l'on tire que tout point de $A \cap \Omega$ est approché par un des points de $Z = \bigcup_{g \in L} g(B)$ à $\varepsilon^2 D(\{x\}, A)$ près. $\square$

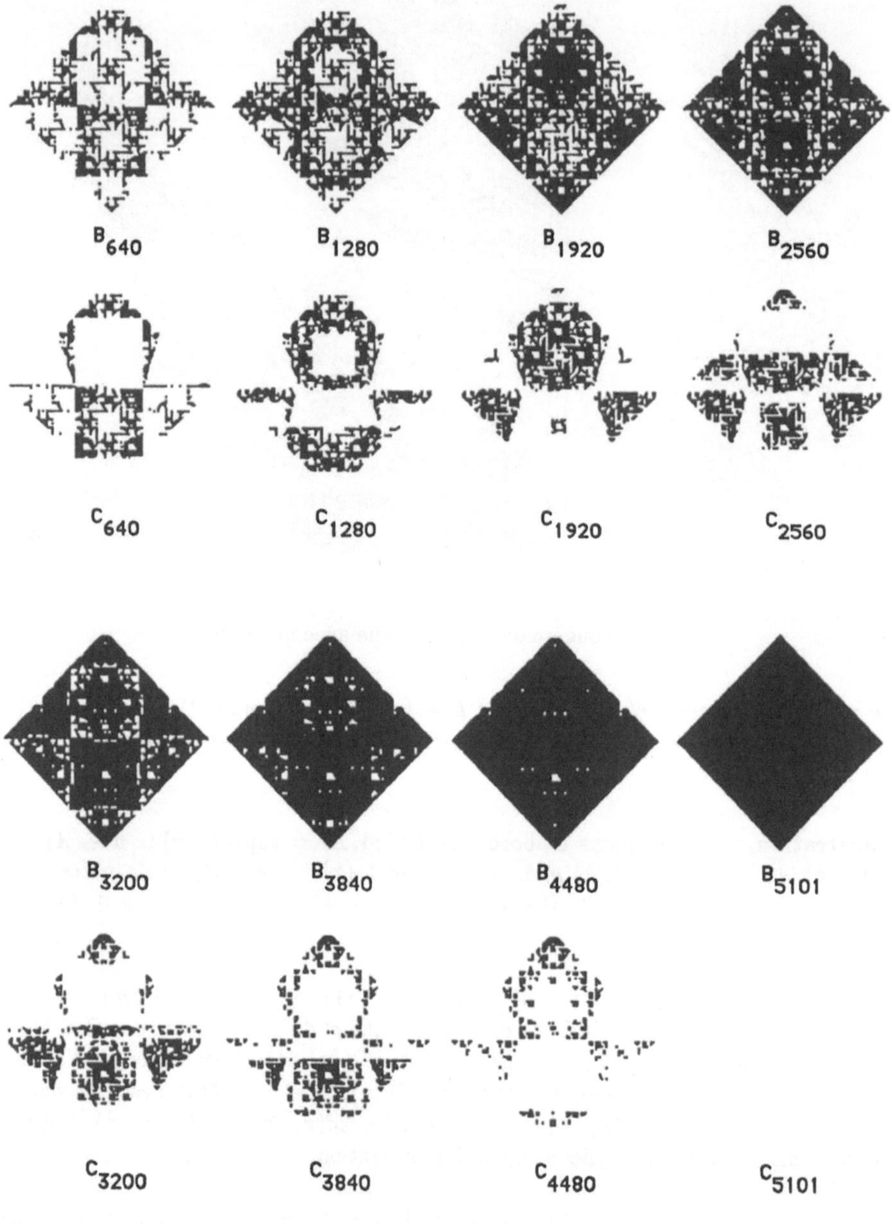

Figure 3:
Échantillon d'images lors de la génération de la courbe de Peano avec son fantôme selon l'algorithme graphique.

Nous illustrons maintenant l'algorithme graphique par la courbe de Peano. La courbe de Peano remplit un carré qui peut être choisi comme l'attracteur du système suivant de contractions. L'espace métrique est le plan $\mathbf{R}^2$. Les contractions sont les neuf similitudes suivantes

$$(x,y) \mapsto (a+bx+cy, d+ex+fy) \quad : \quad \left(\frac{x}{3},\frac{y}{3}\right), \left(\frac{1-y}{3},\frac{x}{3}\right), \left(\frac{1+x}{3},\frac{1+y}{3}\right),$$
$$\left(\frac{2+y}{3},\frac{1-x}{3}\right), \left(\frac{2-x}{3},\frac{-y}{3}\right), \left(\frac{1+y}{3},\frac{-x}{3}\right),$$
$$\left(\frac{1+x}{3},\frac{-1+y}{3}\right), \left(\frac{2-y}{3},\frac{-1+x}{3}\right), \left(\frac{2+x}{3},\frac{y}{3}\right).$$

L'attracteur a été tracé en prenant pour mailles $M(\delta)$ les points du plan dont les coordonnées sont des multiples entiers de $1/100$. Un échantillon de la suite (B_n, C_n) forme le contenu de la figure 3.

Indiquons comment la règle 1 de l'algorithme graphique a été appliquée.

Règle 1- Un point x_n est choisi dans C_n.

C_0 est un singleton, x_0 est ce point. Pour $n > 0$, le choix x_n fera appel à x_{n-1}. Afin de spécifier ce choix, nous nous servons de l'ordre total suivant sur les points du plan: l'ordre lexicographique, un point P est plus petit qu'un point Q si l'ordonnée de P est plus grande que l'ordonnée de Q ou lorsque les ordonnées de P et de Q sont les mêmes, si l'abscisse de P est plus petite que l'abscisse de Q. Si aucun point de C_n n'est plus grand que x_{n-1} (selon l'ordre total en cause), alors x_n est le plus petit point de C_n (selon le même ordre total). S'il y a au moins un point de C_n plus grand que x_{n-1}, alors x_n est le plus petit des points de C_n plus grands que x_{n-1}.

De façon imagée, l'ordre lexicographique consiste à balayer le plan de haut en bas, ligne après ligne, le parcours de chaque ligne se faisant de gauche à droite. En passant, nous mentionnons que Dekking (1980) a suggéré d'autres variations sur la courbe de Peano.

5 Équation fonctionnelle de Bajraktarević

L'équation fonctionnelle de Bajraktarević est

$$(1) \qquad x(t) = f(t, x(b(t)))$$

où t varie dans T, x varie dans X, $f(t,x)$ est une fonction des deux variables t et x à valeurs dans X et $b(t)$ est une transformation de T, la fonction inconnue est la fonction $x(t)$ de T dans X. Cette équation a déjà été considérée dans la littérature mathématique. Nous verrons que cette équation intervient très naturellement pour généraliser la courbe de von Koch. À notre connaissance, Read a été la première personne à introduire l'équation (1) dans le contexte où T et X était le plan complexe, b et f étaient des fonctions analytiques; il cherchait une solution analytique à l'équation (1). En 1957, Bajraktarević a élargi ce contexte en proposant l'étude de l'équation fonctionnelle

$$x(t) = f(t, x(b_1(t)), x(b_2(t)), \dots, x(b_r(t))) \tag{2}$$

où T et X sont des parties de deux espaces euclidiens, t varie dans T, les fonctions $\{b_i\}_{1\le i\le r}$ sont des transformations de T, la fonction f est une application de $T \times X^r$ dans X. Les équations (1) ou (2) ont attiré l'attention de plusieurs mathématiciens depuis, citons les auteurs polonais suivants: Kuczma, Matkowski, Baron, Kwapisz.

Néanmoins, notre intérêt se limite au cas où $r = 1$, nous revenons à l'équation fonctionnelle (1). Reprenons l'essentiel de l'argumentation de Bajraktarević et de Kwapisz. T est un ensemble arbitraire, t varie dans T, $b(t)$ est une transformation de T. X est un espace métrique complet dont la métrique sera notée par d.

$f(t,x)$ est une application de $T \times X$ dans X. Définissons trois conditions qui peuvent s'appliquer pour f:

(A) Il existe un point x_0 de X tel que $\{f(t,x_0)\colon\ t \in T\}$ est une partie bornée de X.

(B) Il existe un nombre $k < 1$ tel que pour tout t de T et pour tout couple de points de X, x_1 et x_2, $d(f(t,x_1), f(t,x_2)) \le kd(x_1,x_2)$.

(C) Il existe une famille F de fonctions continues de T dans X telles que pour tout $y(t)$ de F, $f(t, y(b(t)))$ appartient à la famille F.

Théorème 12 *Si X est un espace métrique complet et si les conditions A et B sont remplies, l'équation* (1) *admet une et une seule solution bornée. Si $x_0(t)$ est une fonction bornée de T dans X et si $x_n(t)$ est définie par récurrence:*

$$x_{n+1}(t) = f(t, x_n(b(t))), n = 0, 1, 2, \dots\ ,$$

alors la solution de (1) *est la limite de la suite $x_n(t)$ et la convergence est uniforme sur T. Si de plus, la condition C est satisfaite, alors la solution de* (1) *est une fonction continue.*

Démonstration. Il suffit de faire appel au principe des contractions de Banach. Soit E l'espace métrique complet des fonctions bornées de T dans X, on a muni E de la métrique uniforme. On introduit la transformation K: si $x(t) \in E$, alors $K(x(t)) = f(t, x(b(t)))$. Les conditions A et B permettent de dire que K est bien définie et est une contraction. Si $x_0(t)$ est une fonction bornée de T dans X et si $x_n(t)$ est définie par récurrence: $x_{n+1}(t) = f(t, x_n(b(t))), n = 0, 1, 2, \dots$, alors le principe des contractions de Banach montre que la suite de fonctions $x_n(t)$ converge uniformément.

Si la condition C est satisfaite, il suffit de choisir $x_0(t)$ dans la famille F; chacune des fonctions $x_n(t)$ est continue; une limite uniforme de fonctions continues reste continue; la solution de (1) est donc une fonction continue. □

6 Courbes de von Koch

Lévy et d'autres auteurs comme Kahane et Mandelbrot ont proposé d'élargir une construction géométrique due à von Koch et ont suggéré de construire des courbes du plan par une propriété d'autosimilitude. Une telle courbe est formée de m arcs semblables à la courbe elle-même. Soient $P_0, P_1, \ldots, P_m, m+1$ points du plan euclidien $\mathbf{R}^2$, désignons par S_k la similitude du plan qui envoie le segment P_0, P_m sur le segment P_{k-1}, P_k. La courbe (généralisée) de von Koch $\{(x(t), y(t)) \colon t \in [0,1]\}$ est telle que

$$x((k-1+t)/m), y((k-1+t)/m) \quad = \quad S_k(x(t), y(t)) \tag{3}$$

pour tout t de $[0,1]$ et pour $k = 1, 2, \ldots, m$.

La courbe originale de von Koch (1904) est obtenue en prenant $m = 4$:

$$P_0 = (0,0), P_1 = (\frac{1}{3}, 0), P_2 = (\frac{1}{2}, \frac{\sqrt{3}}{6}), P_3 = (\frac{2}{3}, 0) \text{ et } \; P_4 = (1,0).$$

La relation fonctionnelle (3) est un cas particulier de l'équation fonctionnelle de Bajraktarević. Posons $T = [0,1)$, $b(t)$ est la partie fractionnaire de mt et $f(t,x,y) = S_i(x,y)$ si $(i-1)/m \leq t < i/m$. Une solution continue de l'équation fonctionnelle (1) induit une solution pour (3).

Parlons des conditions $A), B)$ et $C)$ de la dernière section. La condition A est toujours satisfaite. Désignons par r_k le quotient $\|P_k - P_{k-1}\|/\|P_m - P_0\|$, r_k est le rapport de similitude de la similitude S_k. Si chacun des nombres r_k est inférieur à 1, alors la condition B est satisfaite. Vérifions que la condition C est satisfaite. Par hypothèse, $S_k P_m = S_{k+1} P_0$ pour $k = 1, 2, \ldots, m-1$. De ce fait, on vérifie que la famille F des fonctions continues $z \colon T \to \mathbf{R}^2$ telles que $z(0) = P_0$ et $z(1) = P_m$ permet de remplir la condition C.

Chaque courbe de von Koch peut être réalisée comme limite uniforme d'une suite de lignes polygonales. Si L_n est la n ème ligne polygonale, la ligne polygonale d'ordre $n+1$ consiste à remplacer chacun des segments $[Q, R]$ de la ligne L_n par m segments $[S_1Q, S_1R], [S_2Q, S_2R], \ldots [S_mQ, S_mR]$. La suite de lignes polygonales correspond à la méthode originale de von Koch. Cette construction est aussi celle qui est citée dans le théorème 12 si $x_0(t), y_0(t)$ est choisi comme le segment $[(0,0),(1,0)]$: $x_0(t) = t, y_0(t) = 0$.

7 Interpolation dans une courbe de von Koch

Nous décrivons comment une propriété d'interpolation se présente dans une courbe de von Koch. Soient $m+1$ points du plan complexe $z_0, z_1, \ldots, z_m$, on suppose que $z_0 = 0$ et $z_m = 1$. Nous allons reprendre l'équation (3) en nous servant de nombres complexes. La courbe de von Koch sera donc une fonction $t \to z(t)$ où t varie sur $[0,1]$. L'équation (3) s'écrit comme

$$(4) \qquad z((k-1+t)/m) = z_{k-1} + (z_k - z_{k-1})z(t)$$

pour tout t de $[0,1]$ et pour $k = 1, 2, \ldots, m$.

La construction de von Koch consiste en une suite de lignes polygonales: $L_n = \{z(j/m^n) : j = 0, 1, 2, \ldots, m^n\}$. Le passage de la ligne L_n à L_{n+1} se fait en remplaçant un segment $[u, v]$ de L_n par la ligne polygonale

$$[u, u + z_1(v-u), u + z_2(v-u), \ldots, u + z_m(v-u)].$$

Si j est un entier donné, si q et r sont respectivement le quotient et le reste lors de la division de j par m, la construction de von Koch donne la formule interpolatoire suivante

$$(5) \qquad z(j/m^{n+1}) = (1 - z_r)z(q/m^n) + z_r z((q+1)/m^n).$$

Les sommets de la nouvelle ligne polygonale L_{n+1} s'obtiennent par une pondération des sommets de la ligne L_n, les poids étant des nombres complexes. La figure 4 illustre la construction de von Koch relative à la courbe de Peano. Pour cette courbe, $m = 9$ et les nombres complexes z_k sont $0, \frac{1}{3}, \frac{1+i}{3}, \frac{2+i}{3}, \frac{2}{3}, \frac{1}{3}, \frac{1-i}{3}, \frac{2-i}{3}, \frac{2}{3}$ et 1.

Notes La majeure partie de ce chapitre provient de l'article de Dubuc-Elqortobi. La section 5 est issue de l'article de Dubuc (1985b). C'est Cesàro (1905) qui a reconnu le premier que la courbe de von Koch était formée de parties semblables au tout. Cette propriété d'autosimilitude a été reprise par plusieurs auteurs comme Lévy (1938), de Rham (1957), Kahane-Salem (1963) en particulier. De façon plus récente, cette approche a été privilégiée en géométrie fractale; Mandelbrot (1981) et Hutchinson (1981) sont les principaux responsables de cette prise de conscience.

Problèmes

(1) Si V est la courbe de von Koch, donnez deux transformations affines contractantes du plan en lui-même, f et g telle que $V = f(V) \cup g(V)$. (Une solution est donnée dans l'article de de Rham 1956-57).

(2) En 1918, Knopp a défini une fonction $K(x)$ qui dépend de deux paramètres a et b: c'est la série $\sum_{n=0}^{\infty} a^n d(b^n x)$, où le paramètre a appartient à $(0,1)$, b est un entier supérieur à un et $d(x)$ est la distance de x au plus proche entier. La fonction de Knopp, qui est partout continue, a cette particularité de n'être nulle part dérivable si $ab \geq 1$. Déterminez un système S de deux transformations affines contractantes telles que le graphe $\{(x, K(x)) : x \in [0,1]\}$ est invariant relativement au système S.

(Suggestion: se servir de l'équation fonctionnelle satisfaite par la fonction de Knopp, $K(x) = d(x) + aK(bx)$.)

(3) Existe-t-il deux fonctions continues réelles $x(t)$ et y $y(t)$, définies sur $[0, 1]$, telles que les points du plan $\{(x(t), y(t)) \colon t \in [0, 1]\}$ forment précisément la fougère de Barnsley?

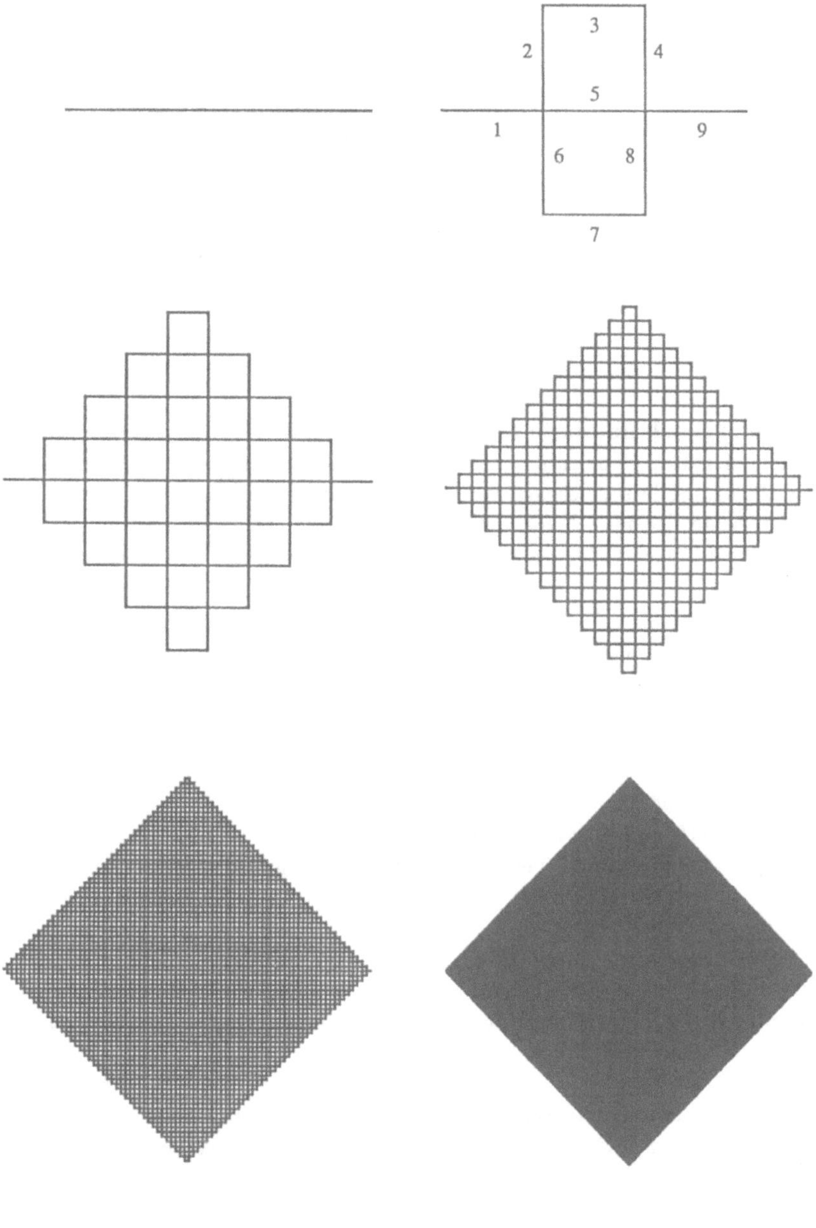

Figure 4: La construction de von Koch à l'occasion de la courbe de Peano.

Chapitre II
Interpolation itérative à une ou plusieurs variables

Sommaire. Nous définissons un procédé d'interpolation itérative pour des données fournies sur un sous-groupe discret G fermé de l'espace euclidien $\mathbf{R}^d$. Ce procédé d'interpolation prolonge une fonction f définie sur G en une fonction g définie sur un sous-groupe plus grand G_∞ de $\mathbf{R}^d$. Après avoir exposé quelques exemples, nous décrirons les principales propriétés algébriques de ce procédé, la linéarité, l'invariance sous les translations par des éléments de G et une invariance d'échelle. Nous déterminerons une condition suffisante pour que le sous-groupe G_∞ soit dense. Si jamais la fonction g est uniformément continue sur ce sous-groupe, le prolongement continu de g à $\mathbf{R}^d$ achève le procédé d'interpolation.

1 Définition du procédé d'interpolation itérative

Une fonction f à valeurs complexes a pour domaine de définition un sous-groupe G de l'espace euclidien $\mathbf{R}^d$. Nous supposons que G est un sous-groupe discret fermé de l'espace euclidien $\mathbf{R}^d$ et que le sous-espace vectoriel engendré par G est tout l'espace $\mathbf{R}^d$. Notre objectif est de décrire un schéma qui permet de prolonger f successivement à une suite croissante de sous-groupes G_n partant de G et telle que l'union des sous-groupes est dense dans $\mathbf{R}^d$.

Les prolongements successifs seront effectués à l'aide d'une transformation linéaire de T de $\mathbf{R}^d$ et d'une fonction de poids p. Les hypothèses suivantes sont retenues pour T et p.

Hypothèse 1 G est contenu dans $T(G)$.
Hypothèse 2 Le domaine de définition de p est $T(G)$.
Hypothèse 3 $p(0) = 1$ et $p(x) = 0$ si x est un élément non-nul de G.
Hypothèse 4 Le support de p est fini: il n'y a qu'un nombre fini de points x pour lesquels $p(x) \neq 0$.

Nous posons $G_0 = G$ et pour chaque entier positif n, $G_n = T^n(G)$. G_∞ est par définition le sous-groupe de $\mathbf{R}^d$ obtenu par la réunion de la suite croissante d'ensembles G_n.

Lemme 1 *Il y a une et une seule fonction g définie sur G_∞ telle que pour tout x de $G, g(x) = f(x)$ et pour tout x de G et pour tout entier non négatif n*

$$(1) \qquad g(T^{n+1}x) = \sum_{y \in G} p(Tx - y) g(T^n y).$$

Démonstration. Une première remarque. Du fait que le sous-espace vectoriel engendré par G est $\mathbf{R}^d$ et que $T(G)$ contient G, il s'ensuit que $T(\mathbf{R}^d) = \mathbf{R}^d$ et que T est une transformation injective.

Par récurrence, on définit une suite de fonctions. f^0 est la fonction de départ, $f(x)$, connue sur le groupe G. Si f_n est définie sur G_n, alors on pose, pour chaque x de G,

$$f_{n+1}(T^{n+1}x) = \sum_{y \in G} p(Tx - y) f_n(T^n y).$$

Puisque T est injective, f_{n+1} est bien définie sur G_{n+1}. Étant donnée l'hypothèse 4, la série apparemment infinie de la dernière formule est de fait une somme finie dès que l'on néglige les termes nuls; si bien que cette formule a bien un sens. L'effet de l'hypothèse 3 (qui est une condition de compatibilité) est d'assurer que pour tout entier positif n, f_{n+1} est un prolongement de f_n : pour tout x de G_n, $f_{n+1}(x)$ est égal à $f_n(x)$. Il est donc possible de définir une fonction g sur G_∞: si x appartient à G_∞, alors il existe un entier n pour lequel x appartient à G_n ; pour un tel indice n, on pose $g(x) = f_n(x)$. Cette fonction g est la fonction cherchée et est unique. □

La fonction g de la formule (1) est par définition l'*interpolation itérative* de f par rapport à T et p.

Dans $\mathbf{R}^d$, un choix possible pour T est $Tx = x/b$ où b est un entier. Cependant, indiquons un moyen différent pour remplir l'hypothèse 1. Pour ce faire, on se sert de d points $\{e_1, e_2, \dots, e_d\}$ de G tels que le sous-groupe engendré par ceux-ci est G et on choisit une matrice carrée régulière d'ordre d dont tous les éléments sont entiers $(n_{i,j})$. Il existe une seule application linéaire H définie sur $\mathbf{R}^d$ telle que $H(e_i) = \sum_j n_{i,j} e_j$. H est inversible et $H(G)$ est contenu dans G. Si $T = H^{-1}$, alors T satisfait l'hypothèse 1.

2 Exemples

Par divers exemples, nous illustrons le procédé d'interpolation itérative. Les cinq premiers exemples décrivent des interpolations unidimensionnelles; dans ces cas, d vaut un et G est le groupe des entiers relatifs $\mathbf{Z}$.

Exemple 1 [Fonction de Hellinger] La fonction f à interpoler a pour domaine de définition le sous-groupe $G = \mathbf{Z}$ de l'axe réel $\mathbf{R}$ alors que $f(n) = 0$ si $n \leq 0$ et $f(n) = 1$ si $n \geq 1$. La transformation linéaire T est $Tx = x/2$. $G_n = T^n(G)$ est formé des nombres rationnels dyadiques $m/2^n$ où m est un entier relatif arbitraire. La fonction de poids p, définie sur G_1, dépend d'un paramètre t compris entre 0 et 1 : $p(-1/2) = t, p(0) = 1, p(1/2) = 1 - t$ et $p(x) = 0$ pour les autres multiples entiers x de $1/2$. Si $g(x)$ est l'interpolation itérative de f, alors la fonction de Hellinger est le prolongement continu de

g restreint à l'intervalle-unité. La détermination successive de g par la formule (1) peut se faire de la façon suivante:

$$g(\frac{m}{2^{n+1}}) = tg(\frac{m-1}{2^{n+1}}) + (1-t)g(\frac{m+1}{2^{n+1}}) \text{ lorsque } m \text{ est un entier impair.}$$

Remarquons bien que le nombre $m/2^{n+1}$ appartient à G_{n+1} alors que les deux nombres $(m-1)/2^{n+1}$ et $(m+1)/2^{n+1}$ appartiennent à G_n.

La figure 1 représente le graphe de la fonction $f(x)$ de Hellinger. Lorsque le paramètre t est différent de 1/2, cette fonction admet la remarquable propriété d'être une fonction à la fois strictement croissante et singulière: presque partout par rapport à la mesure de Lebesgue, $f'(x)$ existe et vaut 0. Sur le plan visuel, ce résultat est surprenant. Une démonstration de ce fait a été donnée par Salem (1943).

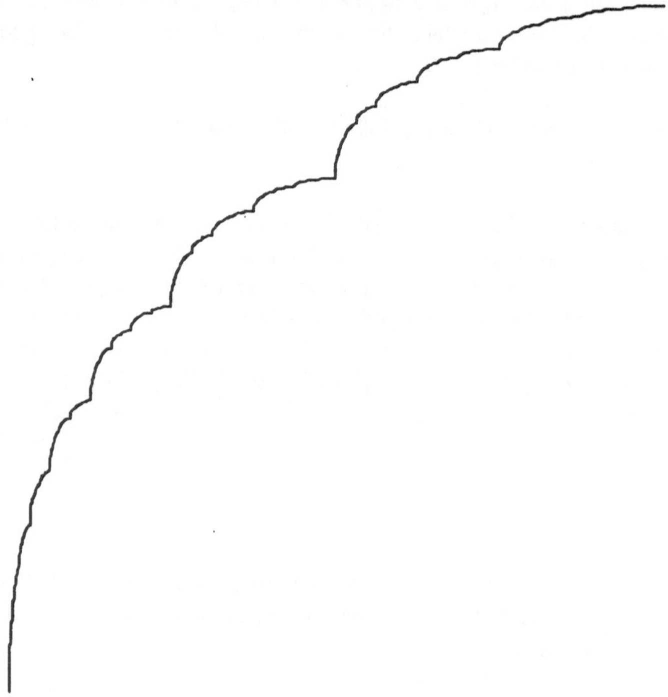

Figure 1: Fonction de Hellinger de paramètre 3/4.

Exemple 2 [Courbes générales de von Koch] Nous décrivons comment le modèle d'interpolation itérative permet de construire les courbes générales de von Koch. On autorise des fonctions et des poids complexes. Le groupe G est encore formé des entiers relatifs $\mathbf{Z}$. Nous posons $f(0) = 0, f(1) = 1$, les autres valeurs $f(n)$ sont choisies de façon arbitraire.

Soient $m+1$ points du plan complexe $z_0, z_1, \ldots, z_m$, on suppose que $z_0 = 0$ et $z_m = 1$. Il sera commode d'avoir défini les coefficients complexes suivants p_n:

$p_n = 1 - z_n$ si n est un entier compris entre 0 et m,
$p_n = z_{m+n}$ si n est un entier compris entre $-m$ et 0,
$p_n = 0$ pour les autres valeurs de n, $|n| > m$.

La transformation linéaire T est $Tx = x/m$. $G_n = T^n(G)$ est formé des nombres rationnels k/m^n où k est un entier relatif arbitraire. La fonction de poids $p(x)$ définie sur $T(G)$ utilise les coefficients p_n: $p(n/m) = p_n$.

Si i est un entier donné, si q et r sont respectivement le quotient et le reste lors de la division de i par m, la relation (1) s'écrit dans ce cas-ci

$$g(i/m^{n+1}) = (1 - z_r)g(q/m^n) + z_r g((q+1)m^n)$$

La fonction $g(t)$ fournie par cette relation et restreinte à l'intervalle $[0,1]$ est une courbe de von Koch observée sur l'ensemble des fractions m-adiques.

La courbe originale de von Koch s'obtient en particulier en prenant $m = 4$ et $z_0 = 0$, $z_1 = 1/3$, $z_2 = 1/2 + i\sqrt{3}/6$, $z_3 = 2/3$, $z_4 = 1$. Si $g(x)$ est l'interpolation itérative de f, alors la courbe de von Koch est produite par le prolongement continu de g à l'intervalle-unité.

La fonction de Hellinger peut se voir comme une courbe de von Koch dégénérée dont tous les poids sont réels avec $m = 2$ et $z_0 = 0$, $z_1 = t$ (le paramètre de la fonction), $z_2 = 1$

Exemple 3 [Interpolation dyadique] Cette méthode d'interpolation, dite *interpolation dyadique*, dépend de quatre paramètres a, b, c et d. Le point de départ est une fonction $f(x)$ définie sur le groupe $G = \mathbf{Z}$ des entiers relatifs, notre désir est de prolonger cette fonction à tout l'axe réel si possible. La transformation linéaire T est $Tx = x/2$. $G_n = T^n(G)$ est l'ensemble des nombres rationnels dyadiques $m/2^n$ où m est un entier relatif arbitraire. Si le prolongement $g(x)$ de f a déjà été complété sur G_n, le prolongement de f à G_{n+1} se fait selon la formule suivante: si x appartient à G_{n+1} mais non à G_n et si $h = 2^{-n-1}$, les nombres $x - 3h$, $x - h$, $x + h$ et $x + 3h$ appartiennent à G_n et on peut poser

$$g(x) = ag(x-3h) + bg(x-h) + cg(x+h) + dg(x+3h)$$

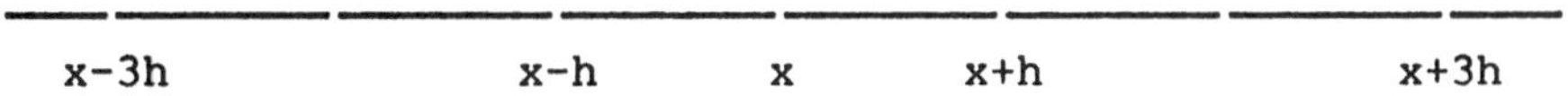

Le prolongement de f se poursuit sur l'ensemble G_∞ des rationnels dyadiques p/q où p est un entier relatif et q est une puissance entière de 2.

Les valeurs de la fonction de poids sont les suivantes: $p(0) = 1, p(3/2) = a, p(1/2) = b, p(-1/2) = c, p(-3/2) = d$ et $p(x) = 0$ pour les autres multiples entiers de 1/2. Ce type d'interpolation dépend donc de quatre paramètres a, b, c et d. Pour obtenir la continuité uniforme de l'interpolation, ces paramètres sont sujets à des restrictions comme l'exigence que $a + b + c + d = 1$.

La figure 2 superpose trois exemples d'interpolations dyadiques avec un décalage vertical entre les courbes pour éviter tout chevauchement entre elles. Dans chacun des cas, la fonction $f(n)$ dont on fait l'interpolation est la même, c'est la fonction qui prend la valeur 1 en $n = 0$ et qui prend la valeur 0 en tout autre entier n. La différence des courbes provient du choix des paramètres a, b, c, d. Ceux-ci sont respectivement

$$(-\frac{1}{16}, \frac{9}{16}, \frac{9}{16}, -\frac{1}{16}), \quad (-\frac{2}{10}, \frac{6}{10}, \frac{7}{10}, -\frac{1}{10}), \quad (\frac{4}{10}, \frac{3}{10}, \frac{2}{10}, \frac{1}{10})$$

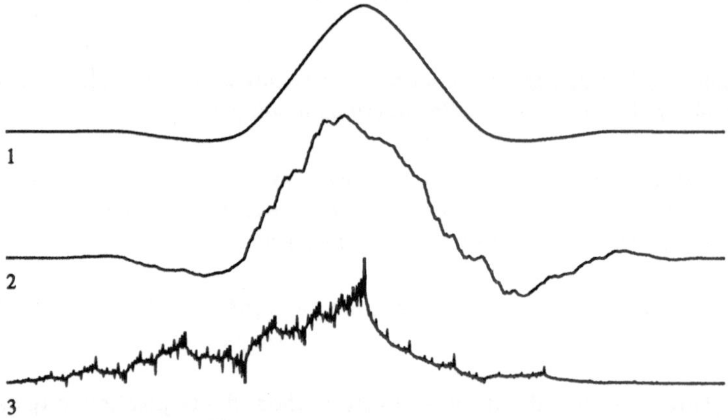

Figure 2: Trois exemples d'interpolations itératives dyadiques.

Le cas particulier où $a = d = -1/16, b = c = 9/16$ est très intéressant en pratique. L'auteur a analysé cette situation en 1986. Dyn, Levin et Gregory ont fait de même en 1987. À ce schéma particulier, nous donnerons le nom d'*interpolation dyadique cubique.* Le qualificatif de cubique provient du fait que si $P(x)$ est un polynôme cubique, l'interpolation itérative de la fonction $f(n) = P(n)$ selon ce schéma va reproduire le polynôme P, l'interpolation itérative g obtenue obéira à l'identité $g(x) = P(x)$.

Exemple 4 [Interpolation itérative selon une base b] Nous partons d'une fonction $f(n)$ définie sur le groupe $G = \mathbf{Z}$. Nous aimerions prolonger f à toute la droite réelle si possible. Premièrement, on se sert d'un entier b supérieur à un; ce sera la base du procédé d'interpolation. Nous posons G_n comme la totalité des nombres rationnels de la forme m/b^n où m est un entier relatif et $n = 0, 1, 2, \ldots$ La fonction f est déjà définie sur $G_0 = G$. Nous prolongerons f successivement à $G_1, G_2, G_3, \ldots$ par récurrence. Pour prolonger f de G à G_1, on se sert d'une suite de poids $\{p_n\}_{-\infty<n<\infty}$. On impose quelques conditions sur cette suite. D'abord $p_0 = 1$ et $p_n = 0$ pour tout autre multiple entier de b: $n = \pm b, \pm 2b, \pm 3b, \ldots$ On suppose de plus qu'un nombre fini des paramètres p_n diffèrent de zéro. Sous ces hypothèses, le prolongement proposé de f noté g est:

$$g(j/b) = \sum_n p_{j-nb} f(n).$$

Le prolongement de f à $G_2, G_3, \ldots$ est effectué de la même façon: si j est un entier, on pose

$$g(j/b^{n+1}) = \sum_n p_{j-nb} g(n/b^n).$$

Sans restriction à imposer sur la suite originale $f(n)$, les propriétés des poids permettent un prolongement consistant. Ce prolongement de la suite $f(n)$ à l'ensemble G_∞ des nombres b-adiques sera appelé *l'interpolation itérative à base b* de la suite. La transformation linéaire T est $Tx = x/b$. La fonction de poids définie sur G_1 est $p(n/b) = p_n$.

Exemple 5 [Interpolation itérative de Lagrange] Nous décrivons une classe d'interpolation itérative inspirée par l'interpolation polynomiale. Le groupe G est formé des entiers relatifs, $\mathbf{Z}$. La transformation linéaire T est $Tx = x/b$ où b est un entier plus grand que un. Si N est une collection finie d'entiers, si $f(n)$ est définie sur le groupe $G = \mathbf{Z}$, il est possible de prolonger f à l'ensemble G_1 par une interpolation polynomiale à partir de translatées de N de la manière suivante.

Si $x = n + r/b$ (où r est un entier de $[1, b-1]$) est un point de G_1 où le prolongement doit se faire, si $P(x)$ est le polynôme d'interpolation tel que $P(n+k) = f(n+k)$ pour tout entier k de N, alors on pose $f(x) = P(x)$.

Si $\{L_k\}_{k\in N}$ sont les polynômes de Lagrange lorsque N sert d'ensemble de noeuds, alors le prolongement est

$$f(n + r/b) = \sum_{k\in N} L_k(r/b) f(n+k)$$

À ce procédé d'interpolation, nous avons donné le nom d'*interpolation itérative de Lagrange*. Si $b = 2$ et si l'ensemble des noeuds N est $\{-1, 0, 1, 2\}$, l'interpolation itérative de Lagrange répète l'interpolation dyadique cubique.

Exemple 6 [Produit cartésien d'interpolations] Si deux procédés d'interpolation itérative sont définis l'un sur $\mathbf{R}^{d_1}$, l'autre sur $\mathbf{R}^{d_2}$, alors il est possible de définir le produit cartésien de ces deux procédés $\mathbf{R}^{d_1} \times \mathbf{R}^{d_2}$. Si le premier procédé se sert du sous-groupe G_1, d'une transformation linéaire T_1 et d'une fonction de poids p_1 et si le second procédé se sert de G_2, T_2 et p_2, alors on pose $G = G_1 \times G_2, T(x,y) = (T_1x, T_2y)$ pour x de $\mathbf{R}^{d_1}$ et y de $\mathbf{R}^{d_2}$ et $p(x,y) = p_1(x)p_2(y)$.

Exemple 7 [Interpolation à quatre poids sur réseau rectangulaire] L'interpolation a lieu dans $\mathbf{R}^2$. $G = \mathbf{Z} \times \mathbf{Z}$, $T(x,y) = ((x-y)/2, (x+y)/2)$. T est une similitude, il s'agit d'une rotation de 45 degrés autour de l'origine, suivie d'une homothétie de rapport $1/\sqrt{2}$ centrée aussi à l'origine. La fonction de poids est déterminée par quatre nombres positifs dont la somme est un, a, b, c, d. $p(0,0) = 1, p(1/2,1/2) = a, p(-1/2,1/2) = b, p(-1/2,-1/2) = c, p(1/2,-1/2) = d$ et $p(x,y) = 0$ autrement.

La figure 3 suggère comment s'opère cette interpolation plane. Les points de G sont représentés par des cercles, les points de G_1 qui ne sont pas dans G sont localisés par des carrés, les points de G_2 qui ne sont pas dans G_1 sont indiqués par des carrés très petits. Pour interpoler en un point de G_1, on indique la distribution des poids chez les voisins de ce point situés dans G. Un autre exemple illustre l'interpolation en un point de G_2.

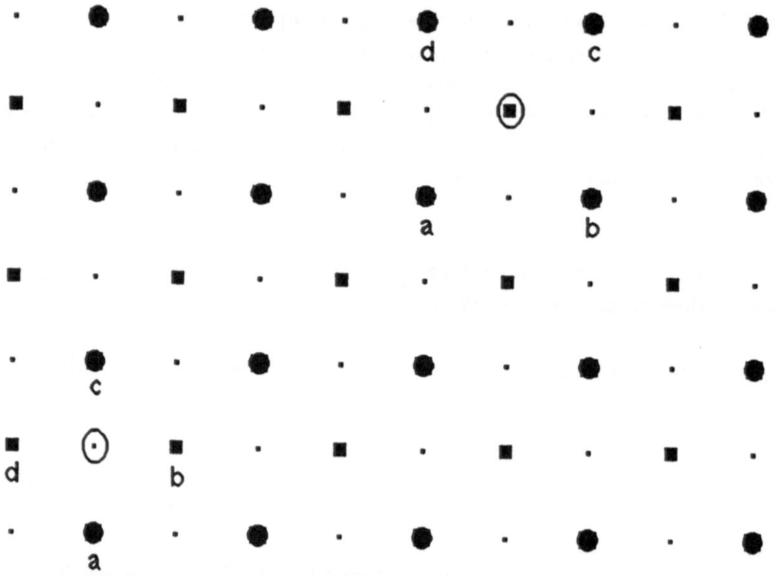

Figure 3:
Interpolation dans le plan à partir d'un réseau rectangulaire à l'aide de quatre poids et par rotation de 45 degrés.

Exemple 8 [Interpolation quadratique sur un réseau triangulaire] L'interpolation a lieu dans $\mathbf{R}^2$. G est le réseau triangulaire $\{m+n/2, n\sqrt{3}/2\}$; $T(x,y) = ((3x-\sqrt{3}y)/6, (\sqrt{3}x+$

$3y)/6)$). Lorsque $(x, y) \in T(G), p(x, y) = 1$ si $(x, y) = (0, 0), p(x, y) = 4/9$ si $x^2 + y^2 = 1/3, p(x, y) = -1/9$ si $x^2 + y^2 = 4/3$ et $p(x, y) = 0$ autrement.

La figure 4 suggère à sa façon cette interpolation plane. Les points de G sont représentés par des cercles, les points G_1 par des carrés, les points de G_2 par des petits carrés. Pour interpoler en un point de G_1, on note la distribution des poids chez les six plus proches voisins de ce point situés dans G. On se sert des trois plus proches voisins et des autres voisins situés deux fois plus loin. Une des propriétés de ce procédé d'interpolation est que si la fonction f à interpoler prend les valeurs d'un polynôme quadratique en deux variables, $f(x, y) = P(x, y)$, alors le résultat de l'interpolation reproduit le polynôme P.

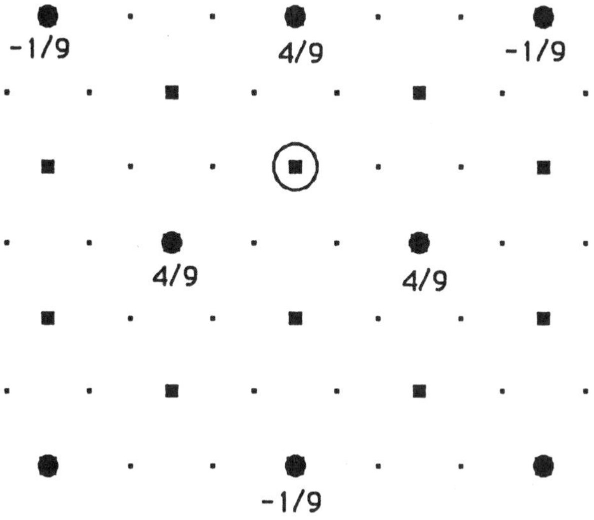

Figure 4: Interpolation quadratique dans le plan à partir d'un réseau triangulaire.

3 Propriétés élémentaires de l'interpolation itérative

Nous exposons trois propriétés élémentaires de tout procédé d'interpolation itérative. Tout procédé d'interpolation est *linéaire* (lemme 2), est *invariant sous les translations* par des éléments de G (lemme 3) et est *invariant sous certains changements d'échelle* (lemme 4). La vérification de ces lemmes ne cause pas de difficultés particulières.

Lemme 2 *Si $f_1(x)$ et $f_2(x)$ sont deux fonctions définies sur G et que $f(x) = f_1(x) + f_2(x)$, alors les interpolations respectives $g_1(x), g_2(x)$ et $g(x)$ définies sur G_∞ satisfont la relation linéaire: $g(x) = g_1(x) + g_2(x)$. Pour tout scalaire c, l'interpolation correspondante à $cf(x)$ est $cg(x)$.*

Lemme 3 *Si $f(x)$ est une fonction définie sur G avec g comme interpolation et si a est un élément de G, l'interpolation de la fonction $f(x+a)$ est la fonction $g(x+a)$.*

Lemme 4 *Si f est définie sur G, si g est l'interpolation itérative de f par rapport à une transformation T et une fonction de poids p, si n est un entier positif, alors l'interpolation de la fonction h définie sur G par $h(x) = g(T^n x)$ est la fonction $g(T^n x)$ déjà définie sur G_∞.*

4 Densité du domaine du prolongement

Dans cette section, nous cherchons des conditions qui nous assurent que le domaine de définition du prolongement selon le procédé d'interpolation itérative est dense dans $\mathbf{R}^d$.

Théorème 5 (Densité de G_∞) *Si G est un sous-groupe de $\mathbf{R}^d$, si le sous-espace vectoriel engendré par G est tout $\mathbf{R}^d$, si T est une transformation linéaire telle que*

(a) *G est contenu dans $T(G)$,*

(b) *le rayon spectral de T est inférieur à un,*

alors le sous-groupe $\cup_{n\geq 1} T^n(G)$ est dense dans $\mathbf{R}^d$.

Démonstration. Nous faisons deux remarques.

(1) Puisque le sous-espace vectoriel engendré par G est l'espace euclidien $\mathbf{R}^d$, il existe un nombre r tel que pour tout point y de $\mathbf{R}^d$, il y a un point x de G dont la distance à y est inférieure à r.

(2) Puisque le rayon spectral de T est inférieur à un, pour tout nombre réel positif ε, on peut trouver un entier n tel que $\|T^n\| < \varepsilon$.

Soit ε un nombre réel positif donné, choisissons un entier n pour lequel $\|T^n\| < \varepsilon$. Si y est un vecteur de $\mathbf{R}^d$, il existe un vecteur z de $\mathbf{R}^d$ tel que $T^n z = y$. Suivant la remarque 1), il y a un point x de G tel que $\|x - z\| < r$. La distance de $T^n x$ à y ne dépasse pas $r\varepsilon$. Si bien que tout point de $\mathbf{R}^d$ est arbitrairement proche de $G_\infty = \cup_{n\geq 1} T^n(G)$. □

5 Interpolation itérative et continuité

Une des questions de base dans un procédé d'interpolation itérative est de savoir si la fonction qui résulte du procédé est uniformément continue. En effet, si $g(x)$ est l'interpolation itérative définie sur G_∞, si g est uniformément continue, alors g se prolonge par continuité à l'adhérence de G_∞. Si l'on savait déjà que G_∞ est dense dans l'espace euclidien total, alors g aurait été prolongée à tout l'espace. Par exemple, est-ce que les fonctions décrites dans les six derniers exemples de la section 3 sont uniformément continues? Il n'est pas facile de répondre à cette question, surtout de justifier sa réponse. Car l'analyse de l'interpolation itérative réclame le développement de techniques propres à ce sujet.

Sous quelles conditions sur la transformation T et sur la fonction de poids peut-on être assuré que la fonction résultant de l'interpolation itérative est uniformément continue? Nous répondrons à cette question dans les deux prochains chapitres.

Notes Le modèle d'interpolation itérative a pris sa source dans divers articles: Dubuc (1986), Deslauriers-Dubuc (1987a, 1987b et 1989) et a pris la forme de ce chapitre lors du récent article Deslauriers-Dubois-Dubuc (1991).

Problèmes

1. Kiesswetter (1966) a défini une fonction K sur $[0,1)$ de la manière suivante: si x appartient à $[0,1)$, on développe le nombre x à la base 4

$$x = \sum_{\nu=1}^{\infty} x_\nu/4^\nu \text{ alors que les } x_\nu \text{ prennent les seules valeurs } 0,1,2, \text{ ou } 3.$$

On pose alors $K(x) = \sum_{n=1}^{\infty}(-1)^{N_\nu}X_\nu/2^\nu$ où $X_\nu = x_\nu - 2$ si $x_\nu > 0$ et $X_\nu = 0$ si $x_\nu = 0$ et N_ν est le nombre de chiffres x_k qui sont nuls de rang $k < \nu$.

(a) Montrez que la fonction de Kiesswetter satisfait l'équation de Bajrakterević:
$K(x) = -K(\{4x\})/2$ si $0 \le x \le 1/4, K(x) = -1/2 + K(\{4x\})/2$ si $1/4 \le x < 1/2$,
$K(x) = K(\{4x\})/2$ si $1/4 \le x < 1/2$ et $K(x) = 1/2 + K(\{4x\})/2$ si $3/4 \le x < 1$.

(b) Si $f(n) = n$ pour $n = 0, \pm1, \pm2, \ldots$ et si G_∞ sont les fractions 4-adiques, $m/4^n$, montrez que la restriction de K à G_∞ est l'interpolation itérative de f selon la base 4.

(Remarque de notation: un nombre réel t mis en accolade, $\{t\}$, désigne la partie fractionnaire de t).

2. Soient A, B et C les sommets d'un triangle du plan, si $P(x,y)$ est un polynôme en deux variables de degré 2, montrez que la relation suivante a lieu:

$$P((A+B+C)/3) = \frac{4}{9}(P(A)+P(B)+P(C)) - \frac{1}{9}(P(A+B-C)+P(A-B+C)+P(-A+B+C).$$

Justifiez l'affirmation du texte: l'interpolation quadratique sur un réseau triangulaire d'une fonction f telle que sur le réseau, $f(x,y)$ prend les valeurs d'un polynôme quadratique $P(x,y)$, reproduit le polynôme P.

Chapitre III
L'interpolante fondamentale

Sommaire. Nous définissons la fonction fondamentale $F(x)$ d'un procédé d'interpolation itérative. Après avoir déterminé sommairement le support de celle-ci, nous relevons comment la fonction fondamentale permet de résoudre tout problème d'interpolation dans le cadre du procédé d'interpolation itérative. Nous identifions les principales équations fonctionnelles dont la fonction fondamentale est la solution. Par la suite, nous entreprenons ce qui pourrait s'appeler l'analyse harmonique de la fonction fondamentale, analyse issue d'un polynôme trigonométrique $P(\theta)$, le polynôme caractéristique du procédé d'interpolation. Sous des hypothèses bénignes, nous introduisons une fonction $\Phi(\xi)$ qui est la transformée de Fourier d'une distribution de Schwartz à support compact et qui se manifeste comme la transformée de Fourier de la fonction fondamentale dans un certain sens. Si $P(\theta) \geq 0$, alors Φ est intégrable. Si Φ est intégrable, alors $F(x)$ est uniformément continue sur le sous-groupe G_∞ et se prolonge donc par continuité à l'espace tout entier. Nous terminons en appliquant les résultats obtenus à l'interpolation dyadique symétrique à quatre poids.

1 L'interpolante fondamentale

Partons de la fonction $F(x)$ définie sur G: $F(0) = 1$ et $F(x)$ est nulle pour les autres valeurs x de G. Cette fonction peut être prolongée à G_∞ par le procédé d'interpolation itérative en une fonction encore notée $F(x)$, ce sera l'*interpolante fondamentale*. Pour désigner F, on se servira aussi de l'expression *fonction fondamentale*. Trois exemples d'interpolante fondamentale se retrouvaient dans la figure 2 du chapitre II.

Remarque Un lien simple unit la fonction de poids et la fonction fondamentale: pour tout x de $T(G)$,

$$F(x) = p(x). \tag{1}$$

Lemme 1 *Si S est le support de la fonction de poids p, si n est un entier positif et si x est un élément de $G_n = T^n(G)$ situé hors de $\sum_{k=0}^{n-1} T^k(S)$ (la somme de Minkowski des ensembles $T^k(S)$), alors $F(x) = 0$.*

Démonstration. Il suffit de raisonner par récurrence sur n. Si x appartient à G et que $F(T^{n+1}x) \neq 0$, alors il existe un y de G tel que $p(Tx - y) \neq 0$ et $F(T^n y) \neq 0$ d'après la formule (1) du chapitre II. Posons $z = Tx - y$, z appartient à S. Par hypothèse d'induction, $T^n y$ appartient à $\sum_{k=0}^{n-1} T^k(S)$. Or $T^{n+1}x = T^n y + T^n z$. D'où $T^{n+1}x$ appartient à $\sum_{k=0}^{n} T^k(S)$. □

Lemme 2 *Si $f(x)$ est une fonction définie sur G, le prolongement g de f à G_∞ suivant le procédé d'interpolation itérative est*

$$g(x) = \sum_{y \in G} f(y)F(x-y).$$

Démonstration. Si x appartient à G_n, alors la série $\sum_{y\in G} f(y)F(x-y)$ est en pratique une somme finie d'après le lemme précédent. Puisque le procédé d'interpolation itérative est linéaire et invariant sous les translations, la dernière série est une solution de la formule (1) du chapitre II. L'unicité de la fonction g, solution de cette formule, témoignée par le lemme 1 du chapitre II montre que la série est égale à $g(x)$. □

2 Équations fonctionnelles satisfaites par l'interpolante fondamentale

Plusieurs équations fonctionnelles sont satisfaites par l'interpolante fondamentale F. Ces équations ont d'autant plus d'importance que plusieurs propriétés de F en sont des conséquences.

Théorème 3 *L'interpolante fondamentale $F(x)$ satisfait l'équation fonctionnelle*

$$(2) \qquad F(T^n x) = \sum_{y \in G} F(T^n y)F(x-y) \text{ où } x \in G_\infty \text{ et } n \in N.$$

Démonstration. Ceci est une conséquence de la linéarité de l'interpolation itérative (lemme 2 du chapitre II) et du lemme précédent (en prenant pour fonction $f(x)$ dans ce lemme la fonction $F(T^n x)$). □

La formule (2) s'avère très utile dans le calcul de la fonction F. Par exemple, si les valeurs de F sont connues sur G_n, en une étape, on peut calculer F en tout autre point de G_{2n}:

$$(3) \qquad F(T^{2n} x) = \sum_{y \in G} F(T^n y)F(T^n x - y).$$

Des deux formules (1) et (2) (avec $n = 1$), on obtient une importante équation fonctionnelle pour F:

$$(4) \qquad F(Tx) = \sum_{y \in G} p(Ty)F(x-y) \text{ où } x \in G_\infty.$$

3 Étude du support de l'interpolante fondamentale

Nous déterminons une condition suffisante fréquemment rencontrée pour que le support de l'interpolante fondamentale soit borné.

Théorème 4 *Si la transformation linéaire T d'un procédé d'interpolation itérative a un rayon spectral inférieur à un, alors le support de l'interpolante fondamentale est borné. De façon plus précise, si R est le rayon de la plus petite boule centrée à l'origine qui recouvre le support de la fonction de poids p, alors $\{x: F(x) \neq 0\}$ est contenu dans la boule centrée à l'origine et dont le rayon est $R\sum_{n\geq 0}\|T^n\|$.*

Démonstration. Il s'agit d'une conséquence directe du lemme 1. □

Il est possible de préciser davantage le support de la fonction fondamentale lorsqu'il s'agit d'interpolation unidimensionnelle.

Lemme 5 *Si dans un procédé d'interpolation itérative, le groupe G est $\mathbf{Z}$, si $Tx = x/b$ où b est un entier, si les poids sont les nombres réels $\{p_n = F(n/b)\}$ et si N est le plus petit indice pour lequel $p_n \neq 0$ et si N^* est le plus grand indice pour lequel $p_n \neq 0$, alors la fonction fondamentale s'annule en dehors de $(N/(b-1), N^*/(b-1))$.*

Démonstration. Après avoir posé $t_0 = 0$, on définit une suite t_n par récurrence: $t_n = t_{n-1} + Nb^{-n}$. La suite t_n est décroissante et converge vers $N/(b-1)$. Selon la relation (4), la restriction de F à G_n s'annule sur $(-\infty, t_n)$, où $n = 1, 2, 3, \ldots$ Pour tout nombre b-adique $t \leq N/(b-1), F(t) = 0$. Un argument semblable montre que $F(t^*)$ est aussi nul si $t^* \geq N^*/(b-1)$. □

4 Transformée de Fourier de l'interpolante fondamentale

Le *polynôme caractéristique* du procédé d'interpolation itérative est le polynôme trigonométrique $P(\theta) = \sum_{x\in G} F(Tx)e^{i<x,\theta>}$ où $< x, \theta >$ est le produit scalaire usuel entre vecteurs de $\mathbf{R}^d$.

L'interpolante fondamentale F est définie sur le sous-groupe G_∞. Supposant pour un instant que cette fonction est uniformément continue et que le rayon spectral de T est inférieur à un, la fonction F se prolongera à tout l'espace euclidien $\mathbf{R}^d$. Si $\Phi(\xi)$ est la transformée de Fourier de $F(x)$: $\Phi(\xi) = \int F(x)e^{-ix\xi}dx$, alors l'équation fonctionnelle (4) induit l'équation fonctionnelle correspondante sur Φ:

$$\Phi(\xi) = |\,\text{dét}\,T|P(-T^*\xi)\Phi(T^*\xi). \tag{5}$$

Il est ici sous-entendu que $\text{dét}\,T$ désigne le déterminant de la matrice représentative de la transformation T et T^* est l'adjoint de l'opérateur T.

Théorème 6 *Si $\sum_{x\in T(G)} p(x) = 1/|\,\text{dét}\,T|$ et si le rayon spectral de T est inférieur à un, alors le produit infini $\prod_{n\geq 1}(P(-T^{*n}\zeta)|\,\text{dét}\,T|)$ est convergent pour tout point ζ de $\mathbf{C}^d$ et la fonction*

$$\Phi(\zeta) = \prod_{n\geq 1}(P(-T^{*n}\zeta)|\,\text{dét}\,T|)$$

est une fonction entière de type exponentiel qui est solution de l'équation (5)*; de façon plus précise, Φ satisfait l'inégalité suivante sur $\mathbf{C}^d$:*

$$|\Phi(\zeta)| \leq C(1+|\zeta|)^N e^{B|Im\zeta|}$$

si les nombres B, C et N sont bien choisis.

Démonstration. Une observation préliminaire. Nous nous servirons à diverses reprises du fait que le rayon spectral de T^* est inférieur à un. On sait que le rayon spectral de T^* coïncide avec le rayon spectral de T. Il est donc possible de trouver un entier L et un nombre $\tau < 1$ tel que $\|T^{*n}\| \leq \tau^n$ pour tout $n \geq L$. Nous retenons donc ces deux valeurs L et τ.

Nous posons $\Delta = |\,\text{dét}\,T|$. Supposons que les variations possibles du point ζ de $\mathbf{C}^d$ ont lieu dans une boule de rayon R. La convergence du produit infini $\prod_{n\geq 1}(P(-T^{*n}\zeta)|\,\text{dét}\,T|)$ proviendra de la finitude de la série $\sum_{n\geq 1}|\Delta P(-T^{*n}\zeta) - 1|$. Par hypothèse $P(0) = 1/\Delta$. P est continûment dérivable, il existe une constante M telle pour tout vecteur ζ de $\mathbf{C}^d$, dans la boule de rayon R,

$$|P(\zeta) - P(0)| \leq M|\zeta|.$$

D'où

$$|P(T^{*n}\zeta) - P(0)| \leq M|T^{*n}\zeta|$$

et

$$\sum_{n\geq 1}|\Delta P(-T^{*n}\zeta) - 1| \leq M\Delta \sum_{n\geq 1}|T^{*n}\zeta| < \infty$$

(vu que le rayon spectral de T est inférieur à un). Selon le critère de Weierstrass, les produits partiels du produit infini $\prod_{n\geq 1}(P(-T^{*n}\zeta)|\,\text{dét}\,T|)$ convergent uniformément sur tout compact $\mathbf{C}^d$. Le produit infini $\prod_{n\geq 1}(P(-T^{*n}\zeta)|\,\text{dét}\,T|)$ est une fonction continue de ζ qui satisfait l'équation fonctionnelle (5).

Nous établirons l'existence de constantes positives C, N et B telles que pour tout vecteur ζ de $\mathbf{C}^d$,

$$|\Phi(\zeta)| \leq C(1+|\zeta|)^N e^{B|Im\,\zeta|}$$

$Im\ \zeta$ est le vecteur réel dont les composantes sont les parties imaginaires des composantes du vecteur complexe ζ).

On peut trouver deux constantes A et R telles que $|P(\zeta)| \leq Ae^{R|Im\ \zeta|}$ pour tout vecteur ζ de $\mathbf{C}^d$. (Un choix possible pour A et R est le suivant: $A = \sum_{x \in G_1} |p(x)|$ et $R = \sup\{|x| : x \in G, F(Tx) \neq 0\}$). La constante N aura pour valeur $-\ell n A/\ell n \tau$.

Pour une valeur entière donnée de n supérieure à L, considérons un point ζ de $\mathbf{C}^d$ tel que $|T^{*n}\zeta| > 1$ et $|T^{*n+1}\zeta| \leq 1$, alors une application répétée de l'équation (5) pour les valeurs suivantes de ξ: $\zeta, T^*\zeta, T^{*2}\zeta, \ldots, T^{*n-1}\zeta$ fait valoir l'inégalité

$$|\Phi(\zeta)| \leq A^n \exp(\sum_{1 \leq k \leq n} R|ImT^{*k}\zeta|)|\Phi(T^{*n}\zeta)|.$$

Parce que le rayon spectral de T est inférieur à 1, il existe une constante B telle que

$$R\sum_{k \geq 1} |ImT^{*k}\zeta| \leq B|Im\ \zeta|.$$

D'où

$$|\Phi(\zeta)| \leq A^n |\Phi(T^{*n}\zeta)| e^{B|Im\ \zeta|}.$$

Nous savons que $|T^{*n}\zeta| \geq 1$ et $n \geq L$. D'où

$$1 \leq |T^{*n}\zeta| \leq \|T^{*n}\|\ |\zeta| \leq \tau^n |\zeta|.$$

Or nous avons que $A = (1/\tau)^N$, d'où $A^n = (1/\tau^n)^N \leq |\zeta|^N$; parce que $|T^{*n+1}\zeta| \leq 1$, alors $|\Phi(T^{*n}\zeta)| \leq M$. D'où $|\Phi(z)| \leq M|z|^N e^{B|Imz|}$.

Remarquons que $E = \{\zeta : (\exists k)k < L$ et $|T^{*k}\zeta| < 1\}$ est un ensemble borné. Nous venons de prouver que si $\zeta \notin E$, alors

$$|\Phi(\zeta)| \leq M|\zeta|^N e^{B|Im\ \zeta|}.$$

Si C est suffisamment grand, alors pour tout ζ de $\mathbf{C}^d$,

$$|\Phi(\zeta)| \leq C(1 + |\zeta|)^N e^{B|Im\ \zeta|}.\ \square$$

Remarque Puisque la fonction $\Phi(\zeta) = \prod_{n \geq 1}(P(-T^{*n}\zeta)|\,\text{dét}\,T|)$ est une fonction entière de type exponentiel, Φ est la transformée de Fourier d'une distribution D à support compact. Ceci provient de la généralisation du théorème de Paley-Wiener telle que proposée par Schwartz et que l'on retrouve dans le volume de Yosida (1966), p. 161. On dira que la fonction Φ est la *transformée de Fourier du procédé d'interpolation itérative* et que D est la *distribution fondamentale associée au procédé d'interpolation itérative.*

5 Distributions de Schwartz et interpolation itérative

Nous venons de voir que sous de faibles conditions, à un procédé d'interpolation itérative est toujours associée une distribution D de Schwartz. Cette distribution peut être considérée comme le prolongement de la fonction fondamentale F à tout l'espace $\mathbf{R}^d$. Cette section expliquera le sens de cette affirmation. Si φ est une fonction indéfiniment dérivable à support compact, on pose $\mu_n(\varphi) = \sum_{x \in G} F(T^n x)\varphi(T^n x)(|\,\text{dét}\, T|)^n$. μ_n est une distribution au sens de Schwartz; de fait, il s'agit d'une combinaison linéaire finie de masses de Dirac disposées sur G_n. Nous verrons que la suite de mesures signées μ_n tendent vers D selon la topologie faible. Auparavant, nous avons besoin de quelques résultats intermédiaires. En particulier, nous aurons besoin d'établir la validité de la formule suivante.

$$\sum_{x \in G} F(T^n x) e^{i<x,\theta>} = \prod_{k=0}^{n-1} P(T^{*-k}\theta). \tag{6}$$

En effet, en partant de l'équation fonctionnelle (4), pour tout $x \in G$ et pour tout entier n, $F(T^{n+1}x) = \sum_{y \in G} p(Ty) F(T^n x - y)$. D'où

$$\begin{aligned}
\sum_{x \in G} F(T^{n+1}x) e^{i<x,\theta>} &= \sum_{x \in G} \sum_{y \in G} F(T^n x - y) F(Ty) e^{i<x,\theta>} \\
&= \sum_{y \in G} \sum_{x \in G} F(T^n(x - T^{-n}y)) F(Ty) e^{i<x,\theta>} \\
&= \sum_{y \in G} \sum_{z \in G} F(T^n z) F(Ty) e^{i<z+T^{-n}y,\theta>} \\
&= \sum_{y \in G} \sum_{z \in G} F(T^n z) F(Ty) e^{i<z,\theta>} e^{i<y,T^{*-n}\theta>} \\
&= \Big(\sum_{x \in G} F(T^n x) e^{i<x,\theta>}\Big) P(T^{*-n}\theta).
\end{aligned}$$

La formule (6) est donc démontrée par récurrence.

Nous reprenons un résultat que l'on peut trouver chez Gel'fand-Chilov (1962) tout en le généralisant aux transformées de Fourier des fonctions à d variables.

Théorème 7 *Si T_n est une suite de distributions tempérées sur $\mathbf{R}^d$ dont les transformées de Fourier $\Phi_n(\xi)$ satisfont les deux conditions suivantes:*

(a) *$\Phi_n(\xi)$ converge ponctuellement vers $\Phi(\xi)$;*

(b) *il existe un polynôme B tel que pour tout n et pour tout y, $|\Phi_n(\xi)| \leq B(\xi)$, alors la suite T_n de distributions converge faiblement vers une distribution tempérée T dont la transformée de Fourier est $\Phi(\xi)$.*

Démonstration. Si φ est une fonction indéfiniment dérivable à support compact et si $f(\xi)$ est la transformée de Fourier de $\varphi(x)$, alors, comme on peut le voir chez Gel'fand-Chilov, $T_n(\varphi) = 1/(2\pi)^d \int_{\mathbf{R}^d} f(\xi)\Phi_n(\xi)d\xi$.

Chacune des fonctions $f(\xi)\Phi_n(\xi)$ est majorée en module par la fonction intégrable $f(\xi)B(\xi)$. Par le théorème de convergence dominée de Lebesgue, la suite $T_n(\varphi)$ converge vers $1/(2\pi)^d \int_{\mathbf{R}^d} f(\xi)d\xi$ lorsque n tend vers l'infini. Selon Schwartz (1966) (cf. le théorème XIII), la fonctionnelle limite $T(\varphi)$, la limite de $T_n(\varphi)$ lorsque n tend vers l'infini, est une distribution. La formule que $T(\varphi) = 1/(2\pi)^d \int_{\mathbf{R}^d} f(\xi)\Phi(\xi)d\xi$ fait également voir que $\Phi(\xi)$ est la transformée de Fourier de la distribution tempérée T. □

Théorème 8 *Si le rayon spectral de T est inférieur à un et si*

$$\sum_{x \in T(\mathbf{G})} p(x) = 1/|\,\text{dét}\, T|,$$

alors la suite μ_n de distributions converge faiblement vers la distribution D.

Démonstration. Étudions la suite des transformées de Fourier $\Phi_n(\xi)$ de μ_n. Nous posons $\Delta = |\,\text{dét}\, T|$. Nous avons que

$$\Phi_n(\xi) = \sum_{x \in G} F(T^n x) e^{-i<T^n x, \xi>} \Delta^n.$$

D'après la formule (6),

$$\Phi_n(\xi) = \prod_{1 \le k \le n} [P(-T^{*k}\xi)\Delta].$$

La convergence de $\Phi_n(\xi)$, c'est la convergence du produit infini $\prod_{n \ge 1}(P(-T^{*n}\xi)\Delta)$, convergence qui a été établie dans le théorème 6.

Démontrons maintenant qu'il existe un polynôme $B(\xi)$ tel que pour tout ξ et pour tout n, $|\Phi_n(\xi)| \le B(\xi)$. Il existe un nombre M tel que pour tout y de la boule-unité de $\mathbf{R}^d$ et pour tout $n, |\Phi_n(\xi)| \le M$. Si C est la valeur maximale de $|P(\xi)\Delta|$, ces paramètres C et M peuvent servir à borner $\Phi_n(\xi)$. Faisons encore appel au fait que le rayon spectral de T est plus petit que un. Il existe un nombre positif τ inférieur à un et un entier L tel que $\|T^{*n}\| \le \tau^n$ pour tout valeur entière de n supérieure à L. Si ξ est un vecteur tel que $1/\tau^L < |\xi| \le 1/\tau^k$, alors $T^{*k}\xi$ est dans la boule-unité et dans ce cas, pour tout n,

$$|\Phi_n(\xi)| = |\Phi_{n-k}(T^{*k}\xi) \prod_{l=1}^{k}(P(T^{*l}\xi)\Delta)| \le MC^k.$$

Par conséquent, il est possible de trouver un nombre E tel que $|\Phi_n(\xi)| \le M(1 + |\xi|)^E$. Ce dernier membre est borné par un polynôme.

On peut donc appliquer le théorème précédent. La suite de distributions μ_n converge faiblement vers une distribution D dont la transformée de Fourier est $\Phi(\xi)$. □

Théorème 9 *Si le rayon spectral de T est inférieur à un, si la fonction de poids est non-négative et si $\sum_{x \in T(G)} p(x) = 1/|\text{dét}.T|$, alors la suite de μ_n mesures converge dans la topologie vague vers une mesure μ. Le support de la mesure μ est une partie invariante pour la famille de transformations affines de $\mathbf{R}^d$: $\{x \to s + Tx\}$ où s parcourt les points de $T(G)$ tels que $p(s) \neq 0$.*

Démonstration. La distribution-limite D est une mesure puisque pour tout $\varphi \geq 0, D(\varphi) = \lim \mu_n(\varphi)$ et pour tout $n, \mu_n(\varphi) \geq 0$. On peut donc désigner D par une mesure μ. Parce que les mesures sont des mesures de probabilités et ont des supports tous contenus dans une même boule, alors μ_n converge vers μ dans la topologie vague.

Avec l'appui du lemme 1, on détermine le support de la mesure μ. Si S est le support de la fonction de poids p, le support de la mesure μ est la somme infinie de Minkowski des ensembles $T^k(S)$: $\sum_{k \geq 0} T^k(S)$. Or ce compact est invariant sous les transformations affines de $\mathbf{R}^d$: $\{x \to a + Tx\}$ où a parcourt les points de $T(G)$ tels que $p(a) \neq 0$. □

6 Condition suffisante de continuité de la fonction fondamentale

Dans cette section, nous donnons des conditions suffisantes pour obtenir la continuité uniforme de la fonction fondamentale.

Théorème 10 *Si le rayon spectral de T est inférieur à un, si $\sum_{x \in T(G)} p(x) = 1/|\text{dét}\, T|$ et si la fonction $\Phi(\xi) = \prod_{n \geq 1}(P(-T^{*n}\xi)|\text{dét}\, T|)$ est une fonction intégrable, alors la fonction fondamentale $F(x)$ se prolonge par continuité à $\mathbf{R}^d$ et la transformée de Fourier de ce prolongement est précisément $\Phi(\xi)$.*

Démonstration. Puisque Φ est intégrable, nous définissons la fonction continue $H(x) = \frac{1}{(2\pi)^d} \int \Phi(\xi) e^{i<x,\xi>} d\xi$. L'étape principale de la démonstration consiste à montrer que $H(y) = F(y)$ pour tout point y de G. Pour tout entier n, nous posons $B_n = \{\xi\colon\ T^{*n}\xi \in [-\pi, \pi]^d\}$. Considérons la suite de fonctions:

$$f_n(\xi) = \prod_{k=1}^{n} [P(-T^{*-k}\xi)|\text{dét}\, T|] \text{ si } \xi \in B_n \text{ , sinon } f_n(\xi) = 0.$$

On peut trouver un voisinage symétrique V de l'origine tel que $|\Phi(\xi)|$ est positive sur V; d'où il existe un nombre A tel que $|f_n(\xi)| \leq A|\Phi(\xi)|$, pour $n = 0, 1, 2, \ldots$, et pour tout ξ de V. Si $T^{*m}\xi$ est dans V et si n plus grand que m, alors

$$\begin{aligned} |f_n(\xi)| &= |\prod_{k=1}^{m} [P(-T^{*k}\xi)|\text{dét}\, T|] f_{n-m}(T^{*m}\xi)| \\ &\leq |\prod_{k=1}^{m} [P(-T^{*k}\xi)|\text{dét}\, T|] A\Phi(T^{*m}\xi)| = A|\Phi(\xi)|. \end{aligned}$$

Si L est un entier pour lequel $T^{*L}([-\pi,\pi]^d)$ est dans V et si M est la valeur maximale de $|P(\theta)(\det T)|$, alors on a l'inégalité suivante:

$$|f_n(\xi)| \leq AM^L|\Phi(\xi)|.$$

La suite $\{f_n\}$ est dominée par la fonction intégrable $AM^L|\Phi|$. Par le théorème de convergence dominée de Lebesgue, $H(y)$ est la limite de $\frac{1}{(2\pi)^d}\int_{B_n} f_n(\xi)e^{i<y,\xi>}d\xi$ quand n tend vers l'infini.

Selon la formule (6) $\prod_{k=0}^{n-1} P(T^{*-k}\theta) = \sum_{x\in G} F(T^n x)e^{i<x,\theta>}$, d'où $\frac{1}{(2\pi)^d}\int_{B_n} df_n(\xi)e^{i<y,\xi>}d\xi$ $= F(y)$ de sorte que $H(y) = F(y)$.

Vu l'équation fonctionnelle (5), $\Phi(\xi) = |\det T|P(-T^*\xi)\Phi(T^*\xi)$, on en tire que $H(Tx) = \sum_{y\in G} p(Ty)H(x-y)$. Par récurrence, $H(x) = F(x)$ pour tout x de G_∞. H est le prolongement continu de F. H a un support compact, est intégrable et sa transformée de Fourier ne peut pas différer de Φ. □

Théorème 11 *Si la transformée de Fourier $\Phi(\xi)$ du procédé d'interpolation ne prend que des valeurs réelles supérieures ou égales à zéro, alors Φ est intégrable.*

Démonstration. On sait que la transformée de Fourier $\Phi(\xi)$ du procédé d'interpolation est $\prod_{n\geq 1}(P(-T^{*n}\xi)|\det T|)$. Si B est la boule-unité de $\mathbf{R}^d$ et si J_n est l'intégrale de Φ sur $T^{*-n}(\bar{B})$, alors

$$J_n = \int_B \Phi(\xi) \prod_{0\leq k\leq n-1} P(-T^{*-k}\xi)d\xi,$$

puisque

$$\Phi(T^{-n}\xi) \;=\; \Phi(\xi) \prod_{0\leq k\leq n-1} [P(-T^{*-k}\xi)/|\det T|].$$

Si M est la valeur maximale de $\Phi(\xi)$ sur B, alors

$$J_n \leq M \int_B \prod_{0\leq k\leq n-1} P(-T^{*-k}\xi)d\xi.$$

Selon la formule (6), la série de Fourier de $\prod_{0\leq k\leq n-1} P(-T^{*-k}\xi)$ est $\sum_{x\in G_n} F(T^n x)e^{i<x,\xi>}$. D'où

$$\int_B \prod_{0\leq k\leq n-1} P(-T^{*-k}\xi)d\xi = F(0) = 1.$$

Chacune des intégrales J_n est bornée par M. Φ est donc intégrable. □

7 Application à l'interpolation itérative dyadique à quatre poids

Nous terminons en appliquant les résultats obtenus dans ce chapitre à l'interpolation dyadique à quatre poids. Ce procédé d'interpolation a été décrit à la section 2 du chapitre II, c'était l'exemple 3. L'interpolation dyadique dépend de quatre paramètres a, b, c et d. Si l'on veut interpoler une fonction $f(x)$ définie sur les entiers relatifs $\mathbf{Z}$, on le fait par une fonction g selon la relation de récurrence $g(x) = ag(x-3h)+bg(x-h)+cg(x+h)+dg(x+3h)$ où $x = m/2^{n+1}$ avec m impair et $h = 2^{-n-1}$. Le prolongement $g(x)$ se poursuit jusqu'à l'ensemble G_∞ des rationnels dyadiques p/q où p est un entier relatif et q est une puissance entière de 2. Le polynôme caractéristique est $P(\theta) = ae^{i3\theta} + be^{i\theta} + 1 + ce^{-i\theta} + de^{-i3\theta}$.

Voici quelques propriétés élémentaires de l'interpolation dyadique.

- L'interpolante fondamentale est nulle hors de l'intervalle $(-3, 3)$.

- L'interpolante fondamentale satisfait l'équation fonctionnelle suivante: $F(t/2) = F(t) + aF(t-3) + bF(t-1) + cF(t+1) + dF(t+3)$.

- Si $a + b + c + d = 1$, la transformée de Fourier Φ du procédé d'interpolation est bien définie: $\Phi(\xi) = \prod_{n\geq 1} [P(-\xi 2^{-n})/2]$. Cette transformée $\Phi(\xi)$ satisfait l'équation fonctionnelle $\Phi(2\xi) = \Phi(\xi)P(-\xi)/2$; de plus $\Phi(0) = 1$.

Propriétés de l'interpolante fondamentale associée à l'interpolation dyadique cubique $(a = d = -1/16, b = c = 9/16)$

Théorème 12 *La transformée de Fourier du procédé d'interpolation dyadique cubique* $(a = d = -1/16, b = c = 9/16)$ *est non-négative et intégrable; dans ce cas, la fonction fondamentale admet un prolongement continu à tout l'axe réel.*

Démonstration. $P(\theta) = 1 + (9/8)\cos\theta - (1/8)\cos 3\theta = 2\cos^4(\theta/2)(2 - \cos\theta) \geq 0$. En se servant des Théorèmes 10 et 11, la conclusion du lemme est atteinte. □

Propriétés de l'interpolante fondamentale associée à l'interpolation dyadique symétrique $(b = c = 1/2 - a$ et $d = a)$

Théorème 13 *Dans le procédé d'interpolation dyadique symétrique à quatre poids* $(b = c = 1/2 - a$ *et* $d = a)$, *si* $-3/16 \leq a \leq 1/16$, *alors la transformée de Fourier* $\Phi(\xi)$ *est intégrable et l'interpolante fondamentale admet un prolongement continu à l'axe réel.*

Démonstration. $P(\theta) = 1 + (1 - 2a)\cos\theta + 2a\cos 3\theta$. Si $t = \cos\theta$, alors $P(\theta) = (1 -$

$8at(1-t))(1+t)$. Si $R(t) = 1 - 8at(1-t)$, alors

$$\Phi(\xi) = [(\sin \xi/2)/(\xi/2))]^2 \prod_{n\geq 1} R(\cos \xi 2^{-n}).$$

Si a est choisi dans l'intervalle $(-3/16, 1/16)$, alors la valeur maximale de $|R|$ est inférieure à 2. Ceci découle du faite que $R(1) = 1, R(-1) = 1 + 16a, R'(1/2) = 0$ et $R(1/2) = 1 - 2a$

Si $-3/16 < a < 1/16$, alors il existe un nombre $r < 1$ tel que lorsque ξ tend vers $\infty, \prod_{n\geq 1} R(\cos \xi 2^{-n})$ est $O(|\xi|^r)$. D'où $\Phi(\xi)$ est $O(|\xi|^{r-2})$.

Sous ces contraintes sur a, la transformée de Fourier $\Phi(\xi)$ est intégrable. D'après le Théorème 10, la fonction fondamentale $F(x)$ se prolonge par continuité à tout l'axe réel. □

Dyn, Levin et Gregory (1987) ont prêté attention aux procédés d'interpolations dyadiques symétriques à quatre poids. Ils ont montré que si $|a| < 1/4$, la fonction d'interpolation produite se prolonge par continuité en une fonction $g(x)$ définie sur l'axe réel. Si de plus $-1/8 < a < 0$, alors $g(x)$ admet en tout point une dérivée continue.

Notes L'interpolante fondamentale d'un procédé d'interpolation itérative a été introduite et étudiée dans divers articles: Dubuc (1986), Deslauriers-Dubuc (1989) et Deslauriers-Dubois-Dubuc (1991). L'analyse de la transformée de Fourier d'un procédé d'interpolation itérative a commencé dans les articles de Deslauriers-Dubuc (1987a et 1987b), elle s'est poursuivie dans l'article de Deslauriers-Dubois-Dubuc (1990) et s'est complétée dans la communication de Deslauriers-Dubuc (1991) au Congrès européen sur la théorie de l'itération à Batschuns en Autriche en 1989. L'application à l'interpolation dyadique à quatre poids exposée à la section 7 était déjà présente dans l'article de Deslauriers-Dubuc (1987a).

L'importante équation fonctionnelle (4) dans le cas de l'interpolation itérative unidimensionnelle a été étudiée de façon très détaillée en 1988 par Daubechies et Lagarias (leurs travaux paraîtront en 1990). Ces auteurs ont recherché des solutions intégrables de cette équation et déterminé des conditions suffisantes pour la régularité des solutions, c'est-à-dire leur continuité ou même leur dérivabilité. Pour mieux atteindre leurs objectifs, ils ont dû abandonner les méthodes de Fourier pour se servir de méthodes plus directes de produits de matrices.

Problèmes

1. Si une fonction définie sur les entiers satisfait la relation $f(n) = P(n)$ pour tout entier n où P est un polynôme du troisième degré, si g est l'interpolation dyadique cubique de f, montrez que pour tout nombre dyadique $x, g(x) = P(x)$.

2. Si Φ est la transformée de Fourier du processus d'interpolation dyadique cubique, montrez que

$$\Phi(\xi) = [(\sin\xi/2)/(\xi/2))]^4 \prod_{n\geq 1}(2 - \cos\xi 2^{-n}).$$

Démontrez que $\Phi(\xi)$ est $O(\xi^{-3})$ lorsque $|\xi|$ tend vers l'infini. Vérifiez que l'interpolante fondamentale F est la restriction aux nombres dyadiques d'une fonction H définie sur la droite réelle non seulement continue, mais aussi continûment dérivable.

3. Si F est l'interpolante fondamentale du procédé d'interpolation dyadique cubique, vérifiez les propriétés suivantes.

(a) Si x est compris entre 2 et 3, alors $F(x) = -F(2x - 3)/16$.

(b) Si $0 < x < 1$, alors $F(x) = (3 - x)(2 - x)(1 - x)/6 - 4F(1 + x) + 10F(2 + x)$.

(c) Pour tout $x, |F(x)| \leq 1$.

Chapitre IV
Interpolations continues

Sommaire. Nous déterminons des conditions nécessaires et suffisantes pour que la fonction fondamentale d'un procédé d'interpolation itérative soit continue. Quelques exemples illustrent la théorie. On s'attarde en particulier à quelques surfaces produites par interpolation fractale.

1 Deux modules de continuité

Un procédé d'interpolation itérative est *continu* par définition si la fonction fondamentale correspondante est uniformément continue. Rappelons ce fait d'analyse (voir par exemple le volume de Topologie générale de Bourbaki) qu'une fonction uniformément continue sur son domaine D de définition de $\mathbf{R}^d$ se prolonge par continuité à l'adhérence de D. Si un procédé d'interpolation itérative est continu et si la transformation linéaire associée T a un rayon spectral inférieur à un, alors l'interpolante fondamentale se prolonge par continuité à tout l'espace euclidien. Dans ce cas, pour éviter de la lourdeur dans la notation, nous désignerons encore par F le prolongement continu de l'interpolante.

Une fonction g définie sur G_∞ est une *fonction d'interpolation* s'il existe une fonction f sur G telle que g est le prolongement de f suivant le procédé d'interpolation itérative.

Deux types de modules de continuité permettront l'étude de la continuité du procédé d'interpolation itérative.

Si h est un nombre positif, si F est l'interpolante fondamentale, si $f(x)$ est une fonction définie sur G_∞ et si nous nous servons d'une norme sur $\mathbf{R}^d$, $x \to |x|$, alors nous posons

$$(1) \quad C_n(h) = \max\{\sum_{z\in G} |F(T^n x - z) - F(T^n y - z)|:\ x \in G, y \in G, |x-y| \le h\}.$$

$$(2) \quad \omega_n(f,h) = \sup\{|f(T^n x) - f(T^n y)|:\ x \in G, y \in G, |x-y| \le h\}.$$

Nous introduisons les supports $S_n = \{x \in G_n:\ F(x) \neq 0\}$. R_n est le rayon de la plus petite boule centrée à l'origine recouvrant S_n. Selon l'usage courant, nous posons $\|T\| = \sup\{|Tx|:\ x \in \mathbf{R}^d, |x| = 1\}$.

Lemme 1 *Supposons que $\|T\| < 1$, si h est un nombre réel supérieur ou égal à $R_1/(1 - \|T\|)$, si a est un point de $\mathbf{R}^d$, $x \in G$ et $|x-a| \le h$ et que f est une fonction d'interpolation, alors*

$$f(Tx) = \sum_{y\in G, |y-Ta|\le h} F(Tx-y)f(y).$$

Démonstration. Nous savons que $f(Tx) = \sum_{y\in G} F(Tx-y)f(y)$. Si x appartient à G, si y est un point de G pour lequel $F(Tx-y) \neq 0$, alors $y - Ta = T(x-a) - z$ où $z = Tx - y$; z appartient à S. D'où $|y - Ta| \le \|T\|h + R_1 \le h$. □

Lemme 2 *Si $f(x)$ est une fonction d'interpolation, si pour tout x de G_1, $\sum_{y\in G} F(x-y) = 1$ et si $\|T\| < 1, h \ge 2R_1/(1-\|T\|)$, alors*

$$\omega_n(f,h) \le [C_1(h)/2]^n \omega_0(f,h).$$

Démonstration. Discutons d'abord du cas où $n = 1$. Considérons deux points x_1 et x_2 de G tels que $|x_1 - x_2| \le h$. Si nous posons $a = (x_1 + x_2)/2$, alors $|x_1 - a| = |x_2 - a| = |x_1 - x_2|/2 \le h/2$. D'après le lemme 1,

$$f(Tx_i) = \sum_{y\in G, |y-Ta|\le h/2} F(Tx_i - y)f(y), i = 1 \text{ et } 2.$$

Si nous posons

$$M = \max\{f(y):\ y \in G, |y-Ta| \le h/2\}, m = \min\{f(y):\ y \in G, |y - Ta| \le h/2\}$$

et

$$c = (M+m)/2,$$

alors nous obtenons que

$$f(Tx_i) - c = \sum_{y\in G, |y-Ta|\le h/2} F(Tx_i - y)(f(y) - c), i = 1 \text{ et } 2.$$

(Nous nous sommes servis ici de l'hypothèse que pour tout x de $G_1, \sum_{y\in G} F(x-y) = 1$.) Il s'ensuit que

$$|f(Tx_1) - f(Tx_2)| \leq \sum_{y\in G, |y-Ta|\leq h/2} |F(Tx_1 - y) - F(Tx_2 - y)|\ |f(y) - c|.$$

$$|f(Tx_1) - f(Tx_2)| \leq \sum_{y\in G, |y-Ta|\leq h/2} |F(Tx_1 - y) - F(Tx_2 - y)|\ (M - m)/2.$$

D'où $\omega_1(f, h) \leq C_1(h)\omega_0(f, h)/2$. L'inégalité du théorème pour le cas $n = 1$ vient d'être démontrée. Le résultat général se vérifie ensuite par récurrence. □

2 Conditions suffisantes de continuité

Lemme 3 *Si le sous-groupe engendré par* $\{x \in G\colon\ |x| \leq h\}$ *est exactement* G *et si* h^* *est un nombre positif, alors on peut trouver une valeur entière positive* L *telle que pour tout entier* n *et pour toute fonction* f *définie sur* G_n, *alors* $\omega_n(f, h^*) \leq L\omega_n(f, h)$ *(*L *est fonction de* h *et de* h^*, *mais non pas de* f *ou de* n.)

Démonstration. Si $B = \{x \in G\colon\ |x| \leq h\}$ et $B^* = \{x \in G\colon\ |x| \leq h^*\}$, tout point de B^* est une somme de points de B. Si N est suffisamment grand, tout point de B^* est la somme de N points de B. Si x et y appartiennent à G et que $|x - y| \leq h^*$, il est alors possible de trouver $N+1$ points de G, $\{x_k\}_{0\leq k\leq N}$, tels que $x_0 = x, x_N = y, |x_k - x_{k-1}| \leq h$ pour $k = 1, 2, \ldots, N$. Puisque

$$f(T^n x) - f(T^n y) = \sum_{k=1}^{N} f(T^n x_{k-1}) - f(T^n x_k),$$

alors

$$|f(T^n x) - f(T^n y)| \leq N\omega_n(f, h).$$

L'inégalité $\omega_n(f, h^*) \leq N\omega_n(f, h)$ est vérifiée. □

Théorème 4 *Si pour tout* x *de* G_1, $\sum_{y\in G} F(x-y) = 1$, *si* $\|T\| < 1$, *s'il existe un nombre positif* h *tel que* $h \geq 2R_1/(1 - \|T\|), C_1(h) < 2$ *et que le sous-groupe engendré par* $\{x \in G\colon$ $|x| \leq h\}$ *est* G, *alors le procédé d'interpolation itérative est continu.*

Démonstration. Si ε est un nombre positif donné, nous choisissons un entier n suffisamment grand pour que

$$2\sum_{k>n} \omega_k(F, \|T^{-1}\|R_1) + \omega_n(F, 4h) < \varepsilon.$$

Ceci est une conséquence des lemmes 2 et 3. Nous choisissons δ comme $2h/\|T^{-1}\|$. Démontrons que si x et y sont deux points de G_∞ tels que $|x - y| < \delta$ alors $|F(x) - F(y)| < \varepsilon$.

Supposons donc que x et y sont deux points de G_∞ tels que $|x-y| < \delta$. Nous pouvons présumer que les points x et y sont dans G_m pour un entier $m \geq n$. Nous construirons une suite de points $\{x_k\}_{n\leq k\leq m}$ tels que $x_m = x$ et alors que k décroît de m à $n+1$, x_{k-1} appartiendra à G_{k-1} et sera trouvé à l'aide de x_k. Puisque x_k appartient à G_k, il existe un point z_k de G tel que $x_k = T^k z_k$. Puisque $\sum_{w\in G} F(Tz_k - w) = 1$, il existe un point w de G tel que $Tz_k - w \in S$. Nous posons $z_{k-1} = w$ et $x_{k-1} = T^{k-1}z_{k-1}$. x_{k-1} appartient à G_{k-1}.

Par définition de $R_1, |Tz_k - z_{k-1}| \leq R_1$. La définition de $\omega_k(F,h)$ montre que

$$|F(x_k) - F(x_{k-1})| = |F(T^k z_k) - F(T^k T^{-1} z_{k-1})| \leq \omega_k(F, \|T^{-1}\|R_1).$$

D'où $|F(x) - F(x_n)| \leq \sum_{k>n} \omega_k(F, \|T^{-1}\|R_1)$.

De la même manière, on peut construire une suite de points $\{y_k\}_{n\leq k\leq m}$ tels que $y_m = y$ et alors que k décroît de m à $n+1$, y_{k-1} appartiendra à G_{k-1} et sera trouvé à l'aide de y_k. Puisque y_k appartient à G_k, il existe un point w_k de G tel que $y_k = T^k w_k$. Puisque $\sum_{w\in G} F(Tw_k - w) = 1$, il existe un point w de G tel que $Tw_k - w \in S$. Nous posons $w_{k-1} = w$ et $y_{k-1} = T^{k-1}w_{k-1}$. y_{k-1} appartient à G_{k-1}. D'où

$$|F(y) - F(y_n)| \leq \sum_{k>n} \omega_k(F, \|T^{-1}\|R_1).$$

Nous remarquons que la distance entre z_n et w_n ne peut pas être si grande:

$$\begin{aligned}
|z_n - w_n| &\leq \sum_{k=0}^{m-n} |T^k z_{n+k} - T^{k+1} z_{n+k+1}| + |T^{m-n} z_m - T^{m-n} w_m| \\
&\quad + \sum_{k=0}^{m-n} |T^k w_{n+k} - T^{k+1} w_{n+k+1}| \\
|z_n - w_n| &\leq \sum_{k=0}^{m-n} \|T^k\| \, |z_{n+k} - T z_{n+k+1}| + |T^{-n}x - T^{-n}y| \\
&\quad + \sum_{k=0}^{m-n} \|T^k\| \, |w_{n+k} - T w_{n+k+1}| \\
|z_n - w_n| &\leq 2\sum_{k=0}^{m-n} \|T^k\| R_1 + |T^{-n}x - T^{-n}y| \\
|z_n - w_n| &\leq 2h + \|T^{-n}\| \, |x-y| \leq 4h
\end{aligned}$$

D'où $|F(x_n) - F(y_n)| \leq \omega_n(F, 4h)$. Ce qui entraîne que

$$|F(x) - F(y)| \leq 2\sum_{k>n} \omega_k(F, \|T^{-1}\|R_1) + \omega_n(F, 4h) < \varepsilon.$$

La fonction F est uniformément continue. □

Corollaire 5 *Si pour tout x de G_1, $\sum_{y \in G} F(x-y) = 1$, s'il existe un entier $n \geq 1$ et un nombre positif h tel que $\|T^n\| < 1, h \geq 2R_n/(1-\|T^n\|), C_n(h) < 2$ et que le sous-groupe engendré par $\{x \in G\colon\ |x| \leq h\}$ est G, alors le procédé d'interpolation itérative est continu.*

Pour se convaincre du bien-fondé du dernier corollaire, on suppose que l'on veuille établir la continuité du processus d'interpolation itérative partant d'un groupe G, selon une transformation linéaire T et une fonction de poids p. Considérons le processus d'interpolation itérative qui se sert du même groupe G, mais dont la transformation linéaire est $T' = T^n$ et la fonction de poids $p'(x) = F(x)$ si $x \in T^n(G)$ (F est l'interpolante fondamentale du premier processus). Si l'on applique le théorème 4 à ce dernier processus d'interpolation, on obtient précisément le corollaire 5.

3 Conditions nécessaires de continuité

Lemme 6 *Si le procédé d'interpolation est continu et si le rayon spectral de T est inférieur à un, alors pour tout x de G_1, $\sum_{y \in G} F(x-y) = 1$.*

Démonstration. Dans la formule (2) du chapitre III, nous prenons x dans G_1 et nous faisons tendre n vers l'infini:

$$1 = \lim_{n\to\infty} F(T^n x) = \lim_{n\to\infty} \sum_{y \in G} F(T^n y) F(x-y) = \sum_{y \in G} F(x-y).$$

Ceci est vrai parce que $T^n x$ converge vers $0, F(0) = 1$ et que la somme sur G est en fait une somme finie dont le nombre de termes est indépendant de k puisque le support F est borné. □

Théorème 7 *Si le procédé d'interpolation itérative est continu et si le rayon spectral de T est inférieur à un, alors il existe un entier n et un nombre positif h tel que $h \geq 2R_n/(1 - \|T^n\|), C_n(h) < 2$ et le sous-groupe engendré par $\{x \in G :\ |x| \leq h\}$ est précisément G.*

Démonstration. Puisque le rayon spectral de T est inférieur à un, le support de F est borné et la suite R_n est bornée. Il est possible de trouver un nombre h tel que le sous-groupe engendré par $\{x \in G\colon\ |x| \leq h\}$ est précisément G et pour tout entier $n \geq 1, h \geq 2R_n/(1-\|T^n\|)$. Nous posons N comme la plus grande cardinalité rencontrée parmi les ensembles: $\{z \in G\colon\ F(a-z) \neq 0\}$ où a est un vecteur de $\mathbf{R}^d$. Puisque F est uniformément continue, il existe un nombre $\delta > 0$ tel que si x et y sont dans G_∞ et que $|x-y| < \delta$, alors $|F(x) - F(y)| < 1/N$. Si n est un entier pour lequel $\|T^n\| h < \delta$, alors $C_n(h) < 2$; puisqu'un tel entier n existe, la démonstration est complète. □

Commentaire À la suite du lemme 6 et du théorème 7, il apparaît que le corollaire 5 fournit une condition nécessaire et suffisante de continuité pour un processus d'interpolation itérative.

4 Exemples de processus continus d'interpolation

Exemple 1 [Interpolation dyadique cubique] Le groupe G est l'ensemble des entiers relatifs $\mathbf{Z}$. La transformation linéaire T est $Tx = x/2$, la fonction de poids est $p(0) = 1, p(\pm 1/2) = 9/16, p(\pm 3/2) = -1/16$ et $p(x) = 0$ pour tout autre multiple entier de $1/2$. Nous faisons appel au corollaire 5 avec sa notation. Nous prenons n égal à 3, $R_3 = 21/8, \|T^3\| = 1/8$, h est choisi comme $2R_3/(1 - \|T^3\|) = 6$; après quelques calculs, on obtient que $C_3(6) = 7/4 < 2$.

Exemple 2 [La courbe de von Koch] $G = \mathbf{Z}$. $Tx = x/4$. $p(0) = 1, p(\pm 3/4) = 1/3, p(-1/2) = 1/2 + i\sqrt{3}/6, p(\pm 1/4) = 2/3, p(1/2) = 1/2 - i\sqrt{3}/6$ et $p(x) = 0$ pour les autres multiples entiers de $1/4$. Avec la notation de la section 2 :

$$R_1 = 3/4, \|T\| = 1/4, n = 1, h = 2R_1/(1 - \|T\|) = 2; C_1(1) = 2/\sqrt{3} < 2.$$

Exemple 3 [Interpolation à quatre poids sur un réseau rectangulaire] Nous considérons la situation: $G = \mathbf{Z} \times \mathbf{Z}, T(x, y) = ((x - y)/2, (x + y)/2)$. $p(0, 0) = 1, p(1/2, 1/2) = a, p(-1/2, 1/2) = b, p(-1/2, -1/2) = c, p(1/2, -1/2) = d$ et $p(x_1, x_2) = 0$ ailleurs. Nous admettons les restrictions suivantes sur les paramètres: chacun d'entre eux est positif et $a + b + c + d = 1$. Sous ces restrictions, l'interpolante fondamentale est uniformément continue.

Théorème 8 *Si les quatre paramètres sont positifs et si leur somme est un, alors la fonction fondamentale F est uniformément continue.*

Démonstration. Esquissons les calculs. Nous nous servons de la norme L^∞ de $\mathbf{R}^2$: $|(x_1, x_2)| = \max(|x_1|, |x_2|)$. Afin d'établir la continuité uniforme de F, nous ferons appel au corollaire 5 avec $n = 6$ et $h = 4$. Parce que $a + b + c + d = 1$, alors pour tout x de G_1, $\sum_{z \in G} F(x - z) = 1$. Le support de F est contenu dans le carré $|x| \leq 2$. G est le sous-groupe engendré par $G \cap \{|x| \leq h\}$. Nous remarquons que $T^6(x_1, x_2) = (x_2/8, -x_1/8)$. On peut montrer que $R_6/(1 - \|T^6\|) = R_2/(1 - \|T^2\|) = 2$; d'où $h \geq 2R_6/(1 - \|T^6\|)$.

Vérifions que $C_6(4) < 2$. $F(x)$ est ici toujours compris entre 0 et 1. Le fait que pour tout x de G, $\sum_{z \in G} F(x - z) = 1$ entraîne que pour tout x de G_∞, $\sum_{z \in G} F(x - z) = 1$. Si bien que pour tout u et v de G_∞, $\sum_{z \in G} |F(u - z) - F(v - z)| \leq 2$. Mais l'égalité à 2 est impossible lorsque $u = T^4 x, v = T^4 y, |x - y| \leq 2$. Dans ce cas $|u - v| \leq 1/2$ et il existe un z de G tel que $F(u - z) > 0$ et $F(v - z) > 0$ (si $|u - z| \leq 1/2$, alors $F(u - z) > 0$); pour ce choix de u, v et z, $|F(u - z) - F(v - z)| < F(u - z) + F(v - z)|$. D'où

$$\sum_{z \in G} |F(u - z) - F(v - z)| \; < \; \sum_{z \in G} F(u - z) + F(v - z) = 2.$$

Le corollaire 5 montre que F est uniformément continue. □

Les figures 1 et 2 (voir la fin du chapitre) illustrent l'exemple 3 dans les deux cas suivants pour les paramètres a, b, c et d. La figure 1 donne les courbes de niveau de la fonction fondamentale alors que chacun des paramètres vaut $1/4$. La figure 2 donne les courbes de niveau de la fonction fondamentale alors que $a = 1/2, b = 1/8, c = 1/4, d = 1/8$. Les niveaux retenus dans chacune des figures sont les multiples entiers de $1/20$.

Les figures 3 à 6 représentent les mêmes surfaces de façon différente. La figure 3 observe la surface de la fonction fondamentale correspondant aux poids égaux à partir d'un point du premier octant sur le plan $x = y$ tandis que la figure 4 fait de même pour la fonction fondamentale correspondant aux poids inégaux $a = 1/2, b = 1/8, c = 1/4$ et $d = 1/8$.

Les figures 5 et 6 tracent les sections de l'une et de l'autre surface selon des plans parallèles au plan vertical $y = 0$.

Exemple 4 [Interpolation quadratique sur un réseau triangulaire] G est le réseau triangulaire: $G = \{m+n/2, n\sqrt{3}/2\}; T(x,y) = ((3x-\sqrt{3}y/6, (\sqrt{3}x+3y)/6))$. $G_1 = T(G) = \{m/2, (m+2n)\sqrt{3}/6\}$. Les valeurs de p sont $p(x,y) = 4/9$ si (x,y) est l'un des six plus proches voisins de $(0,0)$ dans $T(G)$, i.e. $x^2+y^2 = 1/3$,

$$\begin{aligned} p(x,y) &= -1/9 \text{ si } x^2+y^2 = 4/3 \text{ et} \\ p(0,0) &= 1 \text{ et} \\ p(x,y) &= 0 \text{ ailleurs dans } T(G). \end{aligned}$$

On se sert de la norme euclidienne usuelle $\mathbf{R}^2$: $|(x,y)| = \sqrt{(x^2+y^2)}$. Appliquons le corollaire 5 avec $n = 5$ et $h = 2R_5/(1-\|T^5\|)$. La norme de T^5, $\|T^5\|$ est $3^{-5/2} < 1$. Le rayon $R_5 = \max\{|(x,y)|: (x,y) \in G_5, F(x,y) \neq 0\}$ est $(4476)^{1/2}/27$. D'où $h = 5,295$. Comme on peut le vérifier avec un ordinateur, après des milliers d'opérations arithmétiques, la valeur trouvée pour $C_5(h)$ est $1,443$. Selon le corollaire 5, le procédé d'interpolation est continu. Les figures 7, 8 et 9 représentent la surface de cette fonction fondamentale. La figure 7 donne les courbes de niveau de la fonction selon les niveaux multiples entiers de $1/20$; pour le niveau 0, on se limite cependant à l'intérieur du support de la fonction. Le fait que la fonction fondamentale prend aussi des valeurs négatives accentue la visibilité du caractère fractal du support de la fonction. Nous laissons en exercice l'étude ce support. La figure 8 donne une vue de la surface à partir d'un point de l'espace situé dans le plan $x = y$. Dans la figure 9, on a tracé les sections de la surface selon des plans parallèles au plan vertical $y = 0$.

Nous résumons sous forme d'un tableau les calculs qui justifient la continuité des fonctions fondamentales observées.

Nom de l'exemple	n	$\|T^n\|$	R_n	h	$C_n(h)$
Interpolation dyadique	3	1/8	21/8	6	7/4
Courbe de von Koch	1	1/4	3/4	2	$2/\sqrt{3}$

Exemple 3	6	1/8	16/7	4	< 2
Exemple 4	5	$3^{-5/2}$	$\sqrt{4475}/2$	5,3	1,443

Questions ouvertes Quels sont les points de non-dérivabilité des deux surfaces décrites par les figures 3 et 4? Est-ce que la surface de l'interpolation fractale quadratique (figure 8) admet des plans tangents en chacun de ses points? Nous ne connaissons pas la réponse à ces questions.

Notes Ce chapitre est tiré de l'article de Deslauriers-Dubois-Dubuc (1991). Pendant quelques années, un critère pour la continuité des fonctions d'interpolation d'un procédé d'interpolation itérative constituait le chaînon manquant de la théorie. Daubechies et Lagarias ont donné eux-aussi des critères de continuité et de dérivabilité pour des procédés d'interpolation itérative unidimensionelle.

Problèmes

1. Vérifiez que le support de la fonction fondamentale de l'exemple 3 est l'octogone limité par les droites $x = \pm 2, \pm x \pm y = 2\sqrt{2}, y = \pm 2$.

2. Vérifiez que le support de la fonction fondamentale de l'exemple 4 est contenu dans l'attracteur du système des treize similitudes contractantes suivantes (que l'on note avec l'arithmétique des nombres complexes): $z \to a_k + bz$, où $k = 0, 1, 2, \ldots, 12, b = \frac{\sqrt{3}}{3} e^{i\pi/6}, a_0 = 0, a_k = be^{ik\pi/3}$ et $a_{k+6} = 2be^{ik\pi/3}$ où $1 \leq k \leq 6$.

3. (Deslauriers, communication privée) Si un procédé d'interpolation itérative est continu et si la transformation linéaire T du procédé a un rayon spectral inférieur à un, alors $\int_{\mathbf{R}^d} F(x)dx = 1$. Démontrez cette assertion.

Figure 1:
Courbes de niveau aux multiples de 1/20 pour la fonction fondamentale correspondant à des poids égaux.

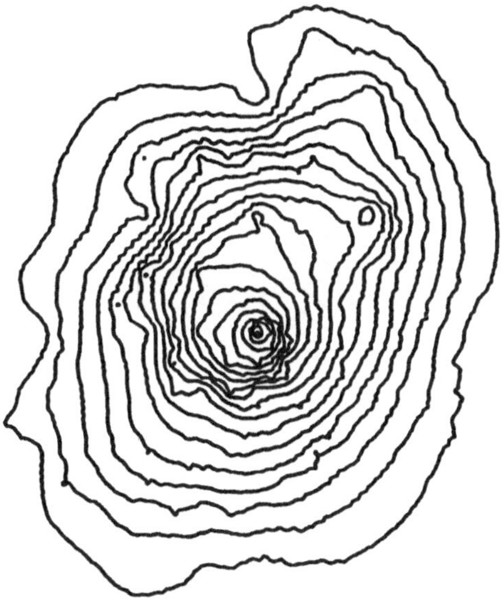

Figure 2:
Courbes de niveau aux multiples de 1/20 pour la fonction fondamentale correspondant à des poids inégaux.

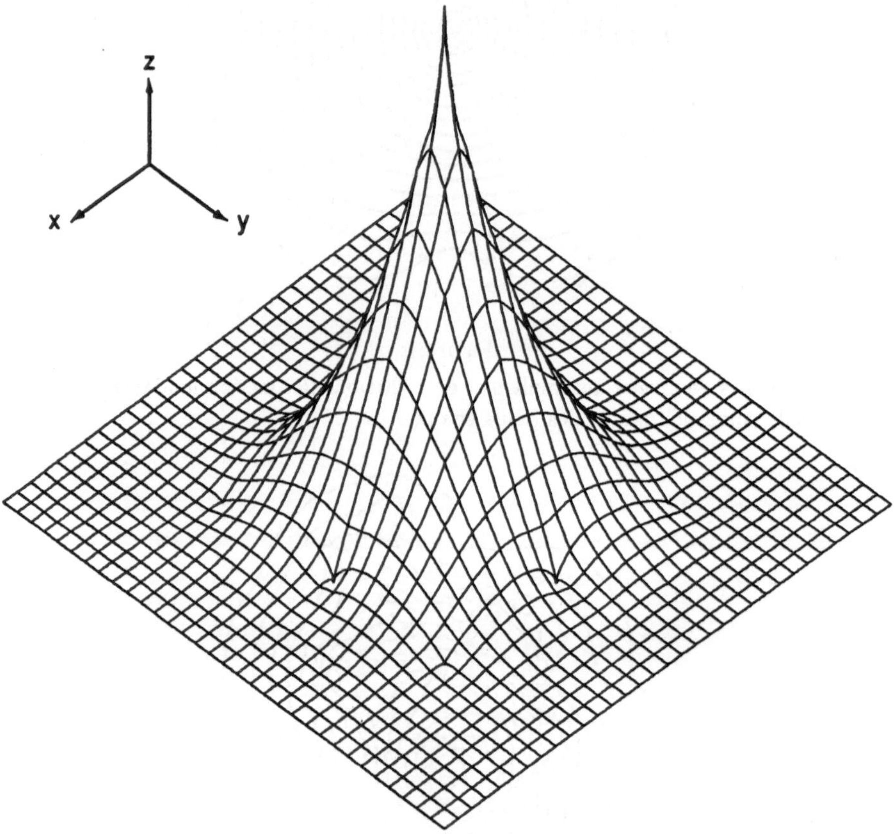

Figure 3:
Projection dimétrique du graphe de la fonction fondamentale correspondant à des poids égaux.

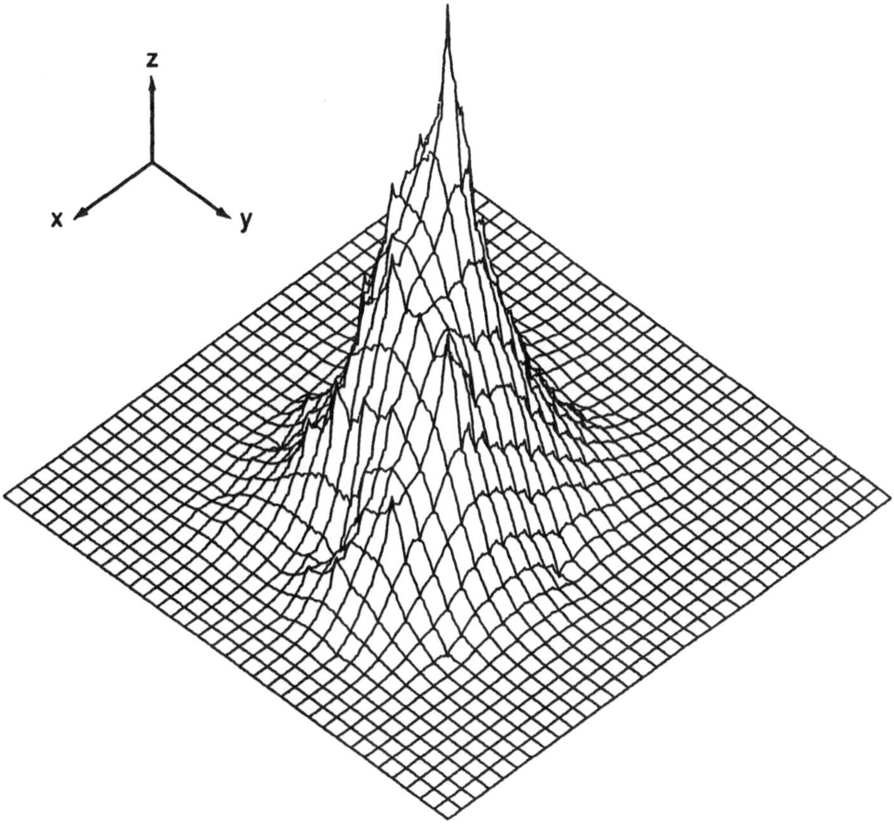

Figure 4:
Projection dimétrique du graphe de la fonction fondamentale correspondant à des poids inégaux.

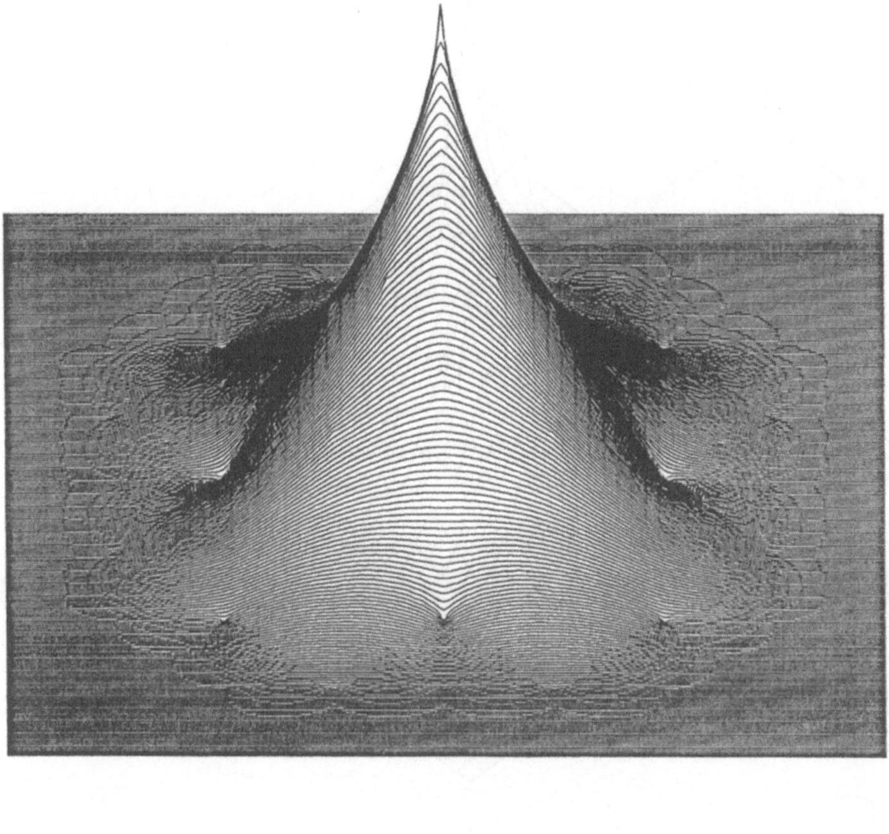

Figure 5: Vues de sections verticales de la surface correspondant à des poids égaux.

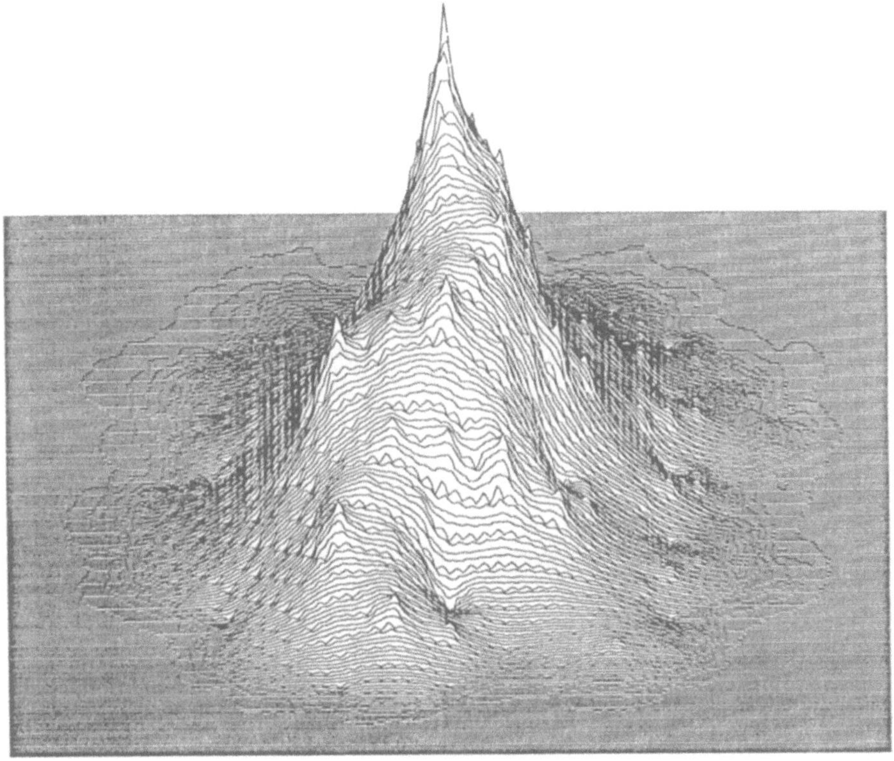

Figure 6: Vues de sections verticales de la surface correspondant à des poids inégaux.

Figure 7:
Courbes de niveau aux multiples de 1/20 pour la fonction fondamentale de l'interpolation quadratique sur un réseau triangulaire.

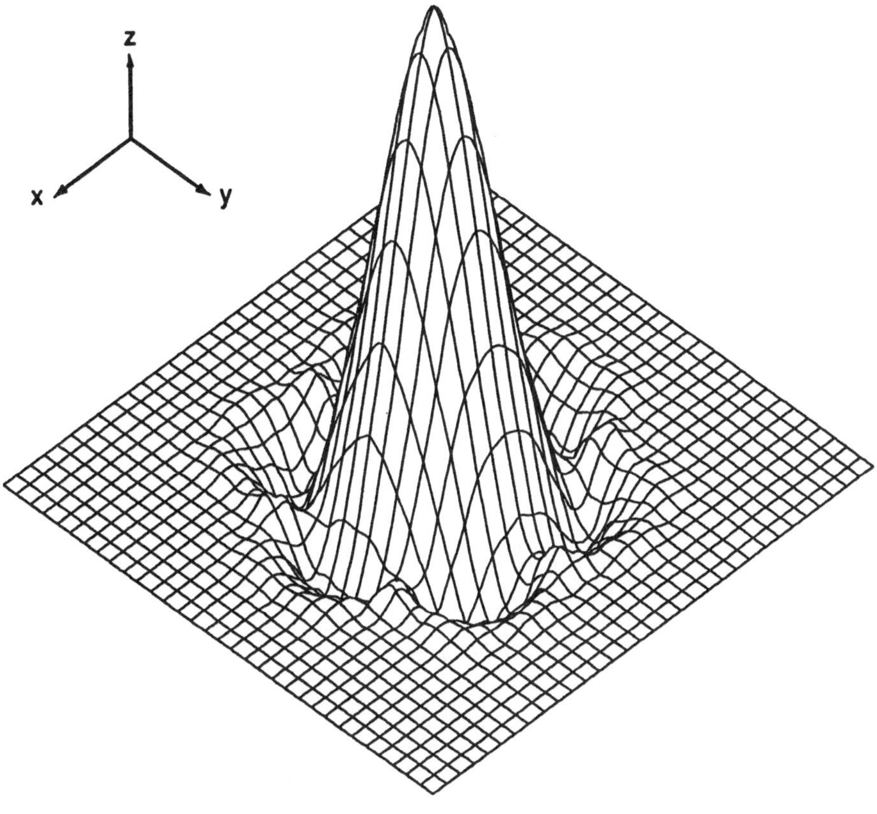

Figure 8:

Projection dimétrique du graphe de la fonction fondamentale correspondant à l'interpolation quadratique sur un réseau triangulaire.

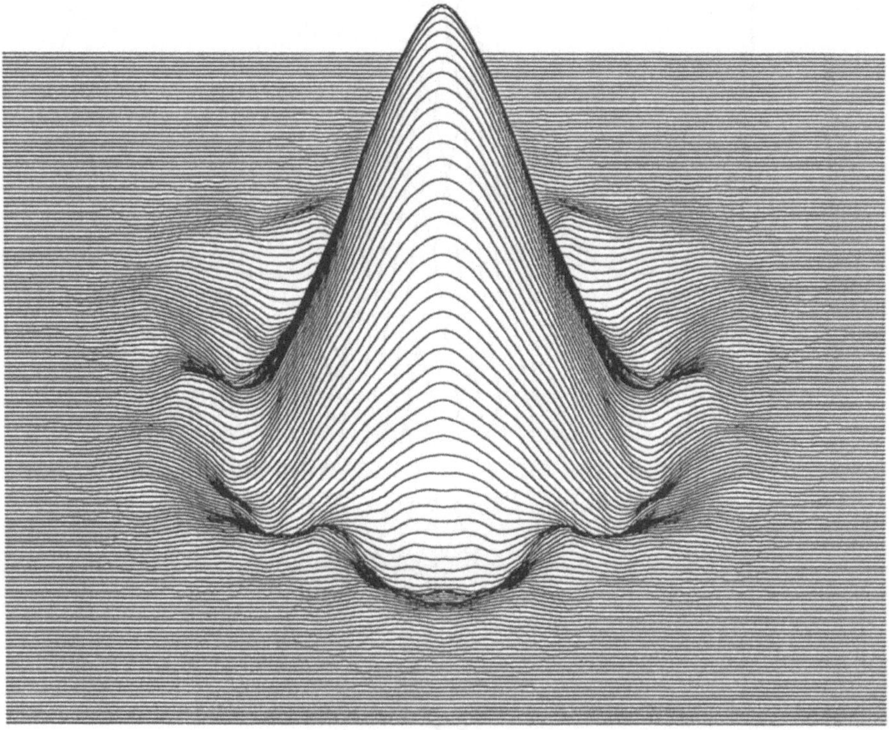

Figure 9:
Vue de sections verticales de la surface de l'interpolation fractale quadratique sur un réseau triangulaire.

Chapitre V
Interpolations de carré localement sommable

Sommaire. Nous nous restreignons à des procédés d'interpolation itérative unidimensionnelle, l'interpolation ayant lieu sur la droite réelle. Nous découvrirons les conditions précises pour que la distribution D associée à l'interpolante fondamentale soit celle d'une fonction de carré sommable. Une première condition est que la transformée de Fourier de D soit de carré sommable. Une autre condition provient de l'examen des valeurs propres d'une matrice. Nous appliquerons ces résultats aux courbes de von Koch pour obtenir un critère simple pour qu'une telle courbe soit de carré sommable et nous aurons un développement de Fourier pour chacune de celles-ci.

1 Interpolation itérative à base b

Nous rappelons ce que nous entendons par l'interpolation itérative selon une base b. Nous partons d'une fonction $f(n)$ définie sur le groupe $G = \mathbf{Z}$. Nous aimerions prolonger f à toute la droite réelle si possible. Premièrement, on se sert d'un entier b supérieur à un; ce sera la base du procédé d'interpolation. Nous posons G_n comme la totalité des nombres rationnels de la forme m/b^n où m est un entier relatif et $n = 0,1,2,\ldots$ La fonction f est déjà définie sur $G_0 = G$. Nous prolongerons f successivement à $G_1, G_2, G_3, \ldots$ par récurrence. Pour prolonger f de G à G_1, on se sert d'une suite de poids $\{p_k\}_{-\infty<k<\infty}$. On impose quelques conditions sur cette suite. D'abord $p_0 = 1$ et $p_k = 0$ pour tout autre multiple entier de b : $k = \pm b, \pm 2b, \pm 3b, \ldots$ On suppose de plus qu'un nombre fini des paramètres p_k diffèrent de zéro. Sous ces hypothèses, le prolongement proposé encore noté f est: $f(j/b) = \sum_k p_{j-kb} f(k)$.

Le prolongement de f à $G_2, G_3, \ldots$ est effectué de la même façon: si j est un entier, on pose

$$f(j/b^{n+1}) = \sum_k p_{j-kb} f(k/b^n) \tag{1}$$

Sans restriction à imposer sur la suite originale $f(n)$, les propriétés des poids permettent un prolongement consistant. Ce prolongement de la suite $f(n)$ à l'ensemble G_∞ des nombres b-adiques sera appelé *l'interpolation itérative à base b* de la suite. On notera ce prolongement $f(x)$.

2 Interpolation unidimensionnelle de carré sommable

Pour le processus itératif (1), nous recherchons des conditions nécessaires et suffisantes pour que la distribution fondamentale soit de carré sommable. Une telle condition est que la norme quadratique de la transformée de Fourier, $\Phi(\xi)$, de la distribution soit finie. Notre objectif est de rendre cette condition plus explicite.

On a déjà défini la suite de distributions au sens de Schwartz $T^n(\varphi) = \sum_m F(m/b^n)\varphi(m/b^n)/b^n$. Introduisons la suite F_n de fonctions: si k est un entier, si $|x - k/b^n| < b^{-n/2}$, alors $F_n(x) = F(k/b^n)$. F_n est une fonction constante par morceaux, la distribution associée $U_n(\varphi) = \int F_n(x)\varphi(x)dx$ est la convolution de la distribution T_n avec la fonction $\varepsilon_n(x)$ qui vaut b^n sur l'intervalle $[-b^{-n}/2, b^{-n}/2]$ et 0 ailleurs. La suite U_n de distributions tend aussi vers la distribution fondamentale T. L'intégrale du carré du module de F_n est $\|F_n\|^2 = \sum_m |F(m/b^n)|^2/b^n$.

Théorème 1 *Si $\sum_k p_k = b$, alors la suite $\|F_n\|^2$ converge au sens propre ou impropre vers $\frac{1}{2\pi}\int |\Phi(\xi)|^2 d\xi$.*

Démonstration. Désignons par $P_n(\theta)$ la série $\sum_k F(kb^{-n})e^{ik\theta}$. $F(kb^{-n})$ est le k ème coefficient de la série de Fourier $P_n(\theta)$. L'identité de Parseval prend ici la forme: $\sum_k |F(kb^{-n})|^2 = \frac{1}{2\pi}\int_{-\pi}^{\pi} |P_n(\theta)|^2 d\theta = \frac{1}{2\pi}\int_{-\pi b^n}^{\pi b^n} |\Phi_n(\xi)|^2 d\xi$ où $\Phi_n(\xi) = P_n(-\xi/b^n)/b^n$. D'après le théorème 6 et la formule (6) du chapitre III, nous avons déjà montré que $\Phi_n(\xi)$ converge vers $\Phi(\xi) = \prod_{k\geq 1}[P(-\xi/b^n)/b]$. Si $\int |\Phi(\xi)|^2 d\xi = \infty$, le lemme de Fatou et le fait que Φ_n converge ponctuellement vers Φ permettent de dire que la limite (lorsque n tend vers l'infini) de $\int_{-\pi b^n}^{\pi b^n} |\Phi_n(\xi)|^2 d\xi$ est infinie; dans ce cas, la limite de la suite $\sum_k |F(kb^{-n})|^2/b^n$ sera infinie.

Désormais, nous supposons que $\int |\Phi(\xi)^2 d\xi < \infty$. Comparons Φ_n et Φ. $\Phi(\xi)/\Phi_n(\xi) = \prod_{k\geq n+1}[P(-\xi/b^k)/b]$. Remarquons que l'on peut trouver un entier N tel que la fonction $P(\theta)$ ne s'annule pas sur $[-\pi b^{-N}, \pi b^{-N}]$. Il est même possible de trouver un nombre ε tel que si $|\xi| \leq \pi b^{-N}$, alors $\prod_{k>0} |P(-\xi/b^k)/b| > \varepsilon$. Si $|\xi| \leq \pi b^{n-N}$, alors $|\Phi(\xi)/\Phi_n(\xi)| > \varepsilon$ ou encore que $|\Phi_n(\xi)| < |\Phi(\xi)|/\varepsilon$. Le théorème de convergence dominée permet de prétendre que la limite de la suite $\int_{-\pi b^{n-N}}^{\pi b^{n-N}} |\Phi_n(\xi)|^2 d\xi$ est $\int |\Phi(\xi)|^2 d\xi$.

Pour terminer la vérification que la suite $\|F_n\|^2$ converge vers le nombre $\frac{1}{2\pi}\int |\Phi(\xi)|^2 d\xi$, il nous reste à montrer que les deux suites d'intégrales $\int_{-\pi b^n}^{-\pi b^{n-N}} |\Phi_n(\xi)|^2 d\xi$ et $\int_{\pi b^{n-N}}^{\pi b^n} |\Phi_n(\xi)|^2 d\xi$ tendent vers zéro. Or $\Phi_n(\xi) = P_N(\xi)\Phi_{n-N}(\xi/b^N)$. La fonction $P_N(\xi)$ est bornée, désignons par M une borne des nombres $|P_N(\xi)|$; si $|\xi| < b^n\pi$, alors $|\Phi_{n-N}(\xi/b^N)| \leq |\Phi(\xi)|/\varepsilon$. La fonction $|\Phi_n(\xi)|$ est donc majorée par $M|\Phi(\xi)|/\varepsilon$ sur $[\pi b^{n-N}, \pi b^n]$. La limite des intégrales $\int_{\pi b^{n-N}}^{\pi b^n} |\Phi_n(\xi)|^2 d\xi$ est donc nulle. □

Théorème 2 *Si $\sum_k p_k = b$ et si la distribution fondamentale est une fonction de carré sommable $F(x)$, alors la suite F_n converge vers F en moyenne quadratique.*

Démonstration. Si la distribution limite T est de carré sommable: $T(\varphi) = \int F(x)\varphi(x)dx$ où $\int |F(x)|^2 dx < \infty$, alors $\int |\Phi(\xi)|^2 d\xi < \infty$. La suite des intégrales $\int |F_n(x)|^2 dx$ est convergente d'après le dernier théorème. La suite F_n est donc bornée dans L^2. Soit f un point d'accumulation de la suite F_n pour la topologie faible, montrons que $f = F$. Pour toute fonction indéfiniment dérivable à support compact $\varphi(x)$, nous avons que

$$\int F(x)\varphi(x)dx = T(\varphi) = \lim T_n(\varphi) = \lim U_n(\varphi) = \lim \int F_n(x)\varphi(x)dx = \int f(x)\varphi(x)dx.$$

La fonction f-F est donc orthogonale à φ. Les fonctions C^∞ à support compact sont denses dans L^2; ceci force f-F à être orthogonale à toute fonction de L^2, ce qui oblige f à se confondre à F.

Nous avons donc montré que F ne peut être que le seul point d'accumulation pour la topologie faible de la suite F_n. La compacité de la boule unité de L^2 pour la topologie faible entraîne que la suite F_n converge faiblement vers F. La convergence faible de F_n vers F et la convergence de $\|F_n\|$ vers $\|F\|$ suffisent à garantir que F_n converge fortement vers F. □

3 Analyse spectrale

Dans cette section, nous verrons comment décider si le schéma d'interpolation (1) est de carré sommable par la décomposition spectrale d'une matrice associée. Nous introduisons une suite de séries trigonométriques: si $Q(\theta) = |P(\theta)|^2, Q_n(\theta) = Q(\theta)Q(b\theta)\dots Q(b^{n-1}\theta)$. Par convention, $Q_0(\theta)$ est le polynôme constant à 1; $Q_1(\theta) = Q(\theta), Q_2(\theta) = Q(\theta)Q(b\theta)$ et ainsi de suite. Nous introduisons aussi les coefficients de Fourier correspondants:

$$Q(\theta) = \sum_k q_k e^{ik\theta} \qquad \text{et} \qquad Q_n(\theta) = \sum_k q_{k,n} e^{ik\theta}.$$

Si $P(\theta) = \sum_k p_k e^{ik\theta}$, les coefficients q_k sont donnés par la formule: $q_k = \sum_j (p_{j+k}\overline{p_j})$. La relation de récurrence $Q_{n+1}(\theta) = Q_n(\theta)Q(b\theta)$ induit la formule

(2)
$$q_{k,n+1} = \sum_j q_{k-jb} q_{j,n}$$

Le nombre $\|F_n\|^2$ est exactement le nombre $q_{0,n}/b^n$. L'étude de la convergence de $\|F_n\|$ revient à déterminer le comportement asymptotique de la suite $q_{0,n}$. Visualisons la suite $q_{k,n}$ comme un vecteur V_n. La relation de récurrence (2) peut s'écrire sous une forme matricielle. Pour ce faire, nous indiquons par $A = (a_{j,k})$ la matrice dont les coefficients sont ainsi formés: $a_{j,k} = q_{j-kb}$.

Lorsque la matrice A s'applique au vecteur V_n, elle produit le vecteur V_{n+1}; ce résultat est consécutif à la relation de récurrence $Q_{n+1}(\theta) = Q_n(b\theta)Q(\theta)$. Si V_0 est le vecteur dont toutes les composantes sont nulles à l'exception de la composante centrale égale à un, alors $V_n = A^n V_0$. Définissons le vecteur ligne $R = \{r_k\}$: $r_0 = 1$ et $r_k = 0$ pour les autres valeurs de k. On obtient que $b^n \|F_n\|^2 = RA^n V_0$.

Lemme 3 *Si l'interpolante fondamentale du processus d'interpolation* (1) *est de carré sommable, alors le vecteur colonne dont toutes les composantes sont les mêmes est un vecteur propre de la matrice A pour la valeur propre b.*

Démonstration. $q_{k,n} = \frac{1}{2\pi}\int_{-\pi}^{\pi} Q_n(\theta)e^{-ik\theta}d\theta$,

$$q_{k,n}/b^n = \frac{1}{2\pi}\int_{-\pi b^n}^{\pi b^n} |\Phi_n(\xi)|^2 \exp(-ik\xi/b^n)d\xi.$$

Par le théorème de convergence dominée, la limite de $q_{k,n}/b^n$ sera $\frac{1}{2\pi}\int |\Phi(\xi)|^2 d\xi$ lorsque n tend vers l'infini. Remarquons que la somme des modules d'une ligne de la matrice A est finie. $\sum_k |a_{j,k}| \leq \sum_k |q_k| \leq (\sum_k |p_k|)^2$. Puisque les vecteurs V_n/b^n ont des composantes uniformément bornées qui convergent vers un vecteur dont toutes les composantes sont identiques et que $V_{n+1} = AV_n$, alors un vecteur dont toutes les composantes sont égales est un vecteur propre de la matrice A pour la valeur propre b. □

Théorème 4 *Supposons que $\sum_k p_k = b$. Le vecteur colonne dont toutes les composantes sont égales à l'unité est un vecteur propre de A pour la valeur propre b si et seulement si pour tout entier j, $\sum_k p_{j+kb} = 1$.*

Démonstration. Supposons que pour tout $j, \sum_k p_{j+kb} = 1$. Il s'ensuit que

$$\sum_k a_{j,k} = \sum_k q_{j-kb} = \sum_k \sum_l p_{l+j-kb}\overline{p}_l = \sum_l p_l = b.$$

Le vecteur colonne dont toutes les composantes valent un est un vecteur propre pour la valeur propre b.

Supposons maintenant que le vecteur colonne dont toutes les composantes sont égales à l'unité est un vecteur propre de A pour la valeur propre b. Posons

$$a_j = \sum_k a_{j,k} \quad \text{et} \quad S_j = \sum_k p_{j+kb}.$$

On obtient les deux identités:

$$a_0 = S_0\,\overline{S}_0 + S_1\,\overline{S}_1 + \ldots + S_{b-1}\,\overline{S}_{b-1} \quad \text{et} \quad a_1 = S_1\,\overline{S}_0 + S_2\,\overline{S}_1 + \ldots + S_0\,\overline{S}_{b-1}.$$

Par hypothèse, $a_0 = a_1 = b$. Appliquons l'inégalité de Cauchy-Schwartz à la quantité a_1 vue comme produit scalaire.

$$b^2 = |a_1|^2 \leq (|S_0|^2 + |S_1|^2 + \ldots + |S_{b-1}|^2)(|S_1|^2 + |S_2|^2 + \ldots + |S_0|^2) = b^2.$$

L'égalité obtenue force les deux vecteurs (S_j) et (S_{j+1}) à être linéairement dépendants. Tous les nombres S_j sont donc égaux. Comme la somme des S_j (pour les j entre 0 et $b-1$) est b, chacun des S_j vaut un. □

Remarque La condition $\sum_k p_{j+kb} = 1$ revient à dire que dans le schéma (1), l'interpolation itérative $f(x)$ de la suite $f(n)$ constante à un est la fonction constante, égale à un.

Faisons l'hypothèse supplémentaire que la fonction caractéristique du processus $P(\theta)$ est un polynôme trigonométrique de degré N. Le degré de $Q(\theta)$ est alors $2N$. La matrice A admet beaucoup de zéros. $a_{j,k} = q_{j-kb}$. Si M est la partie entière de $2N/(b-1)$, si $|j| \leq M$ et si $|k| > M$, alors $a_{j,k} = 0$. Désignons par B la matrice tronquée des $a_{j,k}$ où $|j| \leq M$ et $|k| \leq M$; la matrice B est une matrice carrée d'ordre $2M+1$. Les vecteurs tronqués W_n seront les $2M+1$ composantes centrales des vecteurs V_n. On a la récurrence $W_{n+1} = BW_n$. La décomposition de Jordan de B détermine le comportement asymptotique de la suite $q_{0,n}$. En particulier si b est une valeur propre simple de la matrice B et si toutes les autres valeurs propres de B ont un module inférieur à b, alors les éléments de la suite des matrices B^n sont uniformément bornés; dans ce cas, la suite $q_{0,n}$ sera bornée et le processus d'interpolation (1) sera de carré sommable. Nous énonçons ce résultat sous forme d'un théorème.

Théorème 5 *Si le polynôme caractéristique $P(\theta) = \sum_k p_k e^{ik\theta}$ du processus d'interpolation (1) est un polynôme trigonométrique de degré N, si $\sum_k q_k e^{ik\theta}$ est le polynôme $|P(\theta)|^2$ et si B est la matrice carrée $\{q_{j-kb}\}$ où $|j| \leq 2N/(b-1)$ et $|k| \leq 2N/(b-1)$, alors le processus d'interpolation est de carré sommable lorsque b est une valeur propre simple de la matrice B et lorsque toutes les autres valeurs propres ont un module inférieur à b.*

4 Courbes de von Koch de carré sommable

Soit $\{z_k\}_{0\leq k\leq b}$ une ligne polygonale du plan complexe de b segments et dont les extrémités sont $z_0 = 0$ et $z_b = 1$, considérons l'interpolation itérative à base b telle que si $f(n)$ est une suite de nombres à interpoler, $f(n+k/b) = (1-z_k)f(n) + z_k f(n+1), k = 1, 2, \ldots, b-1$. La question envisagée dans cette section est de déterminer des conditions sur les nombres z_k pour que le schéma d'interpolation soit de carré sommable. Depuis aussi loin que Lévy, il est connu que si chacun des accroissements $z_{k+1} - z_k$ est inférieur à un en module, la courbe de von Koch engendrée par la ligne polygonale est continue. L'interpolante fondamentale correspondante se prolonge continûment à tout l'axe réel avec un support compact. Dans ce cas, il est bien évident que le processus d'interpolation est de carré sommable. Nous examinons des conditions plus générales pour la réalisation de cette propriété.

Le polynôme caractéristique de l'interpolation est

$$P(\theta) = 1 + \sum_{k=1}^{b-1}(1-z_k)e^{ik\theta} + \sum_{k=1}^{b-1} z_{b-k}e^{-ik\theta}.$$

C'est un polynôme de degré $b-1$ et le polynôme $Q(\theta) = |P(\theta)|^2$ sera de degré $2b-2$.

Si $Q(\theta) = \sum_{k=-2b+2}^{2b-2} q_k e^{ik\theta}$, la matrice B de la section précédente est une matrice carré d'ordre 5 dont voici les éléments.

$$\begin{vmatrix} q_{2b-2} & q_{b-1} & q_{-2} & q_{-b-2} & 0 \\ 0 & q_{b-1} & q_{-1} & q_{-b-1} & 0 \\ 0 & q_b & q_0 & q_{-b} & 0 \\ 0 & q_{b+1} & q_1 & q_{-b+1} & 0 \\ 0 & q_{b+2} & q_2 & q_{-b+2} & q_{-2b+2} \end{vmatrix}$$

Entreprenons l'analyse spectrale de B. Deux valeurs propres sont faciles à identifier: q_{2b-2} et q_{-2b+2}. Les trois autres valeurs propres s'obtiennent par les valeurs propres de la matrice carrée C d'ordre trois

$$\begin{vmatrix} q_{b-1} & q_{-1} & q_{-b-1} \\ q_b & q_0 & q_{-b} \\ q_{b+1} & q_1 & q_{-b+1} \end{vmatrix}$$

Grâce au théorème 4, on sait déjà que le vecteur colonne de composantes $1, 1$ et 1 est un vecteur propre de C pour la valeur propre b. Avant d'aller plus loin, introduisons les quantités auxiliaires suivantes:

$$\alpha = \sum_{k=1}^{b-1} |z_k|^2, \beta = \sum_{k=1}^{b-1} z_k \overline{z}_{k+1}, \gamma = \sum_{k=1}^{b-a} z_k.$$

Les quantités q_{b+1}, q_b et q_{b-1} s'expriment simplement en fonction de α, β et γ :

$$q_{b+1} = \overline{\gamma} - \overline{\beta}, q_b = \overline{\gamma} - \alpha \text{ et } q_{b-1} = 1 + \overline{\gamma} - \beta.$$

Le prochain lemme établit la propriété surprenante que le nombre un est toujours une valeur propre de C.

Lemme 6 *Le vecteur ligne dont les composantes sont $\alpha - \overline{\beta}, \overline{\beta} - \beta, \beta - \alpha$ est un vecteur propre de C pour la valeur propre* 1.

Démonstration. Le vecteur ligne donné est orthogonal aux deux vecteurs $(1,1,1)$ et $(\beta, \alpha, \overline{\beta})$. La première colonne et la troisième colonne de la matrice C-I (où I est la matrice identité) sont des combinaisons linéaires de ces deux vecteurs. Parce que la somme de chacune des lignes de C donne la valeur b, on peut vérifier que la deuxième colonne C-I est une combinaison linéaire des deux autres colonnes et du vecteur $(1,1,1)$. La seconde colonne de C-I est également orthogonale au vecteur ligne original. C-I appliquée à ce vecteur ligne produit un vecteur nul. □

Lemme 7 *La quantité $\sum_{k=1}^{b} |z_k - z_{k-1}|^2$ est une valeur propre additionnelle de la matrice C.*

Démonstration. Calculons la trace T de la matrice C. $T = q_{b-1} + q_0 + q_{-b+1} = q_{b-1} + (b - q_b - q_{-b}) + q_{-b+1} = b + 2 + 2\alpha - \beta - \overline{\beta}$. Puisque b et 1 sont des valeurs propres et que la trace de C est la somme de ses trois valeurs propres. La troisième valeur propre sera $1 + 2\alpha - \beta - \overline{\beta}$. Or si l'on développe la quantité $\sum_{k=1}^{b} |z_k - z_{k-1}|^2$ on obtient précisément cette valeur propre. □

Lemme 8 *Les deux valeurs propres q_{2b-2} et q_{2-2b} de la matrice B ont un module majoré par $\frac{1}{2}\sum_{k=1}^{b} |z_k - z_{k-1}|^2$.*

Démonstration. $|q_{2b-2}| = |q_{2b-2}| = |(1 - z_{b-1})z_1| \leq \frac{1}{2}(|1 - z_{b-1}|^2 + |z_1|^2)$, ce qui est majoré par $\frac{1}{2}\sum_{k=1}^{b} |z_k - z_{k-1}|^2$. □

Théorème 9 *La courbe de von Koch est de carré sommable si $\sum_{k=1}^{b} |z_k - z_{k-1}|^2 < b$.*

Démonstration. Supposons que $\sum_{k=1}^{b} |z_k - z_{k-1}|^2 < b$. La valeur propre b est simple et les autres valeurs propres de la matrice B ont un module inférieur à b comme en font foi les derniers lemmes. Le théorème 8 établit que le processus d'interpolation est de carré sommable. □

Lorsque la courbe de von Koch est de carré sommable, $\int |F(x)|^2 dx$ se calcule facilement à partir du vecteur ligne propre associé à la valeur propre b; si (r, s, t) est un tel vecteur, alors

(3) $$\int |F(x)|^2 dx \quad = \quad s/(r + s + t)$$

5 Série de Fourier des courbes de von Koch

Si $z(t)$ est une courbe généralisée de von Koch et si $F(t)$ est l'interpolante fondamentale du processus d'interpolation à base b correspondant, alors on a la relation que $z(t) = F(t - 1)$ et $1 - z(t) = F(t)$ pour $t \in [0, 1]$. La fonction F est nulle hors de $[-1, 1]$, la fonction $\sum_n F(t + 2n - 1)$ sera un prolongement périodique de période 2 de la courbe de von Koch. Si le développement de Fourier de cette fonction est $\sum_k c_k e^{i\pi kt}$, alors

(4) $$c_k \quad = \quad \frac{1}{2}\Phi(k\pi) = \frac{1}{2}\{\prod_{n=1}^{\infty} [P(-k\pi/b^n)/b]\}$$

La série de Fourier $\sum_k c_k e^{i\pi kt}$ est donc le développement de Fourier de F sur l'intervalle $[-1, 1]$. Cette série est convergente lorsque les nombres $\|z_k - z_{k-1}\|$ sont majorés par un. En effet, si les $b + 1$ points du plan complexe qui engendrent la courbe de von Koch sont $z_0, z_1, \ldots, z_b$, si α est le paramètre tel que $(1/b)^\alpha = \max\{|z_k - z_{k-1}| : \ k = 1, 2, \ldots, b\}$, alors $z(t)$ appartient à la classe de Lipschitz d'ordre α. Il en est donc de même pour $F, F \in \text{Lip } \alpha$. Comme le dit le critère de Dini (voir le volume de Zygmund, par exemple), la série de Fourier d'une fonction de la classe Lip α est convergente.

Les figures suivantes illustrent les séries de Fourier (tronquées) de trois courbes: la courbe de Lévy (1938), aussi appelée courbe en C de Gosper, la courbe de von Koch et la courbe de Peano. Les ordres des séries de Fourier sont respectivement: 800, 512 et 729. On a évalué ces polynômes trigonométriques complexes en 4000 points également espacés sur une période, ceci pour chacune des courbes. Remarquons bien que ces calculs se font avec un ordinateur. En situation de calcul automatique, le calcul des coefficients ne pose pas de problème particulier parce que les produits partiels du produit infini dans la formule (4) convergent rapidement; il suffit d'invoquer un produit partiel d'ordre 20 pour obtenir une précision de l'ordre de 10^{-6}.

Dans la figure 1, nous avons tracé les 4000 segments qui relient les évaluations de l'approximation de Fourier d'ordre 800 de la fonction fondamentale correspondant à la courbe de Lévy. Cette courbe est la courbe de von Koch déterminée par les points du plan complexe z_0, z_1, z_2: $z_0 = 0, z_1 = \frac{1+i}{2}$ et $z_2 = 1$. Comme $z(t) = F(t-1)$ et $1 - z(t) = F(t)$ pour $t \in [0, 1]$, l'approximation de Fourier, sur une période complète, approche la courbe de von Koch $z(t)$ et la courbe $1 - z(t)$ symétrique de la première.

Figure 1:
Approximation de Fourier d'ordre 800 de la fonction fondamentale correspondant à la courbe de Lévy.

Puisque le calcul des coefficients de Fourier réclame la connaissance de la transformée de Fourier Φ du procédé d'interpolation itérative, nous avons jugé bon de donner le graphique de Φ. Le long de l'axe des abscisses, on a disposé des traits verticaux, chacun des traits indique un multiple entier de π, le plus grand des traits donne la position 0. Remarquons que $\Phi(0) = 1$, ceci permet d'apprécier l'échelle utilisée pour les ordonnées.

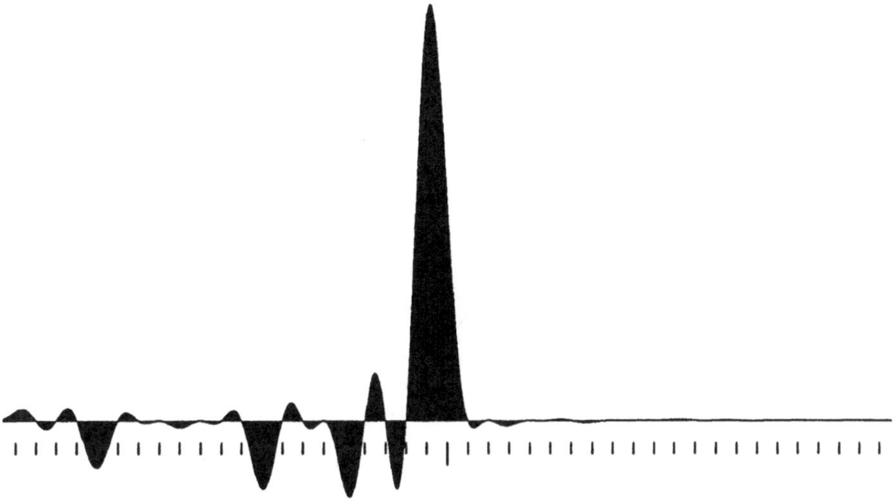

Figure 2:
Transformée de Fourier de la fonction fondamentale correspondant à la courbe de Lévy.

Dans la figure 3, on a tracé les sommes partielles d'ordre $25, 50, 100, 200, 400$ et 800 de la série de Fourier correspondant à la courbe de Lévy. Cependant, on n'a tracé que les évaluations sur une demi-période, ce qui donne des approximations successives de la courbe de Lévy elle-même.

Figure 3:
Sommes partielles d'ordre 25, 50, 100, 200, 400 et 800 de la série de Fourier correspondant à la courbe de Lévy.

Nous donnons dans la figure 4 l'approximation de Fourier d'ordre 512 de la fonction fondamentale correspondant à la courbe originale de von Koch. La partie supérieure de la figure donne déjà une assez bonne approximation de la courbe de von Koch. La figure 5 représente la transformée de Fourier de la fonction fondamentale correspondant à la courbe de von Koch.

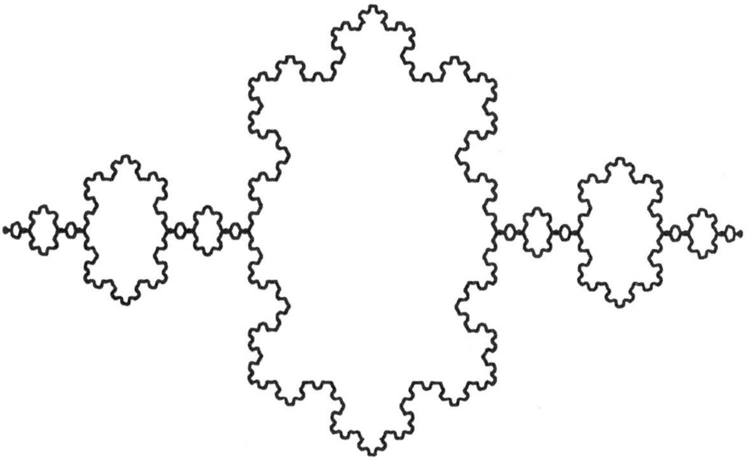

Figure 4:
Approximation de Fourier d'ordre 512 de la fonction fondamentale correspondant à la courbe de von Koch.

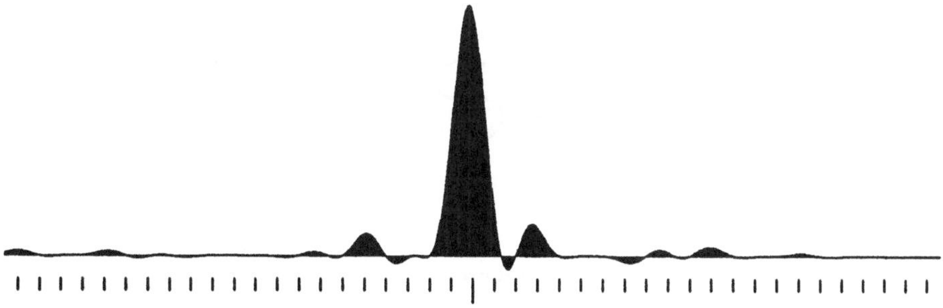

Figure 5:
Transformée de Fourier de la fonction fondamentale correspondant à la courbe de von Koch.

Les figures 6 et 7 sont relatives à la courbe de Peano.

Figure 6:
Approximation de Fourier d'ordre 729 de la fonction fondamentale correspondant à la courbe de Peano.

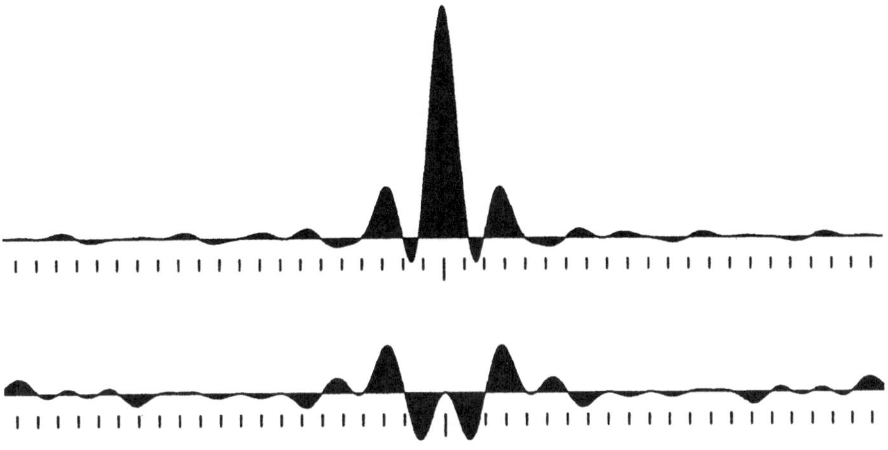

Figure 7:

Partie réelle et complexe de la transformée de Fourier de la fonction fondamentale correspondant à la courbe de Peano.

Notes Ce chapitre provient de l'article de Deslauriers-Dubuc (1987b). Cette recherche a été entreprise pour tirer profit de la transformée de Fourier d'un procédé d'interpolation itérative et pour tâcher d'élargir les outils d'analyse des procédés d'interpolation itérative. La dernière section tient à ce qu'il nous est apparu opportun de nous interroger sur les développements de Fourier de courbes de von Koch. Indiquons que Dekking-van Otterloo ont, eux aussi, prêté intérêt aux développements de Fourier en géométrie fractale.

Problèmes

1. On considère la courbe généralisée de von Koch obtenue par interpolation dyadique en posant $z_0 = 0, z_1 = -1/4$ et $z_2 = 1$. La courbe $z(t)$ est dégénérée en ce sens que les valeurs réalisées sont purement réelles. Démontrez que cette courbe est de carré sommable et que $\int z(t)^2 dt = 8/3$. Vérifiez également que cette fonction restreinte à n'importe quel intervalle ouvert non-vide n'est jamais bornée.

2. Démontrez que dans le développement de Fourier de la fonction fondamentale correspondant à une courbe généralisée de von Koch tous les coefficients de rang pair à l'exception du coefficient de rang zéro sont nuls.

3. Parmi toutes les courbes de von Koch obtenues par interpolation dyadique, engendrées par $z_0 = 0, z_1$ et $z_2 = 1$, déterminez les valeurs de z_1 qui produiront une courbe de carré sommable.

Chapitre VI
Interpolations itératives symétriques de Lagrange

Sommaire. Nous introduisons une famille de procédés d'interpolation. Cette famille est indexée par deux paramètres entiers b, une base, et $2N$ un nombre pair de noeuds mobiles. Étant donnée une fonction f définie sur les entiers relatifs $\mathbf{Z}$, nous prolongeons cette fonction à tous les multiples entiers de $1/b$: si r est un entier entre 0 et b, si n est un entier, $f(n+r/b)$ est définie comme la valeur $P(n+r/b)$ où P est le polynôme de Lagrange de degré inférieur à $2N$ tel que $P(k) = f(k)$ pour tout entier de $[n-N+1, n+N]$. Cette construction peut être répétée pour prolonger f à tous les multiples de $1/b^2$, puis à tous les multiples de $1/b^3$, et ainsi de suite. De cette façon, un prolongement $f(x)$ est effectué. Le domaine de définition du prolongement est formé des fractions dont le dénominateur est une puissance entière de la base b.

Notre objectif principal est de démontrer que le prolongement $f(x)$ est uniformément continu sur tout intervalle fini quels que soient la base b, le nombre pair $2N$ de noeuds ou les valeurs initiales $f(n)$. Les propriétés du procédé d'interpolation proviennent principalement de l'interpolante fondamentale $F(x)$. L'analyse de l'interpolante fondamentale $F(x)$ peut être entreprise avec l'étude de deux compagnons mathématiques du procédé: le polynôme caractéristique: $P(\theta) = \sum_k F(k/b)e^{ik\theta}$ et la matrice infinie $A = (F(k/b-j))_{-\infty<k<\infty, -\infty<j<\infty}$. Nous verrons que pour un procédé d'interpolation itérative symétrique qui se sert de polynômes de Lagrange, $P(\theta)$ est positif et est divisible par le polynôme trigonométrique: $\sin^{2N}(b\theta/2)/\sin^{2N}(\theta/2)$. Nous prouverons que F est une fonction continue définie positive. Nous déterminons presque précisément les classes de Lipschitz auxquelles appartiennent les dérivées successives de F; ces classes dépendent essentiellement des valeurs propres de la matrice A. Par des équations linéaires et par interpolation, nous calculerons les dérivées de F. Nous obtiendrons des bornes sur l'erreur d'une interpolation itérative lorsqu'elle est appliquée à une fonction spécifique.

1 Le modèle de l'interpolation itérative de Lagrange

Nous décrivons une classe d'interpolation itérative inspirée par l'interpolation polynomiale. b un entier plus grand que l'unité sera la base du procédé. Si N est une collection finie d'entiers, si $f(n)$ est définie sur le groupe $G = \mathbf{Z}$, il est possible de prolonger f à l'ensemble $G_1 = \{0, \pm 1/b, \pm 2/b, \ldots\}$ par une interpolation polynomiale à partir de translatées de N de la manière suivante.

Si $x = n + r/b$ (où r est un entier de $[1, b-1]$) est un point de G_1 où le prolongement doit se faire, si $P(x)$ est le polynôme d'interpolation tel que $P(n+k) = f(n+k)$ pour tout entier k de N, alors on pose $f(x) = P(x)$.

Si $\{L_k\}_{k \in N}$ sont les polynômes de Lagrange lorsque N sert d'ensemble de noeuds, alors le prolongement est

$$(1) \qquad f(n + r/b) = \sum_{k \in N} L_k(r/b) f(n + k)$$

Le procédé d'interpolation est appelé une *interpolation itérative de Lagrange.* La formule (1) justifie l'énoncé du prochain lemme.

Lemme 1 *La fonction fondamentale F prend les valeurs suivantes: si $-n$ appartient à N et si r est un entier $[-1, b-1]$, alors $F(n + r/b) = L_{-n}(r/b)$; $F(0) = 1$ et pour les autres valeurs de $G_1, F(x) = 0$.*

La manière dont est définie l'interpolation itérative de Lagrange nous assure de la validité du prochain théorème.

Théorème 2 *Pour tout polynôme P de degré inférieur au nombre de points de N, l'interpolation itérative de Lagrange de la suite $f(n) = P(n)$ est précisément la fonction $f(x) = P(x)$.*

Lorsque les noeuds sont $\mathcal{N} = \{-N+1, -N+2, \ldots, N-1, N\}$ où N est un entier positif, nous parlons d'*interpolation symétrique*; en effet des données symétriques $f(n)$ produisent une fonction $f(x)$ elle-même symétrique. La fonction fondamentale $F(x)$ est paire.

Si $\mathcal{N} = \{-1, 0, 1, 2\}$, on obtient quatre polynôme de Lagrange:

$$L_{-1}(x) = -x(1-x)(2-x)/6, \quad L_0(x) = (x+1)(1-x)(2-x)/2,$$
$$L_1(x) = (x+1)x(2-x)/2, \quad L_2(x) = -(x+1)x(1-x)/6.$$

Quand la base b est 2 ou 3, les valeurs typiques de F sont respectivement:

t	1/2	3/2
$F(t)$	9/16	−1/16

t	1/3	2/3	4/3	5/3
$F(t)$	20/27	10/27	−5/81	−4/81

Si $\mathcal{N} = \{-2, -1, 0, 1, 2,$ et $3\}$, les six polynômes de Lagrange sont:

$$\begin{aligned}
L_{-2}(x) &= (x+1)x(1-x)(2-x)(3-x)/120, \\
L_{-1}(x) &= -(x+2)(1-x)(2-x)(3-x)/24, \\
L_0(x) &= (x+2)(x+1)(1-x)(2-x)(3-x)/12, \\
L_1(x) &= (x+2)(x+1)x(2-x)(3-x)/12, \\
L_2(x) &= -(x+2)(x+1)x(1-x)(3-x)/24, \\
L_3(x) &= (x+2)(x+1)x(1-x)(2-x)/120
\end{aligned}$$

Quand la base b est 2 ou 3, les valeurs typiques de F sont respectivement:

t	1/2	3/2	5/2
$F(t)$	75/128	−25/256	3/256

t	1/3	2/3	4/3	5/3	7/3	8/3
$F(t)$	560/729	280/729	−70/729	−56/729	8/729	7/729

Quand la base est b et que l'ensemble des noeuds $\mathcal{N}$ est $\{-N+1, -N+2, \ldots, N-1, N\}$, nous parlerons de procédé d'***interpolation de type*** (b, N) à la place du procédé d'interpolation itérative de Lagrange. La figure 1 donne le tracé de 6 de ces fonctions fondamentales $F(x)$ pour les valeurs de x allant de -3 à 3. Pour éviter des empiètements entre les graphes des fonctions, on a décalé verticalement chacun des graphes. Les petits traits verticaux indiquent les positions suivantes des abscisses: $-3, -2, -1, 1, 2$ et 3.

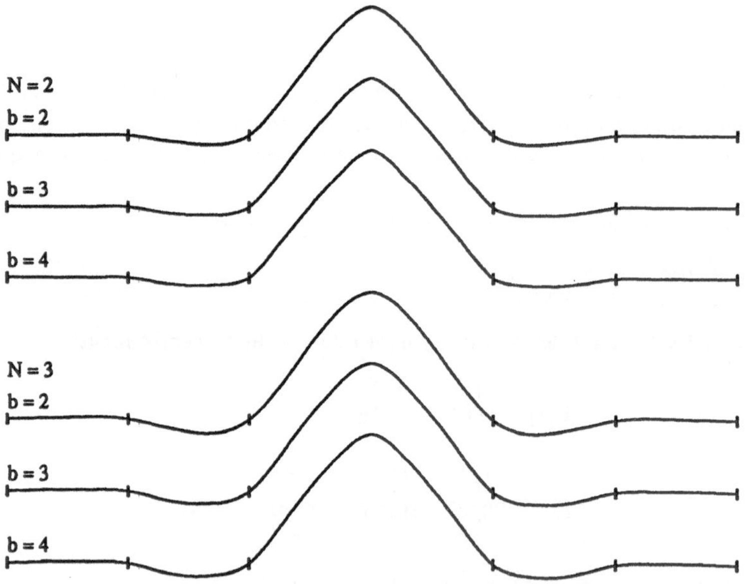

Figure 1:
Tracé des fonctions fondamentales de -3 à 3 pour les interpolations itératives de Lagrange avec 4 noeuds selon les bases 2, 3 et 4, puis avec 6 noeuds selon les mêmes bases 2, 3 et 4.

2 Continuité de l'interpolante fondamentale

Nous considérons un procédé d'interpolation de type (b, N). Nous étudions l'interpolation itérative de Lagrange lorsque les noeuds sont donnés comme $\{-N+1, -N+2, \ldots, 0, 1, \ldots, N-1, N\}$. Nous démontrons que l'interpolante fondamentale est continue. Ceci proviendra essentiellement du fait que le polynôme caractéristique est en tout point non-négatif. La figure 2 trace de 0 à 2π le graphe de six polynômes caractéristiques de procédés d'interpolation de type (b, N). Encore là, nous avons décalé verticalement les divers graphes. Si b est la base, on a indiqué d'un trait vertical les points $(2\pi k/b, 0), k = 1, 2, \ldots, b-1$. L'observation de ces graphiques laissent soupçonner les propriétés suivantes de ces fonctions caractéristiques: les zéros de ces fonctions sont des multiples entiers de $2\pi/b$ et ces fonctions ne prennent jamais des valeurs négatives. Les deux prochains théorèmes vérifient ces propriétés. De celles-ci, nous tirerons d'importantes conséquences.

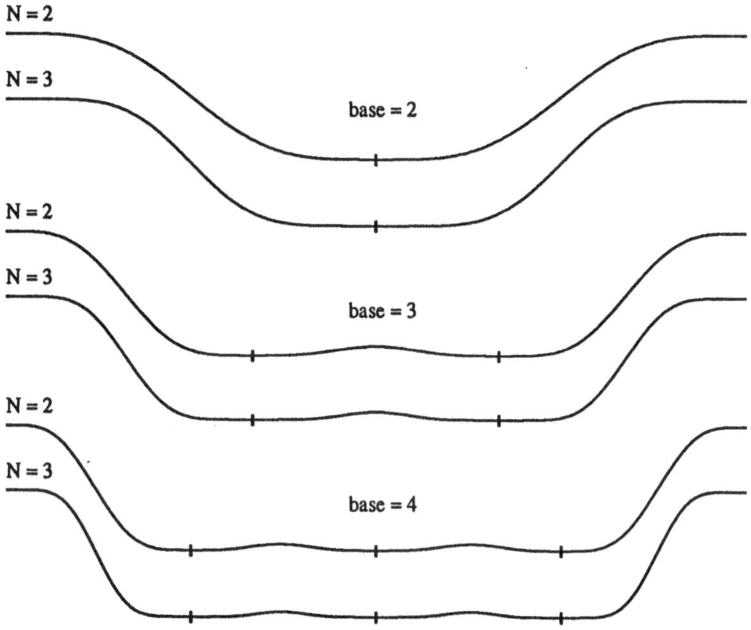

Figure 2:
Tracé avec décalage vertical du graphe des polynômes caractéristiques des procédés d'interpolation de type $(2,2), (2,3), (3,2), (3,3), (4,2)$ et $(4,3)$.

Théorème 3 *Si $P(\theta)$ est le polynôme caractéristique du procédé d'interpolation de type (b, N), alors tout multiple entier de $2\pi/b$ qui n'est pas un multiple entier de 2π est une racine de P de multiplicité au moins égale à $2N$. De plus, les $2N-1$ premières dérivées de P en $\theta = 0$ s'annulent.*

Démonstration. Par définition, $P(\theta) = \sum_k F(k/b)e^{ik\theta}$. Si $\theta = 2\pi r/b$ (où r est un nombre entier positif inférieur à b), si $\omega = e^{i2\pi r/b}$, alors

$$P(\theta) = \sum_k F(k/b)\omega^k = \sum_{s=0}^{b-1}\sum_n F(n+s/b)\omega^{nb+s} = \sum_{s=0}^{b-1}\sum_n F(n+s/b)\omega^s$$

Puisque $1 = \sum_n F(x-n)$ pour tout x (par une application du théorème 2 et du lemme III.2 avec le polynôme égal à la constante 1),

$$P(\theta) = \sum_{s=0}^{b-1}\omega^s = (\omega^b - 1)/(\omega - 1) = 0.$$

Si $\theta = 2\pi r/b$ (avec $0 \leq r < b$), si $\omega = e^{i2\pi r/b}$, si m est un entier positif inférieur à $2N$ et si $P_m = P^{(m)}(\theta)/i^m$, alors

$$\begin{aligned} P_m &= \sum_k k^m F(k/b)\omega^k = \sum_{s=0}^{b-1}\sum_n b^m(n+s/b)^m F(n+s/b)\omega^s \\ P_m &= (-b)^m \sum_{s=0}^{b-1}\sum_n (n-s/b)^m F(s/b-n)\omega^s \end{aligned}$$

Puisque $(x-c)^m = \sum_n (n-c)^m F(x-n)$ pour tout x et pour tout c quand $m < 2N$ (comme le démontrent le théorème 2 et le lemme III.2),

$$P_m = b^m \sum_{s=0}^{b-1}(s/b - s/b)^m \omega^s = 0. \quad \square$$

Théorème 4 *Le polynôme caractéristique d'un procédé d'interpolation de type (b, N) est en tout point non négatif.*

Démonstration. Le degré du polynôme caractéristique $P(\theta)$ est $bN - 1$. La symétrie de $F(t)$ nous assure que $P(\theta)$ ne prend que des valeurs réelles. $P(\theta)$ est nul en chacun des points $2\pi k/b, k = 1, 2, \ldots, b-1$; la multiplicité de chacune de ces racines est au moins égale à $2N$. De plus, les $2N - 1$ premières dérivées de P' s'annulent en zéro. Observons les valeurs critiques de P, les racines du polynôme P'. $P'(\theta)$ s'annule aux endroits suivants: $2\pi k/b, k = 0, 1, \ldots, b-1$; la multiplicité de chacune de ces racines est au moins égale à

$2N - 1$. Suivant le théorème de Rolle, dans chaque intervalle $(2\pi k/b, 2\pi(k+1)/b), k = 1, 2, \ldots, b-2, P'(\theta)$ est nul pour au moins une valeur θ. Le nombre de racines de P' est au moins aussi grand que $2Nb - 2$, le degré de P' comme polynôme trigonométrique est $bN - 1$. Puisque le nombre de racines d'un polynôme trigonométrique ne peut pas dépasser le double de son degré, alors dans chaque intervalle $(2\pi k/b, 2\pi(k+1)/b), k = 1, 2, \ldots, b-2, P'(\theta)$ vaut zéro en un seul point et il n'existe pas de racine de P' dans $(0, 2\pi/b)$ et dans $(-2\pi/b, 0)$. Puisque $P(0) = b$ et $P(2\pi/b) = 0$, la dérivée P' est négative sur $(0, 2\pi/b)$. La dérivée $P^{(2N)}(2\pi/b)$ est positive, sinon $P'(\theta)$ aurait admis trop de racines. Quand la base b est plus grande que deux, on peut démontrer que $P^{(2n)}(4\pi/b)$ est positive; d'abord, $P^{(2n)}(4\pi/b) \neq 0$ vu le nombre de racines de P'; si $P^{(2N)}(4\pi/b) < 0$, P serait positif au voisinage de $2\pi/b$ et négatif au voisinage de $4\pi/b$, ce qui donnerait deux points critiques pour P dans $(2\pi/b, 4\pi/b)$ mais ceci est impossible. Le même argument peut être employé pour tout intervalle $(2\pi k/b, 2\pi(k+1)/b), k = 2, 3, \ldots, b-2$. Il s'ensuit que $P(\theta)$ ne peut jamais prendre des valeurs négatives. □

Remarque Comme le fait voir la démonstration précédente, $P^{(2N)}$ est positive pour toute évaluation en un multiple entier de $2\pi/b$ qui n'est pas un multiple entier de 2π. Si bien que tout multiple entier de $2\pi/b$ qui n'est pas un multiple entier de 2π est une racine de P dont la multiplicité est précisément égale à $2N$.

Corollaire 5 *Dans un procédé d'interpolation de type (b, N), la transformée de Fourier $\Phi(\xi)$ est non négative et l'interpolante fondamentale $F(x)$ d'un procédé d'interpolation de type (b, N) admet un prolongement continu à tout l'axe réel; la transformée de Fourier de ce prolongement est $\Phi(\xi)$.*

3 Régularité de l'interpolante fondamentale

Nous savons maintenant que l'interpolante fondamentale F de tout procédé d'interpolation itérative symétrique de Lagrange est continue. Est-ce que F est dérivable? Existe-t-il des dérivées d'ordre plus élevé? Dans cette section, nous verrons quelques résultats concernant le degré de régularité de F. Les deux instruments de base dans cette étude seront 1) une factorisation du polynôme caractéristique $P(\theta)$ et 2) la recherche des valeurs réelles α pour lesquelles $\int |\xi|^\alpha \Phi(\xi) d\xi$ est finie (où Φ est la transformée de Fourier du procédé d'interpolation itérative). Nous rappelons la définition de *fonction de Lipschitz d'ordre* α; f est une telle fonction s'il existe un nombre L tel que pour tout couple de nombres réels x_1 et x_2, $|f(x_1) - f(x_2)| \leq L|x_1 - x_2|^\alpha$. La classe des fonctions de Lipschitz d'ordre α est habituellement notée par Lip α. Quelques auteurs préfèrent se servir du nom de fonctions de Hölder plutôt que de fonctions de Lipschitz. Le lemme suivant établit une connexion entre la finitude de la transformée de Fourier multipliée par une puissance et les fonctions de Lipschitz. Ce lemme devrait être connu dans la littérature mathématique, mais nous ne connaissons pas de référence sur celui-ci.

Lemme 6 *Soit $f(x)$ une fonction dont la transformée de Fourier est $\phi(\xi)$. On suppose que $|\xi|^{m+\alpha}\phi(\xi)$ est intégrable alors que m est un entier non négatif et α est un nombre réel de $[0,1]$. Si $\alpha = 0$, alors $f^{(m)}$ est continue. Si $\alpha \neq 0$, alors $f^{(m)}$ appartient à la classe Lip α.*

Démonstration. a) Que $f^{(m)}$ est continue quand $|\xi|^m\phi(\xi)$ est intégrable est un résultat classique que l'on peut retrouver dans le volume de Bochner et Chandrasekharan (1949).

b) Supposons que $\int |\xi|^\alpha |\phi(\xi)| < \infty$ où $\alpha \in (0,1)$. Si t_1 et t_2 sont deux nombres tels que $|t_1 - t_2| < \delta$, alors

$$2\pi|f(t_1) - f(t_2)| \leq \int |e^{i\xi t}1 - e^{i\xi t}2| \, |\phi(\xi)|d\xi \leq \int \min(|\xi|\delta, 2)|\phi(\xi)|d\xi.$$

Si $\alpha \in [0,1]$, alors pour toute valeur positive x, $\min(x,1) \leq x^\alpha$. D'où $2\pi|f(t_1) - f(t_2)| \leq 2\int(|\xi|\delta/2)^\alpha|\phi(\xi)|d\xi$.

$|f(t_1) - f(t_2)| \leq \int |\xi|^\alpha|\phi(\xi)|d\xi(\delta/2)^\alpha/\pi$ et f appartient à la classe Lip α.

c) Le cas général est une conséquence directe de a) et de b). □

Théorème 7 *Si $P(\theta)$ est le polynôme caractéristique du procédé d'interpolation de type (b, N), alors il existe un polynôme trigonométrique $S(\theta)$ de degré $N-1$ tel que*

$$P(\theta) = S(\theta)[\sin(b\theta/2)/\sin(\theta/2)]^{2N}. \tag{2}$$

Démonstration. Introduisons le polynôme suivant:

$$\Pi(w) = \sum_{|k|<bN} F(k/b)w^{k+2bN-1}.$$

D'après le théorème 3, chacun des points $2\pi k/b$ ($k = 1, 2, \ldots, b-1$) est une racine de multiplicité $2N$ de l'équation $P(\theta) = 0$, chacune des racines b èmes ω de l'unité autre que un est une racine de multiplicité $2N$ de $\Pi(w)$. $(w-\omega)^{2N}$ est un facteur de $\Pi(w)$. Si $\omega_1, \omega_2, \ldots, \omega_{b-1}$ sont ces racines b èmes de l'unité autre que 1, alors $\prod_{0<k<b}(w-\omega_k)^{2N}$ est un facteur de $\Pi(w)$. $\prod_{0<k<b}(w-\omega_k) = (w^b-1)/(w-1)$. $\Pi(w)[(w-1)/(w^b-1)]^{2N}$ est un polynôme $\Sigma(w)$ de degré $2N-2$. La formule suivante sera valide

$$\begin{aligned} P(\theta) &= e^{-i\theta(bN-1)}\Pi(e^{i\theta}) = e^{-i\theta(bN-1)}[(e^{ib\theta}-1)/(e^{i\theta}-1)]^{2N}\Sigma(e^{i\theta}) \\ P(\theta) &= [\sin(\theta b/2)/\sin(\theta/2)]^{2N}e^{-i\theta(N-1)}\Sigma(e^{i\theta}). \end{aligned}$$

Puisque $e^{-i\theta(N-1)}\Sigma(e^{i\theta})$ est un polynôme trigonométrique de degré $N-1$, $P(\theta)/[\sin(b\theta/2)/\sin(\theta/2)]^{2N}$ est un polynôme trigonométrique du même degré $N-1$. □

Le polynôme $P(\theta)/[\sin(b\theta/2)/\sin(\theta/2)]^{2N}$ sera appelé le *facteur secondaire* de $P(\theta)$. Ce facteur sera noté par $S(\theta)$.

La prochaine étape consiste à étudier la famille d'intégrales: $\int |\xi|^\alpha \Phi(\xi) d\xi$ où α est un paramètre entre 0 et $2N$. Nous commençons par la valeur $\alpha = 2N$.

Lemme 8 *Dans un procédé d'interpolation de type (b, N), il existe une constante C_1 telle que pour tout $n > 0$,*

$$\int_0^{\pi b^n} \xi^{2N} \Phi(\xi) d\xi \leq C_1 b^{2Nn} \int_0^{2\pi} \prod_{0 \leq k \leq n-1} S(\theta b^k) d\theta.$$

Démonstration. $I_n = \int_0^{\pi b^n} \xi^{2N} \Phi(\xi) d\xi = b^{n(2N+1)} \int_0^{\pi} \xi^{2N} \Phi(b^n \xi) d\xi$.

$$\begin{aligned} I_n &= b^{n(2N+1)} \int_0^{\pi} \xi^{2N} \prod_{0 \leq k \leq n-1} [P(\xi b^k)/b] \Phi(\xi) d\xi, \\ I_n &= b^{2Nn} \int_0^{\pi} \xi^{2N} [\sin(b^n \xi/2)/\sin(\xi/2)]^{2N} [\prod_{0 \leq k \leq n-1} S(\xi b^k)] \Phi(\xi) d\xi. \end{aligned}$$

Puisque les fonctions de $\xi, \xi^{2N}[\sin(b^n\xi/2)/sin(\xi/2)]^{2N}\Phi(\xi)$, sont uniformément bornées sur $[0, \pi]$, il existe une borne C_1 telle que $I_n \leq C_1 b^{2Nn} \int_0^{2\pi} \prod_{0 \leq k \leq n-1} S(\theta b^k) d\theta, n = 1, 2, 3, \ldots$
□

Nous avons besoin de connaître le comportement d'une autre suite d'intégrales. Si $S(\theta)$ est un polynôme trigonométrique et si b est un entier, nous posons $J_n = \int_0^{2\pi} \prod_{0 \leq k \leq n-1} S(\theta b^k) d\theta$. Il y a un rapport étroit entre la suite J_n et les valeurs propres d'une matrice infinie B que nous introduisons maintenant.

Si $S(\theta) = \sum_k s_k e^{ik\theta}$, les éléments de la matrice B sont $b_{j,k} = s_{j-bk}$. Voici quelques propriétés de cette matrice.

Lemme 9 *Si $S(\theta) = \sum_k s_k e^{ik\theta}$ est un polynôme trigonométrique, si B est la matrice (s_{j-bk}), si $\sum_k s_k^{(n)} e^{ik\theta}$ est le polynôme $\prod_{0 \leq k \leq n-1} S(b^k\theta)$, alors l'élément à la position (j, k) de la matrice B^n est le nombre $s_{j-b^n k}^{(n)}$.*

Démonstration. L'action de la matrice B est mieux comprise par le biais des séries trigonométriques formelles. Si $V = (v_k)$, alors nous introduisons la série trigonométrique: $Y(\theta) = \sum_k v_k e^{ik\theta}$. On ne porte aucune attention à la convergence de la série. Si $(w_k) = W = BV$ et si $Z(\theta) = \sum_k w_k e^{ik\theta}$, alors $Z(\theta) = S(\theta) Y(b\theta)$. Si $(w_k^{(n)}) = W_n = B^n V$ et si $Z_n(\theta) = \sum_k w_k^{(n)} e^{ik\theta}$, alors $Z_n(\theta) = \prod_{0 \leq k \leq n-1} S(b^k\theta) Y(b^n\theta)$. La k ème colonne de la matrice B^n peut être obtenue du polynôme trigonométrique $Z_n(\theta)$ si $Y(\theta) = e^{ik\theta}$. Avec cette formule, il est possible de voir que l'élément de position (j, k) de la matrice B^n est le nombre $s_{j-b^n k}^{(n)}$ si le polynôme $\prod_{0 \leq k \leq n-1} S(b^k\theta)$ est égal à $\sum_k s_k^{(n)} e^{ik\theta}$. □

Une matrice $B = (b_{j,k})_{-\infty<k<\infty,-\infty<j<\infty}$ est *M-concentrée* si pour tout j de $[-M, M]$ et pour tout k hors de $[-M, M], b_{j,k} = 0$.

La *troncature-M* de la matrice $B = (b_{j,k})_{-\infty<k<\infty,-\infty<j<\infty}$ est la matrice $(b_{j,k})_{-M\leq k\leq M,-M\leq j\leq M}$. Cette matrice sera notée par $[B]_M$.

Lemme 10 *Si $S(\theta) = \sum_k s_k e^{ik\theta}$ est un polynôme trigonométrique de degré N, si B est la matrice (s_{j-bk}) et si M est la partie entière de $N/(b-1)$, alors B est une matrice M-concentrée.*

Démonstration. Si M est la partie entière de $N/(b-1)$, considérons deux indices j et k tels que $|j| \leq M$ et $|k| > M$. Puisque $|j-bk| \geq (M+1)b-M$, alors $|j-bk| \geq M(b-1)+b > N$ et $s_{j-bk} = 0$. B est une matrice M- concentrée. □

Lemme 11 *Si B et C sont deux matrices M-concentrées, BC est une matrice M-concentrée et $[BC]_M = [B]_M[C]_M$.*

Nous laissons la démonstration au lecteur.

Corollaire 12 *Si A est une matrice M-concentrée, alors pour toute valeur entière de n, $[A^n]_M = ([A]_M)^n$.*

Théorème 13 *Dans un procédé d'interpolation itérative de type (b, N), on pose M comme la partie entière de $(N-1)/(b-1)$ et J_n comme $\int_0^{2\pi} \prod_{0\leq k\leq n-1} S(\theta b^k)d\theta$. Si r est le rayon spectral de $[B]_M$, si s est la plus grande multiplicité d'une valeur propre de module égal à r de cette matrice, alors il existe une constante C_2 telle que pour toute valeur entière de n, $|J_n| \leq C_2 n^{s-1} r^n$.*

Démonstration. Si le facteur secondaire d'un procédé d'interpolation itérative de type (b, N) est

$$S(\theta) = P(\theta)/[\sin(b\theta/2)/\sin(\theta/2)]^{2N} = \sum_k s_k e^{ik\theta},$$

la matrice B est $(b_{j,k}) = (s_{j-bk})$. Le nombre J_n est $\int_0^{2\pi} \prod_{0\leq k\leq n-1} S(\theta b^k)d\theta$. J_n est l'élément central (élément correspondant à la rangée 0 et à la colonne 0) de la matrice B^n. Puisque S est un polynôme trigonométrique de degré $N-1$, la matrice B est M-concentrée si M est égal à la partie entière de $(N-1)/(b-1)$. D'après le dernier corollaire, J_n est l'élément central de la matrice tronquée $[B]_{M^n}$. Si $\varphi(x) = \sum_{k\geq 0} c_k x^k$ est le polynôme minimal de $[B]_M$, alors

$$([B]_M)^n \varphi([B]_M) = \sum_{k\geq 0} c_k([B]_M)^{n+k} = 0 \text{ pour } n = 0, 1, 2, \ldots .$$

De cette équation matricielle, il s'ensuit que la suite J_n satisfait la récurrence linéaire: $\sum_{k\geq 0} c_k J_{n+k} = 0$. Comme on peut le voir dans Gel'fond (1971) ou dans d'autres volumes classiques, la suite J_n est une combinaison linéaire finie des suites spécifiques suivantes. Ces suites sont $n^j\lambda^n$ où λ est une racine du polynôme minimal $\varphi(x)$ et j est un entier non négatif inférieur à la multiplicité de la racine λ. Si r est le plus grand module d'une racine de φ, et si s est la plus grande multiplicité d'une racine de φ de module égal à r, alors chaque suite admissible $n^j\lambda^n$ est bornée par un multiple approprié de la suite $n^{s-1}r^n$. Il existe une constante C_2 telle que pour toute valeur entière positive n, $|J_n| \leq C_2 n^{s-1} r^n$. Le nombre r est aussi le rayon spectral de la matrice $[B]_M$. □

Corollaire 14 *Dans un procédé de type* (b, N), *si* r *est le rayon spectral de la matrice* $[B]_M$, *il existe un nombre* C *et un entier* s *tels que pour tout* $n \geq 1$,

$$\int_0^{\pi b^n} \xi^{2N}\Phi(\xi)d\xi \leq Cn^s r^n b^{2Nn}.$$

Théorème 15 *Si* r *est le rayon spectral de la matrice associée* $[B]_M$ *du procédé d'interpolation de type* (b, N), *si* α *est un nombre positif tel que* $rb^\alpha < 1$ *alors* $\int |\xi|^\alpha \Phi(\xi)d\xi < \infty$.

Démonstration.

$$\begin{aligned} K_n &= \int_{\pi b^{n-1}}^{\pi b^n} |\xi|^\alpha \Phi(\xi)d\xi, \\ K_n &\leq (b^{n-1})^{\alpha-2N} \int_{\pi b^{n-1}}^{\pi b^n} \xi^{2N}\Phi(\xi)d\xi, \\ \sum_{n\geq 1} K_n &\leq b^{2N-\alpha} C \sum_{n\geq 1} (b^\alpha r)^n n^s < \infty \end{aligned}$$

puisque par hypothèse $b^\alpha r < 1$. D'où il suit que

$$\int |\xi|^\alpha \Phi(\xi)d\xi = 2\Big(\int_0^\pi |\xi|^\alpha \Phi(\xi)d\xi + \sum_{n\geq 1} K_n\Big) < \infty. \quad \square$$

Dans un procédé d'interpolation itérative de type (b, N), si M est la partie entière de $N/(b-1)$, si r est le rayon spectral de la matrice $[B]_M$, le nombre E tel que $b^E = 1/r$ est appelé l'***exposant critique*** du procédé d'interpolation.

Une application du lemme 6 et du théorème précédent établit le résultat suivant.

Théorème 16 *Soit* E *l'exposant critique d'un procédé d'interpolation itérative de type* (b, N).

Si m *est une valeur entière positive inférieure à* E, *alors* $F^{(m)}(x)$ *est continue.*

Si m est une valeur entière non négative et si α est un nombre réel compris entre 0 et 1 tels que $m+\alpha$ est inférieur à E, alors $F^{(m)}(x)$ appartient à la classe de Lipschitz d'ordre α.

4 Différences finies de la fonction fondamentale

Nous avons trouvé que les dérivées d'un certain ordre de l'interpolante fondamentale appartiennent à une classe de Lipschitz. Nous prouvons dans cette section que ce résultat ne peut pas être subtantiellement amélioré. Si $S(\theta) = \sum_k s_k e^{ik\theta}$ est le facteur secondaire du polynôme caractéristique $P(\theta)$, nous avons déjà défini la matrice $B = (s_{j-bk})$. Nous définissons une autre matrice à partir de la fonction fondamentale F: $A = (F(j/b-k))$. Nous commençons par l'étude des valeurs propres des matrices A et B.

Lemme 17 *Si r est un entier de $[0, 2N-1]$, si V est le vecteur-colonne $(k^r)_{-\infty<k<\infty}$, alors $AV = b^{-r}V$.*

Démonstration. Si r est un entier $[0, 2N-1]$, par le théorème 2 et le lemme III.2, on sait que $\sum_k F(j/b-k)k^r = (j/b)^r$. Si V est le vecteur-colonne $(k^r)_{-\infty<k<\infty}$, alors $AV = b^{-r}V$. □

Théorème 18 *Toute valeur propre de la matrice A autre que les puissances suivantes de $(1/b)$: $1, 1/b, 1/b^2, \ldots, 1/b^{2N-1}$ est une valeur propre de la matrice B. Toute valeur propre de B est une valeur propre de A.*

Démonstration. Supposons que V est un vecteur-colonne $\neq 0$, dont les composantes sont des nombres complexes, et que λ est un nombre complexe tel que $AV = \lambda V$. Nous supposons aussi que λ n'est pas une des puissances entières de $1/b$: $1, 1/b, 1/b^2, \ldots, 1/b^{2N-1}$. Si $V = (v_k)$, alors nous introduisons les séries trigonométriques: $Y(\theta) = \sum_k v_k e^{ik\theta}$. On ne s'inquiète pas de la convergence de cette série, qui est une série trigonométrique formelle. Puisque $AV = \lambda V$, alors $P(\theta)Y(b\theta) = \lambda Y(\theta)$ où $P(\theta)$ est le polynôme caractéristique du procédé d'interpolation. Selon le théorème 7, $P(\theta) = [\sin(b\theta/2)/\sin(\theta/2)]^{2N} S(\theta)$.

Si $Z(\theta) = \sin^{2N}(\theta/2)Y(\theta)$, alors $S(\theta)Z(b\theta) = \lambda Z(\theta)$. Si $Z(\theta) = \sum_k w_k e^{ik\theta}$ et si $W = (w_k)$, alors $BW = \lambda W$. Vérifions que $W \neq 0$. Si $W = 0$, alors nous aurions que $Z(\theta) = 0 = \sin^{2N}(\theta/2)Y(\theta)$. La signification de cette relation $\sin^{2N}(\theta/2)Y(\theta) = 0$ est que chacune des différences finies d'ordre $2N$ de la suite v_k est nulle. Si tel était le cas, alors il y aurait un polynôme $p(t)$ de degré inférieur à $2N$ tel que $v_k = p(k), k = 0, \pm 1, \pm 2, \ldots$ Si $V_r = (k^r)$, alors V serait une combinaison linéaire des vecteurs $\{V_r\}_{0\leq r\leq 2N-1}$. Comme $AV_r = 1/b^r V_r, AV = \lambda V$ et λ n'est pas l'un des nombres $1/b^r (0 \leq r \leq 2N-1)$, alors nous obtiendrons une contradiction.

Discutons maintenant de la seconde partie du théorème. Si λ est une valeur propre de B, nous devons démontrer que λ est une valeur propre de A. Le cas $\lambda = 1/b^r$ (où $r = 0, 1, \ldots, 2N-1$) est traité par le lemme 17. Nous pouvons supposer que λ est une valeur propre qui diffère de $1/b^r$ ($r = 0, 1, \ldots, 2N-1$). Alors il existe un vecteur $W = (w_k)$ tel que $BW = \lambda W$. Nous posons $Z(\theta) = \sum_k w_k e^{ik\theta}$ et la relation $S(\theta)Z(b\theta) = \lambda Z(\theta)$ est induite. Si δ_{2N} est l'opérateur de différences centrales d'ordre $2N$, alors il existe une suite v_k telle que $w_k = \delta_{2N} v_k$. Si nous posons $Y(\theta) = \sum_k v_k e^{ik\theta}$, alors $Z(\theta) = \sin^{2N}(\theta/2)Y(\theta)$. D'où il s'ensuit que

$$S(\theta)\sin^{2N}(b\theta/2)Y(b\theta) = \lambda \sin^{2N}(\theta/2)Y(\theta)$$

et

$$\sin^{2N}(\theta/2)[P(\theta)Y(b\theta) - \lambda Y(\theta)] = 0.$$

Si $V = (v_k)$, alors il existe un polynôme $p(k)$ de degré inférieur à $2N$ tel que $AV - \lambda V = (p(k))$. Il est facile de trouver un polynôme $q(k)$ de degré inférieur à $2N$ tel que $(A - \lambda I)(q(k)) = (p(k))$. Le vecteur-colonne $V - (q(k))$ est alors un vecteur propre pour la matrice A. □

Lemme 19 *Si A est la matrice $(F(j/b - k))_{-\infty<k<\infty, -\infty<j<\infty}$, alors la matrice A^n est $(F(j/b^n - k))$.*

Démonstration. On procède par récurrence sur n. Si $A^{n+1} = (c_{j,k})$, on se sert de la relation $A^{n+1} = AA^n$.

$$c_{j,k} \sum_r F(j/b - r)F(r/b^n - k) = \sum_s F((j - kb^{n+1})/b - s)F(s/b^n)$$

À partir de l'équation (III.2), $c_{j,k} = F(j/b^{n+1} - k)$. □

Lemme 20 *Toute valeur propre de B est dans le disque-unité du plan complexe.*

Démonstration. Si λ est une valeur propre de la matrice B, alors il existe un vecteur-rangée $W = (w_k)_{-M\leq k\leq M}$ qui est un vecteur propre pour la matrice tronquée $[A]_M$: $W[A]_M = \lambda W$. Puisque $W[A]_{M^n} = \lambda^n W$, il s'ensuit que

$$\sum_j w_j F(j/b^n - k) = \lambda^n w_k \text{ (pour } k = -M, -M+1, \ldots, M-1, M).$$

Supposons que $|\lambda|$ est plus grand que un. Le vecteur W est biorthogonal au vecteur propre $(1)_{-M\leq k\leq M}$ correspondant à la valeur propre 1: $\sum_k w_k = 0$. Si $s_k = \sum_{j\leq k} w_j$, alors

$$\sum_j w_j F(j/b^n - k) = -\sum_j s_j[F((j+1)/b^n - k) - F(j/b^n - k].$$

Si k est un indice pour lequel $w_k \neq 0$, il est possible de trouver un nombre positif ε et une suite j_n d'entiers tels que $F((j_n+1)/b^n - k) - F(j_n/b^n - k)$ est plus grand que $\varepsilon|\lambda|^n$ en valeur absolue. Le nombre ε peut être choisi comme $|w_k/[(2M+1)B]|$ où B est le maximum des nombres $|s_j|$. Puisque F est continue, alors $|\lambda| < 1$. Ce qui est une contradiction. □

Théorème 21 *Si E est l'exposant critique du procédé d'interpolation, on pose m comme la partie entière de E et $\alpha = E - m$. Si E n'est pas entier, $F^{(m)}(t)$ ne fait pas partie de la classe de Lipschitz* Lip s *si s plus grand que α. Si E est entier, $F^{(m)}$ n'est pas une fonction continue.*

Démonstration. Discutons d'abord du cas où E n'est pas entier. Si λ est la valeur propre maximale de la matrice B, alors il y a un vecteur rangée $W = (w_k)_{-M \leq k \leq M}$ qui est un vecteur propre pour la matrice tronquée $[A]_M$: $W[A]_M = \lambda W$. Puisque $W[A]_{M^n} = \lambda^n W$, on obtient que

$$\sum_j w_j F(j/b^n - k) = \lambda^n w_k \text{ (pour } k = -M, -M+1, \ldots, M-1, M).$$

Le vecteur W est biorthogonal à chacun des vecteurs-colonnes propres correspondant aux valeurs propres $1/b^r, r = 0, 1, \ldots, m$. W est alors biorthogonal à tout vecteur-colonne de la forme $V = (p(j))_{-M \leq j \leq M}$ où p est un polynôme de degré inférieur ou égal à M. À cause de cela, il s'ensuit que W est une combinaison linéaire d'opérateurs de différences finies d'ordre $m+1$ (tout opérateur de différences finies peut être vu comme un vecteur-rangée). Choisissons un indice k pour lequel $w_k \neq 0$. Il est possible de trouver un nombre positif ε et une suite D_n d'opérateurs de différences finies d'ordre $m+1$ tels que lorsque D_n est appliqué à la suite $V_n = (F(j/b^n - k))_{-M \leq j \leq M}$, un nombre $D_n V_n$ plus grand que $\varepsilon\lambda^n$ en valeur absolue est alors produit. Le nombre ε peut être choisi comme $|w_k/[(2M+1)B]|$ si B est une borne des valeurs absolues des coefficients présents dans la représentation de W comme combinaison linéaire d'opérateurs de différences finies d'ordre $m+1$.

Pour chaque valeur de n, il y a un couple de nombres réels (u_n, t_n) tel que $D_n V_n = [F^{(m)}(t_n) - F^{(m)}(u_n)]b^{-nm}$ et $|u_n - t_n| \leq 2M/b^n$. Avec ces inégalités, il n'est pas possible à la fonction $F^{(m)}$ d'appartenir à la classe de Lipschitz Lip s si s est supérieur à α. La première partie de la démonstration est complète.

Discutons du cas où E est un nombre entier. La valeur propre maximale λ de la matrice B est $1/b^m$.

a) S'il existe un vecteur-rangée $W = (w_k)_{-M \leq k \leq M}$ qui à la fois soit un vecteur propre pour la matrice tronquée $[A]_M$: $W[A]_M = \lambda W$ et soit biorthogonal au vecteur-colonne $(k^m)_{-M \leq k \leq M}$, alors la même argumentation que tantôt montre que $F^{(m)}$ ne peut pas être continue. Une suite de couples de nombres réels (u_n, t_n) peut être trouvée de telle sorte que $F^{(m)}(t_n) - F^{(m)}(u_n)$ ne converge pas vers 0 alors que $|u_n - t_n| \leq 2M/b^n$.

b) Autrement, il existe un vecteur-rangée propre W et un vecteur-rangée W' tels que $W[A]_M = \lambda W$, $W'[A]_M = \lambda W' + W$ et W' est biorthogonal au vecteur-colonne propre $(k^m)_{-M\leq k\leq M}$. Si $W = (w_j)$ et $W' = (w'_j)$, alors $W[A]_{M^n} = \lambda^n W$ et $W'[A]_{M^n} = \lambda^n W' + n\lambda^{n-1}W$.

$$\sum_k w'_j F(j/b^n - k) = \lambda^n w'_k + n\lambda^{n-1} w_k \text{ (pour } k = -M, -M+1, \ldots, M-1, M).$$

Une reprise de l'argumentation antérieure montre que $F^{(m)}$ ne peut pas être continue. On peut trouver un nombre positif ε et une suite de couples de nombres réels (u_n, t_n) tels que $|F^{(m)}(t_n) - F^{(m)}(u_n)| > \varepsilon n$ alors que $|u_n - t_n| \leq 2M/b^n$. □

5 Exemples de facteurs secondaires

Pour illustrer les théorèmes précédents, nous donnons des exemples de facteurs secondaires. Nous revenons au polynôme

$$\Pi(w) = \sum_{k=-bN+1}^{bN-1} F(k/b) w^{k+2Nb-1}.$$

Nous avons vu que $\Pi(w)$ est divisible par $[(w^b - 1)/(2-1)]^{2N}$, le quotient est un polynôme $\Sigma(w)$ de degré $2N-2$. Nous calculons les deux premiers coefficients β_0 et β_1 de $\Sigma(w)$.

Lemme 22 *Dans le procédé d'interpolation de type* (b, N),

$$\beta_0 = (1/b)^{2N-1}[\prod_{1\leq k\leq N-1}(1 - k^2b^2)]/(2N-1)!,$$
$$\beta_1 + 2N\beta_0 = 2(1/b)^{2N-1}[\prod_{1\leq k\leq N-1}(4 - k^2b^2)]/(2N-1)!.$$

Démonstration. On se sert du fait que les deux premiers coefficients de $\Pi(w)$ sont $F(-N+1/b)$ et $F(-N+2/b)$. Selon le lemme 1,

$$F(-N + r/b) = [\prod_{-N+1\leq k\leq N-1}(r/b - k)]/(2N-1)!$$

Les deux premiers coefficients de $[(w^b - 1)/(2-1)]^{2N}$ sont 1 et $2N$. Puisque $(\beta_0 + \beta_1 w + \ldots)(1 + 2Nw + \ldots) = \Pi(w)$, un simple calcul achève la démonstration. □

Lemme 23 *Pour le procédé d'interpolation de type* $(b,2)$, *le facteur secondaire est* $S(\theta) = [(b^2+2)-(b^2-1)\cos\theta]/(3b^3)$.

Lemme 24 *Pour le procédé d'interpolation de type* $(b,3)$, $S(\theta)$ *est*

$$[3(4b^4+5b^2+11)-2(b^2-1)(8b^2+13)\cos\theta+(b^2-1)(4b^2-1)\cos(2\theta)]/(60b^5).$$

Les deux derniers lemmes peuvent se démontrer à l'aide du lemme 22 et d'un peu d'algèbre.

Nous donnons les facteurs secondaires S (quand $N=2$) pour $b=2,3,4$. Si $b=2, S(\theta)=1/4-\cos\theta/8$. Si $b=3, S(\theta)=(11-8\cos\theta)/81$. Si $b=4, S(\theta)=(6-5\cos\theta)/64$.

Observons les valeurs propres de la matrice $[B]_M$.

Si $b=2$, alors $[B]_M$ est une matrice d'ordre 3 dont les valeurs propres sont $1/4$ et $-1/16$. L'exposant critique est dans ce cas $E=2$. Ceci veut dire que la fonction fondamentale est dérivable et que F' appartient à la classe de Lipschitz Lip s pour tout $s\in(0,1)$.

Si $b>2$, alors la matrice $[B]_M$ est d'ordre 1. L'unique valeur propre λ est le coefficient constant du développement de Fourier de $S(\theta), \lambda=(b^2+2)/(3b^3)$. L'exposant critique E est la solution de l'équation $b^E=3b^3/(b^2+2)$. E est compris entre 1 et 2. La fonction F admet toujours une dérivée continue. Par exemple, si $b=3$, alors $E=1,817$ et si $b=4$, alors $E=1,707$.

Nous donnons les facteurs secondaires S (quand $N=3$) pour $b=2,3,4$.

$$\begin{aligned}
S(\theta) &= [19-18\cos\theta+3\cos(2\theta)]/128 \text{ si } b=2.\\
S(\theta) &= [57-68\cos\theta+14\cos(2\theta)]/729 \text{ si } b=3.\\
S(\theta) &= [223-282\cos\theta+63\cos(2\theta)]/4096 \text{ si } b=4.
\end{aligned}$$

Observons les valeurs propres de la matrice $[B]_M$.

Si $b=2$, alors $[B]_M$ est une matrice d'ordre 5 dont après calcul, on trouve les valeurs propres: $3/256, -9/128, 9/64$ et $-1/16$. La plus grande valeur propre de $[B]_M$ est $9/64$. L'exposant critique est dans ce cas $E=\log_2(64/9)=2,830$. La fonction fondamentale est deux fois dérivable et F'' appartient à la classe de Lipschitz Lip s pour tout $s\in(0,\ 0,830)$.

Si $b=3$, alors $[B]_M$ est une matrice d'ordre 3 dont les valeurs propres $7/729$ et $57/729$. L'exposant critique est $2,319$. F'' appartient à la classe de Lipschitz Lip s pour tout $s\in(0,\ 0,319)$. Si $b>3$, alors $[B]_M$ est un scalaire dont la valeur est $(4b^4+5b^2+11)/(20b^5)$. L'exposant critique E est la racine de l'équation $b^E=1/\lambda$. Si $b=4$, alors $E=2,099, F''$ appartient à Lip s pour tout $s\in(0,\ 0,099)$. Si $b\geq 5$, alors $1<E<2$.

Tableau concernant la dérivabilité de $F(x)$

N	b	ordre et	valeurs propres de $[B]_M$	E
2	2	3	$1/4, -1/16$	2
2	3	1	$11/81$	1,817
2	4	1	$3/32$	1,707
3	2	5	$3/256, -9/128, 9/64, -1/16$	2,830
3	3	3	$7/729, 57/729$	2,319
3	4	1	$223/4096$	2,099

6 Calcul des dérivées de l'interpolante fondamentale

Si $F(t)$ est l'interpolante fondamentale et si r est un entier positif inférieur à l'exposant critique E, alors nous indiquerons comment on peut calculer la r ème dérivée de F, $F^{(r)}(t)$, quand t est un nombre rationnel b-adique. Comme nous le verrons immédiatement, le point principal est qu'il suffit de faire le calcul de $F^{(r)}(t)$ pour les valeurs entières de t. Ceci est une conséquence de la formule (2) du chapitre III. Si l'on dérive r fois par rapport à t chaque membre de cette formule, alors nous obtenons

$$F^{(r)}(k/b^m)/b^{rm} = \sum_n F(n/b^m)F^{(r)}(k-n) \tag{3}$$

Nous considérons le cas d'un procédé d'interpolation de type (b, N). $F(t)$ est une fonction paire, $F'(t)$ est une fonction impaire. En particulier, $F'(0) = 0$. $F^{(r)}$ est une fonction paire si r est pair, $F^{(r)}$ est une fonction impaire si r est impair. Nous trouverons une relation entre $F^{(r)}(t)$ aux valeurs entières de t et les vecteurs-rangées propres de la matrice $A = (F(j/b-k))$. Nous calculerons $F'(n)$ pour tout entier n, pour toute base b, pour $N = 2$ et 3.

Théorème 25 *Si r est un entier non négatif plus petit que l'exposant critique d'un procédé d'interpolation de type (b, N), alors le vecteur-rangée $(F^{(r)}(n))_{-\infty<n<\infty}$ est un vecteur propre pour la matrice $A = (a_{j,k}) = (F(j/b-k))$ pour la valeur propre $1/b^r$. De plus, $\sum_n n^r F^{(r)}(n) = (-1)^r r!$.*

Démonstration. Si $V = (F^{(r)}(j))_{-\infty<j<\infty}$ et $(w_j) = VA$, alors

$$w_k = \sum_j F^{(r)}(j)F(j/b-k) = \sum_n F^{(r)}(n+kb)F(n/b).$$

Si dans le dernier membre, n est changé en $-n$ et si la symétrie de F est utilisée, alors $w_k = \sum_n F^{(r)}(kb-n)F(n/b)$. D'après la formule (3), $w_k = F^{(r)}(k)/b^r = v_k/b^r$. W est un vecteur propre pour la valeur propre $1/b^r$.

Afin de démontrer la dernière partie du théorème, on se sert du théorème 2 avec le polynôme $P(t) = t^r$ et du lemme III.2: $t^r = \sum_n n^r F(t-n)$.

Si cette identité est dérivée r fois par rapport à t, alors

$$r! = \sum_n n^r F^{(r)}(t-n) = (-1)^r \sum_n n^r F^{(r)}(t+n).$$

Si t est remplacé par zéro, alors $\sum_n n^r F^{(r)}(n) = (-1)^r r!$. □

Corollaire 26 *Si r est un entier non négatif plus petit que l'exposant critique d'un procédé d'interpolation de type (b, N), si M est le plus grand entier inférieur à $N+(N-1)/(b-1)$ et si $V = (v_j)_{-M\leq j\leq M}$ est un vecteur-rangée de la matrice tronquée $[A]_M = [(F(j/b-k))]_M$ pour la valeur propre $1/b^r$ et si $\sum_n n^r v_n(-1)^r r!$, alors $V = (F^{(r)}(n))$.*

Ceci vient du fait que $1/b^r$ est une valeur propre simple de la matrice $[A]_M$.

Théorème 27 *Dans le procédé d'interpolation de type $(b, 2), F'(1) = -2/3, F'(2) = 1/12$ indépendamment de la base b.*

Démonstration. Vu le lemme III.5, $F(t)$ s'annule sur $[2+1/(b-1), \infty)$. $F(t) = 0$ sur $[3, \infty)$. $F'(3) = 0$. Afin de calculer $F'(2)$, on invoque le théorème 2 et le lemme III.2; si $P(t) = t(t^2-1)$, l'identité suivante a lieu quand $|t| < 1$: $-P(t) = P(2)[F(t+2) - F(t-2)] + P(3)[F(t+3) - F(t-3)]$

Si nous dérivons chaque membre par rapport à t et si nous posons t égal à $0, F'(2) = -P'(0)/2(P(2)) = 1/12$.

Le calcul de $F'(1)$ se fait de façon semblable. Si $Q(t) = t(t^2-4)$ et si $|t| < 1$, alors

$$-Q(t) = Q(1)[F(t+1) - F(t-1)] + Q(3)[F(t+3) - F(t-3)].$$

D'où $F'(1) = -Q'(0)/(2Q(1)) = -2/3$. □

Théorème 28 *Dans le procédé d'interpolation de type $(b, 3)$ avec une base $b \geq 3, F'(1) = -3/4, F'(2) = 3/20, F'(3) = -1/60$.*

Démonstration. Nous raisonnons comme dans le théorème précédent. Selon le corollaire III.5, $F(t)$ s'annule sur $[3+2/(b-1), \infty)$. Puisque $b \geq 3$, alors $F(t) = 0$ sur $[4, \infty)$. $F'(4) = 0$. Pour calculer $F'(3)$, on invoque encore le théorème 2 et le lemme III.2; si $P(t) = t(t^2-1)(t^2-4)$, l'identité suivante a lieu quand $|t| < 1$:

$$-P(t) = P(2)[F(t+2) - F(t-2)] + P(3)[F(t+3) - F(t-3)] + P(4)[F(t+4) - F(t-4)].$$

D'où $F'(3) = -P'(0)/(2P(3)) = -1/60$.

Le calcul de $F'(2)$ se fait à l'aide du polynôme $Q(t) = t(t^2-1)(t^2-9)$. $F'(2) = -Q(0)/(2Q(2)) = 3/20$. Le calcul $F'(1)$ se fait en se servant du polynôme $R(t) = t(t^2-4)(t^2-9)$. $F'(1) = -R'(0)/(2R(1)) = -3/4$. □

Pour le calcul des dérivées de F aux points entiers lorsque l'on est en présence de six noeuds avec la base 2, il y a une petite complication parce que $F'(4) \neq 0$. En se servant des mêmes polynômes P, Q et R comme dans la dernière démonstration, nous obtenons trois équations reliant $F'(1), F'(2), F'(3)$ et $F'(4)$:

$$\begin{aligned} F'(1) + R(4)F'(4)/R(1) &= -R'(0)/(2R(1)), \\ F'(2) + Q(4)F'(4)/Q(2) &= -Q'(0)/(2Q(2)) \text{ et} \\ F'(3) + P(4)F'(4)/P(3) &= -P'(0)/(2P(3)). \end{aligned}$$

Une autre équation est trouvée en se servant du fait que $F(t+4) = 3F(2t+3)/256$ quand t est positif. D'où $F'(4) = 3F'(3)/128$. La solution de ce système d'équations linéaires est le contenu du prochain théorème.

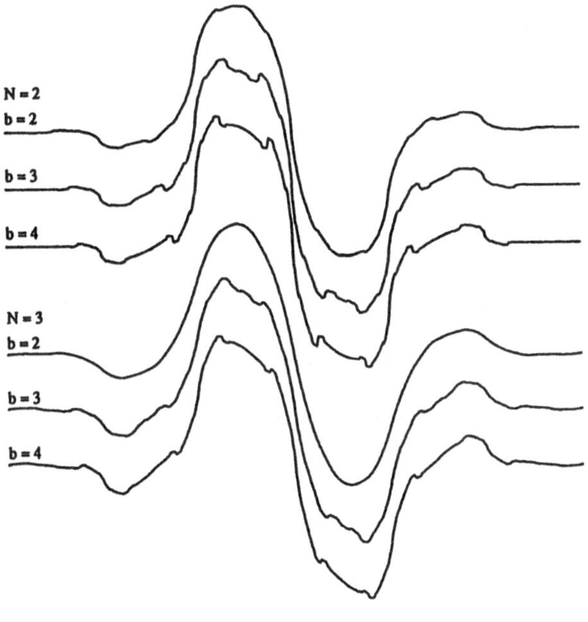

Figure 3:
Tracé des dérivées premières des fonctions fondamentales de -3 à 3 pour les interpolations itératives de Lagrange avec 4 noeuds selon les bases 2, 3 et 4, puis avec 6 noeuds selon les mêmes bases 2, 3 et 4.

Théorème 29 *Dans le procédé d'interpolation de type* $(2,3)$, $F'(1) = -272/365, F'(2) = 53/365, F'(3) = -16/1095$ *et* $F'(4) = -1/2920$.

Les derniers résultats et la formule (3) permettent de tracer le graphe de $F'(x)$ pour les procédés d'interpolations symétriques de type $(b,2)$ et $(b,3)$, pour $2 \leq b \leq 4$.

Un procédé d'interpolation de type (b, N) pour lequel F est deux fois dérivable a lieu dans les cas suivants: $N = 3$ et $b = 2, 3$ ou 4. Nous laissons en exercice le calcul des formules relatives à $F''(n)$.

7 Majoration des erreurs d'interpolation

Soit $f(t)$ une fonction réelle définie sur l'axe réel. Nous considérons la suite: $y(n) = f(n)$ pour les entiers relatifs n. $y(t)$ est l'interpolation de la suite $y(n)$ selon la formule (1) pour une base donnée b et avec $2N$ noeuds. L'erreur de l'interpolation est la fonction $e(t) = f(t) - y(t)$. Nous désirons obtenir des bornes simples pour $e(t)$.

Théorème 30 *Soit $y(t)$ la fonction d'interpolation d'une fonction f selon la formule* (1) *avec une base b et un nombre pair $2N$ de noeuds. Si $t \in [0,1]$ et si $p(t)$ est le polynôme de degré $2N-1$ tel que*

$$p(n) = f(n), n = -N+1, -N+2, \ldots, N-1, N$$

alors

$$|f(t)-y(t)| \leq |f(t)-p(t)| + \sum_{k>0}(|f(N+k)-p(N+k)| + |f(-N+1-k)-p(-N+1-k)|)\varepsilon_k.$$

Le nombre ε_k est la valeur maximale de $|F(t)|$ sur $[k-1,k]$.

Démonstration. D'après l'inégalité triangulaire,

$$|f(t) - y(t)| = |f(t) - p(t) + p(t) - y(t)| \leq |f(t) - p(t)| + |p(t) - y(t)|.$$

Si $t \in [0,1]$, alors le théorème 2 et le lemme III.2 nous assurent que

$$p(t) - y(t) = \sum_k [p(k) - f(k)]F(t-k).$$

Puisque $p(k) = f(k)$ pour $k = -N+1, -N+2, \ldots, N-1, N$ et puisque F est une fonction paire, alors

$$\begin{aligned} p(t) - y(t) &= \sum_{k>N} \{[p(k) - f(k)]F(k-t) + [p(-k+1) - f(-k+1)]F(k-1+t)\}, \\ |p(t) - y(t)| &\leq \sum_{k>N} (|p(k) - f(k)| + |p(-k+1) - f(-k+1)|)\varepsilon_k. \end{aligned}$$

L'inégalité recherchée est alors démontrée. □

Une argumentation semblable permet de borner la r ème dérivée de $y(t) - p(t)$. Si $\varepsilon_k^{(r)}$ est la valeur maximale de $|F^{(r)}(t)|$ sur $[k-1, k]$, alors

$$|f^{(r)}(t) - y^{(r)}(t)| \leq |p^{(r)}(t) - y^{(r)}(t)| + \sum_{k>N} (|p(k) - f(k)| = |p(-k+1) - f(-k+1)|)\varepsilon_k^{(r)}.$$

Nous donnons la table des nombres $\varepsilon_k^{(r)}$ pour $b = 2, 3, 4$ et pour $N = 2$ et 3. Ces nombres ont été trouvés après des expériences numériques.

Cas $N = 2$

b	ε_3	$\varepsilon_3^{(1)}$
2	0,00459	0,08333
3	0,00316	0,08333
4	0,00251	0,08333

Cas $N = 3$

b	ε_4	ε_5	$\varepsilon_4^{(1)}$	$\varepsilon_5^{(1)}$	$\varepsilon_4^{(2)}$	$\varepsilon_5^{(2)}$
2	0,00122	0,00001	0,01461	0,00034	0,11429	0,00536
3	0,00083	0	0,16667	0	0,36111	0
4	0,00063	0	0,16667	0	1,15152	0

Remarque Quand $N = 2$, il semble que la valeur exacte $\varepsilon^{(1)_3}$ soit $1/12$ qui est la valeur de $F'(2)$. Quand $N = 3$, les valeurs exactes de $\varepsilon_4^{(1)}, \varepsilon_5^{(1)}, \varepsilon_4^{(2)}$ et $\varepsilon_5^{(2)}$ sont probablement $|F'(3)|, |F'(4)|, |F''(3)|$ et $|F''(4)|$ respectivement.

Notes Ce chapitre reprend l'article de Deslauriers-Dubuc (1989). Ce sont les résultats de ce chapitre qui pourraient être d'une grande utilité pour l'analyse numérique, lorsque l'on veut réaliser des interpolations plus lisses que l'interpolation linéaire par morceaux. Dyn, Levin et Gregory ont obtenu dans un contexte différent des résultats analogues à la section 6.

Problèmes

1. Si $f(x)$ est l'interpolation dyadique cubique d'une suite $\{f(n)\}$, si $x = m/2^n$ et si $h = 1/2^n$, alors démontrez que

$$f'(x) = (4/3)[f(x+h) - f(x-h)]/(2h) - (1/3)[f(x+2h) - f(x-2h)]/(4h).$$

2. Si $y(t)$ est le résultat de l'interpolation dyadique cubique de la suite n^4 et si $e(t) = t^4 - y(t)$, démontrez que $e(t)$ est une fonction périodique de période un, majorée par un en valeur absolue.

3. Calculez les valeurs des dérivées secondes aux entiers, $F''(n)$ de la fonction fondamentale d'un procédé d'interpolation de type $(b, 3)$ pour chacune des bases $b = 2, b = 3$ et enfin $b = 4$. Tracez le graphe de $F''(x)$.

Chapitre VII
Interpolations anciennes et nouvelles

Sommaire. Nous discutons du problème de la construction d'un arc à tangentes continues passant par une suite donnée de points ou au moins s'approchant suffisamment de ceux-ci. Nous rappelons quelques méthodes classiques ou modernes pour accomplir cette tâche. Nous comparons l'efficacité de ces méthodes avec l'interpolation fractale, l'interpolation dyadique cubique.

1 Courbes liées à des noeuds

Nous décrivons un problème qui se présente fréquemment en infographie. Soient $\{P_k\}_{0\leq k\leq n} = \{(x_k, y_k)\}_{0\leq k\leq n} n+1$ points du plan, on cherche une courbe à tangentes continues qui passe au voisinage de chacun des points et qui respecte l'ordre de l'énumération. Si la courbe passe exactement par chacun des points, on parle d'interpolation et les points P_k sont désignés comme les noeuds de la courbe; sinon, on dira que les points P_k sont les points de contrôle de la courbe. Plusieurs méthodes existent pour résoudre ce problème qui de toute façon admet une infinité de solutions. Nous exposons d'abord les méthodes d'interpolation. Indiquons immédiatement la réduction de ce problème à deux problèmes analogues d'interpolation sur l'axe réel. Nous allons chercher deux fonctions, $f(t)$ et $g(t)$, de classe C^1, définies sur l'intervalle $[0, n]$ telles que $f(k) = x_k$ et $g(k) = y_k$ pour $k = 0, 1, \ldots, n$. La courbe paramétrique $x(t) = f(t)$ et $y(t) = g(t), 0 \leq t \leq n$ donnera une solution du problème.

Il existe plusieurs techniques pour interpoler des données $\{(t_k, x_k)\}$ selon une fonction $x = f(t)$. Nous retiendrons les techniques suivantes: l'interpolation classique de Lagrange (ou de Newton), l'interpolation cubique de Bessel, l'interpolation par spline cubique, en particulier la spline cubique de Swartz et de Varga et l'interpolation dyadique cubique. Les valeurs nodales $t_k = k, 0 \leq k \leq n$, seront les plus souvent utilisées.

Interpolation de Lagrange $f(t) = \sum_{0\leq k\leq n} x_k \prod_{i\neq k} [(t - t_i)/(t_k - t_i)]$, c'est la formule de Lagrange. La formule de Newton permet aussi le calcul de $f(t)$, solution du problème d'interpolation polynomiale.

Interpolation cubique de Bessel Cette technique d'interpolation n'est pas très souvent citée, cependant on la retrouve dans le volume de de Boor (1978). p_k est la pente de la parabole qui passe par les points $(t_{k-1}, x_{k-1})(t_k, x_k)(t_{k+1}, x_{k+1})$ lorsque k varie entre 1 et $n-1$.

- $f(t)$ sur $[t_0, t_1]$ est la parabole solution du problème d'interpolation $f(t_0) = x_0, f(t_1) = x_1$ et $f'(t_1) = p_1$.

- $f(t)$ sur $[t_k, t_{k+1}]$ est le polynôme cubique solution du problème d'interpolation d'Hermite: $f(t_k) = x_k, f'(t_k) = p_k, f(t_{k+1}) = x_{k+1}$ et $f'(t_{k+1}) = p_{k+1}$, k varie de 1 à $n-2$.

-$f(t)$ sur $[t_{n-1}, t_n]$ est la parabole solution du problème d'interpolation $f(t_{n-1})x_{n-1}$, $f'(t_{n-1}) = p_{n-1}$ et $f(t_n) = x_n$.

Lorsque les valeurs nodales sont $t_k = k, 0 \leq k \leq n$, on obtient les formules suivantes pour cette interpolation. Si $t \in [0,1]$,

$$
\begin{aligned}
f(t) &= x_0 + (-3x_0 + 4x_1 - x_2)t/2 + (x_0 - 2x_1 + x_2)t^2/2 \\
f(k+t) &= x_{k-1} + (-x_{k-2} + x_k)t/2 + (2x_{k-2} - 5x_{k-1} + 4x_k - x_{k+1})t^2/2 \\
&\quad +(-x_{k-2} + 3x_{k-1} - 3x_k + 2x_{k+1})t^3/2 \text{ pour } k = 1, 2, \ldots, n-2, \\
\text{et } f(n-1+t) &= x_{n-1} + (-x_{n-2} + x_n)t/2 + (x_{n-1} - 2x_n + x_n)t^2/2
\end{aligned}
$$

Spline cubique On cherche une fonction $f(t)$ définie sur $[t_0, t_n]$ telle que f est de classe C^2 et la restriction de f à chacun des intervalles $[t_k, t_{k+1}]$ est un polynôme cubique. Si l'on pose $p_k = f'(t_k)$, alors l'exigence que la fonction $f(t)$ interpolante soit globalement dérivable deux fois entraîne $n-1$ relations linéaires entres les p_k :

$$\Delta_{k+1}p_{k+1} + 2(\Delta_k + \Delta_{k+1})p_k + \Delta_k p_{k+1} = 3[\Delta_{k+1}(x_k - x_{k-1})/\Delta_k + \Delta_k(x_{k+1} - x_k)\Delta_{k+1}]$$

où les quantités Δ_k représentent les accroissements $t_k - t_{k-1}$.

Les conditions de *Swartz-Varga* sont les suivantes: si $P(t)$ est le polynôme cubique d'interpolation tel que $P(t_0) = x_0, P(t_1) = x_1, P(t_2) = x_2, P(t_3) = x_3$, alors on pose $p_0 = P'(t_0)$; si $Q(t)$ est le polynôme cubique d'interpolation tel que $Q(t_{n-3}) = x_{n-3}, Q(t_{n-2}) = x_{n-2}, Q(t_{n-1}) = x_{n-1}, Q(t_n) = x_n$, alors on pose $p_n = Q'(t_n)$.

Lorsque les pentes p_k sont connues, le calcul de la spline $f(t)$ se complète facilement. Sur chacun des intervalles, $[t_k, t_{k+1}]$, $f(t)$ est le polynôme cubique déterminés par quatre conditions: $f(t_k) = x_k$ $f(t_{k+1}) = x_{k+1}$ $f'(x_k) = p_k$ $f'(x_{k+1}) = p_{k+1}$

$$f(x_k + t) = x_k + p_k t + ct^2 + dt^3$$

où $c = 3(x_{k+1} - x_k) - (p_{k+1} + 2p_k)(t_{k+1} - t_k)$ et $d = -2(x_{k+1} - x_k) + (p_{k+1} + p_k) + (t_{k+1} - t_k)$

Lorsque les valeurs nodales sont $t_k = k, 0 \leq k \leq n$, on obtient les formules suivantes pour cette interpolation:

$$p_0 = -\frac{11}{6}x_0 + 3x_1 - \frac{3}{2}x_2 + \frac{1}{3}x_3$$

$$p_n = \frac{11}{6}x_n - 3x_{n-1} + \frac{3}{2}x_{n-2} - \frac{1}{3}x_{n-2}$$

Les $n-1$ relations linéaires entre les p_k se simplifient:

$$p_{k-1} + 4p_k + p_{k+1} = -3x_{k-1} + 3x_{k+1}$$

Ce système d'équations est tridiagonal et est à diagonale prépondérante. Il est possible de résoudre de façon rapide un tel système.

Interpolation dyadique cubique On cherche une fonction f définie sur les nombres dyadiques compris entre 0 et n telles que $f(k) = x_k$. Pour les autres nombres dyadiques, la fonction f est définie de façon récursive. Si $t = k/2^r$ alors que k est un entier impair, si $h = 1/2^r$, alors

$$\begin{aligned} f(h) &= \frac{5}{16}f(0) + \frac{15}{16}f(2h) - \frac{5}{16}f(4h) + \frac{1}{16}f(6h) \\ f(n-h) &= \frac{5}{16}f(n) + \frac{15}{16}f(n-2h) - \frac{5}{16}f(n-4h) + \frac{1}{16}f(n-6h) \\ f(t) &= -\frac{1}{16}f(t-3h) + \frac{9}{16}f(t-h) + \frac{9}{16}f(t+h) - \frac{1}{16}f(t+3h) \end{aligned}$$

pour les autres valeurs de t compris entre 0 et n.

2 Courbes déterminées par des points de contrôle

Nous maintenons la situation de disposer de $n+1$ points du plan: $\{P_k\}_{0\leq k\leq n} = \{(x_k, y_k)\}_{0\leq k\leq n}$. On cherche une courbe à tangentes continues qui approche chacun des points de contrôle et qui respecte l'ordre de l'énumération. Les courbes les plus fréquemment utilisées sont les courbes de Bézier (1970) et les B-splines. Dans le volume de Schweizer (1987), ces courbes sont définies et certaines de leurs propriétés analysées.

Courbes de Bézier La *courbe de Bézier* d'ordre n déterminée par les points de contrôle $\{P_k\}_{0\leq k\leq n}$ est $P(t) = \sum_{0\leq i\leq n} \frac{n!}{i!(n-i)!} t^i (1-t)^{n-i} P_i$, le paramètre t varie de 0 à 1. Les composantes respectives de $P(t)$ sont des polynômes de Bernstein.

***B*-splines** On se donne une suite de $n+1$ valeurs réelles croissantes $\{t_i\}_{0\leq i\leq n}$. La *B-spline de degré k* est $P(t) = \sum_{0\leq i\leq n} N_{i,k}(t)\ P_i$, les poids $N_{i,k}(t)$ sont définies de façon récursive par rapport à l'indice k:

$$\begin{aligned} N_{i,k}(t) &= 1 \text{ si } t_i \leq t < t_{i+1} \\ N_{i,k}(t) &= 0 \text{ sinon et} \\ N_{i,k}(t) &= \frac{t - t_i}{t_i + k^{-t_i}} N_{i,k-1}(t) + \frac{t_{i+1+k^{-t}}}{t_{i+1+k^{-t}i+1}} N_{i+1,k-1}(t) \end{aligned}$$

Si jamais un dénominateur est nul, on adopte la convention que $0/0 = 0$. Normalement, le paramètre t varie de t_0 à t_{n-k}. On dit que la B-spline est *uniforme* si les valeurs nodales $\{t_i\}_{0 \leq i \leq n}$ sont les entiers qui s'étendent de 0 à n.

La B-spline (uniforme) *quadratique* s'obtient selon la formule

$$P(t) = (t-i)^2/2P_i + (3/4 - (t-i-1/2)^2)P_{i-1} + (1-(t-i)^2)/2P_{i-2}$$

si $i \leq t < i+1$ alors que i peut prendre les valeurs $2, 3, \ldots, n$.

La B-spline (uniforme) *cubique* s'obtient selon la formule

$$\begin{aligned} P(t) &= (t-i)^3/6P_i \\ &+ (2/3 - (t-i-1)^3/2 - (t-i-1)^2)P_{i-1} \\ &+ (2/3 - (t-i)^3/2(t-i)^2)P_{i-2} \\ &+ (1-t+i)^3 P_{i-3} \end{aligned}$$

si $i \leq t < i+1$ alors que i peut prendre les valeurs $3, 4, \ldots, n$.

3 Comparaison théorique entre divers modèles de courbes

Nous venons de décrire plusieurs modèles de courbes engendrées par une ligne polygonale de points $\{P_k\}_{0 \leq k \leq n} = \{(x_k, y_k)\}_{0 \leq k \leq n}$. Dans cette section, nous exposons certaines de leurs propriétés. Puis, dans la section suivante, par plusieurs exemples, nous ferons la comparaison empirique entres les approximations citées. Nous verrons que dans chacun des cas, l'interpolation dyadique cubique supporte bien la comparaison.

Par définition, la ***précision*** d'un procédé de création de courbes est au moins égal à m, si chaque fois que les points $\{(x_k, y_k)\}_{0 \leq k \leq n}$ sont de la forme $x_k = p(k)$ et $y_k = q(k)$ où p et q sont des polynômes de degré m, alors la courbe produite est justement la courbe $x = p(t), y = q(t)$. Dans ce sens donné de précision, les courbes de Bézier sont peu précises: leur précision est un. Les B-splines ont cette même précision limitée: un. La précision de l'interpolation cubique de Bessel est de deux. Les splines cubiques de Swartz-Varga et l'interpolation dyadique cubique ont une précision du troisième ordre, une précision cubique. Quant à l'interpolation de Lagrange, la précision est d'ordre n lorsque les nombres de point de la ligne polygonale est $n+1$.

Énumérons quelques propriétés des polynômes de Bernstein, autant de conséquences en sont tirées pour les courbes de Bézier. Si $f(x)$ est une fonction sur $[0, 1]$, nous rappelons que le *polynôme de Bernstein* de degré n associé à f est $B_n(x) = \sum_{0 \leq i \leq n} \frac{n!}{i!(n-i)!} t^i (1-t)^{n-i} f(\frac{i}{n})$. Les polynômes de Bernstein admettent la propriété de diminution du nombre de changement de signes. Cette propriété a d'abord été reconnue et analysée par Polya et Schoenberg.

Théorème (Polya-Schoenberg) *Si $f(x)$ change de signe r fois sur $[0,1]$ et que $B_n(x)$ n'est pas identiquement nul, alors la somme des multiplicités des racines dans $]0,1]$ de l'équation $B_n(x) = 0$ ne peut dépasser r.*

Plusieurs résultats sur les polynômes de Bernstein se retrouvent dans la monographie de Lorentz. Nous en citons quelques-uns. Si $f(x)$ est non-décroissante, alors $B_n(x)$ est non-décroissante. Si $f(x)$ est convexe, alors $B_n(x)$ est convexe. C'est ainsi que si la ligne polygonale $\{P_k\}$ est une ligne convexe, la courbe de Bézier est convexe. Terminons cette liste de propriétés des polynômes de Bernstein par une formule asymptotique.

Théorème (Voronowskaja) *Soit $f(x)$ une fonction bornée sur $[0,1]$, supposons que la dérivée seconde $f''(x)$ existe en un point donné x de $[0,1]$, alors la limite de $n(f(x)-B_n(x))$, lorsque n tend vers ∞ est*

$$-\frac{1}{2}x(1-x)f''(x).$$

Les B-splines comme les polynômes de Bernstein admettent la propriété de diminution du nombre de changement de signes. Cette propriété des B-splines a été établie par Schoenberg en 1967. Une autre propriété importante des B-splines est le contrôle local de la forme de la courbe engendrée. On dit qu'un procédé qui engendre une courbe à partir d'une ligne polygonale qu'il a un caractère *local* si le déplacement d'un seul sommet de la ligne polygonale ne modifie la courbe que selon un arc restreint de la courbe de départ. Akima (1970) avait relevé l'intérêt de la notion de caractère local et a justement proposé un procédé d'interpolation qui remplissait cette propriété. L'interpolation cubique de Bessel et l'interpolation dyadique cubique sont aussi des procédés locaux de génération de courbes.

4 Comparaison empirique entre modèles de courbes

Les divers procédés d'approximation que nous comparerons sont les suivants:

(1) Interpolation de Lagrange

(2) Interpolation dyadique cubique

(3) Spline cubique de Swartz-Varga

(4) Interpolation cubique de Bessel

(5) Approximation de Bernstein

(6) B-spline quadratique

Pour ce faire, nous ferons appel à une brochette de fonctions $\phi(t)$ définies sur un intervalle $[a,b]$.

(a) e^t sur $[-1, 1]$

(b) $\sin(\pi t)$ sur $[-1, 1]$

(c) e^{5t} sur $[-1, 1]$

(d) $\sin(5\pi t)$ sur $[-1, 1]$

(e) La fonction de Runge $1/(1 + 25t^2)$ sur $[-1, 1]$

(f) La fonction $|t|^3$ sur $[-1, 1]$

(g) La fonction $t\ |t|$ sur $[-1, 1]$

(h) $|t|$ sur [-1,1]

(i) Une fonction en dents de scie, 5 dents de scie sur $[0, 1]$

(j) La fonction de Takagi sur $[0, 1]$

(k) La fonction de Knopp, $a = 1/\sqrt{2}$ et $b = 2$, sur $[0, 1]$

(l) La fonction de Hellinger de paramètre $3/4$ sur $[0, 1]$

(Si $\Psi(t)$ désigne la fonction qui mesure la distance du nombre t au plus proche entier de $\mathbf{Z}$, alors on peut exprimer la fonction (i) comme $\Psi(5t)$. La fonction de Knopp s'exprime aussi en termes de $\Psi, K(t) = \sum_{n=0}^{\infty} a^n \Psi(b^n t)$. La fonction de Takagi, 1903, est un cas particulier de la fonction de Knopp, il suffit de choisir pour a la valeur $1/2$ et pour b la valeur 2.)

Si ϕ est une de ces fonctions définies sur $[a, b]$, nous partageons uniformément en n parties l'intervalle $[a, b]$ selon les $n + 1$ noeuds $\{t_k\}$, $a, a + h, a + 2h, \ldots, b (h = (b - a)/n)$ et nous posons $x_k = \phi(t_k)$. Si $f_n(t)$ est la fonction proposée d'un procédé d'approximation relativement aux points $\{(t_k, x_k)\}$, notre intérêt se porte aux quantités qui mesurent l'écart ou l'erreur de l'approximation de ϕ par f: $E_n = \sup\{|f(t) - \phi(t)| :\ t \in [a, b]\}$.

Pour chacun des procédés d'approximation, le comportement de la suite E_n est grosso modo prévisible si ϕ admet une dérivabilité suffisamment grande. En particulier, voici ce que l'on peut affirmer au sujet de la suite $u_n = -\ln E_n / \ln\ n$.

- Interpolation polynomiale. Si ϕ est de classe C^∞ sans être un polynôme, alors la suite u_n tend vers l'infini lorsque n tend vers l'infini.

- Interpolation dyadique cubique ou par la spline cubique de Swartz-Varga. Si ϕ est de classe C^4 sans être un polynôme du troisième degré, alors la suite u_n tend vers 4 lorsque n tend vers l'infini.

- Interpolation cubique de Bessel. Si ϕ est de classe C^3 sans être un polynôme quadratique, alors la suite u_n tend vers 2 lorsque n tend vers l'infini.

- Approximation de Bernstein ou par la B-spline quadratique. Si ϕ est de classe C^2 sans être une fonction linéaire, alors la suite u_n tend vers 1 lorsque n tend vers l'infini.

Cela est la théorie. Observons maintenant la réalité. À chaque procédé d'approximation et pour chacune des fonctions citées, on trace un diagramme log-log, on trace les points suivants du plan: $(\ln n, -\ln E_n)$ en faisant varier n de 3 à 20. La figure 1 reproduit chacun de ces diagrammes pour les six premières fonctions ϕ, tandis que la figure 2 fait de même avec les autres fonctions. La seule particularité dans le choix des échelles est que la graduation utilisée selon l'axe vertical est quatre fois plus rapprochées que la graduation utilisée selon l'axe horizontal. Un repère vertical est ajouté à chacun des diagrammes. Le trait le plus bas du repère correspond à une erreur d'une unité, les autres traits correspondent à des erreurs respectives d'un dixième, d'un centième et d'un millième.

Voici quelques commentaires à partir de ces deux figures. L'interpolation polynomiale se comporte parfois très bien, parfois très mal, selon le degré de régularité de la fonction à approcher. L'interpolation dyadique cubique et la spline cubique de Swartz-Varga se comportent de la même façon dans chacun des exemples. L'approximation de Bernstein et la B-spline quadratique ont des comportements équivalents dans les exemples traités; elles fournissent de moins bonnes approximations, sauf si la fonction à approcher est très irrégulière. Du point de vue de la qualité de l'approximation, l'interpolation dyadique cubique ou la spline cubique de Swartz-Varga sont à préférer aux autres.

5 Conclusion

Le modèle d'interpolation itérative est un nouveau modèle d'interpolation qui mérite d'être exploré davantage. Le caractère local de ce procédé d'interpolation est probablement un de ses principaux avantages. Ce modèle d'interpolation est capable de se comparer avantageusement à d'autres techniques d'interpolation, l'interpolation de Lagrange ou l'interpolation par fonctions splines. Il permet de construire une très grande famille de fonctions régulières ou irrégulières selon le but cherché. En particulier ce modèle permet de créer des surfaces fractales de façon commode. On pourrait trouver des rapprochements entre l'itération itérative et la théorie des ondelettes qui s'élabore présentement. De plus, la création de courbes ou de surfaces par des schémas itératifs attirent l'attention de chercheurs dans le domaine de la conception géométrique assistée par ordinateurs. En particulier, dans les dernières années, Micchelli et Prautzch ont développé cette approche.

Problème

On considère la suite $\{P_n\}_{-\infty\leq n\leq\infty} = \{(x_n, y_n)\}_{-\infty\leq n\leq\infty}$: où $x_n = \cos(n\pi/2)$ et $y_n = \sin(n\pi/2)$. On définit l'interpolation dyadique périodique suivante: $P(n) = P_n$

Si $t = k/2^r$ alors que k est un entier impair et que $r > 0$, si $h = 1/2^r$, alors $P(t) = -\frac{1}{16}P(t-3h) + \frac{9}{16}P(t-h) + \frac{9}{16}P(t+h) - \frac{1}{16}P(t+3h)$.

Évaluer les deux nombres r et R, r est le rayon du plus grand disque ouvert centré à l'origine qui ne rencontre aucun des points $P(t)$ alors que R est le rayon du plus petit disque fermé centré à l'origine qui contient tous les points $P(t)$.

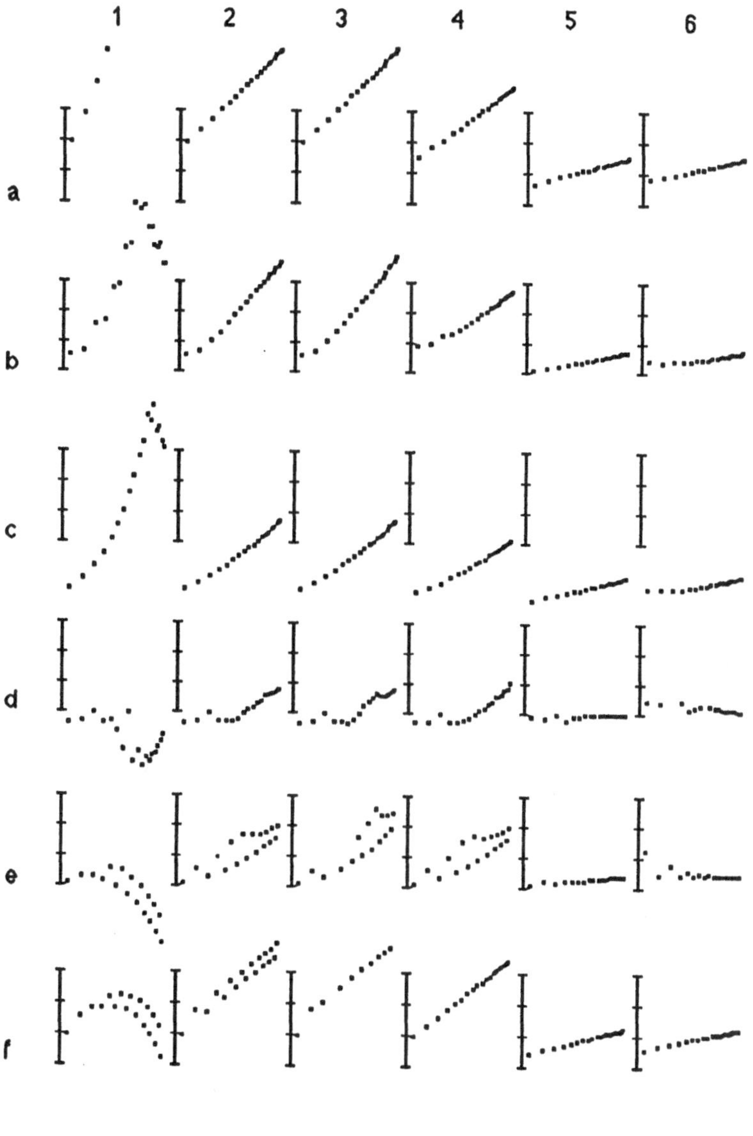

Figure 1: Diagrammes log-log des erreurs pour les fonctions a-f.

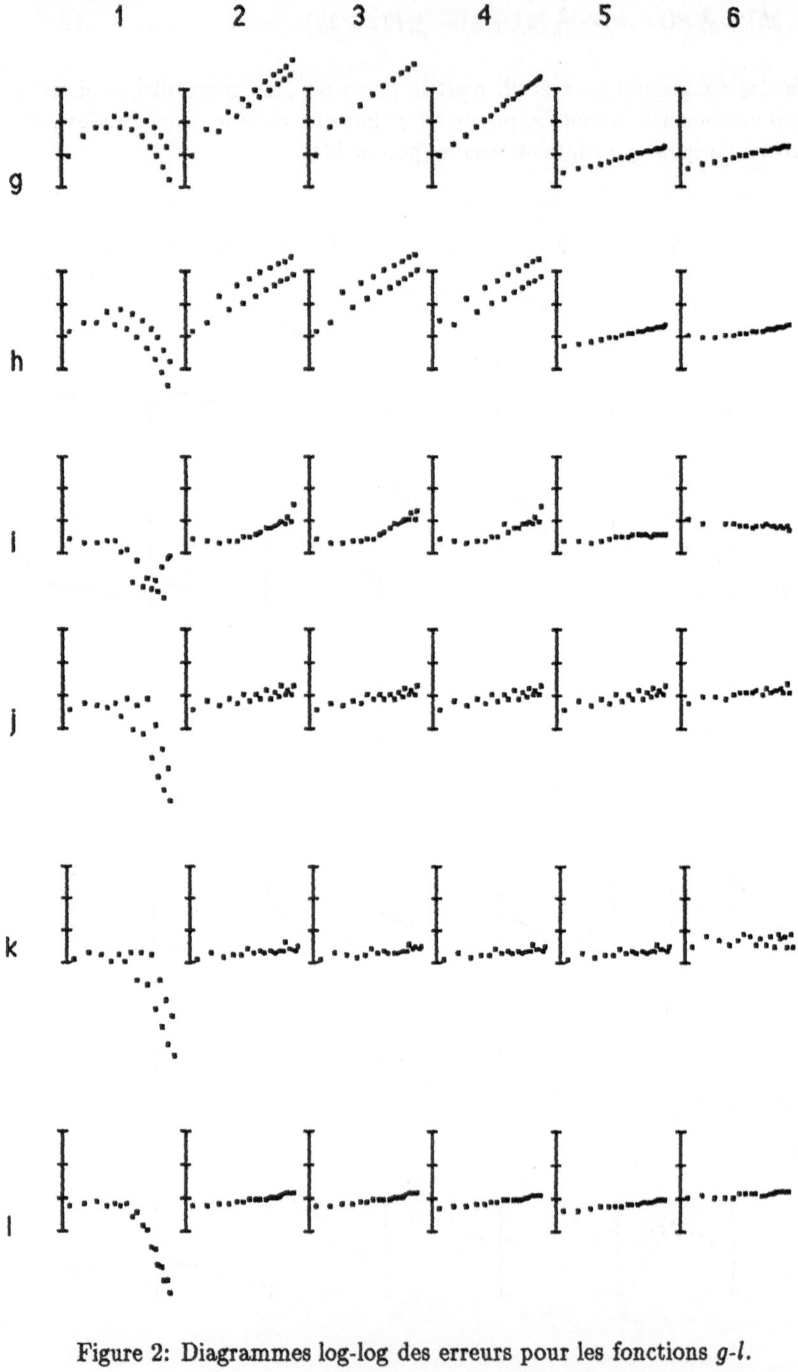

Figure 2: Diagrammes log-log des erreurs pour les fonctions *g*-*l*.

Bibliographie

Akima, H, 1970, A New Method of Interpolation and Smooth Curve Fitting Based on Local Procedures, *J. Assoc. Comp. Mach.* **17**, 589-602.

Bajraktarević, M. 1957, Sur une équation fonctionnelle, *Glasnik Mat.-Fiz. Astr. Ser. II* **12**, 201-205.

Barnsley, M. F. 1986, Fractal Functions and Interpolation, *Constr. Approx.* **2**, 303-329.

Barnsley, M. F. & Demko, S. 1985, Iterated function systems and the global construction of fractals, *Proc. Roy. Soc. London Ser.* **A 399**, 243-275.

Barnsley, M. F., Ervin, V., Hardin, D. et Lancaster J. 1986, Solution of an inverse problem for fractals and other sets, *Proc. Nat. Acad. Sci. U.S.A.* **83**, 1975-1977.

Baron, K. 1974, Note on Continuous Solutions of a Functional Equation, *Aequationes Math.* **11**, 267-269.

Bernstein, S. 1912, Démonstration du théorème de Weierstrass, fondée sur le calcul des probabilités, *Commun. Soc. Math. Kharkow* **13**, 1-2.

Bézier, P. 1970, *Emploi des machines à commande numérique*, Paris: Masson.

Bochner, S. & Chandrasekharan, K. 1949, *Fourier Transforms*, Princeton: Princeton University Press.

Bourbaki, N. 1971, *Éléments de mathématique. Topologie générale*, Paris: Hermann.

Cesàro, E. 1905, Remarques sur la courbe de von Koch. *Atti della R. Accad. Sc. Fis. Mat. (Napoli)* **12**, série 2, no 15, 1-12.

Daubechies, I. 1988, Orthonormal bases of compactly supported wavelets, *Comm. Pure Appl. Math.* **41**, 909-996.

Daubechies, I. et Lagarias, J. 1990, Two-scale difference equations I Global regularity of solutions, *SIAM J. Math. Anal.* (à paraître)

Daubechies, I. et Lagarias, J. 1990, Two-scale difference equations II Local regularity, infinite products of matrices and fractals, *SIAM J. Math. Anal.* (à paraître)

de Boor, C. 1978, *A Practical Guide to Splines*, New York: Springer-Verlag.

Dekking, F. M. 1980, Variations on Peano, *Nieuw Arch. Wisk.* (3), **28**, 275-281.

Dekking, F. M. et van Otterloo, P. J. 1986, Fourier coding and reconstruction of complicated contours, *IEEE Trans. Systems, Man Cybernet.* **16**, 395-404.

de Rham, G. 1956-57, Sur quelques courbes définies par des équations fonctionnelles, *Rend. Sem. Mat. Univ. Politec. Torino* **16**, 101-113.

Deslauriers, G. et Dubuc, S. 1987a, Interpolation dyadique, dans: *Fractals. Dimensions non entières et applications*, Paris: Masson, 44-55.

Deslauriers, G. et Dubuc, S. 1987b, Transformées de Fourier de courbes irrégulières, *Ann. Sci. Math. Québec* **11**, 25-44.

Deslauriers, G. et Dubuc, S. 1989, Symmetric iterative interpolation, *Constr. Approx.* **5**, 49-68.

Deslauriers, G. et Dubuc, S. 1991, Continuous iterative interpolation. (à paraître dans les comptes-rendus, European Conference on Iteration Theory, Batschuns (Austria) (1989))

Deslauriers, G. Dubois, J. et Dubuc, S. 1990, Multidimensional Iterative Interpolation. (à paraître dans *Canad. J. Math.*)

Dubuc, S. 1985a, Functional equations connected with peculiar curves, dans: *Iteration Theory and Its Functional Equations* (Lochau, 1984) Lecture Notes in Math. 1163. Springer-Verlag: Berlin, 33-40.

Dubuc, S. 1985b, Une équation fonctionnelle pour diverses constructions géométriques, *Ann. Sci. Math. Québec* **9**, 151-184.

Dubuc, S. 1986, Interpolation through an iterative scheme, *J. Math. Anal. Appl.* **114**, 185-204.

Dubuc, S. et Elqortobi, A. 1990, Approximations of fractal sets, *J. Comput. Appl. Math.* **29**, 79-89.

Dyn, N., Levin, D. et Gregory, J. A. 1987, A 4-point interpolatory subdivision scheme for curve design, *Comput. Aided Geom. Design* **4**, 257-268.

Gel'fand, I. M. et Chilov, G. E. 1962, *Les distributions*, tome 1, Paris: Dunod.

Gel'fond, A. O. 1963, *Calcul des différences finies*, Paris: Dunod.

Hahn, H. 1948, *Reelle Funktionen*, New York: Chelsea.

Hata, M. 1985, On the structure of self-similar sets, *Japan J. Appl. Math.* **2**, 381-414.

Hausdorff, F. 1962, *Set Theory* (seconde édition), New York: Chelsea.

Hellinger, E. 1907, *Die Orthogonalinvarianten quadratischer Formen von unendlichvielen Variablen*, dissertation, Göttingen.

Hutchinson, J. E. 1981, Fractals and self-similarity, *Indiana Univ. Math. J.* **30**, 713-747.

Kahane, J. P. 1970, Courbes étranges, ensembles minces, *Bull. A.P.M.E.P* **275/276**, 325-339.

Kahane. J. P. et Salem, R. 1963, *Ensembles parfaits et séries trigonométriques*, Paris: Hermann.

Kantorovich, L. V. et Akilov, G. P. 1981, *Analyse fonctionnelle. Tome 2: équations fonctionnelles*, Moscou: Mir.

Kiesswetter, K. 1966, Ein einfaches Beispiel für eine Funktion, welche überall stetig und nicht differenzierbar ist, *Math.-Phys. Semesterber.* **13**, 216-221.

Knopp, K. 1918, Ein einfaches Verfahren zur Bildung stetiger nirgends differenzierbarer Funktionen, *Math. Z.* **2**, 1-26.

Kuczma, M. 1968, *Functional Equation in a Single Variable.* Varsovie: Polish Scientific Publishers.

Kwapisz. M. 1977, On the existence and uniqueness of the solution of a non-linear functional equation of r-th order, *Ann. Polon. Math.* **34**, 35-38.

Lévy, P. 1938, Les courbes planes ou gauches et les surfaces composées de parties semblables au tout, *J. École Poly.* Série III, **7-8**, 227-292.

Lorentz, G. 1953, *Bernstein Polynomials*, Toronto: University of Toronto Press.

Mandelbrot, B. B. 1982, *The Fractal Geometry of Nature*, San Francisco: W. H. Freeman.

Matkowski, J. 1973, On Lipschitzian solutions of a functional equation, *Ann. Polon. Math.* **28**, 135-139.

Micchelli, C. A. et Prautzch, H. 1987, Computing curves invariant under halving, *Comput. Aided Geom. Design* **4**, 133-140.

Micchelli, C. A. et Prautzch, H. 1989, Uniform Refinement of Curves, *Linear Algebra and Applications* **114/115**, 841-870.

Micchelli, C. A. et Prautzch, H. 1987, Computing surfaces invariant under subdivision, *Comput. Aided Geom. Design* **4**, 321-328.

Ostrowski, A. M. 1973, *Solution of Equations in Euclidean and Banach Spaces* (Troisième édition de *Solution of Equations and Systems of Equations*), New York: Academic Press.

Peano, G. 1890, Sur une courbe, qui remplit toute une aire plane, *Math. Annalen* **36**, 157-160.

Polya G. et Schoenberg, I. J. 1958, Remarks on the la Vallée Poussin means and convex conformal map of the circle, *Pac. J. Math.* **8**, 295-334.

Read, A. H. 1951-52, The solution of a functional equation, *Proc, Roy. Soc. Edinburgh Sect. A*, **63**, 336-345.

Salem, R. 1943, On some singular monotonic functions which are strictly increasing, *Trans. Amer. Math. Soc.* **53**, 427-439.

Schwartz, L. 1966, *Théorie des distributions*, Paris: Hermann.

Schweizer P. 1987, *Infographie II*, Lausanne: Presses polytechniques romandes.

Sierpiński, W. 1916, Sur une courbe dont tout point est un point de ramification (en polonais) *Prace Mat.-Fiz.* **27**, 77-86; (en français) dans *Oeuvres chosies. Tome II*, Varsovie: PWN - Éditions scientifiques de Pologne, (1975), 99-106.

Swartz B.K. et Varga R.S. 1972, Error bounds for splines and L-spline interpolation, *J. Approx. Theory* **6**, 6-49.

Takagi, T. 1903, A simple example of the continuous function without derivative, *Proc. Phys. Math. Soc. Japan*, **1**, 176-177; *The Collected Papers of Teiji Takagi*, Iwanami Shoten Publ., Tokyo, 1973, 5-6.

von Koch, H. 1904, Sur une courbe continue sans tangente obtenue par une construction géométrique élémentaire, *Arkiv für Matematik, Astronomie och Fysik* **1**, 681-704.

Voronowskaja, E. 1932, Détermination de la forme asymptotique d'approximation des fonctions par les polynômes de Bernstein, *C. R. Acad. Sci. URSS*, 79-85.

Williams, R. F. 1971, Composition of contractions, *Bol. Soc. Brasil. Mat.* **2**, 55-59.

Yosida, K. 1966, *Functional Analysis*, Berlin: Springer-Verlag.

Zygmund, A. 1968, *Trigonometric Series*, Volume I, London: Cambridge University Press.

Dimensions - their determination and properties

Kenneth J. FALCONER
School of Mathematics
University Walk
University of Bristol
Bristol BS8 1TW
England

Introduction/Abstract

Dimensions of various types are the main mathematical tools for the study of fractals. Dimensions contain much information about the geometry of fractals, such as properties of their projections, intersections or local structure. In these lectures we introduce some of the principal definitions of dimension and discuss their properties and how they may be calculated in certain cases.

These lectures are more concerned with ideas than with mathematical detail. Proofs are included only where they are reasonably straightforward and instructive; most of the topics are covered in more detail in the author's recent book "Fractal Geometry - Mathematical Foundations and Applications" (1990) and the references therein. The figures are from that book, and appear here by permission of John Wiley and Sons, Ltd.

Lecture 1
Introduction to Dimension

In this lecture we introduce some of the principal definitions of dimension. We first define Hausdorff dimension via Hausdorff measures.

Hausdorff Measure

Let F be a subset of $\mathbf{R}^n$, and $0 \leq s \leq n$. For each $\delta > 0$ let

$$H^s_\delta(F) = \inf\{\sum_i |U_i|^s : \ F \text{ is covered by sets } U_i \text{ with } 0 < |U_i| \leq \delta\}. \tag{1.1}$$

J. Bélair and S. Dubuc (eds.), Fractal Geometry and Analysis, 221–254.

(We write $|U| = \sup\{|x-y| : x, y \in U\}$ for the diameter of the set U; thus this infimum is over all coverings of F by a (finite or countable) collection of sets with diameters at most δ.) Decreasing δ imposes increasing restrictions on the allowable coverings of F for the infimum; hence $H^s_\delta(F)$ increases as δ decreases to 0, and we may define

$$H^s(F) = \lim_{\delta \to 0} H^s_\delta(F). \tag{1.2}$$

We call $H^s(F)$ the *s-dimensional Hausdorff measure* of F. It may be shown that H^s is a measure on the Borel subsets of $\mathbf{R}^n$, so in particular

$$H^s\left(\bigcup_{i=1}^{\infty} F_i\right) = \sum_{i=1}^{\infty} H^s(F_i)$$

for a countable collection of disjoint Borel sets $\{F_i\}$. Moreover, if F is, say, a smooth or rectifiable curve, then $H^1(F)$ is the length of F, and if F is a surface then $H^2(F)$ gives the area of F. Thus Hausdorff measures generalise length, area, volume, etc.

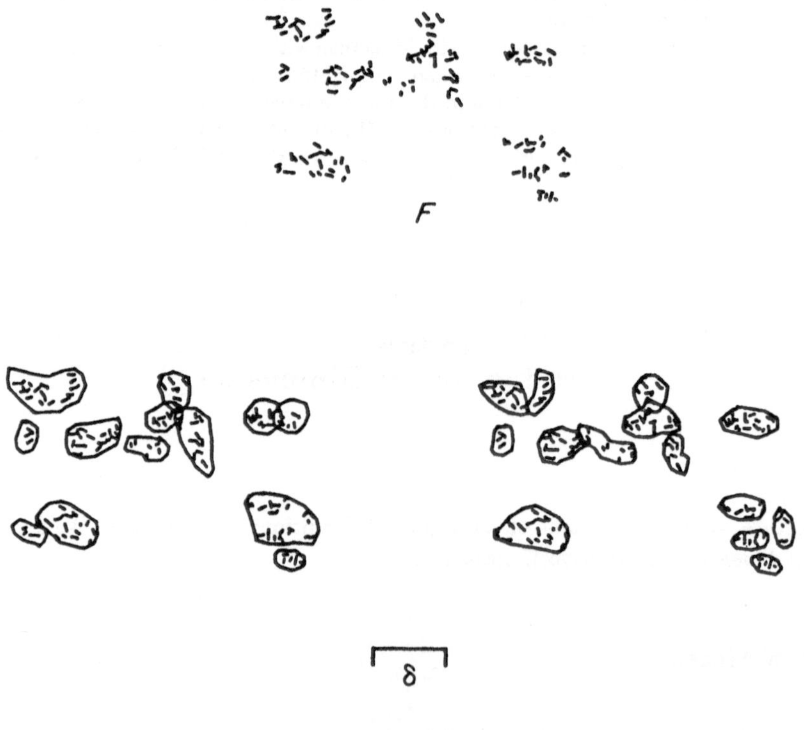

Figure 1:
A set F and two possible covers of F by sets of diameter at most δ.
The infimum of $\sum |U_i|^s$ over all such covers $\{U_i\}$ gives $H^s_\delta(F)$.

The following scaling property, which also generalises familiar properties of length, area, etc., is fundamental in the theory of Hausdorff measures.

Scaling Property *For any set $F \subset \mathbf{R}^n$ and $\lambda > 0$,*

$$H^s(\lambda F) = \lambda^s H^s(F),$$

where $\lambda F \equiv \{\lambda x : x \in F\}$ is F scaled by a factor λ.

Proof. If $F \subset \bigcup U_i$ with $|U_i| \leq \delta$, then $\lambda F \subset \bigcup \lambda U_i$ with $|\lambda U_i| \leq \lambda\delta$. Thus

$$H^s_{\lambda\delta}(\lambda F) \leq \sum_i |\lambda U_i|^s = \lambda^s \sum_i |U_i|^s \leq \lambda^s H^s_\delta(F).$$

Letting $\delta \to 0$ gives the result. □

Hausdorff Dimension

We use Hausdorff measure to define Hausdorff dimension.

Suppose that $t > s$ and $F \subset \bigcup U_i$ with $|U_i| \leq \delta$. Then

$$\sum_i |U_i|^t = \sum_i |U_i|^s |U_i|^{t-s} \leq \delta^{t-s} \sum_i |U_i|^s,$$

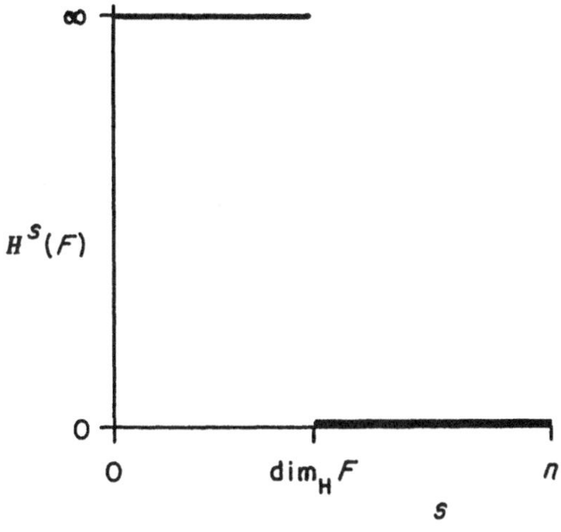

Figure 2:
Graph of $H^s(F)$ against s for a set F. The Hausdorff dimension is the value of s at which the "jump" from ∞ to 0 occurs.

so, taking infima,

$$H^t_\delta(F) \leq \delta^{t-s} H^s_\delta(F).$$

Letting $\delta \to 0$ it follows that if $t > s$ and $H^s(F) < \infty$ then $H^t(F) = 0$. Thus there is a number s at which $H^s(F)$ jumps from ∞ to 0; we call this number the *Hausdorff (or Hausdorff-Besicovitch) dimension of F* which we denote by $\dim_H F$. Thus

$$\dim_H F = \sup\{s: H^s(F) = \infty\} = \inf\{s: H^s(F) = 0\}. \tag{1.3}$$

Note that if $s = \dim_H F$ then $H^s(F)$ can be $0, \infty$, or positive and finite (this last case is by far the most convenient to work with).

Example *The middle third Cantor set F obtained by repeatedly removing the middle third of intervals (see Figure 3) has Hausdorff dimension* $\log 2/\log 3$.

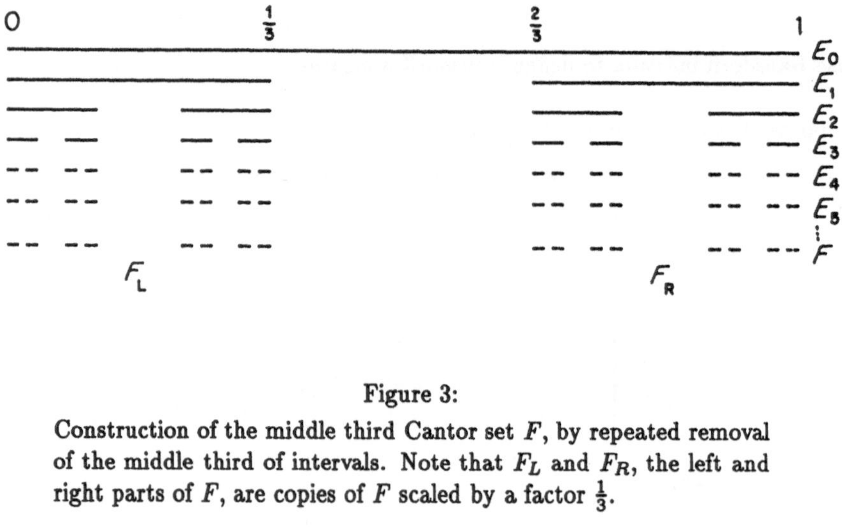

Figure 3:
Construction of the middle third Cantor set F, by repeated removal of the middle third of intervals. Note that F_L and F_R, the left and right parts of F, are copies of F scaled by a factor $\frac{1}{3}$.

Heuristic proof. We have $F = F_L \cup F_R$ where F_L and F_R are the left and right parts of F. Then, for each s,

$$\begin{aligned} H^s(F) &= H^s(F_L) + H^s(F_R) \\ &= (\tfrac{1}{3})^s H^s(F) + (\tfrac{1}{3})^s H^s(F), \end{aligned}$$

noting that F_L and F_R are copies of F scaled by a factor $\frac{1}{3}$ and using the scaling property. Assuming that at the "jump value" $s = \dim_H F$ we have that $0 < H^s(F) < \infty$ (a big assumption, but one that can be justified here), this gives $1 = 2 \times 3^{-s}$ or $s = \log 2/\log 3$. □

A similar heuristic proof shows that the von Koch curve of Figure 4 has Hausdorff dimension $\log 4/\log 3$. We will see further examples of sets of fractional dimension and some rigorous methods of calculation later on.

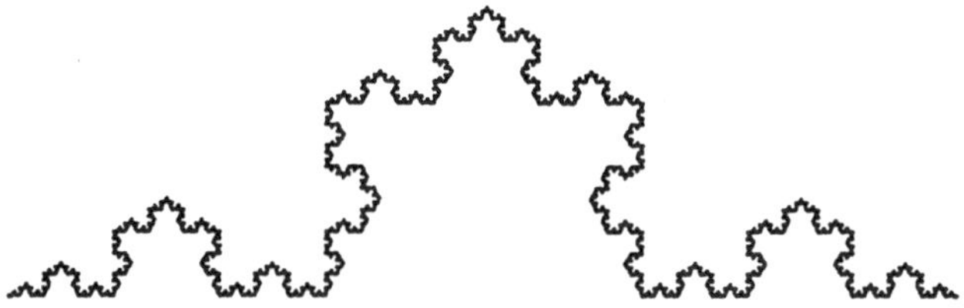

Figure 4: The von Koch curve.

Properties of Hausdorff Dimension

We list some basic properties of Hausdorff dimension. Other definitions of dimension often, but not always, have such properties.

1. *Open sets*: If $F \subset \mathbf{R}^n$ is open (or contains some ball of positive radius) then $\dim_H F = n$.

2. *Smooth sets*: If F is a smooth m-dimensional submanifold (i.e. an m-dimensional "surface") in $\mathbf{R}^n$, then $\dim_H F = m$.

3. *Monotonicity*: If $E \subset F$ then $\dim_H E \leq \dim_H F$.

4. *Countable stability*: Given a sequence of sets $F_1, F_2, \ldots$,

$$\dim_H\left(\bigcup_1^\infty F_i\right) = \sup_{1 \leq i < \infty} \dim_H F_i.$$

5. *Countable sets*: If F is finite or countable then $\dim_H F = 0$.

6. *Effect of Lipschitz mappings*: If $F \subset \mathbf{R}^n$ and $f\colon\ F \to \mathbf{R}^m$ is Lipschitz, i.e.,

$$|f(x) - f(y)| \leq c|x - y| \qquad (x, y \in F), \tag{1.4}$$

where c is a constant, then

$$\dim_H f(F) \leq \dim_H F.$$

[Note: we often have $f\colon\ \mathbf{R}^n \to \mathbf{R}^m$ with F a subset of $\mathbf{R}^n$.]

Proof. Just as in the proof of the scaling property for Hausdorff measures, we get that $H^s_{c\delta}(f(F)) \leq c^s H^s_\delta(F)$, so $H^s(f(F)) \leq c^s H^s(F)$. Thus, if $s > \dim_H F$, then both sides of this last inequality are 0, so that $s \geq \dim_H f(F)$. □

7. *Special cases of 6*:

(a) If $f\colon F \to \mathbf{R}^m$ is bi-Lipschitz, that is

$$c_1|x-y| \le |f(x)-f(y)| \le c_2|x-y| \qquad (x,y \in F) \tag{1.5}$$

with $0 < c_1 \le c_2$, then $\dim_H f(F) = \dim_H F$. Thus *Hausdorff dimension is preserved by bi-Lipschitz mappings.*

(b) If $f\colon \mathbf{R}^n \to \mathbf{R}^n$ is a congruence, similarity or (non-singular) affine transformation, then $\dim_H f(F) = \dim_H F$ (special case of (a)).

(c) Orthogonal projections. Let P be a linear subspace of $\mathbf{R}^n$, and let proj denote orthogonal projection from $\mathbf{R}^n$ onto P. Then

$$\dim_H(\mathrm{proj}\, F) \le \dim_H F, \tag{1.6}$$

since $|\mathrm{proj}\, x - \mathrm{proj}\, y| \le |x-y|$ for $x,y \in \mathbf{R}^n$.

(d) Smooth sets. If F is a C^1diffeomorphic image of an open subset of $\mathbf{R}^n$, then $\dim_H F = n$. (Note that C^1 mappings are Lipschitz on compact sets, by the mean value theorem.)

(e) The classification theorem. By 7(a), sets that may be mapped onto each other by bi-Lipschitz mappings have the same Hausdorff dimension. One approach to fractal geometry is to regard two sets as equivalent if there is a bi-Lipschitz mapping between them, and to seek "invariants", such as Hausdorff dimension, to distinguish non-equivalent sets. This is anologous to topology, where invariants such as homotopy groups distinguish between sets that are not homeomorphic (i.e., not equivalent under bijections that are continuous and have continuous inverses).

8. *Dust-like sets*. It may be shown that if $\dim_H F < 1$ then F is totally disconnected.

Box-counting dimensions

We now discuss another frequently used definition of dimension.

Let F be a bounded subset of $\mathbf{R}^n$. For $\delta > 0$, let $N_\delta(F)$ be the least number of sets of diameter at most δ that can cover F. We define the *lower* and *upper box-counting dimensions* of F as

$$\underline{\dim}_B F = \underline{\lim}_{\delta\to 0} \frac{\log N_\delta(F)}{-\log \delta} \tag{L.B.}$$

$$\overline{\dim}_B F = \overline{\lim}_{\delta\to 0} \frac{\log N_\delta(F)}{-\log \delta}. \tag{U.B.}$$

If these are equal we call the common value the *box-counting dimension*, abbreviated to *box dimension*,

$$\dim_B F = \lim_{\delta\to 0} \frac{\log N_\delta(F)}{-\log \delta}. \tag{B.}$$

[Note that this is the case if $N_\delta(F) \sim \delta^{-\dim_B F}$]. Box-dimension has also been called metric dimension, capacity, logarithmic density, entropy dimension, ...

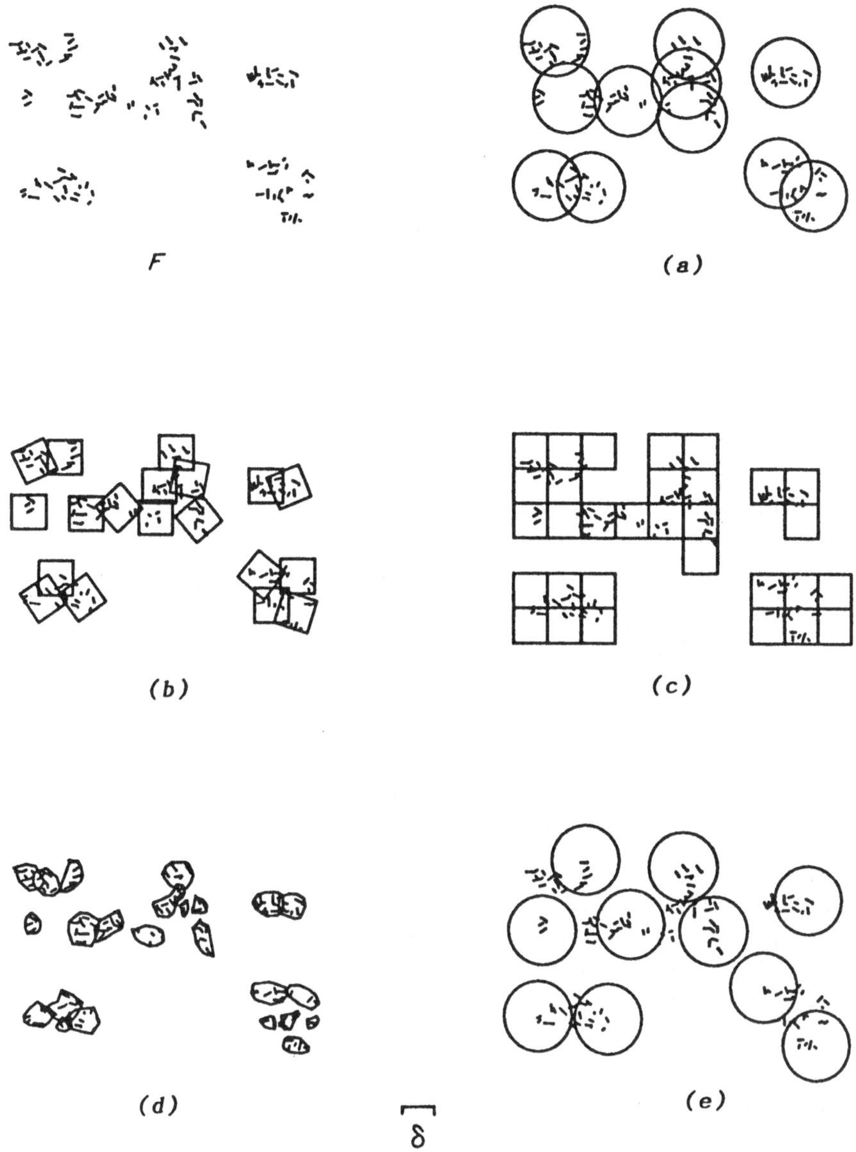

Figure 5:
Five ways of finding the box dimension of F, see the Equivalent Definitions. The number N_δ is taken to be (a) the least number of balls of radius δ that cover F, (b) the least number of cubes of side δ that cover F, (c) the number of δ-mesh cubes that intersect F, (d) the least number of sets of diameter δ that cover F, (e) the greatest number of disjoint balls of radius δ with centres in F.

Equivalent Definitions *We get precisely the same answer in (L.B.), (U.B.) and (B.) if we take $N_\delta(F)$ to be any of the following:*

(a) *the least number of (closed) balls of radius δ that cover F;*

(b) *the least number of sets of diameter at most δ that cover F;*

(c) *the least number of cubes of side δ that cover F;*

(d) *the number of cubes of the lattice of side δ that intersect F;*

(e) *the largest number of disjoint balls of radius δ centred in F.*

Proof. The following proof that (a) and (d) are equivalent is typical of the proofs required. Let N_δ be the least number of sets of diameter at most δ that can cover F, and let N'_δ be the number of cubes of the δ-lattice that intersect F. Then

$$N_{\sqrt{n}\delta} \leq N'_\delta \leq 3^n N_\delta;$$

the left hand inequality holds since the cubes of the δ-lattice have diameter $\sqrt{n}\delta$, and the right hand inequality holds since any set of diameter δ is contained in 3^n δ-lattice cubes (take any cube containing some point of the set together with its immediate neighbours). Thus

$$\frac{\log N_{\sqrt{n}\delta}}{-\log\sqrt{n}\delta + \log\sqrt{n}} \leq \frac{\log N'_\delta}{-\log\delta} \leq \frac{\log 3^n + \log N_\delta}{-\log\delta}.$$

Letting $\delta \to 0$ we see that the limits (L.B.), (U.B.) and (B.) are unchanged when N_δ is replaced by N'_δ. □

Relationship Between Hausdorff and Box-counting Dimensions

Note that, if $F \subset \bigcup_{i=1}^k U_i$, where $|U_i| \leq \delta$, then $\sum_{i=1}^k |U_i|^s \leq k\delta^s$. Hence $H^s_\delta(F) \leq N_\delta(F)\delta^s$, where $N_\delta(F)$ is the least number of sets of diameter δ that can cover F. If $s < \dim_H F$, then $1 < H^s_\delta(F) \leq N_\delta(F)\delta^s$ for sufficiently small δ, so $0 < \log N_\delta(F) + s\log\delta$ or $s < \log N_\delta(F)/-\log\delta$. Thus

$$\dim_H F \leq \underline{\dim}_B F \leq \overline{\dim}_B F.$$

We often, but not always, have equality here. For example, it is not hard to show that the box dimension of the Cantor set is $\log 2/\log 3$, and of the von Koch curve is $\log 4/\log 3$.

The Drawbacks of Box-counting Dimensions

While box dimensions are conceptually simpler and computationally more convenient than Hausdorff dimensions, under certain circumstances they have serious drawbacks. The following example is of a countable set (which one would generally regard as very small) that has large box dimension.

Example *Let F be the set of rational numbers between* 0 *and* 1. *Then*

(a) *F is countable, so* $\dim_H F = 0$;

(b) $\dim_B F = 1$.

Check. For (b), note that F intersects $N_\delta(F) = [1/\delta]$ intervals of the "δ-lattice" (which have the points $\{\ldots, -\delta, 0, \delta, 2\delta, \ldots\}$ as ends; here [ ] denotes the "next integer above"). Thus

$$\frac{\log N_\delta(F)}{-\log\delta} = \frac{\log[1/\delta]}{-\log\delta} \to 1,$$

so $\underline{\dim}_B F = \overline{\dim}_B F = 1$. □

More generally, the same argument shows that $\underline{\dim}_B F = \underline{\dim}_B$ (closure of F) and $\overline{\dim}_B F = \overline{\dim}_B$ (closure of F). One might hope that it would be possible to avoid such problems by restricting attention to closed sets, but the following example shows that even this is not the case.

Example *Let* $F = \{0, 1, \frac{1}{2}, \frac{1}{3}, \frac{1}{4}, \frac{1}{5}, \ldots\}$. *Then*

(a) *F is countable and compact, so* $\dim_H F = 0$,

(b) $\dim_B F = \frac{1}{2}$.

Check. Let U be an interval of length $\delta < \frac{1}{2}$, and let k be the integer satisfying $(k(k+1))^{-1} \le \delta < ((k-1)k)^{-1}$. Then U can cover at most one of the points $\{1, \frac{1}{2}, \frac{1}{3}, \ldots, \frac{1}{k}\}$. Thus, at least k intervals of length δ are required to cover F, so if $N_\delta(F)$ is the least number of intervals of length δ that can cover F,

$$\frac{\log N_\delta(F)}{-\log\delta} \ge \frac{\log k}{\log k(k+1)} \to \frac{1}{2}$$

as $\delta \to 0$ (and $k \to \infty$). Hence $\underline{\dim}_B F \ge \frac{1}{2}$. The similar check that $\overline{\dim}_B F \le \frac{1}{2}$ is left to the reader. □

Despite these examples, box-dimensions can be extremely useful. They are often much easier to calculate than Hausdorff dimensions, and, for many interesting classes of sets, may be shown to equal the Hausdorff dimension.

A word of warning: it is important to be aware which definition of dimension is in use in any discussion. Different definitions can have different properties and different values for the same set.

Lecture 2
Some Fractal Constructions and the Calculation of Dimension

Invariant Sets and Self-Similar Sets

A transformation $S\colon \mathbf{R}^n \to \mathbf{R}^n$ is called a *contraction* if there is a number $0 < c < 1$ such that

$$|S(x) - S(y)| \leq c|x - y| \qquad (x, y \in \mathbf{R}^n).$$

Let $S_1, \ldots, S_k$ be contractions. We call a compact set F *invariant* for the S_i if

$$F = \bigcup_{i=1}^{k} S_i(F).$$

It may be shown that a set of contractions has a unique non-empty compact invariant set. Invariant sets are often fractals, and they may be specified very conveniently by such sets of contractions.

For a simple example, let $S_1, S_2\colon \mathbf{R} \to \mathbf{R}$ be given by

$$S_1(x) = \tfrac{1}{3}x; \qquad S_2(x) = \tfrac{1}{3}x + \tfrac{2}{3}.$$

It is easy to see that the middle third Cantor set (Figure 3) is the invariant set of S_1 and S_2.

For $E \subset \mathbf{R}^n$ we define a transformation ψ by

$$\psi(E) = \bigcup_{i=1}^{k} S_i(E).$$

It is not hard to see that if E is any non-empty compact set with $S_i(E) \subset E$ for all i, then the invariant set F is given by

$$F = \bigcap_{q=1}^{\infty} \psi^q(E) \tag{2.1}$$

where ψ^q denotes the q-th iterate of ψ, so $\psi^1(E) = \psi(E)$ and $\psi^q(E) = \psi(\psi^{q-1}(E))$. Note also that

$$\psi^q(E) = \bigcup S_i(S_{i_2}(\ldots(S_{i_q}(E)))) \tag{2.2}$$

where the union is over all sequences $i_1, \ldots, i_q$ with $1 \leq i_r \leq k$ for each r. [Such families $S_1, \ldots, S_k$ of contractions are sometimes called *iterated function schemes.*]

In the Cantor set example above, if we take $E = [0,1]$, then $\psi^q(E)$ is the q-th stage of the construction in Figure 3, consisting of 2^q intervals of length 3^{-q}.

Now suppose that the S_i are *similarities*, i.e.

$$|S_i(x) - S_i(y)| = c_i|x - y| \quad (x, y \in \mathbf{R}^n),$$

where $0 < c_i < 1$. Let s be the solution of

$$\sum_{i=1}^{k} c_i^s = 1. \tag{2.3}$$

Then $\dim_H F \leq s$. To see this, note that for each q,

$$\begin{aligned}
\sum_{i_1,\ldots,i_q} |S_{i_1}(S_{i_2}(\ldots(S_{i_q}(E))))|^s &= \sum_{i_1,\ldots,i_q} (c_{i_1}c_{i_2}\ldots c_{i_q})^s|E|^s \\
&= (\sum_i c_i^s)^q|E|^s \\
&= |E|^s.
\end{aligned}$$

Using (2.1) and (2.2) with q such that $\delta \geq (\max c_i)^q$ it follows that $H_\delta^s(F) \leq |E|^s$. Thus $H^s(F) \leq |E|^s$ and $\dim_H F \leq s$.

In fact, if the S_i satisfy the *open set condition*, i.e., if there exists an open set U such that

$$U \supset \bigcup_{i=1}^{k} S_i(U) \tag{2.4}$$

with this union disjoint, we have

$$\dim_H F = \underline{\dim}_B F = \overline{\dim}_B F = s$$

where s is given by (2.3) (see Hutchinson (1981)); we show this in a special case shortly. Essentially, the open set condition ensures that the components $S_i(F)$ of F do not overlap "too much".

Finding the Dimension of a Set

We now discuss rigorous methods of calculating the dimensions of a set. In general, it is much easier to find upper bounds for dimensions than lower bounds.

Upper Estimates If F can be covered by n_k sets of diameter at most δ_k, with $\delta_k \to 0$ as $k \to \infty$, then it follows from the definitions that

$$\dim_H F \leq \underline{\dim}_B F \leq \underline{\lim}_{k\to\infty} \frac{\log n_k}{-\log \delta_k}.$$

Lower Estimates Most methods for finding lower bounds for dimensions involve some form of the *mass distribution principle.*

We call μ a *mass distribution* on a set $F \subset \mathbf{R}^n$ if

(a) $\mu(E)$ is defined as a non-negative number for all "reasonable" (i.e. Borel) subsets of $\mathbf{R}$;

(b) $\mu(\bigcup_{i=1}^{\infty} E_i) = \sum_{i=1}^{\infty} \mu(E_i)$ for disjoint E_i;

(c) $0 < \mu(F) < \infty$;

(d) $\mu(\mathbf{R}^n - F) = 0$.

[Intuitively we think of μ as a finite mass or charge spread across the set F in some way.]

Mass Distribution Principle *Suppose that there exists a mass distribution μ on a set F and constants $c > 0$ and $\delta > 0$ such that*

$$\mu(U) \leq c|U|^s$$

for all sets U with $|U| \leq \delta$. Then

$$H^s(F) \geq \mu(F)/c \tag{2.5}$$

so that

$$s \leq \dim_H F \leq \underline{\dim}_B F \leq \overline{\dim}_B F.$$

Proof. If $F \subset \bigcup_i U_i$, then

$$0 < \mu(F) = \mu(\bigcup_i U_i) \leq \sum_i \mu(U_i) \leq c \sum_i |U_i|^s$$

so, taking infima, $H^s_\delta(F) \geq \mu(F)/c$ if δ is small enough, giving (2.5). □

As an illustration, we find the dimension of the set obtained using the iterated construction of Figure 6. At each stage of the construction each square is replaced by 6 squares of a quarter of the side length, always arranged in the same pattern.

Example *If F is the self-similar set of the construction indicated in Figure* 6,

$$\dim_H F = \dim_B F = \log 6/\log 4.$$

Check. Writing E_k for the set obtained at the k-th stage of the construction, E_k consists of 6^k squares of side 4^{-k}. Since $F \subset E_k$ may be covered by these squares,

$$\dim_H F \leq \underline{\dim}_B F \leq \underline{\lim}_{k\to\infty} \frac{\log 6^k}{\log 4^k} = \log 6/\log 4$$

A minor variation of this argument gives $\overline{\dim}_B F \leq \log 6/\log 4$ also.

Now take a unit mass, split it equally between the 6 squares of E_1, split the mass on each of these equally between the 6 sub-squares of E_2, and so on, to get a mass distribution μ on F. Each square in E_k has mass 6^{-k}. Let U be a set with $|U| < 1$, and let k be the integer such that $4^{-k} > |U| \geq 4^{-k-1}$. Then U intersects at most 4 squares of E_k. Hence

$$\begin{aligned} \mu(U) \leq 4 \times 6^{-k} &= 4 \times 4^{-k(\log 6/\log 4)} \\ &\leq 4 \times (4|U|)^{\log 6/\log 4} \\ &= c|U|^{\log 6/\log 4} \end{aligned}$$

for $|U| < 1$. Since $\mu(F) = 1$, the mass distribution principle gives

$$\log 6/\log 4 \leq \dim_H F \leq \underline{\dim}_B F \leq \overline{\dim}_B F. \quad \square$$

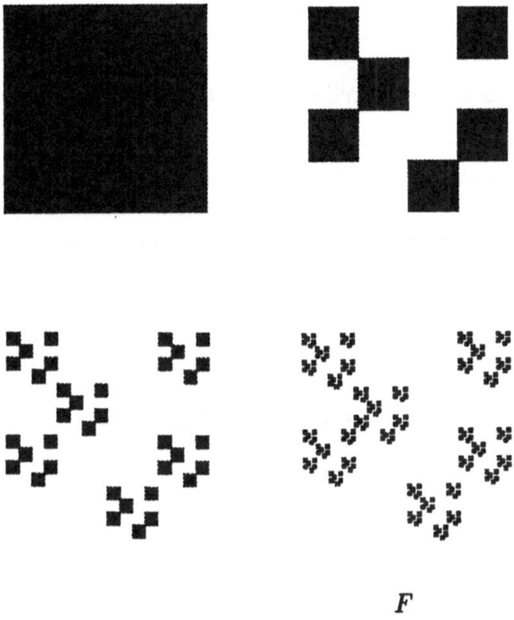

Figure 6:
A fractal F constructed by repeatedly taking an arrangement of sub-squares of each square. The set F has Hausdorff and box dimensions of $\log 6/\log 4$.

Variations

We quote the following result, which is a generalisation of the mass distribution principle. We write $B_r(x)$ for the ball of center x and radius r.

Theorem *Let μ be a mass distribution on $\mathbf{R}^n$. Suppose $F \subset \mathbf{R}^n$ and $0 < c < \infty$.*

(a) *If $\overline{\lim}_{r\to 0}\dfrac{\mu(B_r(x))}{r^s} < c$ for $x \in F$, then $H^s(F) \geq \mu(F)/c$.*

(b) *If $\overline{\lim}_{r\to 0}\dfrac{\mu(B_r(x))}{r^s} > c$ for $x \in F$, then $H^s(F) \leq 2^s\mu(\mathbf{R}^n)/c$.*

The following very useful variant comes essentially from integrating the mass distribution principle. For μ a mass distribution on $\mathbf{R}^n$ we define the *s-energy* of μ to be

$$I_s(\mu) = \int\int \frac{d\mu(x)d\mu(y)}{|x-y|^s}.$$

Theorem *Let F be a subset of $\mathbf{R}^n$.*

(a) *If there exists a mass distribution μ on F with $I_s(\mu) < \infty$, then $H^s(F) = \infty$ and $\dim_H F \geq s$.*

(b) *If $H^s(F) > 0$, then there exists a mass distribution μ on F such that $I_s(\mu) < \infty$.*

We shall use this result to prove the "projection theorems" in the next lecture.

Lecture 3
Projections of Fractals

We first consider projections of subsets of the plane onto lines. We write proj_θ for orthogonal projection onto the straight line ℓ_θ through the origin making angle θ with the x-axis. Note that

$$|\text{proj}_\theta x - \text{proj}_\theta y| \leq |x-y| \qquad (x, y \in \mathbf{R}^2)$$

so that proj_θ is a Lipschitz mapping. It is immediate that

$$\dim_H(\text{proj}_\theta F) \leq \min\{\dim_H F, 1\}$$

for any F and θ.

The following theorem (Marstrand 1954) gives inequalities in the opposite direction. Note that they need only hold for *almost all* θ, i.e., except for a set of θ in $[0, \pi)$ of length 0.

Projection Theorems *Let F be a (Borel) subset of* $\mathbf{R}^2$ *with* $\dim_H F = s$.

(a) *If $s \leq 1$, then* $\dim_H(\mathrm{proj}_\theta F) = s$ *for almost all θ in $[0, \pi)$.*

(b) *If $1 < s \leq 2$, then* length $(\mathrm{proj}_\theta F) > 0$ *for almost all θ in $[0, \pi)$.*

Proof. We give the proof of (a), the proof of (b) is fairly similar. Let $t < s$. Using the final result stated in Lecture 2, we may find a mass distribution μ on F with $I_t(\mu) < \infty$. We may project μ onto each ℓ_θ to get mass distributions μ_θ on $\mathrm{proj}_\theta F$; thus $\int f(u) d\mu_\theta(u) = \int f(x \cdot \boldsymbol{\theta}) d\mu(x)$ for non-negative functions f. (We write $\boldsymbol{\theta}$ for the unit vector along ℓ_θ, "$\cdot$" for the usual dot product and associate x with the position vector that it represents.) Then

$$\begin{aligned} I_t(\mu_\theta) &= \int\int \frac{d\mu_\theta(u) d\mu_\theta(v)}{|u - v|^t} \\ &= \int\int \frac{d\mu(x) d\mu(y)}{|x \cdot \boldsymbol{\theta} - y \cdot \boldsymbol{\theta}|^t} \\ &= \int\int \frac{d\mu(x) d\mu(y)}{|(x - y) \cdot \boldsymbol{\theta}|^t}. \end{aligned}$$

Thus

$$\begin{aligned} \int_0^\pi I_t(\mu_\theta) d\theta &= \int_0^\pi \int\int \frac{d\mu(x) d\mu(y)}{|(x - y) \cdot \boldsymbol{\theta}|^t} d\theta \\ &= \int_0^\pi \frac{d\theta}{|\boldsymbol{\tau} \cdot \boldsymbol{\theta}|^t} \int\int \frac{d\mu(x) d\mu(y)}{|x - y|^t} \end{aligned}$$

where $\boldsymbol{\tau}$ is any unit vector. Since $t < 1$ and $I_t(\mu) < \infty$, this expression is finite, so that $I_t(\mu_\theta) < \infty$ for almost all θ. But $\mu_\theta(\mathrm{proj}_\theta F) = \mu(F)$, so $\dim_H(\mathrm{proj}_\theta F) \geq t$ for almost all θ. This is true for all $t < s$, implying that $\dim_H(\mathrm{proj}_\theta F) \geq t$, as required. □

Note that all the obvious higher dimensional analogues of this result are valid.

The Critical Case

According to the above theorem, length $(\mathrm{proj}_\theta F) = 0$ for all θ if $\dim_H F < 1$ and length $(\mathrm{proj}_\theta F) > 0$ for almost all θ if $\dim_H F > 1$. The critical case of $\dim_H F = 1$ is rather more delicate.

Besicovitch showed that any Borel subset F of $\mathbf{R}^2$ with $0 < H^2(F) < \infty$ has a disjoint decomposition $F = F_R \cup F_I$ where F_R is *regular* or *curve-like* (that is consists of a countable union of pieces of rectifiable curve) and F_I is *irregular* or *curve-free* (that is, intersects every rectifiable curve in length 0.) These two types of set have contrasting projection properties.

Theorem *Suppose* $0 < H^1(F) < \infty$.

(a) *If F is regular, then* length$(\text{proj}_\theta F) > 0$ *for almost all* θ.

(b) *If F is irregular, then* length$(\text{proj}_\theta F) = 0$ *for almost all* θ.

We are left with the case of $\dim_H F = 1$ and $H^1(F) = \infty$. In this case almost anything can happen. The following theorem says that there exists such a set with more or less any projections desired.

Theorem *Let G_θ be a given subset of ℓ_θ for each θ. (We ignore a technical measurability condition on the G_θ.) Then there exists $F \subset \mathbf{R}^2$ such that for almost all θ, we have $\text{proj}_\theta F = G_\theta$ to within zero length (i.e.,* $\text{proj}_\theta F \supset G_\theta$ *and* length$(\text{proj}_\theta F \setminus G_\theta) = 0$.*)*

The proof of this is quite complicated. An underlying idea is the "iterated venetian blind" construction indicated in Figure 7 - the set shown has projections similar to that of a line segment in some directions, and very small projections in other directions.

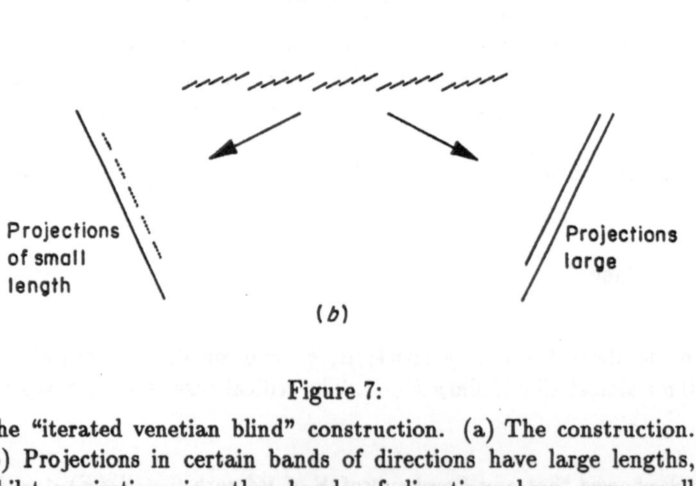

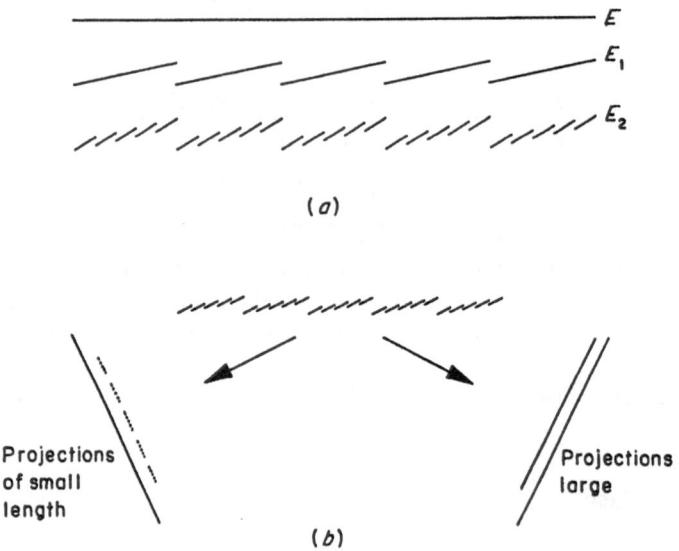

Figure 7:
The "iterated venetian blind" construction. (a) The construction. (b) Projections in certain bands of directions have large lengths, whilst projections in other bands of directions have very small lengths.

The higher dimensional analogues of this result are also true. In particular, there exists a subset of $\mathbf{R}^3$ with any prescribed projections (to within area 0) onto planes through the origin. One way of thinking of this is that there exists a digital sundial (Figure 8). show the thickened digits of the time.

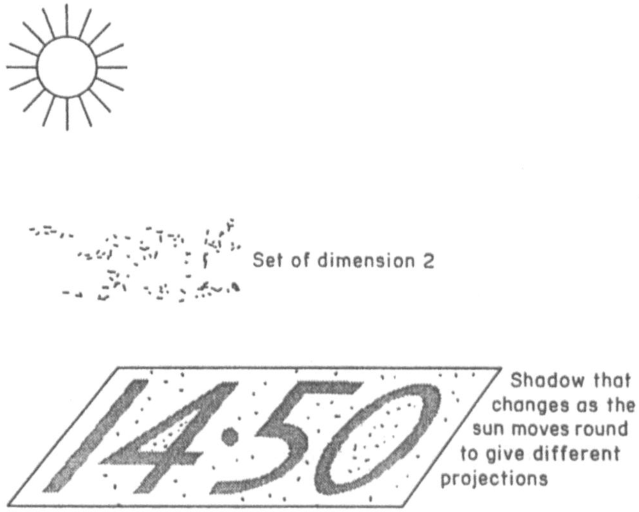

Figure 8: A digital sundial.

Vitushkin's Conjecture in Complex Variable Theory

As an aside, we mention another recent use of an "iterated venetian blind" construction in complex variable theory.

We call a compact set $F \subset \mathbf{C}$ *removable* if, for any bounded open set $U \supset F$, any bounded analytic function on $U \setminus F$ has an analytic extension to U. It has been known for a long time that if $\dim_H F < 1$ or $0 < H^1(F) < \infty$ with F irregular then F is removable, and if $\dim_H > 1$ or $0 < H^1(F) < \infty$ with F regular then F is not removable. This, and much further evidence, led to Vitushkin's conjecture: A compact set F is removable if and only if $\text{length}\,(\text{proj}_\theta F) = 0$ for almost all θ.

Let $f\colon\ U \to f(U)$ be an analytic bijection that is "bending", i.e., maps some straight lines onto curves. Mattila (1986) used a version of the iterated venetian blind construction to obtain a set E such that $\text{length}\,(\text{proj}_\theta E) = 0$ for almost all θ, but with $\text{length}\,(\text{proj}_\theta f(E)) > 0$ for almost all θ. This immediately implies that Vitushkin's conjecture is false, since "length $(\text{proj}_\theta F) = 0$ for almost all θ" is not invariant under conformal transformations, but "F is removable" obviously is; thus these conditions cannot be equivalent. This argument gives no indication of which of the implications fails; however Jones and Murai (1988) have very recently constructed a set with almost all projections of zero length that is not removable.

Lecture 4
Intersections of Fractals

We consider the dimension of the intersection of sets $E, F \subset \mathbf{R}^n$ as they are moved relative to each other. In particular, what is the relationship between $\dim_H(E \cap \sigma(F))$ and $\dim_H E$ and $\dim_H F$ as σ ranges over a group of transformations, such as the group of translations, similarities or congruences?

In the classical situation, where E, F are smooth curves, surfaces, etc. in $\mathbf{R}^n$ we have that:

$$\dim(E \cap \sigma(F)) \leq \max\{0, \dim E + \dim F - n\} \qquad \text{"in general"} \qquad \text{(A)}$$

$$\dim(E \cap \sigma(F)) \geq \dim E + \dim F - n \qquad \text{"often".} \qquad \text{(B)}$$

Of course, by "in general" we mean "except for a set of transformations σ of measure zero" and by "often" we mean "for a set of σ of positive measure", taking natural measures on groups of transformations σ. We seek analogues of (A) and (B) in the fractal situation.

The following proposition, concerning the intersection of a set with a translate of a straight line, leads to the best result that can be obtained in the direction of (A).

Proposition *Let $E \subset \mathbf{R}^2$. Then if $\dim_H E \geq 1$, we have*

$$\dim_H(E \cap L_x) \leq \dim_H E - 1,$$

for almost all x, where L_w is the line $x = w$ in the (x, y)-plane.

Proof. Let $E \subset \bigcup U_i$ with $0 < |U_i| \leq \delta$. For each i let $S_i \supset U_i$ be a square with sides of length $|U_i|$ parallel to the coordinate axes. We write χ_i for the indicator function of S_i, so $\chi_i(x, y) = 1$ if $(x, y) \in S_i$ and 0 otherwise. Then, for $s \geq 1$,

$$\int_{L_x} \sum \chi_i(x, y)|U_i|^{s-2} dy \;=\; \sum_i |S_i \cap L_x||U_i|^{s-2}$$

$$= \sum\{|S_i|^{s-1} : \ L_x \text{ cuts } S_i\}$$
$$\geq H_\delta^{s-1}(E \cap L_x).$$

Thus

$$\int H_\delta^{s-1}(E \cap L_x)dx \leq \sum_i \int\int \chi_i(x,y)|U_i|^{s-2}dx\,dy = \sum_i |U_i|^s.$$

Taking infima

$$\int H_\delta^{s-1}(E \cap L_x)dx \leq H_\delta^s(E),$$

so letting $\delta \to 0$,

$$\int H^{s-1}(E \cap L_x)dx \leq H^s(E),$$

and the result follows. □

Theorem *Let $E, F \subset \mathbf{R}^n$. Then for almost all $\sigma \in G$ we have*

$$\dim_H(E \cap \sigma(F)) \leq \max\{0, \dim_H(E \times F) - 1\}, \tag{4.1}$$

where G is the group of translations, congruences or similarities.

Proof. We give the proof in the case $n = 1$; a similar proof works for larger n. We work in $\mathbf{R}^2$ and write L_t for the line $y = x + t$. Observe that

$$(x, x+t) \in E \times F \text{ if and only if } x \in E \cap (F - t).$$

Hence, for each t, $(E \times F) \cap L_t$ and $E \cap (F - t)$ are similar sets. Thus

$$\dim_H(E \cap (F - t)) = \dim_H((E \times F) \cap L_t) \leq \max\{\dim_H(E \times F) - 1, 0\}$$

for almost all t, by the previous proposition. Thus (4.1) holds for the group of translations, and so for the larger groups of congruences and similarities. □

This theorem is the closest we can get to (A), even for the group of similarities. Of course, (A) will follow if $\dim_H(E \times F) = \dim_H E + \dim_H F$; this is very often the case, for example, whenever $\dim_H E = \overline{\dim}_B E$, see Falconer (1990).

The proof of the following version of (B) is rather hard, see Kahane (1986) and Mattila (1975, 1985).

Theorem *Let $E, F \subset \mathbf{R}^n$. Then*

$$\dim_H(E \cap \sigma(F)) \geq \dim_H E + \dim_H F - n$$

for a set of σ of positive measure, in the following cases.

(a) *G is the group of congruences and E is a regular set, e.g. a smooth or rectifiable curve, surface, etc.;*

(b) *G is the group of similarities and E and F are arbitrary;*

(c) *G is the group of congruences, E and F are arbitrary and* $\dim_H E \geq \frac{1}{2}(n+1)$.

It is not known whether (c) holds without the restriction on $\dim_H E$.

Sets with Large Intersection

It is possible to construct classes of sets of dimension s such that the intersection of any countable collection of sets from the class also has dimension s; thus (A) is as far from true as it can be. We indicate one such example here.

Fix $\alpha > 2$ and let

$$F = \{x \in \mathbf{R}: \text{ for infinitely many positive integers } q, \text{ there is an integer } p \text{ such that } |x - \frac{p}{q}| \leq \frac{1}{q^\alpha}\}.$$

(Such sets of numbers occur in the theory of Diophantine approximation, and in the KAM theorem of Hamiltonian dynamics.)

Jarník's theorem states that $\dim_H F = 2/\alpha$. By translation, $\dim_H(F+t) = 2/\alpha$ for any real number t. However, it may be shown that, if $t_1, t_2, \ldots$ is any sequence of real numbers,

$$\dim_H \bigcap_{i=1}^{\infty}(F + t_i) = 2/\alpha,$$

thus, these sets have intersection of larger dimension than might be expected.

Lecture 5
Self-Affine Sets

In Lecture 2 we pointed out that, if $S_1, \ldots, S_k\colon \mathbf{R}^n \to \mathbf{R}^n$ are contractions there is a unique (compact non-empty) invariant set F such that

$$F = \bigcup S_i(F).$$

In the simplest case where $S_1, \ldots, S_k$ are similarities (i.e., $|S_i(x) - S_i(y)| = c_i|x - y|$) we have $\dim_H F = \dim_B F = s$, where

$$\sum_{i=1}^{k} c_i^s = 1, \tag{5.1}$$

provided that the open set condition (2.4) holds.

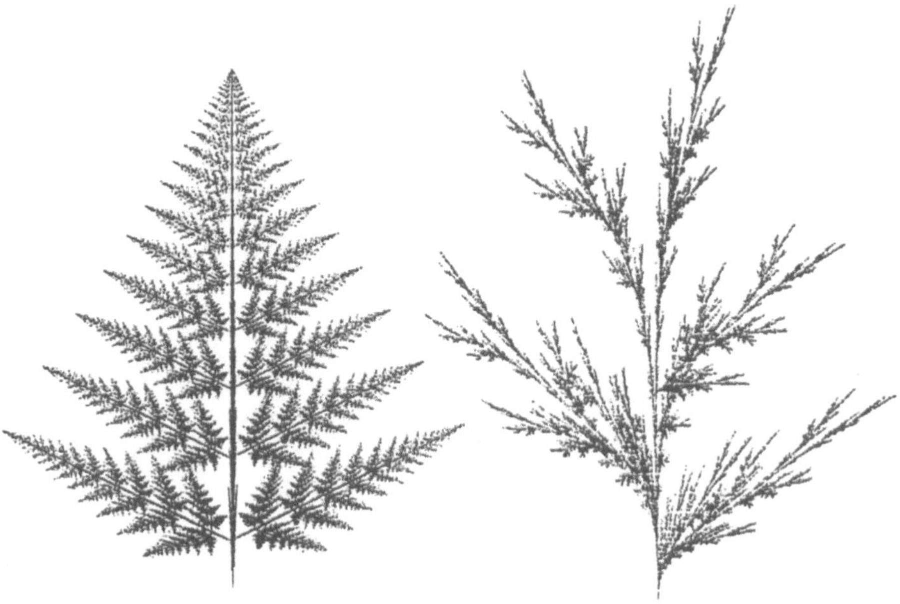

Figure 9:
The fern is the invariant set of just 4 affine transformations, and the grass is the invariant set of 6 affine transformations.

It is natural to try to extend this to the case where the S_i are *affine* contractions, i.e., where

$$S_i(x) = T_i(x) + a_i,$$

T_i being linear and $a_i \in \mathbf{R}^n$. Invariant sets for such families of transformations are called *self-affine* sets, examples, which often provide realistic pictures of natural objects, are shown in Figure 9. One might hope for an expression for the dimension of self-affine sets in terms of the linear parts of the transformations, T_i, generalising (5.1). Unfortunately the situation is far more complicated than for self-similar sets.

A very special case which has been completely analysed (McMullen (1984)) is shown in Figure 10. The square is divided into a p by q array of rectangles of side p^{-1} by q^{-1} (we take $p < q$). A subcollection of these rectangles is selected, with N_j rectangles selected from the j-th column; we call the union of the selected rectangles E_1. Each rectangle of E_1 is replaced by an affine copy of E_1 to give a set E_2, and the process is iterated in the usual way to give a fractal F. It may be shown that

$$\dim_H F = \log(\sum_{j=1}^{p} N_j^{\log p/\log q})/\log p \tag{5.2}$$

$$\dim_B F = \log p_1/\log p + \log((\sum_{i=1}^{p} N_j)/p_1)/\log q \tag{5.3}$$

(p_1 is the number of columns containing at least one rectangle.)

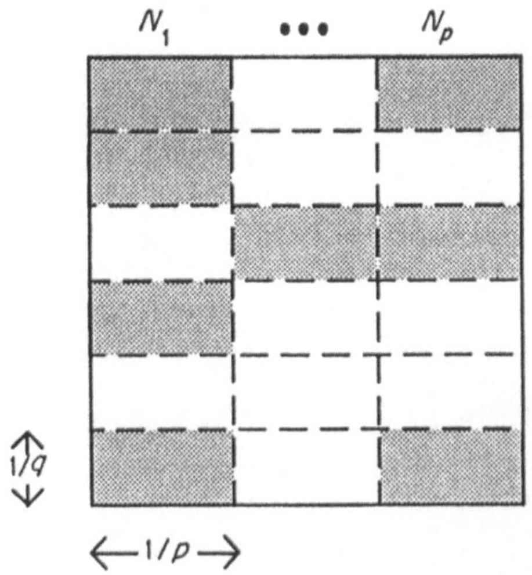

Figure 10a:

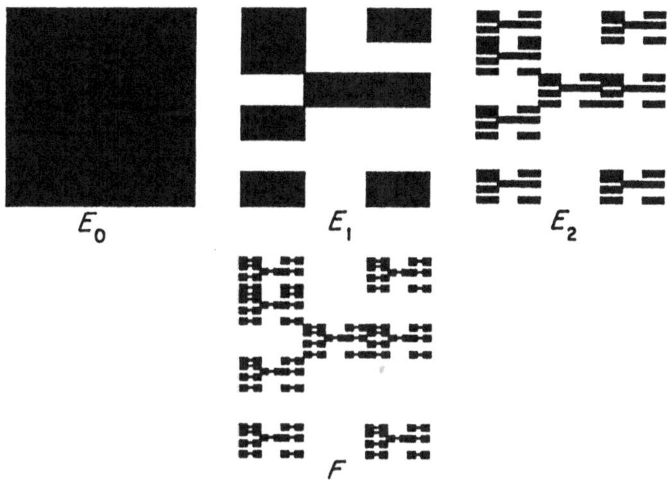

Figure 10b:
Construction of a self-affine set: (a) data for the construction, (b) the first few stages of construction.

The main things to note about these formulae are that the box dimension and Hausdorff dimension are not, in general, equal, and that the dimensions depend crucially on the position of the rectangles selected, as well as their size and number. This suggests that the problem of finding the dimensions of self-affine sets is likely to be difficult.

Returning to the general situation, we recall that, if E is a convenient set (which we usually choose to be a square) such that $S_i(E) \subset E$ for all i, then the sequence of sets

$$\psi^q(E) = \bigcup S_{i_1}(S_{i_2}(\ldots(S_{i_q}(E)))) \tag{5.4}$$

(where the union is over sequences $i_1, \ldots, i_q$ with $1 \leq i_j \leq k$) decreases to the self-affine set F. Such sequences often provide the "usual" constructions of F.

The examples shown in Figure 11 provide another illustration of the difficulties. We consider affine transformations S_1, S_2 that map the unit square E onto the rectangles of sides $\frac{1}{2}$ and $\epsilon < \frac{1}{2}$ indicated. Note that $S_1(E)$ abuts the y-axis, and $S_2(E)$ is distance λ from the y-axis. Iterating the construction, it is easy to see that the projection of $\psi^q(E)$ onto the x-axis contains the interval $[0, 2\lambda]$ for all q; thus the same is true of the invariant set F. It follows that, if $\lambda > 0$, then $\dim_H F \geq 1$, since a set has dimension at least that of any projection. If $\lambda = 0$, however, (Figure 11 (b)), the set $\psi^q(E)$ lies within 2^{-q} of the y-axis, so that F is a subset of the y-axis which may be obtained by a "Cantor-like" construction, repeatedly removing the middle proportion $1 - 2\epsilon$ of intervals. A simple calculation gives that $\dim_H F = \log 2 / -\log \epsilon < 1$. Thus the dimension of F does not vary continuously with

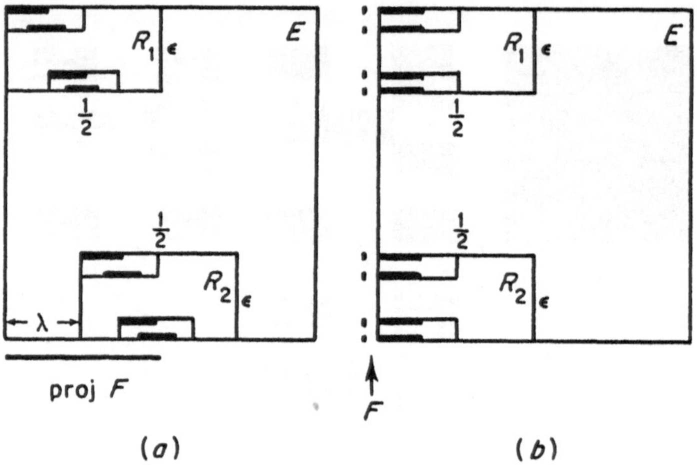

Figure 11:
Discontinuity of the dimension of self-affine sets. The affine mappings S_1 and S_2 map the unit square E onto rectangles R_1 and R_2. In (a) $\lambda > 0$ and $\dim_H F \geq \dim_H \text{ proj } F = 1$, but in (b) $\lambda = 0$, and $\dim_H F = \log 2 / -\log \epsilon < 1$.

λ, which suggests that any formula for the dimension of self-affine sets is likely to be very awkward.

It is natural to use the approximations to an invariant set F given by (5.4) in analysing dimensions. While this works perfectly well for self similar sets, two problems are encountered when using the "basic sets" $S_{i_1}(S_{i_2}(\ldots(S_{i_q}(E))))$ as covering sets in the case of self-affine sets. First, these sets tend to become "long and thin" when q is large. Rather than taking the set $S_{i_1}(S_{i_2}(\ldots(S_{i_q}(E))))$ itself as a covering set, we may be able to get a smaller value for $\sum |U_i|^s$ by cutting the set into a large number of cube like pieces and using these. For example, if $U \equiv S_{i_1}(S_{i_2}(\ldots(S_{i_q}(E))))$ is a rectangle of sides $\alpha_1 \gg \alpha_2$, then $|U|^s \simeq \alpha_1^s$. However, dividing U into about α_1/α_2 squares U_i of sides α_2 we get $\sum |U_i|^s \simeq \alpha_1 \alpha_2^{s-1}$, which will tend to be smaller than $|U|^s$ if $s > 1$.

The second difficulty is that it is possible for several sets $S_{i_1}(S_{i_2}(\ldots(S_{i_q}(E))))$ to lie nearly side by side for different $i_1, \ldots, i_q$. It might then be possible to find a more efficient cover by using sets that straddle several of these basic sets, than by covering each basic set individually. However, it turns out that this situation only affects the dimension in "exceptional" cases.

To formalise this, we define the *singular values* $1 > \alpha_1 \geq \alpha_2 \geq \ldots \geq \alpha_n > 0$ of a linear contraction T on $\mathbf{R}^n$ to be the positive square roots of the eigenvalues of T^*T, equivalently the lengths of the semi-axes of the ellipsoid $T(B)$ where B is the unit ball in $\mathbf{R}^n$. For

$0 \le s \le n$ we define the *singular value function* $\phi^s(T)$ by

$$\phi^s(T) = \alpha_1\alpha_2 \dots \alpha_{m-1}\alpha_m^{s-m+1}$$

where m is the integer such that $m-1 < s \le m$. Then $\phi^s(T)$ is continuous and strictly decreasing in s. Also, $\psi^s(T)$ is submultiplicative for each s, i.e.

$$\phi^s(TT') \le \phi^s(T)\phi^s(T'). \tag{5.5}$$

Write

$$\Sigma_q^s = \sum_{1 \le i_1,\dots,i_q \le k} \phi^s(T_{i_1}T_{i_2}\dots T_{i_q}).$$

Then, if $F = \bigcup_{i=1}^k S_i(F)$ is the invariant set of the affinities $S_i = T_i + a_i$, we have, for each δ,

$$H_\delta^s(F) \le c\Sigma_q^s \tag{5.6}$$

if q is large enough, where c is a constant. This follows because $S_{i_1}(S_{i_2}(\dots(S_{i_q}(E))))$ may be covered by sets U_i with $\sum |U_i|^s \le c\phi^s(T_{i_1}T_{i_2}\dots T_{i_q})$ by dividing it up into a large number of pieces, depending on the value of s, as indicated above.

Since Σ_q^s is a submultiplicative sequence of numbers (i.e. $\Sigma_{q_1+q_2}^s \le \Sigma_{q_1}^s\Sigma_{q_2}^s$), we have that $(\Sigma_q^s)^{1/q}$ converges to a function that is strictly decreasing in s. We define $d(T_1,\dots,T_k)$ to be the unique s (assumed to exist) such that

$$\lim_{q\to\infty}(\Sigma_q^s)^{1/q} = 1, \tag{5.7}$$

equivalently

$$d(T_1,\dots,T_k) = \inf\{s\colon \sum_{q=1}^{\infty}\sum_{i_1,\dots,i_q} \phi^s(T_{i_1}T_{i_2}\dots T_{i_k}) < \infty\}.$$

Theorem *Let F be the invariant set satisfying*

$$F = \bigcup_{i=1}^{k}(T_i(F) + a_i).$$

Then $\dim_H F \le d(T_1,\dots,T_k)$. Moreover, for fixed T_i,

$$\dim_H F = \dim_B F = d(T_1,\dots,T_k)$$

for almost all $(a_1,\dots,a_k) \in \mathbf{R}^{nk}$ (i.e., the exceptional points have zero nk-dimensional volume).

Note on proof. The upper bound follows from (5.6) and (5.7) indicated above. The "almost sure" lower bound is much harder (see Falconer (1988)) and is obtained using the potential theoretic method at the end of Lecture 2. □

It is worth mentioning some special cases of this result. If $S_1, \ldots, S_k$ are similarities with ratios $c_1, \ldots, c_k$, the singular values of S_i are $\alpha_1 = \alpha_2 = \ldots = \alpha_n = c_i$, giving

$$\Sigma_r^s = \sum_{1 \le i_1, \ldots, i_r \le k} c_{i_1}^s \ldots c_{i_k}^s = (\sum_{i=1}^{k} c_i^s)^r,$$

Thus, in general, $\dim_H F = \dim_B F$ is given by the solution of (5.1). Note that this holds for the similarities $S_i = T_i + a_i$ for almost all $(a_1, \ldots, a_k)$, even without the open set condition.

If $T_1 = T_2 = \ldots = T_k = T$ and the eigenvectors of T are mutually perpendicular, then

$$\phi^s(T_{i_1} \ldots T_{i_r}) = \phi^s(T^r) = [\phi^s(T)]^r,$$

so that the invariant set of the transformations $S_i = T + a_i$ has box and Hausdorff dimensions given by $k\phi^s(T) = 1$ for almost all $(a_1, \ldots, a_k)$. This does not contradict (5.2) and (5.3) - the $(a_1, \ldots, a_k)$ corresponding to the highly regular arrangement of rectangles in that case are, in general, "exceptional" as far as this result is concerned.

Lecture 6
Lipschitz Equivalence of Julia Sets

We have seen that if two sets are equivalent under a bi-Lipschitz mapping, then they have equal Hausdorff dimensions. In this lecture we look at some situations where the converse of this is true. Our principal examples are Julia sets, though the ideas apply to much larger classes of sets.

Julia sets of Conformal Transformations

We give a very brief, intuitive introduction to Julia sets, which are covered much more fully in other lectures, see Blanchard (1984, 1990). We consider iterates of a point z in the complex plane under an analytic function $f\colon \mathbf{C} \to \mathbf{C}$. We write f^k for the k-th iterate of f. In the case $f(z) = z^2$, it is easy to see that, if $|z| < 1$, then $f^k(z) \to 0$ (the fixed point of f), and if $|z| > 1$ then $f^k(z) \to \infty$ as $k \to \infty$. Thus the circle $|z| = 1$, which is mapped onto itself by f, is a boundary between these two very different types of behaviour. If c is now a small complex number, we again have that $f^k(z) \to \infty$ if z is reasonably large, and that $f^k(z)$ converges to a fixed point of f near 0 if z is small. Thus, it is not unreasonable to expect that there is a closed curve J that is the boundary between the z with xthese two types of behaviour. This is indeed the case if c is small enough, but for $c \neq 0$, J is a fractal curve (Figure 12).

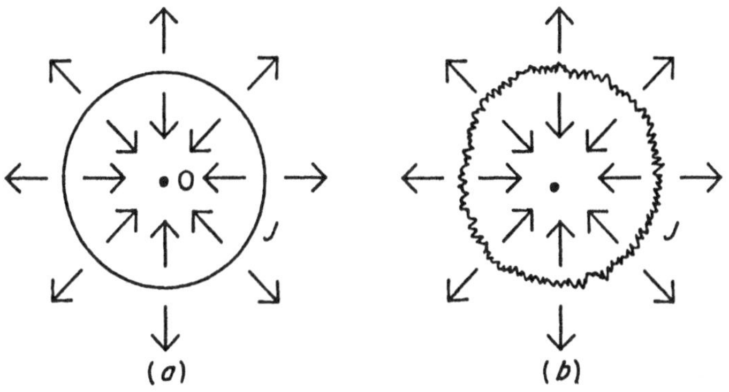

Figure 12:
(a) The Julia set of $f(z) = z^2$ is the circle $|z| = 1$, with $f^k(z) \to 0$ if z is inside J, and $|f^k(z)| \to \infty$ if z is outside J. (b) If f is perturbed to the function $f(z) = z^2 + c$ for small c this picture distorts slightly with a curve J separating those points z for which $f^k(z)$ converges to the fixed point p of f near 0 from those points z with $|f^k(z)| \to \infty$. The curve J is now a fractal.

This boundary J is the Julia set of f. More generally, we define the *Julia set* of f to be the closure of the repelling periodic points of f. (An equivalent definition, involving normal families of functions, is often given.)

Some examples of Julia sets are shown in Figure 13. The form of the Julia set J of $f(z) = z^2 + c$ depends on the position of the complex number c relative to the Mandelbrot set M, Figure 14. (The *Mandelbrot set* is defined to be the set of c such that the Julia set of $f(z) = z^2 + c$ is connected; equivalently, it is the set of c such that $f^k(0)$ is a bounded sequence of numbers.) It may be shown that if c lies in the main cardioid of M, then J is homeomorphic to a circle, and if c is outside M, then J is totally disconnected. If c lies in any of the circular buds of M, then J is topologically rather more complicated, containing infinitely many loops.

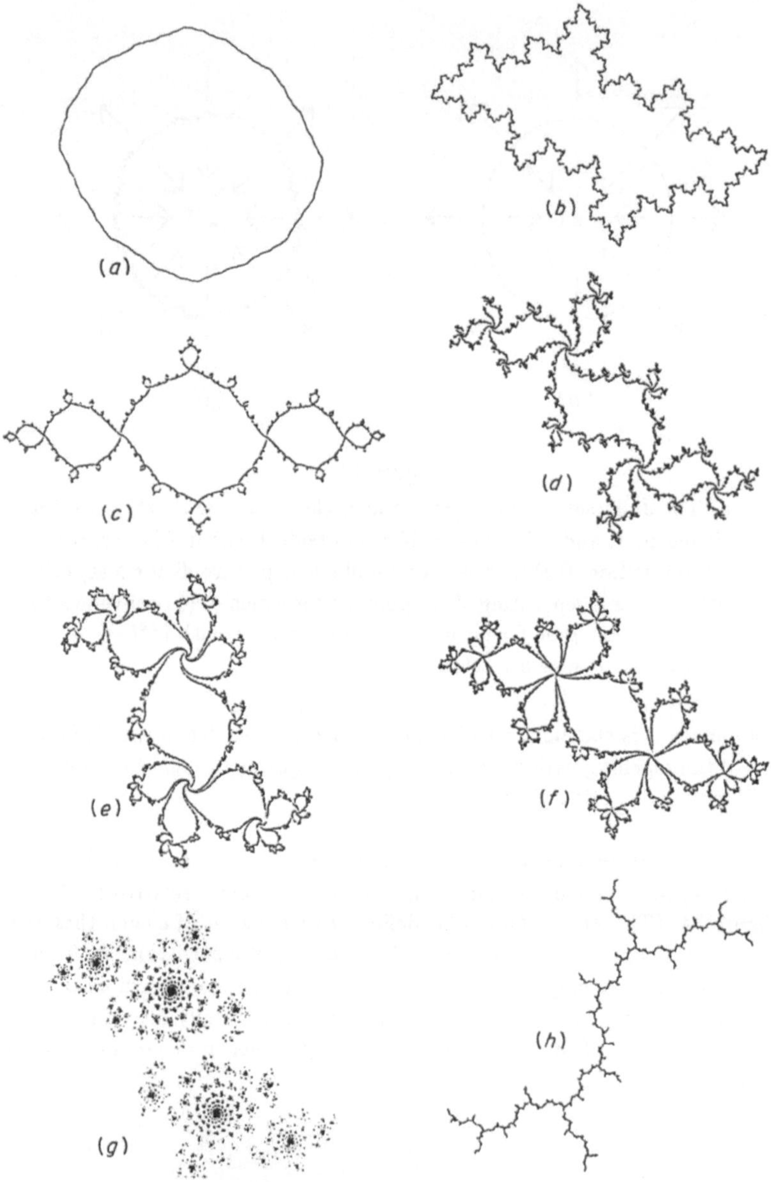

Figure 13:
A selection of Julia sets of the quadratic function $f_c(z) = z^2 + c$. (a) $c = -0.1 + 0.1i$, J a quasi-circle; (b) $c = -0.5 + 0.5i$, J a quasi-circle; (c) $c = -1 + 0.05i$; (d) $c = -0.2 + 0.75i$; (e) $c = 0.25 + 0.52i$; (f) $c = -0.5 + 0.55i$; (g) $c = 0.66i$, J is totally disconnected; (h) $c = -i$.

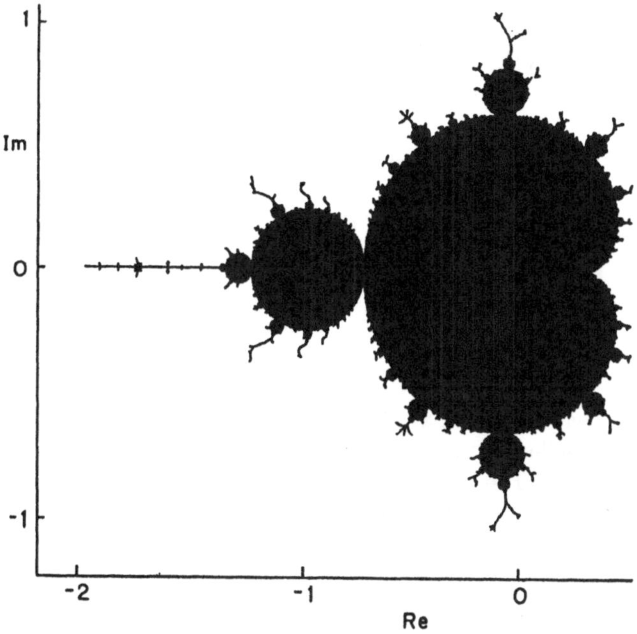

Figure 14: The Mandelbrot set in the complex plane.

Dimensions of Julia Sets

Ruelle (1982) proved that the Hausdorff dimension of the Julia set of z^2+c is a real analytic function of c, and that

$$\dim_H J = 1 + \frac{|c|^2}{4\log 2} + O(c^3)$$

for small c. On the other hand, for large c,

$$\dim_H J \simeq 2\log 2/\log|c|.$$

In all cases $\dim_H J = \underline{\dim}_B J = \overline{\dim}_B J$.

Quasi-Self-Similarity of Julia Sets

Whilst Julia sets are not strictly self-similar in the sense that the middle third Cantor set and von Koch curve are, they are "quasi-self-similar" in that arbitrairly small parts may be enlarged and distorted by a bounded amount to coincide with large parts of the whole set.

Let J be the Julia set of any "reasonable" analytic function $f\colon\ \mathbf{C} \to \mathbf{C}$. Then J is *quasi-self-similar* in the sense that there exist positive constants a, b and λ_0 such that for any neighbourhood $N \subset J$ with $|N| \leq \lambda_0$ there exists a mapping $\phi\colon\ N \to J$ such that

$$a \leq |N|\frac{|\phi(x)-\phi(y)|}{|x-y|} \leq b \qquad (x, y \in N,\ x \neq y) \tag{6.1}$$

i.e. small neighbourhoods of J are "roughly similar" to large parts of J. It may be shown that this holds on taking $\phi = f^k$ for a suitable value of k.

In a similar way, it may be shown that there are constants c and r_0 such that, for any disc or ball B of radius $r \leq r_0$, there is a mapping $\psi\colon\ J \to B \cap J$ such that

$$rc \leq \frac{|\psi(x)-\psi(y)|}{|x-y|} \qquad (x, y \in J,\ x \neq y) \tag{6.2}$$

It turns out that conditions such as (6.1) and (6.2) have implications for the dimensions of J. The proof of the following result is of interest in that it does not involve calculating an expression for $\dim_H J$. The part of the proof given is due to McLaughlin (1987), the rest of the result may be found in Falconer (1989).

Proposition *Let J be any subset of $\mathbf{C}$ or $\mathbf{R}^n$ satisfying* (6.1) *and* (6.2). *Then if* $\dim_H J = s$, *we have* $0 < H^s(J) < \infty$ *and* $\dim_H J = \underline{\dim}_B J = \overline{\dim}_B J$.

Partial Proof. We use the left hand inequality of (6.1) to show that $H^s(J) \geq a^s > 0$. Suppose, to the contrary, that $H^s(J) < a^s$, and take $\delta < \frac{1}{2}a$ to be sufficiently small. Then we may find a cover $\bigcup U_i$ of J with $|U_i| < \delta$ and $\sum_i |U_i|^s < a^s$; thus we may choose $t < s$ such that

$$\sum_i |U_i|^t < a^t. \tag{6.3}$$

Using (6.1) there exist $\phi_i\colon\ U_i \cap J \to J$ with

$$a|x-y| \leq |U_i||\phi_i(x)-\phi_i(y)| \qquad (x, y \in U_i \cap J) \tag{6.4}$$

for each i. Then

$$J \subset \bigcup_i \bigcup_j \phi_i^{-1}(U_j \cap J). \tag{6.5}$$

Moreover, from (6.4)

$$|\phi_i^{-1}(U_j \cap J)| \leq a^{-1}|U_i||U_j| < \tfrac{1}{2}\delta \tag{6.6}$$

so that

$$\begin{aligned} \sum_i \sum_j |\phi_i^{-1}(U_j \cap J)|^t &\leq \sum_i \sum_j a^{-t}|U_i|^t|U_j|^t \\ &\leq a^{-t}(\sum_i |U_i|^t)(\sum_j |U_j|^t) \\ &< a^t \end{aligned} \tag{6.7}$$

by (6.3). Thus (6.5) gives a cover of J by sets of diameters at most $\frac{1}{2}\delta$ with t-th powers of diameters summing to less than a^t. Replacing (6.3) by (6.7) and carrying out this process repeatedly, we get covers of J by sets of diameters at most $2^{-k}\delta$ satisfying (6.3), for each k. It follows that $H^t(J) \leq a^t$, giving $\dim_H J \leq t < s$, contradicting that $\dim_H J = s$. We conclude that $H^s(J) \geq a^s$. □

Lipschitz Equivalence of Quasi-Circles

We call a set F a *quasi-self-similar circle* or *quasi-circle* if

(a) F is homeomorphic to a circle;

(b) F satisfies (6.1) and (6.2);

(c) $0 < H^s(F) < \infty$, where $s = \dim_H F$.

(We include (c) for convenience; it may be deduced from (b) using the previous proposition.) Our principal examples of quasi-circles are Julia sets that are homeomorphic to circles. We now show that quasi-circles are Lipschitz equivalent if and only if they have the same Hausdorff dimension.

Theorem *There exists a bi-Lipschitz bijection between quasi-circles E and F if and only if* $\dim_H E = \dim_H F$.

Sketch of proof. (see Falconer and Marsh (1989)). We already know that if $\dim_H E \neq \dim_H F$ then E and F cannot be Lipschitz equivalent.

Suppose that $\dim_H E = \dim_H F = s$. Let $F(x,y)$ denote the part of F between x and y taken in a clockwise sense. Then there are constants $0 < c_1, c_2$ such that

$$c_1 \leq \frac{H^s(F(x,y))}{|x-y|^s} \leq c_2 \tag{6.8}$$

for $x \neq y$, provided that $F(x,y)$ is the "short arc", say $H^s(F(x,y)) \leq H^s(F(y,x))$. This is true by a continuity and compactness argument for $|x-y| \geq \epsilon$. By choosing ϵ small enough, if $|x-y| < \epsilon$ we may use condition (6.1) for quasi-circles to map $F(x,y)$ into F in such a way that the ratio in (6.8) changes by at most a bounded amount, so (6.8) holds for all relevant x, y.

We now use s-dimensional Hausdorff measure to "measure round" the quasi-circles. We choose "base points" $p \in F$ and $q \in E$, and define ϕ: $F \to E$ by requiring that

$$H^s(F(p,x)) = \frac{H^s(F)}{H^s(E)} H^s(E(q, \phi(x))).$$

Then ϕ is a bijection (note that as x completes a circuit of F, $\phi(x)$ completes a circuit of E). Also

$$H^s(F(x,y)) = \frac{H^s(F)}{H^s(E)} H^s(E(\phi(x),\phi(y))) \qquad (x,y \in F).$$

By (6.8) applied to F and E

$$c_1 \le \frac{H^s(F(x,y))}{|x-y|^s} \le c_2; \qquad c_3 \le \frac{H^s(E(\phi(x),\phi(y)))}{|\phi(x)-\phi(y)|^s} \le c_4,$$

so we get

$$c_5 \le \frac{|\phi(x)-\phi(y)|}{|x-y|} \le c_6 \qquad (x \neq y)$$

for $0 < c_5, c_6$, that is ϕ is bi-Lipschitz. □

In particular, it follows that, if E and F are Julia sets that are homeomorphic to circles, then E and F are Lipschitz equivalent if and only if $\dim_H E = \dim_H F$. For example, if c and d are small complex numbers such that the Julia sets of $z^2 + c$ and $z^3 + d$ are quasi-circles and have the same dimension, then they are Lipschitz equivalent. Note, however, that the dynamics of the two functions on the Julia sets cannot be conjugate, since one of the functions is 2 to 1 and the other 3 to 1.

Other Cases

It is natural to ask whether theorems such as the above hold for quasi-self-similar sets or Julia sets of other topological forms. The next simplest case is of totally disconnected sets, for example the Julia sets of $z^2 + c$ with c outside the Mandelbrot set. Curiously, it may be shown that equality of dimensions of such E and F does not guarantee a bi-Lipschitz mapping between the sets. However, sets of equal dimension must be "nearly Lipschitz equivalent" in the sense that, for any $\epsilon > 0$, there is a bijection ϕ: $F \to E$ and constants $0 < c_1, c_2 < \infty$ such that

$$c_1|x-y|^{1+\epsilon} \le |\phi(x)-\phi(y)| \le c_2|x-y|^{1-\epsilon}.$$

Postscript

With these results our discussion of dimensions has come full-circle. In the first lecture we pointed out that dimensions are invariant under Lipschitz transformations; we have ended by presenting a situation in which dimension actually determines sets to within Lipschitz equivalence. On the way, we have shown that dimension tells us much about geometrical properties, such as the projection or intersection of sets, and we have seen something of the problems associated with calculating the dimensions of fractals. Clearly, dimensions are intimately bound up with the study of fractal geometry.

One often hears phrases such as "the beauty of fractals". It is a beauty that is manifested in many different ways, that, itself, is part of the beauty. In my mind, a major part of the beauty of fractals lies in their mathematics. This goes back many years, to Cantor and Hausdorff, through Besicovitch and his students, and more recently, has received tremendous impetus from the ideas of Benoit Mandelbrot. Mathematicians continue to seek the mathematical truth underlying the beauty of fractals. In the words of John Keats

"Beauty is truth, truth beauty - that is all
Ye know on earth, and all you need to know."

References

Blanchard, P. (1984), Complex analytic dynamics on the Riemann sphere, *Bull. Amer. Math. Soc.* **11**, 85 - 141.

Blanchard, P. and Chiu, A., Complex dynamics: an informal discussion, *this volume*.

Falconer, K.J. (1985), *The Geometry of Fractal Sets*, Cambridge University Press, Cambridge.

Falconer, K.J. (1990), *Fractal Geometry - Mathematical Foundations and Applications*, John Wiley, Chichester.

Falconer, K.J. (1989), Dimensions and measures of quasi-self-similar sets, *Proc. Amer. Math. Soc.* **106**, 543 - 554.

Falconer, K.J. and Marsh D.T. (1989), Classification of quasi-circles by Hausdorff dimension, *Nonlinearity* **2**, 489 - 493.

Hutchinson, J.E. (1981), Fractals and self-similarity, *Indiana Univ. Math. J.* **30**, 713 - 747.

Jones, P.W. and Murai, T. (1988), Positive analytic capacity but zero Buffon needle probability, *Pacific J. Math.* **133**, 99 - 114.

Kahane, J.-P. (1986), Sur la dimension des intersections, in: *Aspects of Mathematics and its Applications* (J.A. Barroso, ed.), North-Holland, Amsterdam, 419 - 430.

Marstrand, J.M. (1954), Some fundamental geometrical properties of plane sets of fractional dimensions, *Proc. London Math. Soc. (3)* **4**, 257 - 302.

Mattila, P. (1984), Hausdorff dimension and capacities of intersections of sets in n-space, *Acta Math.* **152**, 77 - 105.

Mattila, P. (1985), On the Hausdorff dimension and capacities of intersections, *Mathematika* **32**, 213 - 217.

Mattila, P. (1986), Smooth maps, null-sets for integral geometric measures and analytic capacity, *Ann. of Math.* **123**, 303 - 309.

McLaughlin, J. (1987), A note on Hausdorff measures of quasi-self-similar sets, *Proc. Amer. Math. Soc.* **100**, 183 - 186.

McMullen, C. (1984), The Hausdorff dimension of general Sierpinski carpets, *Nagoya Math. J.* **96**, 1-9.

Ruelle, D. (1982), Repellers for real analytic maps, *Ergodic Theory Dynamical Systems* **2**, 99 - 108.

Topological aspects of self-similar sets and singular functions

Masayoshi HATA
Department of Mathematics
Faculty of Science
Kyoto University
Kyoto 606
Japan

Abstract

Let $f_1, f_2, ..., f_m$ be a family of contractions defined on some (separable) complete metric space X. Then the self-similar set $K \equiv K(f_1, \ldots, f_m)$ with respect to $\{f_1, \ldots, f_m\}$ is defined as the unique non-empty compact solution of the set-equation:

$$K = f_1(K) \cup f_2(K) \cup \ldots \cup f_m(K).$$

We survey some fundamental results concerning self-similar sets from a general topological point of view rather than a measure-theoretical point of view; in particular, Williams' problem on the structure of self-affine sets defined in Euclidean space.

Secondly some recent topics concerning Weierstrass's and Takagi's nowhere differentiable functions will be discussed. Such a singular function is not only an instructive counter-example, but also provides a good stimulus to Mathematics; indeed, in the theory of Fractals.

1 Introduction

Let X be a (separable) complete metric space with a metric d. A mapping $f\colon\ X \to X$ is said to be a contraction provided that its Lipschitz constant $\mathrm{Lip}(f)$ satisfies

$$\mathrm{Lip}(f) \equiv \sup_{x \neq y \in X} \frac{d(f(x), f(y))}{d(x, y)} < 1.$$

As is well-known, any contraction f has a unique fixed point $\mathrm{Fix}(f)$ in X.

J. Bélair and S. Dubuc (eds.), Fractal Geometry and Analysis, 255–276.

Definition 1.1 Let $f_1, f_2, \ldots, f_m$ be contractions on $X (m \geq 2)$. We say that a non-empty compact subset $K \subset X$ is *self-similar* with respect to $\{f_1, \ldots, f_m\}$ if K satisfies the set-equation:

$$K = f_1(K) \cup f_2(K) \cup \ldots \cup f_m(K). \tag{1.1}$$

The word "self-similar" is used here in a very weak sense. Indeed, Hutchinson's [13] definition of self-similarity requires some strong separation condition on the Hausdorff measure. Here our terminology will be justified merely in the sense that the set K is a finite union of its miniatures $f_j(K)$.

Self-similar sets defined in Euclidean space often give typical examples of Mandelbrot's "fractal sets" having fractional Hausdorff dimension. For example, the well-known Koch curve and the Sierpinski gasket are typical self-similar sets with respect to similitudes. It is a classical problem to ask for measurements of such sets and the most popular mathematical tool to measure them is no doubt Hausdorff measure.

On the other hand, it may be interesting to consider them from a general topological point of view; that is, what is the shape of self-similar sets? For example, under what condition $K(f_1, \ldots, f_m)$ becomes a connected set? Of course, this approach is closely connected with the following "inverse problem": For which non-empty compact subset $E \subset X$ does there exist a finite family of contractions satisfying $E = K(f_1, \ldots, f_m)$? This inverse problem is still open and we will give some partial answers in the following sections. Note that this problem depends on the number of contractions. For example, the $p-1$ dimensional unit sphere S^{p-1} in $\mathbf{R}^p$ cannot be self-similar for any p contractions (this follows from the Lusternik-Schnirelman-Borsuk theorem); however S^{p-1} is certainly self-similar for some $p+1$ contractions.

2 Fundamental results

The first fundamental result concerns the existence and uniqueness of a solution of the set equation (1.1), which was proven by J.E. Hutchinson [13] in 1981. The author [7] extends this result for weak contractions (see the remark below).

Theorem 2.1 *The set-equation* (1.1) *has a unique nom-empty compact solution* $K \equiv K(f_1, \ldots, f_m)$.

On the other hand, in 1971, R.F. Williams [22] studied the following set:

$$K_0 = \text{closure of} \bigcup_{\substack{1 \leq i_1, \ldots, i_n \leq m \\ n \geq 1}} \text{Fix}(f_{i_1} \circ \ldots \circ f_{i_n}) \tag{2.1}$$

toward a study of generic properties of the action of free (non-abelian) groups on manifolds. Since it is easily verified that K_0 satisfies the set-equation (1.1), we have $K_0 = K(f_1, \ldots, f_m)$. So Williams' formula (2.1) represents a unique compact solution of (1.1).

Someone may think that this formula resembles the definition of Julia sets, because the Julia set is defined as the closure of all repulsive periodic points.

To investigate the structure of $K(f_1, \ldots, f_m)$ it is convenient to introduce the one-sided symbol space on m symbols:

$$\Sigma_m = \{1, 2, \ldots, m\}^{\mathbf{N}}.$$

Equipped with the metric $d_\Sigma(\omega, \omega') = \sum_{n\geq 1} 2^{-n}|\omega_n - \omega'_n|$ for $\omega = (\omega_n), \omega' = (\omega'_n)$, Σ_m becomes a totally disconnected compact metric space. Then we have

Theorem 2.2 *There exists a continuous onto mapping ψ: $\Sigma_m \to K(f_1, \ldots, f_m)$ such that the following diagram:*

$$\begin{array}{ccc} \Sigma_m & \xrightarrow{\sigma_j} & \Sigma_m \\ \psi\downarrow & & \downarrow\psi \\ K(f_1, \ldots, f_m) & \xrightarrow[f_j]{} & K(f_1, \ldots, f_m) \end{array}$$

is commutative for $1 \leq j \leq m$; that is, $f_j \circ \psi = \psi \circ \sigma_j$, where σ_j: $\Sigma_m \to \Sigma_m$ is the right-shift operator: $\sigma_j(\omega) = j\omega$.

Remark Theorems 2.1, 2.2 and Williams' formula (2.1) hold even for weak contractions defined as follows: We say that f: $X \to X$ is a weak contraction if there exists a non-negative right-continuous function Ω_f on $[0, \infty)$ such that

$$d(f(x), f(y)) \leq \Omega_f(d(x, y))$$

for any $x, y \in X$ and that $\Omega_f(s) < s$ for $s > 0$. (Ω_f is the modulus of continuity of f.) For example, $f(x) = x/(1 + x)$ defined on $X = [0, \infty)$ is a weak contraction with $\Omega_f(s) = f(s)$ but f is not a contraction. Every weak contraction f has a unique fixed point Fix(f) in X. Let $C(X)$ be the set of all non-empty compact subsets of X; then $C(X)$ becomes a complete metric space with Hausdorff metric d_H. ($C(X)$ is compact if X is compact.) Let $f_1, f_2, \ldots, f_m$ be weak contractions on X and define the mapping F: $C(X) \to C(X)$ by

$$F(A) = f_1(A) \cup f_2(A) \cup \ldots \cup f_m(A)$$

for any $A \in C(X)$. Then the author [8] showed that F becomes a weak contraction on $C(X)$ satisfying

$$\Omega_F(s) \leq \max_j \Omega_{f_j}(s) < s$$

for any $s > 0$. This fact immediately implies the existence and uniqueness of the solution of (1.1) for weak contractions, since Fix(F) in $C(X)$ is clearly the self-similar set with respect to $\{f_1, \ldots, f_m\}$.

Throughout this survey we assume that $\mathrm{Fix}(f_i) \neq \mathrm{Fix}(f_j)$ for some $i \neq j$; otherwise $K(f_1, \ldots, f_m)$ consists of a single point (trivial case).

3 Connectedness

In this section we survey some results concerning "the shape" of self-similar sets. For the proofs of the theorems mentioned below see the author's paper [7]. The first theorem gives a necessary and sufficient condition for the connectedness of self-similar sets.

Theorem 3.1 *The set $K(f_1, \ldots, f_m)$ is connected if and only if, for any $i \neq j \in \{1, 2, \ldots, m\}$, there exists a subset $\{r_1, \ldots, r_k\} \subset \{1, 2, \ldots, m\}$ such that*

$$K_i \cap K_{r_1} \neq \emptyset,\ K_{r_1} \cap K_{r_2} \neq \emptyset,\ \ldots, K_{r_k} \cap K_j \neq \emptyset,$$

where $K_\ell = f_\ell(K(f_1, \ldots, f_m))$ for $1 \leq \ell \leq m$.

Example Let X be a complex plane and consider the following two contractions on X:

$$f_1(z) = \frac{1+i}{2}\bar{z} \quad \text{and} \quad f_2(z) = \frac{i-1}{2}\bar{z} + \frac{3-i}{2}.$$

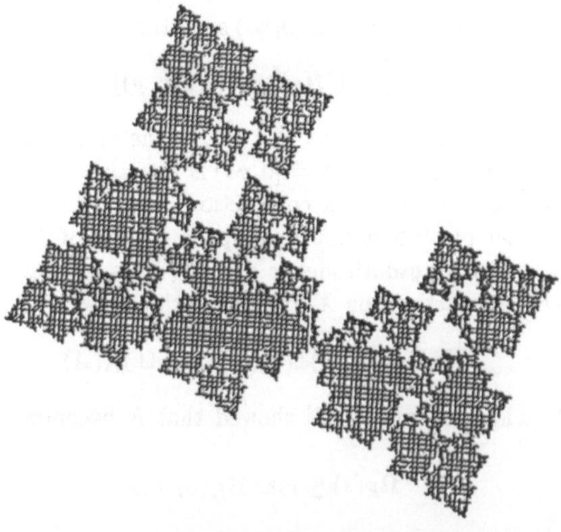

Figure 1.

Then in order to show the connectedness of $K(f_1, f_2)$ illustrated in Figure 1, it is sufficient to check that $K_1 \cap K_2 \neq \emptyset$ from the above theorem. This is easily verified since $0 = \text{Fix}(f_1), 1 = \text{Fix}(f_2) \in K(f_1, f_2)$ (this follows from Williams' formula) and

$$f_1 \circ f_1(\text{Fix}(f_2)) = \frac{1}{2} = f_2 \circ f_2(\text{Fix}(f_1)) \in K_1 \cap K_2.$$

In this case we have $\psi(11222\ldots) = \psi(22111\ldots)$, where $\psi\colon \Sigma_2 \to K(f_1, f_2)$ is the continuous mapping stated in Theorem 2.2. Hence $\{f_1, f_2\}$ defines an equivalence relation $\sim$ on Σ_2 such that $\omega \sim \omega'$ if and only if $\psi(\omega) = \psi(\omega')$. Then $K(f_1, f_2)$ is homeomorphic to the quotient space $\Sigma_2/\sim$.

We say that a subset $Q \subset X$ is locally connected provided that, for any point $p \in Q$ and for any neighborhood U of p, there exists a neighborhood V of p such that $Q \cap V$ lies in a single component of $Q \cap U$ containing p. Then we have

Theorem 3.2 *The set $K(f_1, \ldots, f_m)$ is a locally connected continuum if $K(f_1, \ldots, f_m)$ is connected.*

Hence it follows from the Hahn-Mazurkiewicz theorem that, if $K(f_1, \ldots, f_m)$ is connected, there exists a continuous onto mapping $\alpha\colon [0,1] \to K(f_1, \ldots, f_m)$; in particular $K(f_1, \ldots, f_m)$ is arcwise-connected.

Theorem 3.2 implies immediately the following partial answer for the "inverse problem":

Corollary 3.3 *No non-locally connected continuum can be self-similar for any finite number of weak contractions.*

Using the continuous mapping $\psi\colon \Sigma_m \to K(f_1, \ldots, f_m)$ stated in Theorem 2.2, we can get the following:

Theorem 3.4 *Suppose that all f_j's are injective and that the sets $\{K_j\}_{1 \leq j \leq m}$ are pairwise disjoint. Then the set $K(f_1, \ldots, f_m)$ is totally disconnected and perfect. In particular, the topological dimension* $\dim_T$ *(in the sense of Menger-Urysohn) of $K(f_1, \ldots, f_m)$ is zero and $K(f_1, \ldots, f_m)$ is uncountable.*

The assumption of the above theorem implies that ψ is one to one; therefore ψ becomes a homeomorphism and the conclusion follows immediately. Combining Theorems 3.1 and 3.4 we have the following alternative theorem in the case $m = 2$:

Theorem 3.5 *Suppose that both f_1 and f_2 are injective weak contractions. Then the set $K(f_1, f_2)$ is either a totally disconnected perfect set or a locally connected continuum according as $K_1 \cap K_2$ is empty or not.*

This phenomenon on self-similar sets is similar to that for Julia sets, because the Julia set for a rational function is either a totally disconnected set (Cantor-type set) or a connected set.

Now it is an interesting fact that the shapes of self-similar sets are varied like those of Julia sets.

For example, the self-similar set illustrated in Figure 2(a) is a tree-like set and the curve illustrated in Figure 2(b) studied by P. Lévy [18] in 1939 has a positive 2-dimensional Lebesgue measure, and the set of multiple points is uncountable and dense in itself. Such sets are particular cases of the self-similar sets for the following four-parameter family of contractions on $\mathbf{R}^2$:

$$\begin{cases} f_1(z) = az + b\overline{z} \\ f_2(z) = c(z-1) + d(\overline{z}-1) + 1. \end{cases}$$

In fact, Figures 2(a) and (b) correspond to the following choice of parameters: $(0, \frac{1}{2} + \frac{\sqrt{3}}{6}i, 0, \frac{2}{3})$ and $(\frac{1+i}{2}, 0, \frac{1-i}{2}, 0)$ respectively.

Finally, concerning the shape of connected self-similar sets, we have

Theorem 3.6 *Suppose that every f_j is an injective weak contraction. If the set $K(f_1, \ldots, f_m)$ is connected, then one of the following must hold:*

(a) $K(f_1, \ldots, f_m)$ *is a simple arc (Jordan arc);*

(b) $K(f_1, \ldots, f_m)$ *contains a simple link;*

(c) $K(f_1, \ldots, f_m)$ *has an infinite number of endpoints.*

For example, Figures 2(a) and (b) are typical examples of (c) and (b) of the above theorem respectively. This theorem implies that if $E \subset X$ is a tree-shaped locally connected continuum having a finite number ≥ 3 of endpoints, then E cannot be self-similar for any finite number of injective weak contractions.

4 Williams' problem

In 1971, R.F. Williams [22] studied some fundamental relations between the topological structure of self-similar sets and the Lipschitz constants of contractions. Recall that the

Figure 2 (a).

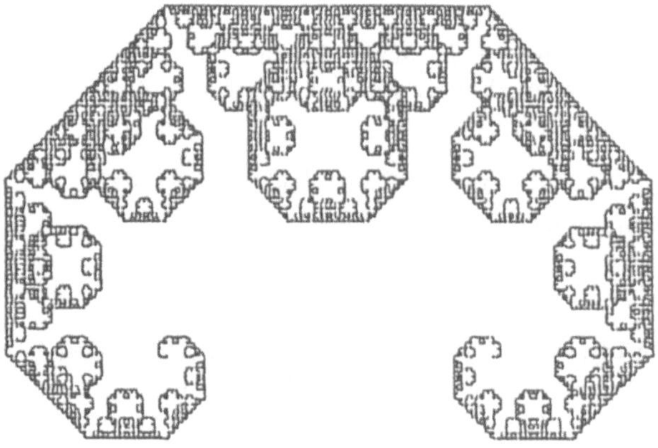

Figure 2 (b).

Hausdorff dimension of particular self-similar sets defined by similitudes in Euclidean space satisfying the so-called "open set condition" is determined merely by its contraction ratios (= Lipschitz constants). Similarly, concerning the topological structure of self-similar sets, some fundamental properties like connectedness are closely related with the Lipschitz constants.

The first result in this direction was obtained by Williams:

Theorem 4.1 *Suppose that*

$$\sum_{j=1}^{m} \mathrm{Lip}(f_j) < 1. \tag{4.1}$$

Then the set $K(f_1, \ldots, f_m)$ *is totally disconnected. In particular, the topological dimension of* $K(f_1, \ldots, f_m)$ *is zero.*

Note that the converse of the above theorem is not true in general. However, in the special case in which $m = 2$ and each f_j is a similitude, the converse is also true.

The above theorem can be easily generalized as follows:

Theorem 4.2 *Let* $\delta \equiv \delta(f_1, \ldots, f_m)$ *be a unique root of the equation:*

$$\sum_{j=1}^{m} (\mathrm{Lip}(f_j))^{\delta} = 1.$$

Then we have $\dim_T K(f_1, \ldots, f_m) \leq \dim_H K(f_1, \ldots, f_m) \leq \delta$.

Proof. Since $\dim_T Y \leq \dim_H Y$ for any Y, it suffices to show the second inequality. For brevity, let W_n^m be the set of all finite words of length $n \in \mathbf{N}$ on m symbols $\{1, 2, \ldots, m\}$. For any $w = (w_1 \ldots w_n) \in W_n^m$ let

$$K_w = f_{w_1} \circ f_{w_2} \circ \ldots \circ f_{w_n}(K(f_1, \ldots, f_m)).$$

Then it easily follows from (1.1) that

$$K(f_1, \ldots, f_m) = \bigcup_{w \in W_n^m} K_w. \tag{4.2}$$

Moreover, denoting by $|E|$ the diameter of E, we have $|K_w| \leq \lambda^n |K(f_1, \ldots, f_m)| \equiv \varepsilon_n$, say, where $\lambda = \max_j \mathrm{Lip}(f_j) < 1$. Since $\{K_w\}$ is a finite ε_n-covering of $K(f_1, \ldots, f_m)$ by (4.2), it follows that

$$\begin{aligned} H_{\varepsilon_n}^{\delta}(K(f_1, \ldots, f_m)) &\leq \sum_{w \in W_n^m} |K_w|^{\delta} \\ &\leq |K(f_1, \ldots, f_m)|^{\delta} \left(\sum_{j=1}^{m} (\mathrm{Lip}(f_j))^{\delta} \right)^n \\ &= |K(f_1, \ldots, f_m)|^{\delta} < \infty; \end{aligned}$$

thus, letting $n \to \infty$, we have $H^\delta(K(f_1, \ldots, f_m)) < \infty$, where H^δ is the Hausdorff δ-dimensional outer measure. This completes the proof. □

Note that this theorem contains Theorem 4.1 since the condition (4.1) implies $\delta(f_1, \ldots, f_m) < 1$; therefore $\dim_T K(f_1, \ldots, f_m) = 0$.

In particular, Theorem 4.2 means the finiteness of the topological dimension of self-similar sets for contractions (*not* weak contractions). Thus we have the following partial answer for the "inverse problem":

Corollary 4.3 *No infinite dimensional compact set is self-similar for a finite number of contractions.*

For example,

$$Q = \{x = (x_1 x_2 \ldots);\ |x_n| \leq n^{-2} \text{ for any } \ n \geq 1\}$$

known as the Hilbert cube, being an infinite dimensional compact subset of the Hilbert space $X = \ell^2$, cannot be self-similar for any finite number of contractions.

To find a condition on the Lipschitz constants of $f_1, \ldots, f_m$ under which the self-similar set $K(f_1, \ldots, f_m)$ becomes a connected set, will be a more interesting problem. In this respect, Williams gave the following result for two injective contractions defined on $X = \mathbf{R}^1$ and raised the question of a higher-dimensional analogue for this:

Theorem 4.4 *Let $X = \mathbf{R}^1$ and f_1, f_2 be two injective contractions defined on X satisfying*

$$(\mathrm{Lip}(f_1^{-1}))^{-1} + (\mathrm{Lip}(f_2^{-1}))^{-1} \geq 1.$$

Then the set $K(f_1, f_2)$ is a closed interval.

Since $\mathrm{Lip}(f) = (\mathrm{Lip}(f^{-1}))^{-1}$ for any similitude f, we have immediately:

Corollary 4.5 *Let $X = \mathbf{R}^1$ and f_1, f_2 be similitudes (contractions). Then the set $K(f_1, f_2)$ is either a totally disconnected set (Cantor set) or a closed line interval according as $\mathrm{Lip}(f_1) + \mathrm{Lip}(f_2) < 1$ or not.*

Compare with Theorem 3.5. The above theorem can now be generalized to higher-dimensional Euclidean space as follows:

Theorem 4.6 *Let $X = \mathbf{R}^p (p \in \mathbf{N})$ and let f_1, f_2 be two injective contractions defined on X satisfying*

$$(\mathrm{Lip}(f_1^{-1}))^{-p} + (\mathrm{Lip}(f_2^{-1}))^{-p} \geq 1.$$

Then the set $K(f_1, f_2)$ is connected.

Proof. Suppose, on the contrary, that $K(f_1, f_2)$ is not connected. Then we have $K_1 \cap K_2 = \emptyset$ by Theorem 3.5. So put $\rho = \mathrm{dist}\,(K_1, K_2) > 0$. Let τ be a positive number satisfying

$$\mathrm{Lip}(f_1) + \mathrm{Lip}(f_2) < \frac{\rho}{\tau}.$$

We also put $U = N_\tau(K(f_1, f_2))$, the τ-neighborhood of $K(f_1, f_2)$; that is,

$$U = \{x \in \mathbf{R}^p; \| x - y \| < \tau \text{ for some } y \in K(f_1, f_2)\}.$$

For brevity, put $U_j = f_j(U)$ for $j = 1, 2$.

We first claim that $U \supset \overline{U}_1 \cup \overline{U}_2$. To see this, for any $x \in \overline{U}_1$, there exists a $y \in \overline{U}$ with $f_1(y) = x$. Then there exists a $z \in K(f_1, f_2)$ such that $\| y - z \| \le \tau$; therefore

$$\| x - f_1(z) \| = \| f_1(y) - f_1(z) \| \le \mathrm{Lip}(f_1) \| y - z \| < \tau.$$

Since $f_1(z) \in K_1 \subset K(f_1, f_2)$, we have $x \in N_\tau(K(f_1, f_2)) = U$; thus $\overline{U}_1$ is contained by U. Similarly $\overline{U}_2 \subset U$, as required.

Secondly we claim that $\overline{U}_1 \cap \overline{U}_2 = \emptyset$. To see this, suppose that $x \in \overline{U}_1 \cap \overline{U}_2$. Then it follows in the same way as above that there exist $y_i \in \overline{U}$ and $z_i \in K(f_1, f_2)$ satisfying $f_i(y_i) = x$ and $\| y_i - z_i \| \le \tau$ for $i = 1, 2$. Then

$$\begin{aligned}
\rho = \mathrm{dist}\,(K_1, K_2) &\le \| f_1(z_1) - f_2(z_2) \| \\
&\le \| x - f_1(z_1) \| + \| x - f_2(z_2) \| \\
&\le \| f_1(y_1) - f_1(z_1) \| + \| f_2(y_2) - f_2(z_2) \| \\
&\le \mathrm{Lip}(f_1) \| y_1 - z_1 \| + \mathrm{Lip}(f_2) \| y_2 - z_2 \| \\
&\le (\mathrm{Lip}(f_1) + \mathrm{Lip}(f_2))\tau < \rho.
\end{aligned}$$

This is a contradiction, as required.

We are now ready to show the theorem. We denote by $\mu(E)$ the usual p-dimensional Lebesgue measure of E. Then, since U is open, we have

$$\mu(U) > \mu(U_1) + \mu(U_2) > 0$$

from the first and the second claim. On the other hand, since the mapping f_i^{-1} maps U_i onto U, we have

$$\mu(U) \le (\mathrm{Lip}(f_i^{-1}))^p \mu(U_i)$$

for $i = 1, 2$. Thus

$$\begin{aligned}
\mu(U) &> \mu(U_1) + \mu(U_2) \\
&\ge \{(\mathrm{Lip}(f_1^{-1}))^{-p} + (\mathrm{Lip}(f_2^{-1}))^{-p}\}\mu(U) \ge \mu(U).
\end{aligned}$$

This contradiction completes the proof. □

Using the above theorem we can obtain a more general result as follows:

Theorem 4.7 *Let $X = \mathbf{R}^p (p \in \mathbf{N})$ and let $f_1, f_2, \ldots, f_m$ be injective contractions on X satisfying*

$$\sum_{j=1}^{m} (\mathrm{Lip}(f_j^{-1}))^{-p} \geq m - 1. \tag{4.3}$$

Then the set $K(f_1, \ldots, f_m)$ is connected.

Proof. Suppose, on the contrary, that $K(f_1, \ldots, f_m)$ is not connected. Then by Theorem 3.1 the symbols $\{1, 2, \ldots, m\}$ can be divided into two groups A, B such that $A \neq \emptyset \neq B$, $A \cup B = \{1, 2, \ldots, m\}$ and $K_i \cap K_j = \emptyset$ for any $i \in A$ and $j \in B$. Renumbering if necessary, we can assume that $A = \{1, 2, \ldots, r\}$ and $B = \{r+1, \ldots, r+s\}$ where $r, s \geq 1$ and $r + s = m$.

For brevity, let $a_j = (\mathrm{Lip}(f_j^{-1}))^{-p}$ for $1 \leq j \leq m$. Suppose now that $a_i + a_j \geq 1$ for some $i \in A$ and $j \in B$. Then the set $K(f_i, f_j)$ is connected by Theorem 4.6. Since the set $K(f_1, \ldots, f_m)$ contains $K(f_i, f_j)$ from Williams' formula (2.1), we have

$$\begin{aligned} K_i \cap K_j &= f_i(K(f_1, \ldots, f_m)) \cap f_j(K(f_1, \ldots, f_m)) \\ &\supset f_i(K(f_i, f_j)) \cap f_j(K(f_i, f_j)). \end{aligned}$$

However the right-hand side is not empty from Theorem 3.5, a contradiction. Thus we have $a_i + a_j < 1$ for any $i \in A$ and $j \in B$. Adding up these rs inequalities, we obtain

$$\begin{aligned} rs &> s\sum_{i=1}^{r} a_i + r\sum_{j=1}^{s} a_{r+j} \\ &\geq \min(r, s)\sum_{i=1}^{m} a_i \\ &\geq \min(r, s) \cdot (m-1) \geq rs. \end{aligned}$$

This contradiction completes the proof. □

Remark The constant $m - 1$ on the right-hand side of (4.3) is best possible in general.

Using Hölder's inequality we have immediately:

Corollary 4.8 *The set $K(f_1, \ldots, f_m)$ is connected if*

$$\frac{1}{m}\sum_{j=1}^{m} (\mathrm{Lip}(f_j^{-1}))^{-1} \geq \left(1 - \frac{1}{m}\right)^{1/p}.$$

In an infinite dimensional complete metric space like Hilbert space, one will not be able to control the connectedness of self-similar sets by Lipschitz constants of the inverse mappings f_j^{-1}.

5 De Rham's functional equations

In 1957, G. de Rham [3] studied some functional equations, which are described by using two contractions on Euclidean space, and discussed the differentiability of the solutions, which contain the parametrizations of the well-known Koch curve, the Lévy curve, etc. We begin with the following theorem of de Rham:

Theorem 5.1 *Let f_1, f_2 be two contractions defined on $X = \mathbf{R}^p (p \in \mathbf{N})$. Then the following functional equation:*

$$M(t) = \begin{cases} f_1(M(2t)) & \text{for } t \in [0, \frac{1}{2}], \\ f_2(M(2t-1)) & \text{for } t \in [\frac{1}{2}, 1], \end{cases}$$

has a unique continuous solution $M(t)$ if and only if

$$f_1(\mathrm{Fix}\,(f_2)) = f_2(\mathrm{Fix}\,(f_1)).$$

Note that the proof of his theorem holds even for two weak contractions in a general complete metric space. The continuous solution $M(t)$ stated in the above theorem gives a parametrization of the self-similar set $K(f_1, f_2)$, since the compact set $E = M([0,1])$ satisfies

$$E = M([0, \tfrac{1}{2}]) \cup M([\tfrac{1}{2}, 1]) = f_1(E) \cup f_2(E);$$

therefore $E = M([0,1]) = K(f_1, f_2)$ by the uniqueness theorem 2.1.

De Rham's theorem is easily generalized to m contractions $f_1, \ldots, f_m$ as follows:

Theorem 5.2 *Let $f_1, \ldots, f_m$ be weak contractions defined on a complete metric space X. Then the equation*

$$M(t) = \begin{cases} f_1(M(mt)) & \text{for } t \in [0, 1/m], \\ f_2(M(mt-1)) & \text{for } t \in [1/m, 2/m], \\ \quad\vdots & \qquad\vdots \\ f_m(M(mt-m+1)) & \text{for } t \in [1-\frac{1}{m}, 1], \end{cases} \tag{5.1}$$

has a unique continuous solution $M(t)$ if and only if

$$f_2(\mathrm{Fix}\,(f_1)) = f_1(\mathrm{Fix}\,(f_m)), \ldots, f_m(\mathrm{Fix}\,(f_1)) = f_{m-1}(\mathrm{Fix}\,(f_m)). \tag{5.2}$$

Of course, for the same reason as above, the continuous mapping $M\colon [0,1] \to K(f_1, \ldots, f_m)$ is onto. For a more general extension of de Rham's theorem, see the author's paper [7].

In this section we survey some results concerning the regularity of the continuous solution $M(t)$ of (5.1); in particular, we will generalize P. Lax's result [17] concerning Pólya's triangle-filling curves. In the following three theorems we assume that X is a closed subset of a Banach space.

Theorem 5.3 *Let $f_1, \ldots, f_m$ be contractions satisfying* (5.2) *defined on X. Suppose that*

$$\prod_{j=1}^{m} \mathrm{Lip}(f_j) < m^{-m}.$$

Then the Fréchet derivative of $M(t)$ is zero for almost every t; that is,

$$\lim_{s \to t} \frac{\| M(s) - M(t) \|}{s - t} = 0.$$

For example, in the one-dimensional case, we consider two contractions: $f_1(x) = ax$ and $f_2(x) = (1-a)x + a$ for $a \in (0,1)$. Then it is easily seen that $\{f_1, f_2\}$ satisfies the conditions of the above theorem except for $a = 1/2$; so $M(t)$ becomes a *purely singular* function in the sense of Lebesgue. Although Cantor's function is known as such a singular function, this function $M(t)$ is strictly monotone increasing.

The proof of the above theorem is essentially due to P. Lax. It depends on the fact that almost every number is *normal* in the scale of m.

Concerning the non-differentiability of $M(t)$ we have

Theorem 5.4 *Let $f_1, \ldots, f_m$ be injective contractions satisfying* (5.2) *defined on X. Suppose that*

$$\prod_{j=1}^{m} \mathrm{Lip}(f_j^{-1}) < m^{m}.$$

Then the continuous solution $M(t)$ of (5.1) *is not Fréchet differentiable at almost every t; more precisely,*

$$\limsup_{s \to t} \frac{\| M(s) - M(t) \|}{|s - t|} = \infty.$$

Theorem 5.5 *Let $f_1, \ldots, f_m$ be the same contractions as above. Suppose that*

$$\mathrm{Lip}(f_j^{-1}) < m$$

for $1 \leq j \leq m$. Then $M(t)$ is nowhere differentiable.

Example Let X be the complex plane and consider two contractions: $f_1(z) = \alpha\bar{z}$ and $f_2 = (1-\alpha)\bar{z} + \alpha$, where α is a complex parameter satisfying $\max\{|\alpha|, |1-\alpha|\} < 1$. (This reduces to the one-dimensional example discussed above if we restrict ourselves to the case $\alpha \in (0,1)$.) Clearly $\{f_1, f_2\}$ satisfies the conditions (5.2); so let $M_\alpha(t)$ be a unique continuous solution of (5.1). Then de Rham pointed out that $M_\alpha(t)$ at $\alpha = \frac{1}{2} + \frac{\sqrt{3}}{6}i$ gives the well-known Koch curve and $M_\alpha(t)$ at $\alpha = \frac{1}{2} + \frac{1}{2}e^{i\theta}$ gives Pólya's triangle-filling curve. It is easily verified that $M_\alpha(t)$ is a simple arc (Jordan curve) if $|\alpha - \frac{1}{2}| < \frac{1}{2}$ and $M_\alpha(t)$ becomes a Pénao curve (triangle-filling curve) if $|\alpha - \frac{1}{2}| \geq \frac{1}{2}$. In 1973, Lax [17] studied the regularity of Pólya's triangle-filling curve ($|\alpha - \frac{1}{2}| = \frac{1}{2}$). His results are now generalized by Theorems 5.3, 5.4 and 5.5. Indeed $M_\alpha(t)$ has zero derivative or is not differentiable according as $|\alpha(1-\alpha)| < \frac{1}{4}$ or $> \frac{1}{4}$. Moreover, $M_\alpha(t)$ is nowhere differentiable if $\min\{|\alpha|, |1-\alpha|\} > \frac{1}{2}$.

Example Let $X = [0,1]$ and define $f_1(x) = ax, f_2(x) = (1-a-b)x + a$ and $f_3(x) = b(x-1)+1$ for $a > 0, b > 0, a + b < 2$. Then the contractions f_1, f_2, f_3 satisfy the conditions (5.2). So let $M_{a,b}(t)$ be a unique continuous solution of (5.1). Then, applying Theorem 5.5, the function $M_{a,b}(t)$ is nowhere differentiable if $a > 1/3, b > 1/3$ and $a + b > 4/3$.

In the following two sections we deal with the famous nowhere differentiable functions known as Weierstrass and Takagi functions.

6 Weierstrass's non-differentiable function

The question of the existence of a continuous nowhere differentiable function was settled affirmatively by K. Weierstrass in 1872. Indeed he showed that the function

$$W_{a,b}(x) = \sum_{n=1}^{\infty} a^n \cos(b^n \pi x),$$

$0 < a < 1$, does not possess a finite or infinite derivative at any point if b is an odd integer with $ab > 1 + \frac{3\pi}{2}$. Later, in 1916, G.H. Hardy [5] succeeded in proving that $W_{a,b}(x)$ does not possess a finite derivative at any point if $ab \geq 1$ (b not necessarily an integer).

We now know that almost all continuous functions are nowhere differentiable in the sense of Baire category. However, such functions are not only instructive counter-examples, but also provide us with a good stimulus.

One of the most fundamental problems concerning Weierstrass's function $W_{a,b}(x)$, is to determine the Hausdorff dimension of the graph of $W_{a,b}(x)$. In this respect we have the following

Conjecture

$$\dim_H(\textit{Graph of } W_{a,b}) = 2 + \frac{\log a}{\log b}.$$

This conjecture is quite reasonable since Hardy showed that the local Lipschitz order of $W_{a,b}(x)$ is exactly equal to $\xi = \log(1/a)/\log b$ at every point x if $ab > 1$. Thus we have at least

$$\dim_H(\textit{Graph of } W_{a,b}) \leq 2 + \frac{\log a}{\log b}$$

from the theorem of Besicovitch-Ursell [2].

Secondly, as is discussed by Mandelbrot, the following slightly modified function

$$D(x) = \sum_{n=-\infty}^{\infty} a^n(1 - \cos(b^n \pi x)),$$

which can be expressed as the sum of Weierstrass's function $W_{a,b}(x)$ and some continuously differentiable function, satisfies the simple functional equation: $D(x) = aD(bx)$. This scaling property also suggests our conjecture.

Remark Historically this argument was first discussed by Hardy and Littlewood [6] in 1916. Indeed the scaling property $D(x) = aD(bx)$ reduces to merely the periodic property:

$$d(x) = d(x + \log b)$$

under the transformation $d(x) = e^{-\xi x} D(e^x)$. Thus $d(x)$ can be expressed as a Fourier series and they obtained the following "Hardy-Littlewood identity":

$$\sum_{n=0}^{\infty} a^n e^{-b^n z} + \sum_{n=0}^{\infty} \frac{(-z)^n}{n!(b^{n-\xi} - 1)} = \frac{1}{\log b} \sum_{n=-\infty}^{\infty} \Gamma\left(-\xi + \frac{2n\pi i}{\log b}\right) z^{\xi - 2n\pi i/\log b}.$$

Recently Mauldin and Williams [19] obtained a non-trivial lower estimate of the Hausdorff dimension of the graph of $W_{a,b}(x)$ as follows:

Theorem 6.1 *There exists a positive constant C such that*

$$\dim_H(\textit{Graph of } W_{a,b}) \geq 2 + \frac{\log a - C}{\log b}$$

for sufficiently large b.

On the other hand, the box-counting dimension $\dim_B$ of the graph of $W_{a,b}(x)$ was determined by Kaplan, Mallet-Paret and Yorke [14] as follows:

Theorem 6.2

$$\dim_B(\textit{Graph of } W_{a,b}) = 2 + \frac{\log a}{\log b}.$$

The essential part of their proof is to show the existence of positive constants C_1, C_2 such that

$$C_1|J|^{\xi} \leq \max_{x\in J} W_{a,b}(x) - \min_{x\in J} W_{a,b}(x) \leq C_2|J|^{\xi}$$

for every sufficiently small interval $J(|J|$ denotes the length of $J)$.

We say that x is a knot point of a real-valued continuous function $u(x)$ provided that its four Dini derivatives satisfy

$$D^+u(x) = D^-u(x) = -D_+u(x) = -D_-u(x) = \infty.$$

Intuitively the graph of $u(x)$ is highly oscillating in a neighborhood of the knot point. The author [9] obtained the following

Theorem 6.3 *Let $ab > 1$. Then every point of* $\mathbf{R}$ *except for a null set* e *is a knot point of*

$$W_{a,b,\theta}(x) = \sum_{n=0}^{\infty} a^n \cos(b^n \pi x + \theta),$$

where the set e *depends only on b.*

Concerning the function $W_{a,b,\theta}(x)$, let κ be the least number such that the condition $ab > \kappa$ implies that $W_{a,b,\theta}(x)$ does not possess a finite or infinite derivative at any point x for any θ (b is not necessarily an integer.) Then the author [10] succeeded in determining the exact value of κ as follows:

Theorem 6.4 *We have*

$$\kappa = 1 + \frac{1}{\cos\theta^*},$$

where θ^* is the unique root of the equation $\tan\theta = \pi + \theta$ *in* $(0, \pi/2)$.

On the other hand, A.S. Besicovitch [1] gave the first example of a continuous function $B(x)$ which has nowhere a finite or infinite unilateral derivative by a geometrical process. The graph of $B(x)$ is illustrated in Figure 3. The difficulty of finding such functions may be explained by the fact that the set of such functions is of only the first category in the space of continuous functions (S. Saks [20]). In this respect the author [11] obtained the following

Theorem 6.5 *For any $\alpha \in [0, 1)$ and $\varepsilon \in (0, 1)$, there exists a continuous function $B_{\alpha,\varepsilon}(x)$ defined on $[0, 1]$ satisfying the following three conditions:*

(a) *$B_{\alpha,\varepsilon}(x)$ has nowhere a unilateral derivative finite or infinite;*

(b) *the Lebesgue measure of the set of all knot points of* $B_{\alpha,\varepsilon}(x)$ *is equal to* α;

(c) $B_{\alpha,\varepsilon}(x)$ *satisfies Hölder's condition of order* $1-\varepsilon$.

The function $B_{\alpha,\varepsilon}(x)$ of the above theorem is constructed as the unique continuous solution of generalized de Rham type functional equations.

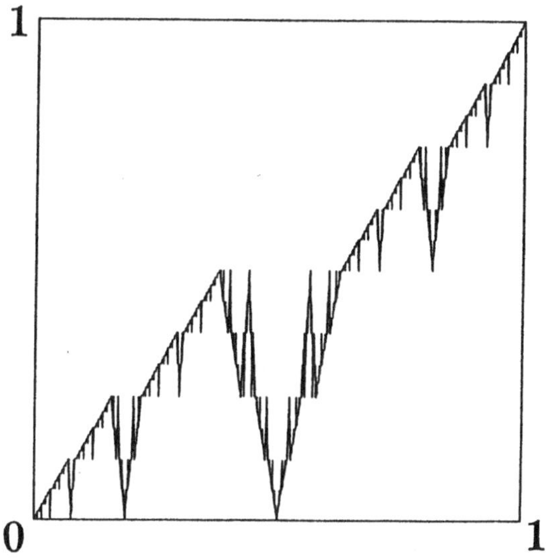

Figure 3.

7 The Takagi function

In 1903, T. Takagi [21] gave an example of a nowhere differentiable continuous function, which is simpler than Weierstrass's function discussed in the previous section. This function $T(x)$, which we call "Takagi function", is defined as follows:

$$T(x)=\sum_{n=1}^{\infty}\frac{1}{2^n}\psi(2^{n-1}x),$$

where $\psi(x)=|2x-2[x+\frac{1}{2}]|$ is a periodic piecewise-linear function with period 1 (so-called "saw function" or "zig-zag function"). The graph of $T(x)$ is illustrated in Figure 4. One can easily find infinitely many nested structures on the graph.

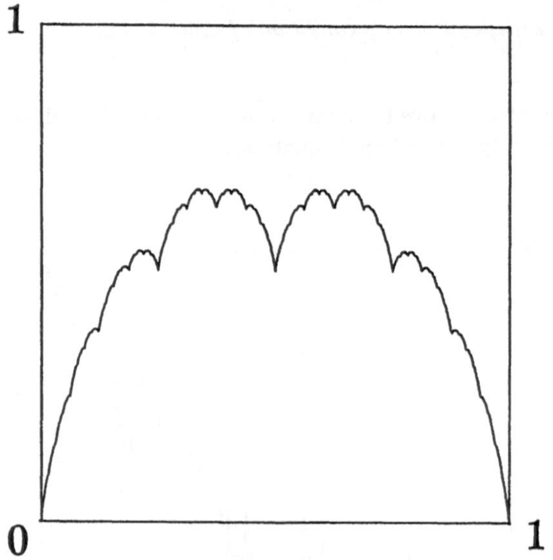

Figure 4.

Historically, nowhere differentiable functions of this kind were rediscovered by many mathematicians; for example, by Knopp (1918), Hobson (1926), van der Waerden (1930), Hildebrandt (1933), and by de Rham (1957), etc.

Remark The Takagi function $T(x)$ can be defined as the generating function of a chaotic dynamical system $\psi|_{[0,1]}$ as follows:

$$T(x) = \sum_{n=1}^{\infty} \frac{1}{2^n}\psi^n(x),$$

where $\psi^n(x) = \psi(\psi^{n-1}(x))$ denotes n-fold iteration of ψ. Similarly, if we replace $\psi|_{[0,1]}$ by another chaotic mapping $4x(1-x)$, then the corresponding generating function becomes essentially Weierstrass's function $W_{1/2,2}(x)$. So one may think that the non-differentiability of such functions follows from the sensitive dependence of initial values for chaotic dynamical systems. For details, see Yamaguti and Hata [23].

Now it is easily seen that

$$T(x+h) - T(x) = O\left(|h|\log\frac{1}{|h|}\right)$$

as $h \to 0$ for every point x. This implies that the Hausdorff dimension of the graph of $T(x)$ is 1. However Mauldin and Williams [19] showed that the graph of $T(X)$ has σ-finite measure with respect to

$$h(s) = \frac{s}{\log(1/s)}$$

and asked whether it has σ-finite linear measure or not.

As a generalization of the Takagi function, we consider the following series:

$$g(x) = \sum_{n=1}^{\infty} a_n \psi^n(x), \tag{7.1}$$

where $\{a_n\}$ is a given real sequence. In this respect, Hata and Yamaguti [12] obtained the following

Theorem 7.1 *The series* (7.1) *converges at every point* $x \in [0,1]$ *if and only if* $\{a_n\} \in \ell^1$.

Next, concerning the non-differentiability of $g(x)$, G. Faber [4] and N. Kono [16] obtained the following

Theorem 7.2 *Suppose* $\{a_n\} \in \ell^1$. *Then* $g(x)$ *is nowhere differentiable if and only if*

$$\limsup_{n\to\infty} 2^n |a_n| > 0.$$

Of course, the above theorem contains the non-differentiability of the Takagi function $T(x)$ since $a_n = 2^{-n}$.

On the other hand, the series (7.1) contains a smooth function. For example, if we take $a_n = 4^{-n}$, we have

$$g(x) = \sum_{n=1}^{\infty} \frac{1}{4^n} \psi^n(x) = x(1-x).$$

However this is the only C^2-function among the series (7.1).

Exercise Show that

$$\sum_{n=1}^{\infty} \frac{1}{2^n} (\psi^n(x))^2 = 2x(1-x).$$

Now, it is well-known that every continuous function $u(x)$ on $[0,1]$ can be uniquely expanded as follows:

$$u(x) = u(0)(1-x) + u(1)x + \sum_{j=0}^{\infty} \sum_{i=0}^{2^j-1} c_{i,j}(u) S_{i,j}(x),$$

where

$$\begin{aligned} c_{i,j}(u) &= u\left(\frac{2i+1}{2^{j+1}}\right) - \frac{1}{2}\left\{u\left(\frac{i}{2^j}\right) + u\left(\frac{i+1}{2^j}\right)\right\}, \\ S_{i,j}(x) &= 2^j\left\{\left|x - \frac{i}{2^j}\right| + \left|x - \frac{i+1}{2^j}\right| - \left|2x - \frac{2i+1}{2^j}\right|\right\}. \end{aligned}$$

This is known as "Schauder expansion" . Note that $c_{i,j}(u)$ is a central difference scheme.

Example Suppose $\{a_n\} \in \ell^1$. Then the series (7.1) is the unique continuous solution of the following infinite number of difference equations:

$$c_{i,j}(g) = a_{j+1}$$

for $0 \leq i \leq 2^j - 1, j \geq 0$ with the boundary condition $g(0) = g(1) = 0$. In particular, the Takagi function $T(x)$ is the unique continuous solution of the difference equations

$$c_{i,j}(g) = 2^{-j-1}$$

for $0 \leq i \leq 2^j - 1, j \geq 0$ with $g(0) = g(1) = 0$.

Example For a given continuous function $w(x)$ defined on $[0,1]$ consider the Dirichlet problem

$$\Delta g(x) = w(x)$$

with the boundary condition $g(0) = \alpha$ and $g(1) = \beta$. Then the solution $g(x)$ is given as the unique continuous solution of

$$c_{i,j}(g) = -2^{j+2}\int_0^1 S_{i,j}(s)w(s)ds$$

for $0 \leq i \leq 2^j - 1, j \geq 0$ with the same boundary condition. J. Kigami [15] recently succeeded in generalizing this argument to $(\log 3/\log 2)$-dimensional case. Indeed, he constructed a quite reasonable Laplacian operator on the Sierpinski gasket using difference schemes.

References

[1] A.S. Besicovitch, Diskussion der stetigen Funktionen in Zusammenhang mit der Frage über ihre Differenzierbarkeit, *Bull. Acad. Sci. Russie* (1925).

[2] A.S. Besicovitch and H.D. Ursell, Sets of fractional dimensions V: On dimensional numbers of some continuous curves, *J. London Math. Soc.* **32** (1937), 142-153.

[3] G. de Rham, Sur quelques courbes définies par des équations fonctionnelles, *Rend. Sem. Mat. Torino* **16** (1957), 101-113.

[4] G. Faber, Über stetige Funktionen, *Math. Ann.* **69** (1910), 372-443.

[5] G.H. Hardy, Weierstrass's non-differentiable function, *Trans. Amer. Math. Soc.* **17** (1916), 301-325.

[6] G.H. Hardy and J.E. Littlewood, Some problems of Diophantine approximation: remarkable trigonometrical series, *Proc. Nat. Acad. Sci. U.S.A.* **2** (1916), 583-586.

[7] M. Hata, On the structure of self-similar sets, *Japan J. Appl. Math.* **2** (1985), 381-414.

[8] M. Hata, On some properties of set-dynamical systems, *Proc. Japan Acad.* **61** (1985), Ser. A, 99-102.

[9] M. Hata, Singularities of the Weierstrass type functions, *J. d'analyse Math.* **51** (1988), 62-90.

[10] M. Hata, On Weierstrass's non-differentiable function, *C.R. Acad. Sci. Paris* **307** (1988), 119-123.

[11] M. Hata, On continuous functions with no unilateral derivatives, *Ann. Inst. Fourier*, Grenoble **38** (1988), 43-62.

[12] M. Hata and M. Yamaguti, The Takagi function and its generalization, *Japan J. Appl. Math.* **1** (1984), 183-199.

[13] J.E. Hutchinson, Fractals and self-similarity, *Indiana Univ. Math. J.* **30** (1981), 713-747.

[14] J.L. Kaplan, J. Mallet-Paret and J. A. Yorke, The Lyapunov dimension of a nowhere differentiable attracting torus, *Ergodic Theory and Systems* 4 (1984), 261-281.

[15] J. Kigami, A harmonic calculus on the Sierpinski spaces, *Japan J. Appl. Math.* **6** (1989), 259-290.

[16] N. Kono, On generalized Tagaki functions, *Acta Math. Hung.* **49** (1987), 315-324.

[17] P.D. Lax, The differentiability of Pólya's function, *Adv. in Math.* **10** (1973), 456-464.

[18] P. Lévy, Les courbes planes ou gauches et les surfaces composées de parties semblables au tout, *J. École Poly.* 1939, 227-292.

[19] R.D. Mauldin and S.C. Williams, On the Hausdorff dimension of some graphs, *Trans. Amer. Math. Soc.* **298** (1986), 793-803.

[20] S. Saks, On the functions of Besicovitch in the space of continuous functions, *Fund. Math.* **19** (1932), 211-219.

[21] T. Takagi, A simple example of the continuous function without derivative, *The Collected Papers of Teiji Takagi*, Iwanami Shoten Publ., Tokyo 1973, 5-6.

[22] R.F. Williams, Composition of contractions, *Bol. Soc. Brasil. Mat.* **2** (1971), 55-59.

[23] M. Yamaguti and M. Hata, Weierstrass's function and chaos, *Hokkaido Math. J.* **12** (1983), 333-342.

Produits de poids aléatoires indépendants et applications

Jean-Pierre KAHANE
Université de Paris-Sud
Mathématiques
Bâtiment 425
F-91405 Orsay Cedex
France

Résumé

Le recouvrement par des arcs disposés au hasard, les cascades self-similaires de Benoît Mandelbrot, et le chaos multiplicatif sont des cas particuliers d'une théorie générale des produits de poids aléatoires indépendants. Un tel produit définit un opérateur Q qui transforme les mesures en mesures aléatoires. Les résultats les plus précis sont obtenus pour les recouvrements. La caractérisation, en termes de capacité nulle par rapport à un noyau convenable, des ensembles presque sûrement recouverts, est exposée sous une forme plus compacte et plus claire que précédemment. Elle s'explique d'ailleurs par un théorème de Fitzsimmons, Fristedt et Shepp dont une démonstration rapide est donnée à la fin du cours.

Le cours comprend six leçons: 1) trois problèmes (recouvrement aléatoire, cascades self-similaires, exponentiation des processus gaussiens); 2) martingales, opérateurs Q, mesures Q–régulières et Q–singulières, mesures α–régulières et α–singulières; 3) recouvrement de $\mathcal{R}$ ou $\mathcal{T}$ par des intervalles aléatoires, application à la forêt hyperbolique; 4) processus de naissance et de mort, cascades self-similaires, opérateurs Q_h, analyse multifractale et multimesures; 5) le chaos multiplicatif gaussien (théorie résumée); 6) problèmes ouverts liés aux recouvrements aléatoires, problèmes ouverts liés aux multimesures et multifractales, compléments sur les recouvrements poissonniens de $\mathcal{R}$, recouvrements poissonniens de $\mathcal{R}^+$ et processus de Lévy.

Abstract

Covering by random arcs, self-similar cascades (B. Mandelbrot) and multiplicative chaos are included in a general theory of products of independent random weights. The operator Q defined by such a product transforms measures into random measures. Most precise results are obtained for coverings. The almost-surely covered sets are exactly the sets of vanishing capacity with respect to a certain kernel; this is explained rapidly and, hopefully, in a clearer way than before. However, the best explanation relies on a theorem of Fitzsimmons, Fristed and Shepp and a short proof of this is given at the end of the course.

The course comprehends six lectures: 1) three problems (random coverings, self-similar cascades, exponentiation of Gaussian processes); 2) martingales, operators

J. Bélair and S. Dubuc (eds.), Fractal Geometry and Analysis, 277–324.

Q, Q–regular and Q–singular measures, α–regular and α–singular measures; 3) coverings of $\mathcal{R}$ and $\mathcal{T}$, application to the hyperbolic forest; 4) birth and death processes, self-similar cascades, operators Q_h, multifractals and multimeasures; 5) summary of the theory of Gaussian multiplicative chaos; 5) open problems on coverings and multifractals, more on Poisson coverings on $\mathcal{R}$, Poisson coverings on $\mathcal{R}^+$ and their relation with Lévy processes.

Première leçon

Il s'agit d'une introduction aux leçons à venir. On va considérer trois séries de problèmes, apparemment sans grand rapport entre eux, et voir qu'ils conduisent à des produits de poids aléatoires indépendants.

Premier problème: recouvrement aléatoire

Il s'agit, en gros, des questions suivantes. Dans quelles conditions un objet donné est-il recouvert par des petits objets aléatoires? Quand il n'est pas recouvert, que peut-on dire de l'ensemble non recouvert?

Signalons quelques travaux récents sur ces questions. Svante Janson [J1983, J1986] étudie le nombre (aléatoire) des objets aléatoires, supposés de forme donnée, nécessaire pour recouvrir l'objet donné. Sa méthode de temps d'arrêt s'avère fructueuse pour d'autres usages [K1987(1)]. La théorie du potentiel intervient naturellement dans ces questions. L'explication la plus profonde en est apportée par le travail de Fitzsimmons, Fristedt et Shepp [FFS1985], justifiant une intuition de B. Mandelbrot [M1972(1)]: pour des intervalles poissonniens, l'ensemble non recouvert a la même loi que la fermeture de l'ensemble des valeurs d'un processus de Lévy croissant (appelé aussi subordinateur). Fan Ai hua [F1989(3)] étudie des recouvrements par des intervalles de longueur aléatoire, et découvre le rôle de la capacité dans les processus de naissance et de mort. Russel Lyons [L1990] montre la relation entre ce problème, les promenades au hasard sur les arbres, et la percolation.

L'origine est un problème d'A. Dvoretzky (1956). Sur le cercle $\mathcal{T}$, de longueur 1, on jette au hasard des arcs I_n de longueurs données $\ell_n(\ell_1 \geq \ell_2 \geq \cdots)$. L'ensemble $\limsup I_n$ (constitué des points recouverts une infinité de fois) est de mesure nulle si $\sum_1^\infty \ell_n < \infty$, et presque sûrement de mesure pleine si $\sum_1^\infty \ell_n = \infty$. À quelle condition a-t-on recouvrement du cercle, c'est-à-dire $\limsup I_n = \mathcal{T}$ p.s.?

Outre Dvoretzky, ce problème a été considéré dans les années suivantes par Erdös, Billard et moi. Voici le point tel qu'il apparaît en 1968 dans la première édition de mon livre *Some Random Series of Functions.* Observons que les propriétés de $\limsup I_n$ que nous considérons sont justiciables de la loi du $0-1$: leur probabilité est 0 ou 1. Les propriétés qui suivent sont presque-sûres.

Si $\ell_n = \frac{1+\epsilon}{n}, \epsilon > 0$, le recouvrement a lieu. Au contraire, si $\ell_n = \frac{1-\epsilon}{n}$, le recouvrement n'a pas lieu; les sous-ensembles de $\mathcal{T}$ de dimension $D < 1-\epsilon$ sont recouvert, ceux de dimension $D > 1-\epsilon$ ne sont pas recouverts. Lorsqu'il n'y a pas recouvrement, l'ensemble

$$E = \mathcal{T} \backslash \overline{\lim} I_n$$

a pour dimension $1 - \overline{\lim} \frac{1}{\log n}(\ell_1 + \cdots + \ell_n)$.

Le cas $\ell_n = \frac{1}{n}$ restait en suspens. C'est un cas de recouvrement, comme l'ont montré B. Mandelbrot (1972) et S. Orey (non publié). En 1972, L. Shepp obtint une réponse définitive à la question de Dvoretzky: le recouvrement a lieu si et seulement si

$$\sum_1^\infty \frac{1}{n^2} \exp(\ell_1 + \cdots + \ell_n) = \infty$$

ou, de manière équivalente, si

$$\int_0^1 \exp \sum_1^\infty (\ell_n - t)^+ dt = \infty$$

(a^+ désigne la partie positive de a) [S1972(1)]. La méthode de Shepp et des résultats concernant $\mathcal{T}^n$ sont exposés dans la seconde édition de mon livre [K1985(1)].

Il est intéressant de considérer

$$k(t) = \exp \sum_1^\infty (\ell_n - |t|)^+ \quad (|t| \leq \frac{1}{2})$$

comme un noyau sur $\mathcal{T}$, identifié à l'intervalle $[-\frac{1}{2}, \frac{1}{2}[$. Pour tout borélien A de $\mathcal{T}$ on peut définir la capacité relativement à $k(t)$, $\mathrm{Cap}_k A$ (nous reviendrons sur cette notion). Le problème du recouvrement de A admet, lui aussi, une réponse complète:

$$A \subset \overline{\lim} \quad I_n \quad \text{p. s.} \Leftrightarrow \mathrm{Cap}_k A = 0$$

et, dans le cas $\lambda(A) > 0$ (λ est la mesure de Lebesgue), $\mathrm{Cap}_k A = 0$ doit s'interpréter comme $k \notin L^1(\mathcal{T})$: on retrouve la condition de Shepp, sous la seconde forme.

Si k' est un autre noyau, de la même forme que k (avec des ℓ_n différents), on a

$$\mathrm{Cap}_{k'} E = 0 \Leftrightarrow kk' \notin L^1(\mathcal{T}).$$

Cela précise beaucoup les résultats de 1968, et se trouve dans la note [K1987(1)].

En 1972, B. Mandelbrot a considéré le problème analogue sur la droite $\mathcal{R}$ au lieu du cercle $\mathcal{T}$ [M1972(1)]. L'idée fondamentale est que ce qui remplace un point aléatoire ω sur

$\mathcal{T}$, c'est un processus de Poisson ponctuel sur $\mathcal{R}$, d'intensité λ (la mesure de Lebesgue). Ce qui remplace l'intervalle $I_n = (\omega_n, \omega_n + \ell_n)$, c'est une réunion d'intervalles $(x, x + \ell_n)$, où x parcourt un tel processus de Poisson. Ce qui remplace la réunion des I_n, c'est une réunion d'intervalles $(x, x+y)$ correspondants aux points (x, y) d'un processus de Poisson ponctuel dans $\mathcal{R} \times \mathcal{R}^+$, dont l'intensité est $\lambda \otimes \mu$ avec $\mu = \sum_1^\infty \delta_{\ell_n}$. Rappelons qu'un processus de Poisson ponctuel d'intensité σ, sur un espace mesuré (X, σ), est une partie aléatoire de X telle que le cardinal de son intersection avec toute partie A mesurable de X soit une variable aléatoire de Poisson de paramètre $\sigma(A)$.

Le problème général, dans ce cadre, est le suivant. On donne une mesure positive sur $\mathcal{R}^+, \mu$, et on considère l'ensemble fermé aléatoire

$$E = \mathcal{R} \setminus \cup (x, x + y)$$

la réunion étant prise pour tous les points (x, y) du processus de Poisson ponctuel d'intensité $\lambda \otimes \mu$ sur $\mathcal{R} \times \mathcal{R}^+$. À quelle condition a-t-on $E = \emptyset$ p.s.? Sinon, quelles propriétés a l'ensemble E?

La réponse à la première question a été également apportée par L. Shepp en 1972 [S1972(2)]. On a recouvrement (c'est-à-dire $E = \emptyset$ p.s.) si et seulement si

$$\int_0^\infty \exp\left(\int_s^\infty \mu(y, \infty) dy\right) e^{-s} ds = \infty.$$

Losque $1 = \sum_1^\infty \delta_{\ell_n}$ avec $0 < \ell_n < \ell 1$, on retrouve la condition

$$\int_0^1 \exp \sum_1^\infty (\ell_n - s)^+ ds = \infty.$$

La réponse à la seconde question se trouve dans l'article déjà cité de Fitzsimmons, Fristedt et Shepp. La loi de l'ensemble

$$E^+ = \mathcal{R}^+ \setminus \cup (x, x + y)$$

(la réunion étant maintenant prise pour tous les points (x, y) du processus de Poisson ponctuel d'intensité $\lambda \otimes \mu$ sur $\mathcal{R}^+ \times \mathcal{R}^+$) coïncide avec la loi de la fermeture de l'ensemble des valeurs d'un certain processus de Lévy croissant, partant de 0. On comprend ainsi que la condition que A soit contenu dans la réunion des $(x, x + y)$, c'est-à-dire que A soit disjointe de E^+, s'exprime en termes de polarité de A par rapport au processus de Lévy, donc (si l'on connaît le lien entre probabilité et potentiel) en termes de capacité nulle de A par rapport à un certain noyau. Ceux que le sujet intéresse peuvent, outre l'article de FFS et la partie de la sixième leçon qui y est relative, consulter le très éclairant travail de J. Hawkes sur les processus de Lévy et le potentiel [H1979].

Considérons enfin un problème de recouvrement beaucoup plus simple: on considère tous les sous-intervalles dyadiques de l'intervalle $[0, 1[$, soit $[0, \frac{1}{2}[, [\frac{1}{2}, 1[, [0, \frac{1}{4}[, [\frac{1}{4}, \frac{1}{2}[, \cdots$

on les garde avec probabilité p, on les tue avec probabilité $1-p$. Que peut-on dire de l'ensemble E qui survit? De manière équivalente, on considère un processus de naissance et de mort dans lequel, à chaque génération, chaque individu vivant donne naissance à deux enfants dont la probabilité de survie après la naissance est p. En supposant que chaque individu meurt au moment de la naissance des enfants, on sait que la population disparaît p.s. lorsque $p \leq \frac{1}{2}$, et survit avec probabilité positive lorsque $p > \frac{1}{2}$. Nous verrons que l'ensemble E (qui dépeint les branches infinies de l'arbre généalogique) est disjoint p.s. d'un ensenble donné A si et seulement si la capacité de A par rapport à un noyau convenable est nulle (Fan Ai hua).

Second problème: cascades self-similaires

C'est un modèle présenté par B. Mandelbrot en 1974 pour rendre compte de certains aspects de la turbulence [M1974(1), M1974(2)]. La théorie en a été développée par J. Peyrière et moi [P1974, K1974, KP1976].

On part de l'ensemble $T = \{1, 2, \cdots c\}^{\mathcal{N}}$. La première coordonnée le décompose en c cellules d'ordre 1, les deux premières en c^2 cellules d'ordre 2, et ainsi de suite. On désigne par C une cellule quelconque.

On considère une variable aléatoire $W \geq 0$ telle que $EW = 1$. Les problèmes à venir vont seulement impliquer c et la distribution de W. On considère une suite de v.a. indépendantes, indexées par les cellules C, et distribuées comme W: soit (W_C).

On définit à partir de là une suite de mesures aléatoires: $\mu_0 = \lambda$ est la meusre de probabilité naturelle sur T (qui donne la masse c^{-n} aux cellules d'ordre n), μ_n est de densité constante par rapport à λ sur chaque cellule C d'ordre n, et cette densité est le produit de la densité de μ_{n-1} par W_C:

$$\text{densité } \mu_n \text{ sur } C = W_C \times \text{ densité } \mu_{n-1} \text{ sur } C.$$

Ainsi, si $C_1 \supset C_2 \supset \cdots \supset C_n$ est une suite de cellules d'ordre $1, 2, \cdots n$ emboîtées,

$$\text{densité } \mu_n \text{ sur } C_n = W_{C_1} W_{C_2} \cdots W_{C_n}.$$

La suite $\mu_n(T)$ est une martingale positive (on rappellera le sens de ce terme, et les résultats à connaître). Elle converge p.s. vers une limite Z. La suite des mesures μ_n converge elle-même p.s. faiblement vers une mesure aléatoire μ:

$$\mu_n \xrightarrow{*} \mu \quad \text{p.s.},$$

et on a $Z = \mu(T)$.

On s'intéresse au cas de dégénérescence ($Z = 0$), au problème des moments ($EZ^h < \infty$?), et à la dimension des boréliens sur lesquels μ est concentrée. Voici les principaux résultats.

1. $EZ > 0 \Leftrightarrow EZ = 1 \Leftrightarrow E(W \log W) < \log c$
2. $0 < EZ^h < \infty \Leftrightarrow EW^h < c^{h-1}$
3. Si $D = 1 - \dfrac{E(W\log W)}{\log c} > 0$, alors $\dim \mu = D$.

Ici, $\dim \mu = D$ signifie qu'il existe un borélien (aléatoire) B, de dimension de Hausdorff D, tel que $\mu(B) = 1$, et que pour tout borélien B de dimension de Hausdorff $< D$ on a $\mu(B) = 0$.

Dans le cas $c = 2$ et $P(W = 0) = 1 - p, P(W = \frac{1}{p}) = p$, l'énoncé 1 redonne $p \leq \frac{1}{2}$ comme condition nécessaire et suffisante de dégénérenscence; mais il est moins précis que le résultat mentionné plus haut, c'est-à-dire la disparition de la population (autrement dit, $\mu_n = 0$) au bout d'un temps fini.

Troisième problème: exponentiation des processus gaussiens

Prenons un exemple. Considérons la série de Fourier aléatoire

$$\sum_1^\infty a_n(\xi_n \cos nt + \xi'_n \sin nt) \tag{1}$$

où les a_n sont donnés ($a_n \geq 0$), et les ξ_n, ξ'_n forment un échantillon de loi $N(0,1)$ (c'est-à-dire qu'elles sont indépendantes, et de loi $N(0,1)$). On sait que pour

$$\sum_1^\infty a_n^2 < \infty \tag{2}$$

la série converge presque sûrement presque partout vers une fonction aléatoire $f(t,\omega)$ qui, pour presque tout ω, appartient à $L^2(dt)$. Au contraire, lorsque (2) n'a pas lieu, la série ne représente aucune fonction; p.s. ce n'est pas une série de Fourier-Lebesgue.

Lorsque (2) a lieu, on peut exponentier la série, et on obtient

$$\exp f(t,\omega).$$

Peut-on encore exponentier quand (2) n'a pas lieu?

On peut avoir l'idée de la solution en considérant une question voisine. Soit

$$\sum_1^\infty b_j \cos 3^j t \tag{3}$$

une série lacunaire, avec $-1 \leq b_j \leq 1$. Pour

$$\sum_1^\infty b_j^2 < \infty \tag{4}$$

la série converge presque partout vers une fonction $g(t) \in L^2(dt)$, qu'on peut exponentier. Si (4) est violée la série représente une distribution de Schwartz qui n'est pas une fonction, et qu'on ne sait donc pas exponentier. Cependant, au lieu de $\exp g(t)$, considérons le produit de Riesz

$$\prod_1^\infty (1 + b_j \cos 3^j t)$$

qui, dans le cas (4), n'en diffère que par un facteur positif borné. Le produit de Riesz conserve un sens lorsque $-1 \leq b_j \leq 1$, ce que nous avons supposé.

On peut donc espérer une sorte d'exponentiation pour la série (1), lorsque l'énergie dans des bandes de largeur logarithmique constante n'est pas trop grande. En fait, nous verrons que l'exponentiation est possible lorsque $a_n = \frac{\alpha}{n}$ et que $\alpha < \alpha_0$, et qu'elle est impossible (par le procédé que nous donnerons) quand $\alpha > \alpha_0$, pour une constante α_0 facile à calculer.

Dans ces trois problèmes il apparaît naturellement des poids aléatoires indépendants.

Dans le premier (intervalles $I_n = (\omega_n, \omega_n + \ell_n)$ sur le cercle T), l'ensemble non recouvert par $I_1 \cup I_2 \cdots \cup I_n$ est le support de la fonction

$$\prod_{m=1}^{n} \frac{1 - \chi_m(t - \omega_m)}{1 - \ell_m}.$$

Dans le second, la densité de μ_n par rapport à μ_{n-1} est le poids $\sum_C W_C 1_C$, la somme étant prise sur les cellules d'ordre n. En désignant ce poids par P_n, la densité de μ_n par rapport à λ est $P_1 P_2 \cdots P_n$.

Dans le troisième, il s'agit de produits d'exponentielles de processus gaussiens.

On se propose, dès la prochaine leçon, d'étudier quelques propriétés générales de ces produits de poids aléatoires indépendants.

Deuxième leçon

On va voir, au cours de cette leçon, deux façon de décomposer des mesures en une partie singulière et une partie régulière. La première est associée aux produits de poids aléatoires indépendants, la seconde aux noyaux de potentiel. Le lien entre ces deux points de vue apparaîtra dans les leçons suivantes.

Voici d'abord quelques résultats classiques sur les martingales

On donne un espace de probabilité $(\Omega, \mathcal{A}, \mathcal{P})$, une suite croissante de sous-tribus de $\mathcal{A}$, $(\mathcal{A}_n)_{n\in\mathcal{N}}$, et une suite de variables aléatoires $(Z_n)_{n\in\mathcal{N}}$ telle que chaque Z_n soit $\mathcal{A}_n$-mesurable. On dit que (Z_n) est une martingale (resp. sous martingale, resp. surmartingale) si

$$Z_n = E(Z_{n+1}|\mathcal{A}_n) \quad (n \in \mathcal{N})$$

(resp. $\leq$, resp. $\geq$ au lieu de $=$).

La définition implique $Z_n \in L^1(\Omega)$ pour chaque n, et $EZ_n = EZ_0$. De plus

$$Z_n = E(Z_{n+p}|\mathcal{A}_n)$$

pour tout $p \geq 1$.

L'exemple le plus important est celui-ci: on donne $Z \in L^1(\Omega)$ et on considère

$$Z_n = E(Z|\mathcal{A}_n).$$

Si, pour fixer les idées, $\Omega = [0,1[$ muni de la mesure de Lebesgue, et si $\mathcal{A}_n$ est la tribu engendrée par les intervalles dyadiques d'ordre n $[0,2^{-n}[, [2^{-n}, 2\times 2^{-n}[, \cdots [1-2^{-n}, 1[$, Z_n est obtenue en remplaçant Z, sur chacun de ces intervalles, par la valeur moyenne de Z sur l'intervalle.

Le contre exemple typique, avec les mêmes tribus $\mathcal{A}_n$, est $Z_n = 2^n 1_{[0,2^{-n}[}$ (c'est le gain au temps n du joueur qui joue à chaque instant toute sa fortune à pile ou face; il est presque sûr de tout perdre!).

Théorème de convergence–L^1 *Les propositions suivantes sont équivalentes (sauf avis contraire, on suppose toujours que Z_n est une martingale):*

a) $Z_n = E(Z|\mathcal{A}_n)$ *pour une* $Z \in L^1(\Omega)$;

b) Z_n *converge dans* $L^1(\Omega)$;

c) Z_n *est équiintégrable, c'est-à-dire qu'on a*

$$E\varphi(|Z_n|) = O(1)$$

pour une certaine fonction $\varphi(t) \geq 0$ *telle que* $\lim\limits_{t\to\infty} \dfrac{\varphi(t)}{t} = \infty$.

Remarquons que la condition

$$E|Z_n| = O(1)$$

est nécessaire, mais elle n'est pas suffisante (cf. contre exemple). Remarquons aussi que, lorsque Z_n converge dans $L^1(\Omega)$, on a a) avec $Z = \lim\limits_{L^1} Z_n$.

Théorème de convergence–L^p $(1 < p < \infty)$ *Les propositions suivantes sont équivalentes:*

a) $Z_n = E(Z|\mathcal{A}_n)$ *pour une* $Z \in L^p(\Omega)$;

b) Z_n *converge dans* $L^p(\Omega)$;

c) Z_n *est bornée dans* $L^p(\Omega)$.

Remarquons encore que, quand b) a lieu, on a a) avec $Z = \lim_{L^p} Z_n$. La condition c) est commode, particulièrement dans le cas $p = 2$.

Théorème de convergence presque sûre

1) *Si* Z_n *est bornée dans* $L^1(\Omega), Z_n$ *admet une limite presque sûre.*

2) *Il en est ainsi si* $Z_n \geq 0$.

3) *Lorsque* $Z_n \geq 0$, *soit* $Z = \lim Z_n$ *(presque sûre). Pour que* Z_n *converge vers* Z *dans* $L^1(\Omega)$, *il faut et suffit que*
$$EZ = EZ_n = EZ_0.$$

4) *Si* Z_n *est une surmartingale* ≥ 0, *elle admet une limite presque sûre.*

Quand $Z_0 \neq 0$ et que Z_n converge vers 0 p.s., on dira que Z_n est dégénérée. Le contre-exemple ci-dessus est un exemple de martingale dégénérée. Il est très facile de voir que toute martingale ≥ 0 est la somme d'une martingale dégénérée et d'une martingale vérifiant la condition 3.

Considérons maintenant des martingales positives (au sens ≥ 0) indexées par un point d'un espace T métrique localement compact: on les appellera T–martingales. Une T–martingale est donc une suite
$$Q_n(t,\omega) \quad (n \in \mathcal{N}, t \in T, \omega \in \Omega)$$
telle que, pour chaque $t \in T$, ce soit une martingale positive, et pour chaque $\omega \in \Omega$ (à l'exception peut-être d'un évènement de probabilité nulle) une suite de fonctions boréliennes. Au lieu de $Q_n(t,\omega)$, on écrit couramment $Q_n(t)$, et on pose
$$q(t) = EQ_n(t).$$

Le cas le plus intéressant est celui où toutes les martingales $Q_n(t)(t \in T)$ sont dégénérées; sans que ce soit une hypothèse dans la suite, c'est ce cas qu'il convient d'avoir à l'esprit.

Lorsque $\sigma \in M^+(T)$ (mesure positive portée par T), $Q_n\sigma$ est une suite de mesures aléatoires.

Théorème (= Proposition 1) *Supposons $q \in L^1(\sigma)$. Alors*

$$Q_n\sigma \overset{*}{\to} S \quad p.s.$$

ce qui veut dire que $Q_n\sigma$ converge faiblement (sur $C_0(T)$) vers une mesure aléatoire S, avec probabilité 1.

Notation: $Q : \sigma \to S$.

L'opérateur Q transforme donc des mesures en mesures aléatoires.

Il y a deux cas extrêmes. Le premier est $Q\sigma = 0$; on dit que Q est dégénéré sur σ, ou meurt sur σ. Le second est $E(Q\sigma) = \sigma$; on dit que Q agit pleinement sur σ, ou vit sur σ. Désormais on suppose toujours $q \in L^1(\sigma)$.

Proposition 2 *On peut écrire, de manière unique, Q_n comme somme de deux martingales Q'_n et Q''_n, dont la première vit et la seconde meurt sur σ.*

Notation: $q' = EQ'_N\ (\leq q)$.

Les propositions suivantes donnent des conditions suffisantes pour l'action pleine (la vie) ou la dégénérescence (la mort).

Proposition 3 *Si $\varphi : \mathcal{R}^+ \to \mathcal{R}$ vérifie $\lim_{t\to\infty} \frac{\varphi(t)}{t} = \infty$ et si $E\varphi(|\int Q_n d\sigma|) = O(1), Q$ vit sur σ.*

Cas particulier: on peut choisir $\varphi(t) = t^p (1 < p < \infty)$, et le cas $p = 2$ est simple à expliciter. En effet,

$$E(\int Q_n d\sigma)^2 = \int\int EQ_n(t)Q_n(s)d\sigma(t)d\sigma(s)$$

ce qui apparaît comme une intégrale d'énergie par rapport à un certain noyau. On pressent ainsi la relation possible avec la théorie du potentiel.

Proposition 4 *Si $\psi : \mathcal{R}^+ \to \mathcal{R}^+$ est concave et $\lim_{t\to\infty} \frac{\psi(t)}{t} = 0$, et si $E\psi(|\int Q_n d\sigma|) = o(1)$ (tend vers 0 quand $n \to \infty$), Q meurt sur σ.*

Cas particulier: $\psi(t) = t^h$ avec $0 < h < 1$. En application de la proposition 4, on a:

Proposition 5 *Supposons* $\text{mes}_\alpha T < \infty$ *(c'est la mesure de Hausdorff en dimension* α*) et supposons que pour toute boule* B *il existe* $n = n(B)$ *tel que*

$$E(\sup_{t\in B} Q_n(t))^h \leq C(\text{diam} \quad B)^{\alpha(1-h)}$$

où C *est une constante (indépendante de* B*). Alors* Q *meurt sur toutes les mesures* σ.

Définition Dans ce cas, on dit que Q est *complètement dégénéré.*

Exercice Tester la proposition 5 sur le processus de naissance et de mort décrit dans la leçon 1 (on retrouve que, pour la probabilité de décès égale à $\frac{1}{2}$, la population ne croît pas exponentiellement; mais on ne retrouve pas le résultat plus précis qu'est l'extinction de la population).

La proposition 5 est facile à démontrer sous l'hypothèse plus forte $\text{mes}_\alpha T = 0$. Les autres propositions sont faciles, et les preuves se trouvent détaillées dans mon cours d'Urbana de 1987 [K1989] et dans l'article [K1987(2)].

Le cas le plus important de T–martingales est celui des poids aléatoires indépendants. On considère une suite

$$P_n(t, \omega) \quad (n \in \mathcal{N}, t \in T, \omega \in \Omega)$$

où presque sûrement, les $P_n(t, \omega)$ sont des fonctions boréliennes positives, où ces fonctions sont indépendantes sur Ω (par exemple, quand Ω est un espace produit $\prod \Omega_n$, et que $P_n(t, \omega)$ ne fait intervenir que la n-ième coordonnée ω_n), et où de plus

$$EP_n(t) = 1 \quad (n \in \mathcal{N}, t \in T).$$

Si on pose

$$Q_n(t) = P_0(t)P_1(t) \cdots P_n(t),$$

on obtient bien une T–martingale. Soit Q l'opérateur correspondant. La décomposition donnée par la proposition 2 prend alors une forme beaucoup plus intéressante. On peut démontrer que dans ce cas $q' = 1_B$ pour un certain borélien B, et qu'on a donc

$$\begin{aligned} EQ'\sigma &= 1_B\sigma = \sigma' \\ EQ''\sigma &= (1 - 1_B)\sigma = \sigma'' \end{aligned}$$

(cf. les références ci-dessus), et on en déduit:

Théorème (Proposition 6) *L'opérateur* EQ *qui à* σ *fait correspondre* $E(Q\sigma)$ *est une projection. Ainsi* σ *se décompose de manière unique sous la forme*

$$\sigma = \sigma' + \sigma''$$

de façon que Q *vive sur* σ' *et meure sur* σ''.

Définition On dira que σ' (élément de l'image de EQ) est une mesure *Q–régulière* et que σ'' (élément du noyau de EQ) est une mesure *Q–singulière*.

Dans les cas intéressants, les mesures de Dirac δ_t sont Q–singulières (cela veut dire que les $Q_n(t)$ sont martingales dégénérées).

Retenons que Q définit une décomposition des mesures en une partie régulière et une partie singulière.

Supposons que σ est une mesure de probabilité Q–régulière. Un bon outil pour l'étude de $S = Q\sigma$ est le *probabilité de Peyrière*, ainsi définie: c'est l'unique mesure de probabilité $\mathcal{Q}$ sur la tribu engendrée par les $B \times A$ (B: borélien dans T, A: évènement dans Ω) qui vérifie

$$\int_{T\times\Omega} f(t,\omega)d\mathcal{Q}(t,\omega) = E\int_T f(t,\omega)dS(t)$$

pour toutes les fonctions positives mesurables $f(t,\omega)$. Écrivons aussi $E_q(f)$ pour le premier membre.

Proposition 7 *Outre l'hypothèse que σ est une mesure de probabilité Q–régulière, supposons que, pour chaque n, la distribution de $P_n(t)$ ne dépend pas de $t(t \in T)$. Alors les $P_n(= P_n(t,\omega))$ sont $\mathcal{Q}$–indépendants.*

Les preuves des propositions 6 et 7 ne sont pas difficiles, et sont détaillées dans l'article [K1987(2)].

Nous venons de voir que l'opérateur Q associé à un produit de poids aléatoires indépendants définit une décomposition des mesures en une partie régulière et une partie singulière. Il en est de même dans un cadre tout différent, celui de la théorie du potentiel.

Pour fixer les idées, supposons T compact, et choisissons $\alpha > 0$. On définit le potentiel d'ordre α et l'énergie d'ordre α d'une mesure $\sigma \in M^+(T)$ par les formules:

$$U_\alpha^\sigma(t) = \int \frac{\sigma(ds)}{(\mathrm{dist}(t,s))^\alpha}$$

$$I_\alpha^\sigma = \int U_\alpha^\sigma d\sigma = \int\int \frac{\sigma(ds)\sigma(dt)}{(\mathrm{dist}(t,s))^\alpha}.$$

On dit que σ est α–singulière, ce qu'on écrit $\sigma \in \mathcal{S}_\alpha$, si $U_\alpha^\sigma(t) = \infty$ σ–presque partout.

On dit que σ est α–régulière, ce qu'on écrit $\sigma \in \mathcal{R}_\alpha$, si $\sigma = \sum_1^\infty \sigma_n$ avec $I_\alpha^{\sigma_n} < \infty$ (somme de mesures d'énergie finie).

Exercice Démontrer (c'est très facile)

$$U^\sigma_\alpha < \infty \quad \text{p.p.} \quad \Rightarrow \sigma \in \mathcal{R}_\alpha. \tag{1}$$

Soit A un borélien dans T. On dit que la capacité d'ordre α de A est nulle, et on écrit $\mathrm{Cap}_\alpha A = 0$, si $\sigma(A) = 0$ pour toute $\sigma \in \mathcal{R}_\alpha$. Désignons par $M^+(A)$ l'ensemble des σ telles que $\sigma(A) = \sigma(T)$.

Exercice Démontrer

$$\mathrm{Cap}_\alpha A = 0 \Rightarrow M^+(A) \subset \mathcal{S}_\alpha. \tag{2}$$

Théorème (Proposition 8) (1) *et* (2) *admettent des réciproques, donc*

$$U^\sigma_\alpha < \infty \quad \sigma - p.p. \quad \Leftrightarrow \sigma \in \mathcal{R}_\alpha$$

$$\mathrm{Cap}_\alpha A = 0 \Leftrightarrow M^+(A) \subset \mathcal{S}_\alpha,$$

et les mesures α–singulières sont caractérisées par le fait qu'elles sont concentrées sur des boréliens de α–capacité nulle. On a

$$M^+(T) = \mathcal{S}_\alpha \oplus \mathcal{R}_\alpha$$

c'est-à-dire que toute mesure $\sigma \in M^+(T)$ se décompose de manière unique en somme d'une partie α–régulière et d'une partie α–singulière.

Remarquons que, lorsque α croît, $\mathcal{S}_\alpha$ croît et $\mathcal{R}_\alpha$ décroît. La masse $\sigma_\alpha(T)$ est la premitive d'une mesure sur $\mathcal{R}^+$ qu'on peut appeler le spectre de dimension de la mesure. Si le support du spectre de dimension est réduit à un point, d, on dira que la mesure σ est unidimensionnelle et de dimension d.

La démonstration du théorème repose sur le fait suivant: si U^σ_α est borné sur le support de σ, U^σ_α est borné partout (principe du potentiel borné). Les définitions et le théorème se transcrivent donc pour d'autres noyaux que $(\mathrm{dist}(t,s))^{-\alpha}$. Dès que le noyau $k(t,s)$ vérifie le principe du potentiel borné, on peut parler de $U^\sigma_k, I^\sigma_k, \mathrm{Cap}_k A, \mathcal{S}_k, \mathcal{R}_k$, et le théorème est valable en remplaçant partout α par k.

Le cas particulier le plus important est celui d'un noyau $k(t,s) = k(t-s)$, où k est paire, positive, décroissante et convexe sur $\mathcal{R}^+$ et nulle à l'infini [KS1963].

Troisième leçon

On va démontrer les résultats annoncés dans la première leçon sur le recouvrement de $\mathcal{T}$ ou de $\mathcal{R}$ par des intervalles aléatoires.

Dans le cas du cercle $\mathcal{T}$, on donne une suite positive ℓ_n tendant vers 0, et on considère les intervalles aléatoires

$$I_n = (\ell_n, \ell_n + \omega_n)$$

et leur réunion; on va montrer que

$$\begin{array}{ll} (1) & P(\mathcal{T} \subset \cup I_n) = 1 \Leftrightarrow k \notin L^1 \\ (2) & P(A \subset \cup I_n) = 1 \Leftrightarrow \ \mathrm{Cap}_k A = 0 \\ (3) & P(\mathrm{Cap}_{k'}(\mathcal{T} \backslash \cup I_n) = 0) = 1 \Leftrightarrow kk' \notin L^1 \end{array}$$

où le noyau k $(= k_\ell)$ est

$$k(t) = \exp \sum (\ell_n - |t|)^+,$$

et k' est un noyau analogue. Dans le cas de la droite $\mathcal{R}$, on donne une mesure μ localement bornée sur $\mathcal{R}^+$, et on considère les intervalles aléatoires.

$$J_n = (X_n, X_n + Y_n)$$

où (X_n, Y_n) est le processus de Poisson ponctuel associé à la mesure $\lambda \otimes \mu$ sur $\mathcal{R} \times \mathcal{R}^+$. On va montrer que

$$\begin{array}{ll} (1') & P(\mathcal{R} \subset \cup J_n) = 1 \Leftrightarrow k \notin L^1(\mathcal{R}^+, e^{-x}dx) \\ (2') & P(A \subset \cup J_n) = 1 \Leftrightarrow \ \mathrm{Cap}_k A = 0 \\ (3') & P(\mathrm{Cap}_{k'}(\mathcal{R} \backslash \cup J_n) = 0) = 1 \Leftrightarrow kk' \notin L^1(\mathcal{R}^+, e^{-x}dx) \end{array}$$

où le noyau k $(= k_\mu)$ est

$$k(x) = \exp \int_{|x|}^{\infty} \mu(y, \infty) dy$$

et k' est un noyau analogue.

Lorsque $\ell_n = \frac{a}{n}$, on vérifie facilement que $k(t) \simeq |t|^{-a}$, donc T est recouvert quand $a \geq 1$, et $A \subset \mathcal{T}$ est recouvert quand $\mathrm{Cap}_\alpha A = 0$. Lorsque $\mu(dy) = a\frac{dy}{y^2}$ sur $[0, 1[$ et $\mu(dy) = 0$ sur $[1, \infty[$ on a $k(x) \simeq |x|^{-a}$, donc $\mathcal{R}$ est recouvert quand $a \geq 1$, et $A \subset \mathcal{R}$ est recouvert quand $\mathrm{Cap}_\alpha A = 0$.

Montrons comment on démontre (3) à partir de (1) et (2). Soit (ℓ'_m) une suite positive, et

$$k'(t) = \exp \sum (\ell_m - |t|)^+.$$

On considère des intervalles

$$I'_m = (\omega'_m, \omega'_m + \ell'_m)$$

les ω'_m étant indépendants entre eux, et indépendants des ω_n. La condition $kk' \notin L^1$ est, d'après (1), nécessaire et suffisante pour que les I_n et les I'_m recouvrent $\mathcal{T}$, donc pour que les I'_m recouvrent $\mathcal{T}\setminus \cup I_n$, donc, d'après (2), pour que $\mathrm{Cap}_{k'}(\mathcal{T}\setminus \cup I_n) = 0$ p.s.

De même (3') découle de (1') et (2').

Il s'agit maintenant de démontrer (1), (2), (1'), (2'). La première étape consistera à établir

(4) $k \in L^1(\mathcal{T}) \Rightarrow P(\mathcal{T} \not\subset I_n) > 0$
(5) $\mathrm{Cap}_k A > 0 \Rightarrow P(A \not\subset \cup I_n) > 0$
(4') $k \in L^1(e^{-|x|}dx) \Rightarrow P(\mathcal{R} \not\subset \cup J_n) > 0$
(5') $\mathrm{Cap}_k A > 0 \Rightarrow P(A \not\subset \cup J_n) > 0$

et pour cela nous utiliserons des martingales convenables.

La seconde étape consistera à établir les réciproques de (4') et de (5'); à ce stade, on aura donc (1') et (2').

La troisième étape consistera à déduire de ces réciproques celles de (4) et (5), ce qui établira (1) et (2).

Première étape

Considérons d'abord le cas de $\mathcal{T}$. Posons

$$P_n(t) = \frac{1 - \chi_n(t - \omega_n)}{1 - \ell_n}$$

($\chi_n = 1_{(0,\ell_n)}$) et $Q_n = P_1 P_2 \dots P_n$. C'est une $\mathcal{T}$–martingale, définissant un opérateur Q. Étant donné $\sigma \in M^+(\mathcal{T}), Q$ vit sur σ dès que

$$\int\int EQ_n(t)Q_n(s)d\sigma(t)d\sigma(s) = O(1).$$

Explicitons cette condition. Pour cela, écartons le cas $\sum \ell_n^2 = \infty$, qui est un cas de couverture de $\mathcal{T}$ très facile à traiter, et supposons donc $\sum \ell_n^2 < \infty$. Posons $\xi_n = \chi_n * \chi_n$, c'est-à-dire

$$\xi_n(t) = (\ell_n - |t|)^+.$$

On a successivement

$$\begin{aligned} EP_n(t)P_n(s) &= \frac{1 - 2\ell_n + \xi_n(t-s)}{(1-\ell_n)^2} \\ &= \exp(\xi_n(t-s) + O(\ell_n^2)) \\ EQ_n(t)Q_n(s) &= \exp(\sum \xi_n(t-s) + O(1)) \end{aligned}$$

donc Q vit sur σ dès que $\sigma \in \mathcal{R}_k$. Si $k \in L^1$, on peut choisir $\sigma = \lambda$; comme $P(Q\lambda \neq 0) \neq 0$, on a bien (4). Si $\mathrm{Cap}_k A > 0$, on peut choisir $\sigma \in \mathcal{R}_k \cap M^+(A)$, et on a bien (5).

Considérons maintenant le cas de $\mathcal{R}$. Posons

$$Q_\epsilon(x) = \frac{1_{x \notin G_\epsilon}}{P(x \notin G_\epsilon)}$$

où G_ϵ est la réunion des intervalles J_n de longueur $\geq \epsilon$. Comme les intervalles J_n correspondant à des bandes horizontales disjointes forment des familles indépendantes (parce que les processus de Poisson restreints à des domaines disjoints sont des processus indépendants), la martingale $Q_\epsilon(x)(\epsilon \downarrow 0)$ peut être obtenue par produit de poids aléatoires indépendants. Elle définit donc un opérateur Q, qui va vivre sur toute mesure σ telle que

$$\int\int EQ_\epsilon(t)Q_\epsilon(s)d\sigma(t)d\sigma(s) = O(1).$$

Cela signifie

$$\int\int EQ_0(t)Q_0(s)d\sigma(t)d\sigma(s) < \infty.$$

Posons $G_0 = G$. Dire que $t \notin G$, c'est dire que l'angle

$$A_t = \{(x,y) : x \leq t, y \geq t - x\}$$

ne contient pas de point (X_n, Y_n), donc

$$P(t \notin G) = \exp(-(\lambda \otimes \mu)(A_t)).$$

De même

$$E(1_{t \notin G} 1_{s \notin G}) = \exp(-(\lambda \otimes \mu)(A_t \cup A_s)).$$

Donc (en supposant d'abord les seconds membres ci-dessus finis)

$$\begin{aligned} EQ_0(t)Q_0(s) &= \exp(-(\lambda \otimes \mu)(A_t \cup A_s) + (\lambda \otimes \mu)A_t + (\lambda \otimes \mu)A_s) \\ &= \exp(\lambda \otimes \mu)(A_t \cap A_s) \\ &= \exp \int_{|s-t|}^{\infty} \mu(y,\infty)dy \\ &= k(t-s) \end{aligned}$$

et le résultat est valable, par passage à la limite, quand $P(t \notin G) = 0$. Si $k \in L^1$, on peut choisir $\sigma = e^{-t}1_{\mathcal{R}^+}$, et on a bien (4'). Si $\mathrm{Cap}_k A > 0$, on choisit $\sigma \in \mathcal{R}_k \cap M^+(A)$ et on a (5').

Deuxième étape

Soit encore

$$k(t) = \int_{|t|}^{\infty} \mu(y,\infty)dy.$$

Si $k(t) = \infty$ pour un $t > 0$, tout point est recouvert p.s. par des J_n arbitrairement grands (voir le calcul de $P(t \notin G)$); nous pouvons éliminer ce cas et admettre désormais $k(t) < \infty$ pour tout t. Supposons pour un instant que $\mu(0,\epsilon) = 0$ et posons $G_\epsilon = G$; G est l'ouvert réunion de tous les intervalles J_n, puisqu'ils ont tous des longueurs $\geq \epsilon$.

Soit τ le premier point non recouvert par G à droite de 0:

$$\tau = \inf\{t \geq 0, t \notin G\}.$$

On va calculer de deux manières l'intégrale

$$I = E\int_0^\infty e^{-t} 1_{t \notin G} dt,$$

et en déduire $E(e^{-\tau})$.

D'abord

$$I = \int_0^\infty e^{-t} P(t \notin G) dt$$

avec

$$\begin{aligned} P(t \notin G) &= \exp(-(\lambda \otimes \mu)(A_t)) \\ &= \exp(-\int_0^\infty \mu(y,\infty)dy) \end{aligned}$$

donc

$$I = \exp(-\int_0^\infty \mu(y,\infty)dy.$$

Ensuite, en faisant intervenir τ, et en observant que $1_{t \notin G} = 0$ pour $t \leq \tau$,

$$\begin{aligned} I &= E\left(e^{-\tau}\int_0^\infty e^{-s} 1_{\tau+s \notin G} ds\right) \\ &= \left(e^{-\tau} E\left(\int_0^\infty e^{-s} 1_{\tau+s \notin G} ds | \tau\right)\right) \\ &= E\left(e^{-\tau}\int_0^\infty e^{-s} P(\tau + s \notin G|\tau) ds\right). \end{aligned}$$

Or $P(\tau + s \notin G|\tau)$ est le probabilité qu'il n'y ait pas de point (X_n, Y_n) dans l'angle $A_{\tau+s}$ sachant qu'il n'y en a pas dans l'angle A_τ. C'est donc

$$\begin{aligned} P(\tau + s \notin G|\tau) &= \exp -(\lambda \otimes \mu(A_{\tau+s} \backslash A_\tau)) \\ &= \exp(-\int_0^s \mu(y,\infty)dy) \end{aligned}$$

donc

$$I = E(e^{-\tau})\int_0^\infty e^{-s} \exp(-\int_0^s \mu(y,\infty)dy) ds$$

et, en divisant par la première expression de I,

$$\begin{aligned} 1 &= (Ee^{-\tau})\int_0^\infty e^{-s}\exp(\int_s^\infty \mu(y,\infty)dy)ds \\ &= (Ee^{-\tau})\int_0^\infty e^{-s}k(s)ds. \end{aligned}$$

Par passage à la limite la formule reste valable sans la restriction $\mu(0,\epsilon) = 0$. En particulier, sous l'hypothèse

$$\int_0^\infty e^{-s}k(s)ds = \infty,$$

on a $\tau = \infty$ p.s., c'est-à-dire que $\mathcal{R}^+ \subset G$ p.s. Donc aussi $\mathcal{R} \subset G$ p.s. On a donc établi l'implication inverse de (4').

Pour établir l'inverse de (5') il faut un tout peu plus de travail. On peut supposer A compact, et d'abord $\mu(0,\epsilon) = 0$, de façon que k est borné. Il existe une mesure de probabilité $\sigma \in M^+(A)$ dont le k–potentiel U_k^σ a la plus petite borne supérieure, et on vérifie facilement que U_k^σ est constante sur A, soit

$$U_k^\sigma(t) = C \quad (t \in A).$$

A est la réunion de deux fermés A^+ et A^- définis par

$$\begin{aligned} A^+ &= \{t \in A : \int_0^\infty k(s)\sigma(t+ds) \geq \frac{1}{2}C\} \\ A^- &= \{t \in A : \int_{-\infty}^0 k(s)\sigma(t+ds) \geq \frac{1}{2}C\}. \end{aligned}$$

Soit

$$\begin{aligned} \tau &= \begin{cases} \inf\{t \in A^+, t \notin G\} & \text{si} \quad A^+ \not\subset G \\ \infty & \text{si} \quad A^+ \subset G \end{cases} \\ I &= E\int 1_{t\notin G}\sigma(dt). \end{aligned}$$

D'abord, comme ci-dessus,

$$I = \exp(-\int_0^\infty \mu(y,\infty)dy).$$

Ensuite,

$$I = E\int_{\mathcal{R}^+} \exp(-\int_s^\infty \mu(y,\infty)dy)\sigma(\tau+ds)$$

d'où, en divisant membre à membre

$$1 = E\int_{\mathcal{R}^+} k(s)\sigma(\tau+ds)$$

et, d'après la définition de A^+,

$$1 \geq \frac{C}{2} P(\tau < \infty) = \frac{C}{2} P(A^+ \not\subset G).$$

On a la même inégalité pour A^- donc

$$P(A \not\subset G) \leq \frac{4}{C}.$$

La même inégalité subsiste, avec $C = (\mathrm{Cap}_k A)^{-1}$, par passage à la limite, en supprimant la restriction $\mu(0, \epsilon) = 0$. En particulier, si $\mathrm{Cap}_k A = 0$ on a $A \subset G$ p.s., c'est-à-dire la réciproque de (5').

À ce stade, nous avons donc établi (1'), (2') et (3').

Trosième étape

Montrons comment la réciproque de (5) se déduit de celle de (5'). Supposons donc $A \subset T$ et $\mathrm{Cap}_k A = 0$ avec $k = k_\ell$, associé aux intervalles $\ell_1, \ell_2, \cdots \ell_n, \cdots$ On peut sans restriction supposer diam $A \leq \frac{1}{2}$. Désormais on considèrera A comme une partie de $[0, 1/2]$. La stratégie consiste à modifier la suite ℓ_n en diminuant les longueurs pour obtenir une suite ℓ'_n non strictement monotone, telle que A soit presque sûrement recouvert par les intervalles I'_n correspondants.

Supposons $\frac{1}{2} \geq \ell_1 \geq \ell_2 \geq \ell_3 \cdots$ et posons

$$\begin{aligned} &\ell'_1 = \ell_1 = \lambda_1 \\ &\ell'_2 = \ell'_3 = \ell_3 = \lambda_2 \\ &\ell'_4 = \ell'_5 = \ell'_6 = \ell_6 = \lambda_3 \quad \text{etc.} \end{aligned}$$

La suite ℓ'_n prend m fois la valeur λ_m. Sous la condition que $\sum_m^{\ell^{1+\epsilon}} < \infty$ pour tout $\epsilon > 0$ (sinon, il est facile de vérifier $T \subset \cup I_n$ p.s.; cf. K1985, p. 145), on a $\ell_n^{1+\epsilon} = o\left(\frac{1}{n}\right)$ et

$$\sum(\ell_n - \ell'_n) \leq \sum m(\lambda_m - \lambda_{m-1}) = \sum \lambda_m = \sum \ell_{\frac{m(m+1)}{2}}$$

donc $\sum(\ell_n - \ell'_n) < \infty$, donc $k(t) \leq C^{\mathrm{te}} k'(t)$, et

$$\mathrm{Cap}_{k'} A = 0.$$

Retirons maintenant, pour chaque m, $[m^{2/3}]$ termes égaux à λ_m. Il reste une suite ℓ''_n, et on vérifie que $\sum(\ell'_n - \ell''_n) < \infty$, donc

$$\mathrm{Cap}_{k''} A = 0.$$

On interprète maintenant k'' comme k_μ avec

$$\mu = \sum \delta_{\ell''_n},$$

et on considère les intervalles J_n associés au processus de Poisson d'intensité $\lambda \otimes \mu$. On sait que $A \subset \cup J_n$ p.s. (seconde étape). Or, pour chaque m, le nombre N_m d'intervalles J_n dont la longueur est λ_m et dont l'origine est dans $[-1/2, 1/2]$ est une v.a. de Poisson de paramètre $m - [m^{2/3}]$. Observons qu'on a p.s. $N_m \leq m$ à partir d'un m assez grand. Pour recouvrir A, on peut donc choisir au hasard des v.a. de Poisson indépendantes, N_m, de paramètres $m - [m^{2/3}]$, puis pour chaque m, N_m points au hasard selon la probabilité naturelle sur $[-1/2, 1/2]$ et les intervalles ayant ces points comme origines et comme longueur λ_m. On ne fait qu'améliorer le recouvrement si on remplace N_m par m; on a donc recouvrement de A par les I'_n. A fortiori, le recouvrement a lieu avec les I_n.

On a donc établi la réciproque de (5); celle de (4) s'obtient exactement de la même façon. En conséquence, on a établi (1), (2), (3).

Comme intermédiaire entre le recouvrement poissonnien sur $\mathcal{R}$ et le recouvrement par des intervalles de longueurs fixées sur $\mathcal{T}$, on peut considérer un recouvrement poissonnien sur $\mathcal{T}$. On considère $\mathcal{T}$ comme la frontière du disque unité $|z| < 1 (z = re^{2\pi i t})$, on donne une mesure radiale ν localement bornée sur le disque unité ouvert et on pose

$$\mu[r, r + dr[= \nu(r \leq |z| < r + dr).$$

Le processus de Poisson ponctuel d'intensité ν est constitué de points $z_n = r_n e^{2\pi i t_n}$, auxquels on associe les intervalles $[t_n - 1/2(1 - r_n), t_n + 1/2(1 - r_n)]$. On pose

$$k(t) = \exp \int_{|t|}^{1} \mu[0, 1 - s] ds$$

et on a les résultats analogues à (1), (2), (3). Résumons

Théorème *À chacun des problèmes de recouvrement considérés (intervalles aléatoires de longueurs données sur I, intervalles poissonniens sur $\mathcal{R}$, intervalles poissonniens sur $\mathcal{T}$) correspond un noyau $k(.)$. Le recouvrement presque sûr (de $\mathcal{T}, \mathcal{R}, \mathcal{T}$) a lieu si et seulement si $k \notin L^1(L^1(\mathcal{T}, dt), L^1(\mathcal{R}^+, e^{-x}dx), L^1(\mathcal{T}, dt))$. Le recouvrement presque sûr d'une partie borélienne A (de $\mathcal{T}, \mathcal{R}, \mathcal{T}$) a lieu si et seulement si* $\mathrm{Cap}_k A = 0$. *Pour un noyau k' de la même forme la capacité d'ordre k' de l'ensemble non recouvert est p.s. nulle si et seulement si $kk' \notin l^1$. Suivant le cas, on a*

$$k(t) = \exp \sum (\ell_n - |t|)^+$$

$$k(x) = \exp \int_{|x|}^{\infty} \mu(y, \infty) dy$$

$$k(x) = \exp \int_{|t|}^{1} \mu(0, 1 - s) ds.$$

Voici une application amusante, suggérée par un problème de R. Lyons. Considérons le demi-plan de Poincaré ($y = \mathcal{R}e\ z > 0$) muni de la métrique hyperbolique $ds = \frac{\sqrt{dx^2+dy^2}}{y}$ et de la meusre d'aire correspondante

$$\alpha(dxdy) = \frac{dxdy}{y^2}.$$

Considérons le processus de Poisson ponctuel associé, et la réunion de tous les disques hyperboliques de rayon r centrés aux points de ce processus. Restreignons nous aux disques situés dans une bande horizontale $0 < y < y_0$. Enfin, projetons ces disques orthogonalement sur la frontière $y = 0$. On obtient les intervalles $(x - \rho y, x + \rho y)$ avec $\rho = r\sqrt{1 + \frac{r^2}{4}}((x, y) \in$ processus), c'est-à-dire des intervalles poissonniens associés à une mesure $y(dy)$ nulle sur $[y_0, \infty]$, et égale au voisinage de 0 à 2 $\rho\frac{dy}{y^2}$. Pour $\rho < \frac{1}{2}$, c'est-à-dire $r < r_0 = (2 + \sqrt{5})^{-1/2} = 0,4858\ldots$ ils ne recouvrent pas la frontière et on a une bonne information sur l'ensemble non recouvert; pour $r \geq r_0$, ils la recouvrent. Si, au lieu de projeter les disques orthogonalement (c'est-à-dire, en géométrie hyperbolique, en suivant les géodésiques issues du point frontière $0 + i\infty$), on projette à partir d'un autre point frontière en suivant les géodésiques, on a le même résultat. Il est facile de vérifier que le résultat subsiste si on projette à partir d'un point intérieur au demi-plan hyperbolique.

On peut considérer le processus de Poisson comme une plantation d'arbres, et r comme le rayon atteint par chaque arbre au temps r, donc la réunion des disque de rayon r centrés sur le processus comme l'état de la forêt au temps r. Pour $\rho < \frac{1}{2}$, un observateur donné voit p.s. à l'infini à travers la forêt si on ménage une clairière assez vaste autour de lui, et l'ensemble qu'il voit à l'infini est de dimension $1 - 2\rho$ et de $(1 - 2\rho)$–capacité nulle. Pour $\rho \geq \frac{1}{2}$, p.s. l'observateur donné ne voit plus à travers la forêt: les arbres lui cachent la forêt.

Le même problème se pose, avec les mêmes résultats, pour le disque hyperbolique $|z| < 1$, muni de la métrique $ds = \frac{2\sqrt{dx^2+dy^2}}{1-(x^2+y^2)}$; le cas de l'observateur placé en 0 donne le cas général pour l'observateur placé à l'interieur, et il est très facile à traiter à partir des résultats concernant le recouvrement poissonnien de $\mathcal{T}$.

Quatrième leçon

On va étudier de près les processus de naissance et de mort décrits dans la première leçon, puis les cascades self-similaires de B. Mandelbrot, et donner dans ce cadre l'interprétation des multifractals par des multimesures.

Considérons $T = \{0, 1\}^{\mathcal{N}}$, muni de la distance ultramétrique; ainsi $\mathrm{dist}(t, s) = 2^{-n}$ signifie que les n premières coordonnées de t et de s coïncident, et que les $(n + 1)$–ièmes diffèrent. Le processus de naissance au temps n consiste à passer d'un individu vivant

$(t_1, t_2, \ldots, t_n)$ aux deux individus

$$(t_1, t_2, \ldots, t_n, t_{n+1}) \quad (t_{n+1} = 0 \quad \text{ou} \quad 1),$$

et le processus de mort est aléatoire: il consiste à conserver chaque rejeton avec probabilité $p = 2^{-\alpha}$, et à le tuer avec probabilité $1-p$, toutes ces morts aléatoires sont indépendantes.

Problème On donne $A \subset \mathcal{T}$. À quelle condition A est-il tué presque sûrement?

Théorème 1 *La condition nécessaire et suffisante est* $\mathrm{Cap}_\alpha A = 0$.

Rappelons que ce théorème est dû à Fan Ai hua [F1989(2),(3)]. La preuve que nous allons donner va consister à nous ramener au cas du recouvrement d'un ensemble B et $\mathcal{T}$ par des intervalles aléatoires, ce qui fait l'objet de la leçon 3. Cependant, avant de procéder à la preuve selon cette méthode, indiquons que la condition est presque évidemment nécessaire. En effet, soit

$$P_n(t) = \sum_{C:\ \text{cellule d'ordre } n} W_C 1_C$$

où (W_C) est un échantillon (= suite de variables aléatoires de même distribution) dont la loi est $P(W = 0) = 1 - 2^{-\alpha}, P(W = 2^\alpha) = 2^{-\alpha}$. Avec les notations de la leçon 2,

$$\int\int E(Q_n(t)Q_n(s))d\sigma(t)d\sigma(s) = O(1)$$

suffit pour que Q vive sur σ (proposition 3). Or (calcule rapide)

$$E(Q_n(t)Q_n(s)) = (\mathrm{dist}(t,s))^\alpha.$$

Donc, si $\mathrm{Cap}_\alpha A > 0$ et $\sigma \in M^+(A) \cap \mathcal{R}_\alpha$, on a $Q\sigma \neq 0$, donc

$$P(A \cap E \neq \emptyset) = 0,$$

E étant l'ensemble des $(t_1, t_2, \ldots, t_n, \ldots)$ dont toutes les sections $(t_1, t_2, \ldots, t_n)$ sont vivantes.

Pour nous ramener au cas traité dans la leçon 3, considérons un ensemble de Cantor symétrique K, intersection d'ensembles K_j qui sont la réunion de 2^j intervalles fermés de longueur commune d_j, avec la règle de passage de K_j à K_{j+1} définie par la figure:

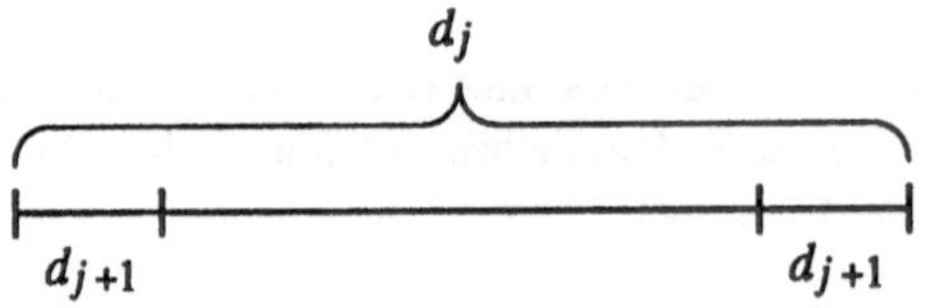

On appellera D_j l'un quelconque des intervalles constituant K_j. On supposera que d_{j+1}/d_j

tend vers 0 très rapidement; disons, assez rapidement pour que les approximations que nous ferons ci-dessous soient négligeables.

Les intervalles D_j se numérotent naturellement $D_{t_1,\ldots,t_j}$, avec la règle

$$D_{t_1,\ldots,t_j,t_{j+1}} \subset D_{t_1,\ldots,t_j}.$$

Ainsi à chaque $t \in T$ correspond un unique $x \in K$, à savoir

$$x = \bigcap_j D_{t_1,\ldots t_j}.$$

Soit B l'image de A dans cette correspondance. Remarquons que si $k(t)$ est un noyau défini sur $[-1/2, 1/2]$, pair, convexe et décroissant sur $]0, 1/2]$, tel que

$$k(d_j) = 2^{-j\alpha}$$

on a

$$\mathrm{Cap}_\alpha A = 0 \Leftrightarrow \quad \mathrm{Cap}_k B = 0;$$

en effet, le rapport $I_\alpha^\sigma / I_k^\tau$ est borné dans les deux sens, quand la mesure $\tau(\in M(B))$ est l'image de la mesure $\sigma(\in M^+(A))$.

Les intervalles aléatoires I_n sont associés à une suite (ℓ_n) qui, par définition, est obtenue en répétant ν_j fois $d_j (j = 1, 2, \ldots)$, et ν_j sera précisé dans un instant. Observons que le noyau correspondant est

$$k(t) = \exp \sum (\ell_n - t)^+ = \exp \sum \nu_j (d_j - t)^+.$$

Appelons G_j la réunion des I_n de longueur d_j. Étant donnés des points $x, y, \ldots$ à des distances mutuelles $\geq d_j$, on a

$$P(x \notin G_j) = (1 - d_j)^{\nu_j} \simeq e^{-\nu_j d_j}$$

$$P(x \notin G_j, y \notin G_j) = (1 - 2d_j)^{\nu_j} \simeq e^{-2\nu_j d_j}$$

et ainsi de suite. Si d_{j+1} est très petit par rapport à d_j, on peut presque identifier un intervalle D_{j+1} à son centre x, et écrire

$$P(D_{j+1} \subset G_j) \simeq P(D_{j+1} \cap G_j \neq \emptyset) \simeq P(x \in G_j)$$

et de même pour plusieurs intervalles D_{j+1}. Choisissons

$$\nu_j d_j \simeq b, \quad e^b = 2^\alpha.$$

Considérons comme vivants les intervalles D_{j+1} disjoints de G_j, et comme morts ceux qui touchent G_j. Pratiquement, cela consiste à laisser vivre chaque D_{j+1} avec probabilité

$2^{-\alpha}$, et à le tuer avec probabilité $1-2^{-\alpha}$, les choix étant indépendants pour les différents D_{j+1}. Il revient donc au même de dire que B est recouvert p.s. par $\bigcup_1^\infty I_n$, ou que A est tué p.s. par le processus donné. Or

$$k(d_j) = \exp \sum_{i<j} \nu_i(d_i - d_j) \simeq \exp(j-1)b.$$

Comme nous l'avons remarqué, cela entraîne

$$\mathrm{Cap}_\alpha A = 0 \Leftrightarrow \mathrm{Cap}_k B = 0.$$

Comme $\mathrm{Cap}_k B = 0$ est la condition nécessaire et suffisante pour que B soit recouvert p.s. par les I_n, $\mathrm{Cap}_\alpha A = 0$ est la condition nécessaire et suffisante pour que A soit tué p.s., CQFD.

Nous venons de considérer le cas $T = \{0,1\}^{\mathcal{N}}$. Le cas $T = \{0,1,\ldots c-1\}^{\mathcal{N}}$ se traite exactement de la même façon, en prenant sur T la distance c–adique

$$\mathrm{dist}(t,s) = c^{-n} \Leftrightarrow (t_1 = s_1, \ldots t_n = s_n, t_{n+1} \neq s_{n+1}),$$

et en définissant la probabilité de survie p comme $c^{-\alpha}$. *Le théorème garde alors la même forme : A est tué si et seulement si* $\mathrm{Cap}_\alpha A = 0$.

Désignons par W_α une variable aléatoire telle que $P(W_\alpha = c^\alpha) = c^{-\alpha}$ et $P(W_\alpha = 0) = 1 - c^{-\alpha}$, et par Q_α l'opérateur correspondant. *Le théorème exprime que Q_α vit sur $\mathcal{R}_\alpha$ et meurt sur $\mathcal{S}_\alpha$.* Nous allons tout de suite l'utiliser sous cette forme.

Soit encore $T = \{0,1,\ldots c-1\}^{\mathcal{N}}$, muni de la distance c–adique et de la probabilité naturelle λ, et soit maintenant W une variable aléatoire ≥ 0 soumise à la seule condition $EW = 1$. Désignons par Q l'opérateur correspondant (rappelons la définition des P_n:

$$P_n(t) = \sum_{C \text{ d'ordre } n} W_C 1_C(t)).$$

Nous allons établir tout à l'heure le théorème suivant.

Théorème 2 *Si $EW \log_c W > 1, Q$ est complètement dégénéré, c'est-à-dire meurt sur toutes les mesures $\sigma \in M^+(T)$. Si $EW \log_c W < 1, Q$ vit sur λ.*

(En fait, quand $EW \log_c W = 1$ on a $Q\lambda = 0$ (théorème 1 de [KP1976], et il ne doit pas être trop difficile de montrer que Q est complètement dégénéré).

Admettons le théorème 2 pour le moment, et supposons

$$1 - EW \log_c W = d > 0.$$

Considérons W_α comme ci-dessus, un échantillon $(W_{\alpha,C})$ indépendant des (W_C), et l'opérateur Q_α correspondant. Il y a deux façons de réaliser l'opérateur produit $Q_\alpha Q$. La première est de faire opérer Q puis Q_α. La seconde est de prendre la suite des poids $(W_{\alpha,C}W_C)$, et l'opérateur correspondant. Observons que

$$\begin{aligned} E(W_\alpha W \log_c(W_\alpha W)) &= E(W_\alpha \log_c W_\alpha) + E(W \log_c W) \\ &= \alpha + E(W \log_c W) \\ &= \alpha - d + 1. \end{aligned}$$

Si $\alpha < d, Q_\alpha Q$ vit sur λ, donc $P(Q\lambda \neq 0) > 0$ et, lorsque $Q\lambda \neq 0$, Q_α vit sur $Q\lambda$, donc $Q\lambda \in \mathcal{R}_\alpha$. Si $\alpha > d, Q_\alpha Q$ meurt sur λ, donc Q_α sur $Q\lambda$, donc $Q\lambda \in \mathcal{S}_\alpha$ p.s. Énonçons le résultat.

Théorème 3 *En supposant* $1 - E(W \log_c W) = d > 0$ *on a* $P(Q\lambda \neq 0) > 0$ *et presque sûrement pour tout* $\epsilon > 0$

$$Q\lambda \in \mathcal{S}_{d+\epsilon} \cap \mathcal{R}_{d-\epsilon},$$

donc $Q\lambda$ *est unidimensionnelle et de dimension* d *dès que* $Q\lambda \neq 0$.

Remarquons que le cas $P(Q\lambda = 0) > 0$ n'est pas exclu: c'est ce qui arrive quand $W = W_\beta$. Remarquons aussi qu'en fait,

$$Q\lambda \in \mathcal{S}_d \cap \mathcal{R}_{d-\epsilon}.$$

Établissons maintenant le théorème 2. La proposition 5 du chapitre 2 s'applique facilement et donne que, si

$$EW^h \leq c^{h-1}$$

avec $0 < h < 1, Q$ est complètement dégénéré. La proposition 3 s'applique également (pour le détail, consulter [KP1976]) et montre que, si

$$EW^h < c^{h-1}$$

avec $h > 1, Q$ vit sur λ et $E(Q\lambda(T))^h < \infty$. Posons, comme Benoît Mandelbrot [M1974(1),(2)],

$$\varphi(h) = \log_c EW^h - (h-1).$$

C'est toujours défini pour $0 \leq h \leq 1$, et

$$\varphi'(1-0) = EW \log_c W - 1 (= -d).$$

Si $d < 0$, on applique le premier résultat: Q est complètement dégénéré. Si $d > 0$, on applique le second: Q vit sur λ (il est commode de supposer que $\varphi(h)$ existe au voisinage de 1, mais on peut se passer de cette hypothèse, cf. [KP1976]). C'est le théorème 2.

Il est intéressant de calculer $\varphi(h)$ dans des cas particuliers. Si $W = W_\alpha, EW_\alpha^h = c^{-\alpha}c^{\alpha h}$ donc

$$\varphi(h) = -(1-\alpha)(h-1), \quad d = 1 - \alpha.$$

Si $W = e^{\tau\xi - (1/2)\tau^2}$, où ξ est une v.a. normale standard, et $\tau > 0$,

$$EW^h = \exp\frac{1}{2}(\tau^2h^2 - \tau^2h)$$

$$\varphi(h) = \left(\frac{\tau^2}{2\log c}h - 1\right)(h-1)$$

$$d = 1 - \frac{\tau^2}{2\log c}.$$

Dans le cas général, $\varphi(h)$ est une fonction convexe, dérivable à l'intérieur du plus grand intervalle où elle est finie, nulle en $h = 1$, et d s'interprête soit comme l'opposé de la pente du graphe en ce point, soit comme l'ordonnée de l'intersection de la tangente au graphe en (1,0) avec l'axe vertical:

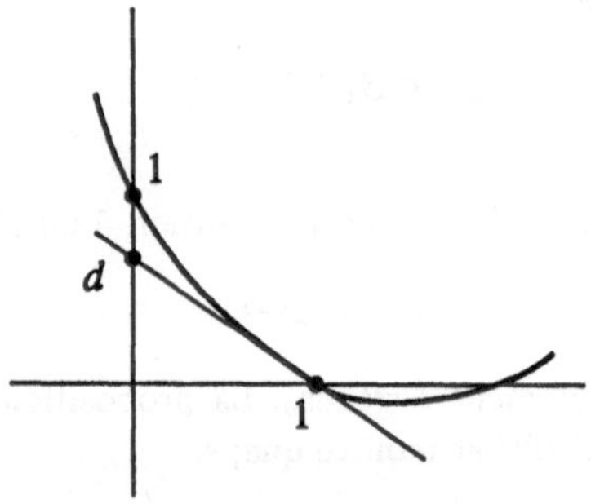

Nous allons maintenant associer à W non pas un, mais une famille d'opérateurs Q^q, correspondant aux poids $\frac{W^q}{EW^q}$. À la mesure σ correspond la famille de mesures aléatoires $S_q = Q^q\sigma$, et $S_1 = S$. Les fonctions $\varphi_q(h)$ correspondantes sont

$$\begin{aligned}\varphi_a(h) &= \log_c \frac{EW^{qh}}{(EW^q)^h} - (h-1)\\ &= \varphi(qh) - h\varphi(q)\end{aligned}$$

et les dimensions des S_q sont donc

$$d_q = -\varphi_q'(1) = \varphi(q) - q\varphi'(q)$$

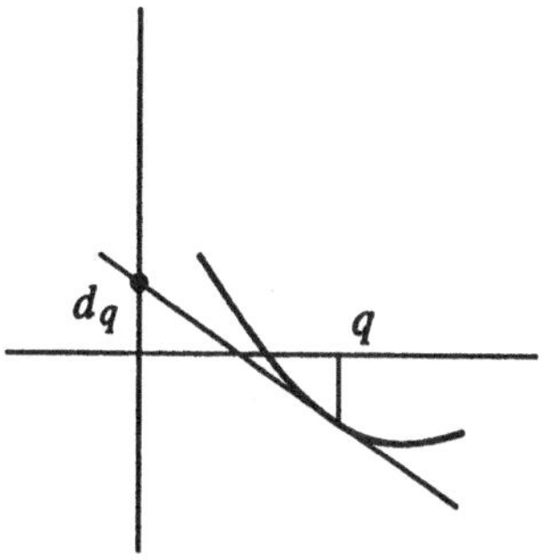

lorsque ces quantités sont positives. Remarquons que d_q est l'ordonnée de l'intersection avec l'axe vertical de la tangente au graphe de $\varphi(h)$ au point $(q, \varphi(q))$. Ici comme dans la suite, q est un point intérieur à l'intervalle de finitude de φ.

Introduisons les probabilités de Peyrière. On suppose $\sigma(T) = 1$. Sur l'espace $T \times \Omega$, $\mathcal{Q}_q$ est la probabilité définie par

$$\int f(t,\omega) d\mathcal{Q}_q(t,\omega) = E \int f(t,\omega) dS_q(t).$$

On sait (voir leçon 2) que les poids $P_n(t)$ sont $\mathcal{Q}_q$–indépendants.

On vérifie aisément, en distinguant dans S_q ce qui vient de $P_n(t)$ et ce qui vient des autres $P_m(t)$, la formule

$$E \int f(P_n(t)) dS_q(t) = E\left(\frac{W^q}{EW^q} f(W)\right).$$

En conséquence, la loi des grands nombres s'applique à

$$\frac{1}{n}(\log P_1(t) + \cdots + \log P_n(t))$$

et donne

$$\frac{1}{n} \log Q_n(t) \to E\left(\frac{W^q}{EW^q} \log W\right) \qquad \mathcal{Q}_q - \text{ p.s.} \tag{1}$$

Désignons par $I_n(t)$ la cellule d'ordre n contenant t. Nous allons montrer que, moyennant une condition sur W, on a

$$\lim_{n\to\infty} \frac{\log S(I_n(t))}{\log \lambda(I_n(t))} = \alpha_q \qquad \mathcal{Q}_q - \text{ p.s.} \tag{2}$$

avec $\alpha_p = -\varphi'(q)$. Rappelons que $S = Q\lambda$. Écrivons

$$S = Q_n S^{(n)}$$

et observons que, pour chaque $t, c^n S^{(n)}(I_n(t))$ a la même distribution que $S(T)$. Comme

$$\frac{\log S(I_n(t))}{\log \lambda(I_n(t))} = \frac{1}{-n \log c}(\log Q_n(t) + \log S^{(n)}(I_n(t))).$$

(2) résulte de (1) et de

$$\lim_{n\to\infty} \frac{1}{n} \log(c^n S^{(n)}(I_n(t))) = 0 \quad \mathcal{Q}_q - \text{ p.s.} \tag{3}$$

qui reste à démontrer; on aura bien

$$\alpha_p = -E\left(\frac{W^q}{EW^q} \log_c W\right) + 1 = -\varphi'(q).$$

Pour avoir (3), il suffit que, pour un $N > 1$, la série

$$\sum |\frac{1}{n} \log(c^n S^{(n)}(I_n(t)))|^N$$

converge $\mathcal{Q}_q$ p.s., donc il suffit que

$$E \int \sum |\frac{1}{n} \log(c^n S^{(n)}(I_n(t)))|^N dS_q(t) < \infty,$$

donc il suffit que

$$E \int |\log(c^n S^{(n)}(I_n(t)))|^N dS_q(t) = O(1)$$

et le premier membre est la valeur commune (indépendante de t) des

$$c^n E((\log |c^n S^{(n)}(I_n(t))|)^N S_q(I_n(t)))$$

c'est-à-dire

$$E(|\log S(T)|^N S_q(T)).$$

Il suffit donc que cette dernière espérance soit finie. Il suffit donc (Hölder) que pour un $h > 1$ on ait simultanément

$$E(S_q(T))^h < \infty$$

$$E|\log S(T)|^{\frac{Nh}{h-1}} < \infty.$$

La première condition est satisfaite dès que $\varphi_q(h) < 0$ (voir [KP1976]), ce qui est garanti lorsque la fonction φ est définie au voisinage de q et que $d_q > 0$ (rappelons que $d_q = \varphi_q'(1)$). La seconde exprime que la distribution de $S(T)$ n'est pas trop concentrée au voisinage de 0. Il est facile de vérifier qu'elle est satisfaite pour tout $h > 1$ si et seulement si

$$P(S(T) < e^{-x}) = O(x^{-A}) \quad \text{pour tout} \quad A > 0 (x \to \infty). \tag{4}$$

Résumons à ce stade.

Théorème 4 *Soit q un point intérieur à un intervalle où la fonction φ est finie et dérivable. Supposons $d_q = \varphi(q) - q\varphi'(q) > 0$, et supposons de plus la condition (4). Alors*

$$\lim_{n\to\infty} \frac{\log S(I_n(t))}{\log \lambda(I_n(t))} = \alpha_q = -\varphi'(q) \quad \mathcal{Q}_q - \ p.s.$$

et en conséquence

$$\lim_{n\to\infty} \frac{\log S(I_n(t))}{\log \lambda(I_n(t))} = \alpha_q \quad S_q - \ p.p.$$

presque sûrement.

Corollaire *L'ensemble des points t tels que*

$$\lim_{n\to\infty} \frac{\log S(I_n(t))}{\log \lambda(I_n(t))} = \alpha_q$$

a une dimension $\geq d_q$ presque sûrement.

Voici une formule permettant de calculer d_q en fonction de α_q; c'est

$$d_q = \inf_x(\varphi(x) + \alpha_q x)$$

(évident sur l'interprétation géométrique).

Reste à exprimer la condition (4) en fonction de la distribution de W. Posons $Y = S(T)$ et partons de l'égalité

$$cY = W_1Y_1 + \cdots + W_cY_c,$$

où les variables aléatoires du second membre sont toutes indépendantes, les W_j ayant la distribution de W et les Y_j celle de Y. Posons

$$\mu(w) = P(W < w) \quad (w > 0)$$
$$\ell(t) = E(e^{-tY}) \quad (t > 0).$$

L'égalité ci-dessus équivaut à

$$E(e^{-ctY}) = (E(e^{-tWY}))^c$$

soit

$$\ell(ct) = \left(\int \ell(wt)d\mu(w)\right)^c. \tag{5}$$

À partir de cette égalité des majorations sur $\mu(w)$ permettent d'obtenir des majorations sur $\ell(t)$, qui donnent à leur tour des majorations sur $P(Y < y)$ en vertu de l'inégalité

$$P(Y < y) \leq e\ell(\frac{1}{y})$$

(cas particulier de

$$e^{-ty}P(Y < y) \leq E(e^{-tY}) \quad (t > 0)).$$

Énonçons comme théorème deux cas simples.

Théorème 5 *Soit $\alpha > 0$.*

(1) *si $\mu(w) = O(|\log w|^{-\alpha})(w \to 0)$, on a $P(Y < y) = O((\log \frac{1}{y})^{-c\alpha})(y \to 0)$.*
(2) *si $\mu(w) = O(w^{\alpha})(w \to 0)$, on a $P(Y < y) = O(y^{c\alpha})(y \to 0)$.*

Pour démontrer (1), choisissons dans (5) $t = c^{2k-1}$ et intégrons sur les deux intervalles $w \leq c^{1-k}$ et $w > c^{1-k}$. On obtient

$$\ell(c^{2k}) \leq (\mu(c^{1-k}) + \ell(c^k))^c.$$

L'hypothèse est $\mu(c^{1-k}) = O(k^{-\alpha})$. On vérifie que $\ell(c^k) = O(k^{-\alpha c})$ (c'est-à-dire la conclusion) en se restreignant à $k = 2^j$, par induction sur j.

Pour démontrer (2), choisissons dans (5) $t = c^k$ et intégrons sur les intervalles $[0, c^{-k}]$, $[c^{-k}, c^{1-k}]$, $[c^{1-k}, c^{2-k}], \cdots, [c^{-1}, 1]$, $[1, \infty]$. On obtient

$$\ell(c^{k+1}) \leq (\mu(c^{-k}) + \sum_{j=1}^{k} \mu(c^{j-k})\ell(c^{j-1}) + \ell(c^k))^c.$$

L'hypothèse est $\mu(c^{-k}) = O(c^{-\alpha k})$, et on établit la conclusion, $\ell(c^k) = O(c^{-\alpha ck})$, par induction sur k.

En corollaire du théorème 5, on peut remplacer la condition (4) *dans l'énoncé du théorème 4, par*

$$P(W < e^{-x}) = O(x^{-A}) \quad \text{pour tout} \quad A > 0 \quad (x \to \infty). \tag{6}$$

Pour terminer ce chapitre, il est bon d'introduire quelques formules et d'en rappeler d'autres.

Supposons d'abord c (donc T) fixé. À l'opérateur Q associé au poids W on associe sont *entropie*

$$\epsilon(Q) = E(W \log_c W)$$

et sa fonction φ

$$\varphi(h) = \log_c EW^h - (h-1)$$

définie pour $0 \leq h \leq 1$, et peut-être au delà. Donc

$$\epsilon(Q) = 1 + \varphi'(1-0).$$

Pour l'opérateur Q_α (cas où $P(W = c^\alpha) = c^{-\alpha}$ et $P(W = 0) = 1 - c^{-\alpha}$),

$$\epsilon(Q) = \alpha.$$

Quand on compose deux opérateurs Q' et Q'' (en les supposant indépendants), on ajoute leurs entropies:

$$\epsilon(Q'Q'') = \epsilon(Q') + \epsilon(Q'').$$

L'opérateur Q_α vit sur $\mathcal{R}_\alpha$ et meurt sur $\mathcal{S}_\alpha$. En d'autres termes, une mesure tuée par Q_α appartient à $\mathcal{S}_\alpha$, et réciproquement. Partons de $\sigma \in \mathcal{S}_{\alpha+\beta}$, et considérons deux opérateurs Q_α et Q_β indépendants: comme $Q_\beta Q_\alpha$ a la loi de $Q_{\alpha+\beta}$, $Q_\alpha\sigma$ est tuée par Q_β, donc $Q_\alpha\sigma \in \mathcal{S}_\beta$ p.s. De même, lorsque $\sigma \in \mathcal{R}_{\alpha+\beta}, Q_\alpha\sigma \in \mathcal{R}_\beta$ p.s. L'entropie de Q_α mesure précisément la perte de régularité quand on passe de σ à $Q_\alpha\sigma$.

Dans le cas général il en est presque de même: le spectre de dimension de $Q\sigma$ s'obtient en translatant à gauche le spectre de dimension de σ par $\epsilon(Q)$, et en supprimant la partie à gauche de 0. C'est la signification du théorème 3 (quand $\sigma = \lambda$, c'est-à-dire que le spectre de dimension de σ est δ_1).

Quand $W = e^{\tau\xi-(1/2)\tau^2}$ (ξ normale standard, $\tau > 0$),

$$\begin{aligned}\epsilon(Q) &= \frac{\tau^2}{2\log c}\\ \varphi(h) &= (h\epsilon(Q)-1)(h-1).\end{aligned}$$

Pour les opérateurs Q^q, correspondant aux poids $\frac{W^q}{EW^q}$, on a la règle d'itération

$$(Q^q)^{q'} = Q^{qq'}$$

(l'égalité signifiant l'égalité des lois), et les formules

$$\begin{aligned}\varphi_q(h) &= \varphi(qh) - h\varphi(q)\\ \epsilon(Q^q) &= 1 + q\varphi'(q) - \varphi(q).\end{aligned}$$

Lorsque W vérifie la condition (6), et que p, q, r sont intérieurs à l'intervalle $I = \{h : \epsilon(Q^h) < 1\}$, on a

$$\lim_{n\to\infty} \frac{\log S_p(I_n(t))}{\log S_q(I_n(t))} = \frac{p\varphi'(r) - \varphi(p)}{q\varphi'(r) - \varphi(q)} \quad S_r - \text{ p.p.}$$

donc sur un ensemble de dimension $\varphi(r) - r\varphi'(r)$ (en supposant pour simplifier φ dérivable

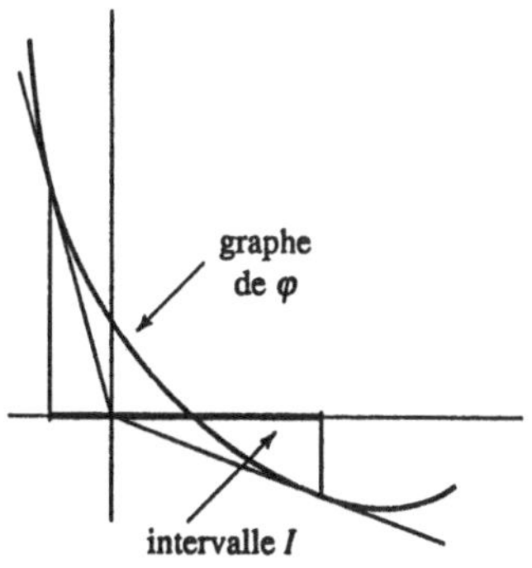

sur I; on part de

$$\lim_{n\to\infty} \frac{\log S_p(I_n(t))}{\log S_q(I_n(t))} = -\varphi_p'(\frac{r}{p}) \quad S_r - \text{ p.p.}$$

Pour élargir le champ de ces formules il est bon de laisser c indéterminé. Supposons pour fixer les idées que la fonction

$$\psi(h) = \log EW^h$$

est définie et dérivable pour tout h (c'est le cas quand $W = e^{\xi - 1/2}$; alors $\psi(h) = \frac{1}{2}h(h-1)$), et que $\gamma = \frac{1}{\log c}$ est petit. On a

$$\varphi(h) = \gamma\psi(h) - (h-1)$$

$$\varphi(h) - h\varphi'(h) = \gamma(\psi(h) - h\psi'(h)) + 1$$

et l'intervalle I ci-dessus est défini par

$$\psi(h) - h\psi'(h) \geq -\frac{1}{\gamma}$$

(dans le cas ci-dessus, $-\frac{1}{2}h^2 \geq -\frac{1}{\gamma}$).

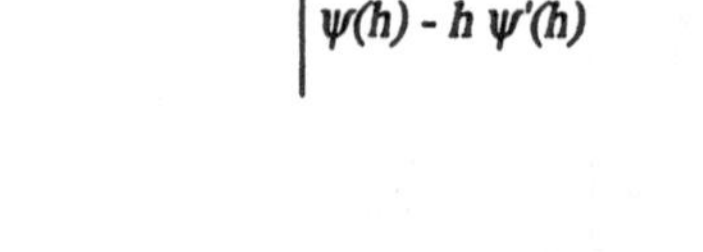

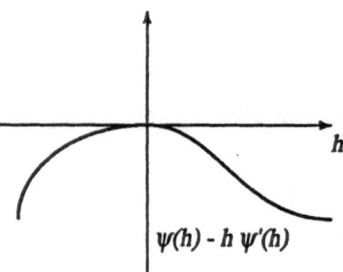

Au facteur α près, la fonction positive $h\psi'(h) - \psi(h)$ est égale à l'entropie $\epsilon(Q^h)$.

Indiquons pour finir l'allure des graphes des fonctions

$$\alpha(h) = -\varphi'(h)$$

$$d(h) = \varphi(h) - h\varphi'(h)$$

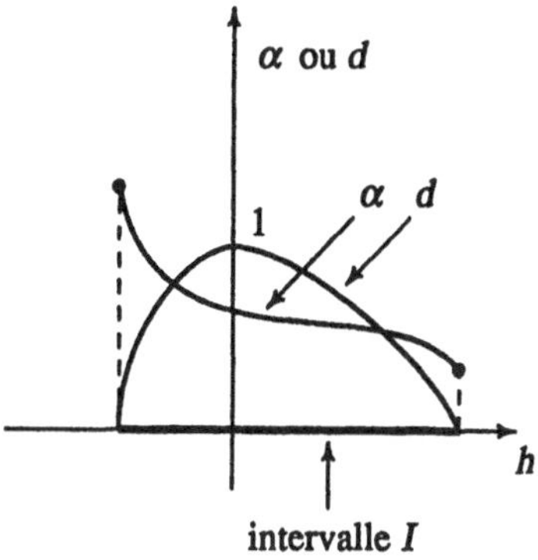

ainsi que la variation de d en fonction de α.

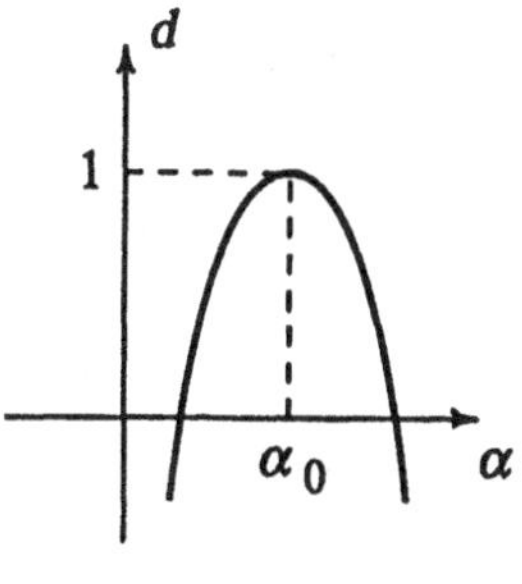

Cinquième leçon

Nous donnerons très brièvement ici la théorie du chaos multiplicatif gaussien: c'est l'opérateur Q correspondant à des poids P_n log-normaux. Le terme de chaos a été utilisé par N. Wiener en 1938 pour désigner des processus indexés par des ensembles, comme

$$E \to \int_E dW(t)$$

($W(t)$ étant la fonction de Wiener du mouvement brownien) qu'il appelle le chaos pur sur $\mathcal{R}$, et à l'aide duquel il construit des chaos plus compliqués.

En fait, comme l'a noté Kakutani, le chaos pur est plus facile à définir que le mouvement brownien, et permet de construire très rapidement la fonction de Wiener. On part d'une mesure μ localement finie définie sur un espace localement compact, et de l'espace de Hilbert réel $L^2(\mu)$. On prend $(\Omega, \mathcal{A}, P)$ un espace de probabilité assez riche pour qu'il existe un échantillon infini de loi $N(0,1)$; il engendre alors dans $L^2(\Omega)$ réel un espace de Hilbert séparable $\mathcal{H}$ que nous appelerons espace de Hilbert gaussien. Soit W un isomorphisme isométrique de $L^2(\mu)$ dans $\mathcal{H}$: c'est, par définition, une version du chaos additif de Wiener. En particulier, si on prend pour μ la mesure de Lebesgue sur $\mathcal{R}$, on peut définir le processus de Wiener comme l'image par W des fonctions indicatrices $1_{[0,t]}$ (ou $1_{[t,0]}$ si $t < 0$); en posant

$$W_t = W(1_{[0,t]}),$$

W_t décrit dans $\mathcal{H}$ l'hélice brownienne, qui satisfait

$$||W_t - W_s||^2 = |t - s|$$

et toutes les bonnes propriétés (effet des translations du paramètre, dimension 2, mesure dans la dimension 2, indépendance du futur par rapport au passé) qui s'ensuivent.

Par analogie avec le chaos additif de Wiener, Fan Ai hua a récemment défini les chaos additifs de Lévy. Il s'agit pour chaque $\alpha (0 < \alpha < 2$ et $\alpha \neq 1)$, d'un isomorphisme isométrique entre l'espace $L^\alpha(\mu)$ et un espace de v.a. α–stables. Et son exponentiation peut se réaliser comme on va le faire dans le cas gaussien [F1989(1)].

La première motivation du chaos multiplicatif gaussien se trouve dans les modèles probabilistes de la turbulence (A.N. Kolmogorof 1962, B. Mandelbrot 1974). Je renvoie pour l'historique à mon article [K1985] qui contient aussi les preuves développées (à l'exception de celles qui se trouvent dans [K1987(2)].

Sur un espace métrique et localement compact T est défini un processus $X(t) (\in \mathcal{H}, t \in T)$ de sorte que les distributions jointes sont aussi gaussiennes. La loi du processus X est donc bien définie par sa covariance, un noyau de type positif

$$p(t,s) = \mathcal{E}X(t)X(s).$$

Ayant un tel processus, on l'exponentie et puis le normalise pour obtenir un poids

$$P(t) = \exp(X(t) - \frac{1}{2}\mathcal{E}(X(t))^2), \qquad \mathcal{E}P(t) = 1.$$

Les relations suivantes sont faciles à établir:

$$\mathcal{E}P(t)P(s) = \exp p(t,s)$$

$$\mathcal{E}P(t_1)\cdots P(t_n) = \exp p(t_1,\cdots,t_n) \quad \text{avec} \quad p(t_1,\cdots,t_n) = \sum_{1\leq j<k\leq n} p(t_j,t_k)$$

$$\mathcal{E}(P(t))^h = \exp(\frac{1}{2}(h^2 - h)p(t,t))$$

$$\mathcal{E}P(t)\log P(t) = \frac{1}{2}p(t,t).$$

Remarquons que si P se décompose en produit de deux poids indépendants, sa covariance est la somme de celles de ses facteurs

$$p(t,s) = p_1(t,s) + p_2(t,s).$$

Inversement, si un noyau se décompose en somme de deux noyaux, on a une égalité en loi entre les poids correspondants

$$P(t) = P_1(t)P_2(t).$$

Supposons que l'on ait une suite de processus indépendants X_n. Par exponentiation, on a une suite de poids indépendants (P_n). Formons alors les produits

$$Q_n = P_1 \cdots P_n.$$

D'après la remarque précédente, le noyau correspondant à Q_n est

$$q_n = p_1 + \cdots + p_n.$$

Posons

$$q(t,s) = \sum_{n=1}^{\infty} p_n(t,s).$$

Si les noyaux p_n sont positifs, on dit que $q(t,s)$ est un noyau de type σ–positif.

Étant donné une mesure $\sigma \in M^+(T)$; d'après un théorème général (proposition 1 de la leçon 2), la limite vague

$$Q_n\sigma \xrightarrow{*} Q\sigma$$

existe. Cette limite est une mesure aléatoire qui est l'objet de nos études.

Posons maintenant les problèmes.

Problème 1 (dégénérescence) Quand a-t-on $Q\sigma = 0$ p.s.?

Problème 2 (action pleine) Quand a-t-on $\mathcal{E}Q\sigma = \sigma$? Autrement dit, quand la martingale $Q_n\sigma(T)$ converge-t-elle dans $L^1(\Omega)$? ou quand

$$\mathcal{E}f(Q_n\sigma(T)) = O(1)$$

pour une certaine fonction convexe f telle que $\lim_{x\to\infty} x^{-1}f(x) = \infty$?

Problème 3 (existence de moments) Pour $p > 1$, quand la martingale $Q_n\sigma(T)$ converge-t-elle dans $L^p(\Omega)$? Ceci est équivalent à l'action pleine sur σ et l'existence du moment d'ordre p.

Problème 4 (diffusion, concentration) Quand $Q\sigma \in \mathcal{R}_\alpha$? Quand $Q_\sigma \in \mathcal{S}_\alpha$?

Problème 5 (unicité) Est-il vrai que la loi du chaos multiplicatif ne dépend que de la somme $q(t, s)$?

Problème 6 (comparaison) Améliore-t-on les réponses aux problèmes 1 à 3 si l'on diminue q? Si l'on contracte σ?

Problème 7 (dépendance du paramètre) Supposons que T est $\mathcal{R}^d$ euclidien et que

$$q(t, s) = u \log^+ \frac{1}{d(t, s)} + O(1).$$

Décider, autant que possible, des réponses aux problèmes 1 à 4 en fonction de u. Dans ce cas-là, l'opérateur est désigné par Q_u. Son entropie est définie par

$$\epsilon(Q_u) = \frac{u}{2}.$$

Comme dans le cas des cascades, on a

$$\epsilon(Q_u Q_{u'}) = \epsilon(Q_u) + \epsilon(Q_{u'}).$$

Dans ce qui suit, on va essayer de répondre à ces problèmes successivement.

Nous n'avons pas de réponses complètes aux problèmes 1 et 2. Mais certaines réponses partielles existent. Elles sont attachées à la réponse au problème 4 dans le cas décrit au problème 7. C'est-à-dire si $\epsilon(Q_u) > \dim \sigma$, on a la dégénérescence; si $\epsilon(Q_u) < \dim \sigma$, on a la non-dégénérescence, et même l'opérateur Q_u agit pleinement sur toute mesure appartenant à $\bigcap_{\alpha > \epsilon(Q_u)} \mathcal{R}_\alpha$.

Le problème 3 est relativement simple. Quand $p = m$, un entier ≥ 2, on a une réponse complète: pour avoir la convergence dans L^m il faut et suffit que l'on ait

$$\int \cdots \int \exp q(t_1, \cdots, t_m) d\sigma(t_1, \cdots, t_m) < \infty$$

comme c'est une intégrale m-uple, il peut être utile de trouver des conditions plus parlantes. En voici une suffisante

$$\int \left(\int \exp \frac{m}{2} q(t, s) d\sigma(t) \right)^{m-1} d\sigma(s) < \infty.$$

En particulier, il suffit que

$$\sup_{s \in \operatorname{supp} \sigma} \int \exp \frac{m}{2} q(t, s) d\sigma(t) < \infty$$

(voir [F1986]). Inspiré par cela, on hasarde ici une conjecture au problème 2:

$$\int\int \exp \frac{1}{2} q(t,s) d\sigma(t) d\sigma(s) < \infty \Rightarrow \mathcal{E}Q\sigma = \sigma.$$

Quant au problème 5, la réponse est oui sous réserve de la positivité des noyaux p_n. La démonstration repose sur le lemme suivant, qui est dans l'esprit du lemme de Slepian concernant les processus gaussiens: si $p_0 \leq p_1$, on a l'inégalité

$$Ef \int p_0 \sigma \leq Ef \int p_1 \sigma$$

pour toute fonction convexe f.

Comme conséquence de la résolution du problème de l'unicité, on peut dire que quand on diminue le noyau q, on améliore les réponses; quand on contracte σ, on empire les réponses. Ceci est une réponse au problème de comparaison.

Le théorème de comparaison permet d'aborder le problème 7 à partir du cas des cascades log-normales, étudiées dans la leçon précédente. En effet, les cascades log-normales (cas $W = e^{\tau\xi - \frac{1}{2}\tau^2}$) définissent un chaos multiplicatif gaussien sur $T = \{0, 1, \cdots c-1\}^{\mathcal{N}}$ dont la fonction q est

$$q_c(t,s) = \tau^2 \log_c \frac{1}{d_c(t,s)}$$

$d_c(t,s)$ désignant la distance ultramétrique sur T. Si K est un ensemble de Cantor dans $\mathcal{R}^n$, image de T par une application qui ne modifie pas trop les distances, on connaîtra les propriétés du chaos multiplicatif $(K, d, u \log \frac{1}{d(t,s)})$ à partir de celles du chaos multiplicatif $(T, d_c, \tau^2 \log_c \frac{1}{d_c(t,s)})$. Avec la terminologie introduite dans ce cours voici le résultat le plus saillant relatif au problème 7.

Théorème *L'opérateur Q_u transforme toute mesure $\sigma \in M^+(\mathcal{R}^n)$ en une mesure aléatoire S dont le spectre de dimension s'obtient en translatant à gauche par $\epsilon(Q_u)(= \frac{u}{2})$ le spectre de dimension de σ, et en supprimant la partie de ce translaté qui est à gauche de* 0.

La méthode étant clairement indiquée, et développée ailleurs [K1985(2), K1989], la preuve du théorème est laissée en exercice au lecteur.

Naturellement on peut travailler sur $\mathcal{T}^n$ au lieu de $\mathcal{R}^n$. Restreignons nous au cas $q(t,s) = q(t-s)$. Pour $n = 1$, il est commode de définir $q(t)$ par sa série de Fourier

$$\sum_{m \in \mathcal{Z}} \hat{q}_m e^{2\pi i m t}.$$

Les plus commodes des fonctions de type σ-positif sont les fonctions de Polya: $\hat{q}_m = \hat{q}_{-m}$, et la suite $\hat{q}_m (m \geq 0)$ est décroissante, convexe, et tend vers 0 à l'infini. En effet, toute suite $\hat{q}_m$ de ce type peut s'écrire

$$\hat{q}_m = \sum_{j=1}^{\infty} \alpha_j (1 - \frac{m}{j})^+, \alpha_j \geq 0$$

donc

$$q(t) = \sum_{j=0}^{\infty} \alpha_j K_j(t)$$

$K_j(t)$ désignant le j-ième noyau de Fejér (qui est positif et de type positif), donc q est de type σ–positif. Dans ce cadre, le problème 7 se pose avec

$$q_u(t) = u \left(1 + \sum_{m=1}^{\infty} \frac{1}{m} \cos 2\pi m t\right)$$

et l'opérateur Q_u est, formellement, la multiplication par

$$\exp u \left(\xi_0 + \sum_{m=1}^{\infty} \frac{\xi_m}{\sqrt{m}} \cos 2\pi m t - 1 - \sum_{m=1}^{\infty} \frac{1}{m}\right).$$

On voit le rôle critique de $u = 1$ dans le problème d'exponentiation des séries aléatoires gaussiennes posé lors de la première leçon.

Sixième leçon

L'exposé oral comprenait une rétrospective sur le mouvement brownien et sur les processus de Lévy, que nous laisserons ici de côté. Par contre nous allons expliciter quelques problèmes relatifs aux recouvrements aléatoires d'une part, aux multimesures et multifractales liées au chaos multiplicatif d'autre part. Nous conclurons par quelques compléments sur les recouvrements aléatoires dans le cas poissonnien, et sur la relation entre recouvrements poissonniens et processus de Lévy.

Problèmes liés aux recouvrements aléatoires

Limitons nous à un cas simple: T^d est le tore de dimension d, K un compact dans T^d, $1 > v_1 \geq v_2 \geq \cdots \geq v_n \geq \cdots > 0$ une suite décroissante, $B_n(= B_n(\omega))$ une suite de boules dans T^d, de volumes v_n, dont les centres sont disposés aléatoirement et indépendamment sur T^d. On cherche à quelle condition $K \subset \bigcup_1^\infty B_n$ p.s.

1. La méthode de martingale–L^2 exposée dans le cas $d = 1$ donne une condition nécessaire, sous la forme $\mathrm{Cap}_k K = 0, k(.)$ étant un noyau lié de façon simple à la suite (v_n). Est-ce aussi une condition suffisante?

2. Est-ce une condition suffisante dans le cas $K = \mathcal{T}^d$? La condition $\mathrm{Cap}_k \mathcal{T}^d = 0$ doit se lire comme $k \notin L^1(\mathcal{T}^d)$.

3. La condition $k \notin L^1(\mathcal{T}^d)$ est réalisée quand $v_n = \frac{1}{n+1}$. A-t-on recouvrement de $\mathcal{T}^d$ dans ce cas?

4. Quand $v_n = \frac{1}{n+1}$ on sait que $\mathcal{T}^d \setminus \bigcup_1^\infty B_n$ est au plus dénombrable p.s. Est-il possible de choisir les v_n de façon que $\mathcal{T}^d \setminus \bigcup_1^\infty B_n$ soit au plus dénombrable p.s., et non vide avec probabilité positive (ce serait bien étonnant)?

5. Quand $v_n = \frac{a}{n} (0 < a < 1)$ on sait que $K \setminus \bigcup_1^\infty B_n$ est vide p.s. lorsque $ad > \dim K$ (dimension de Hausdorff), et a pour dimension $\dim K - ad$ avec probabilité positive lorsque $ad < \dim K$ (l'alternative p.s. étant $\dim((K \setminus \bigcup_1^\infty B_n)) = \dim K - ad$ ou $K \setminus \bigcup_1^\infty B_n = \emptyset$). En particulier, la dimension de Hausdorff de $\mathcal{T}^d \setminus \bigcup_1^\infty B_n$ est $d - ad$ avec probabilité positive. Qu'en est-il de la dimension topologique?

6. (Complément à la question posée par Russell Lyons) Considérons le disque $|z| < 1$ muni de la métrique hyperbolique $\left(ds = \frac{2|dz|}{1-|z|^2}\right)$, le processus de Poisson F ponctuel associé à l'aire hyperbolique, et la réunion F_r des disques hyperboliques de rayon r centrés sur F. À quelle condition la réunion des demi-droites issues de 0 et rencontrant F est-elle p.s. le disque entier? Nous connaissons la réponse: c'est $r \geq r_0 (= (2 + \sqrt{5})^{-1/2})$. À quelle condition est-il presque sûr que, pour tout point P du disque, la réunion des demi-droites hyperboliques issues de P et rencontrant F_r est-elle le disque entier? Pour $r = r_0$, les arbres cachent p.s. la forêt à tout observateur donné; est-il vrai que p.s. les arbres cachent la forêt à tout observateur?

Problèmes liés aux multimesures et multifractales

Comme on l'a vu sur deux exemples (cascades, et chaos multiplicatif gaussien), on peut associer à des poids aléatoires indépendants P_n sur un espase T non pas seulement *un* opérateur Q, mais toute une famille $Q^{(r)} (r \in \mathcal{R})$: on définit $S^{(r)} = Q^{(r)}\sigma$ comme limite vague presque sûre des mesures aléatoires $Q_n^{(r)}\sigma$, où $Q_n^{(r)} = \frac{Q_n^r}{EQ_n^r} (Q_n = P_1 P_2 \cdots P_n)$.

1. Fixons $\sigma \in M^+(T)$. Pour tout r, $S^{(r)}$ est définie p.s. Donc $S^{(r)}$ est définie pour presque tout r. Peut-on définir p.s. $S^{(r)}$ pour tout r?

   À défaut de savoir résoudre cette question, restreignons-nous à une famille dénombrable de r, contenant 0 et 1, par exemple $\mathbb{Q}$. La famille $S^{(r)}$ contient σ (pour $r = 0$) et S (pour $r = 1$), et, dans certaines conditions (nous en avons donné dans la leçon 4), on a p.s.

$$\lim_{\rho \to 0} \frac{\log S(B_\rho(t))}{\log \sigma(B_\rho(t))} = \alpha_r \qquad S^r - \text{p.p.}$$

   ($B_\rho(t)$ désigne la boule de rayon ρ centrée en t).

2. Donner d'autres exemples et, si possible, des conditions générales sur Q et σ pour qu'il en soit ainsi.

3. Pour le chaos multiplicatif gaussien, seul le cas $T = \{0,1,\cdots c-1\}^{\mathcal{N}}, q(t,s) = C\log\frac{1}{d_c(t,s)}, \sigma = \lambda =$ mesure de Haar sur T a été traité. Traiter le cas $T = \mathcal{T}^n, \sigma = \lambda =$ mesure de Haar sur T, et $q(t,s) = C\log\frac{1}{d(t,s)}$, où $d(t,s)$ est la distance euclidienne. Mutatis mutandis, traiter le cas $T = \mathcal{R}^n$ euclidien.

4. Les $S^{(r)}$ sont les analogues des *mesures de Gibbs* associées à une mesure S et une mesure de base σ. Préciser cette analogie. La référence la plus utile pour les mesures de Gibbs est un article à paraître, de Michon et Peyrière.

5. Prenons maintenant le cadre déterministe. Considérons deux mesures σ et S de même support dans T (disons pour fixer les idées, $T = \mathcal{R}^n$). Une multimesure associée à σ et S est une famille $S^{(r)}$ de mesures ayant le même support que σ et S, telle que $S^{(0)} = \sigma$ et $S^{(1)} = S$, et que pour tout r et S^r-presque tout t on ait, pour un α_r réel convenable,

$$\lim_{\rho\to 0}\frac{\log S(B_\rho(t))}{\log\sigma(B_\rho(t))} = \alpha_r. \tag{1}$$

La condition, écrite pour $r = 1$, entraîne que, dans le cas où σ est la mesure de Lebesgue sur $\mathcal{R}^n, S$ est unidimensionnelle et de dimension $\alpha_1 n$. Dans tous les cas connus, les $S^{(r)}$ sont aussi des mesures unidimensionnelles, de dimension d_r, et il s'ensuit que l'ensemble des t tels que (1) a lieu est de dimension d_r. La loi de d_r en fonction de α est l'objet de l'analyse multifractale de la mesure S. Le cours de Benoît Mandelbrot donne beaucoup d'information à cet égard, et de très bons problèmes (par exemple, l'analyse multifractale de la mesure harmonique relative à un point et à un ouvert du plan contenant ce point, et homéomorphe au disque; d'après N. Makarov, on sait que c'est une mesure unidimensionnelle et de dimension 1).

Compléments sur les recouvrements poissonniens de $\mathcal{R}$

Rappelons que les intervalles du recouvrement sont, par définition, les intervalles ouverts $J = (x, x+y)$ de $\mathcal{R}$, associés aux points (x,y) du processus de Poisson ponctuel sur $\mathcal{R}\times\mathcal{R}^+$ dont l'intensité est $\lambda\otimes\mu, \lambda$ étant la mesure de Lebesgue sur $\mathcal{R}$ et μ la mesure donnée sur $\mathcal{R}^+$. Le noyau associé au problème est

$$k(x) = \exp\left(\int_{|x|}^{\infty}\mu(y,\infty)dy\right).$$

Supposons $k\in L^1(\mathcal{R}, e^{-x}dx)$, c'est-à-dire que la réunion G des intervalles J diffère de $\mathcal{R}$ avec probabilité positive. Remarquons que, si l'on exclut le recouvrement de $\mathcal{R}$ par un nombre fini d'intervalles J (par exemple, si $\mu(1,\infty) = 0$), on a alors $G\neq\mathcal{R}$ p.s. Nous continuous à parler des intervalles de recouvrement même lorsque $\mathcal{R}$ n'est pas recouvert.

On établi dans la troisième leçon la formule

$$E(e^{-\tau}) = \left(\int_0^\infty e^{-s}k(s)ds\right)^{-1},$$

où τ est le premier point non recouvert à droite de 0, soit

$$\tau = \inf\{t \geq 0, t \notin G\}.$$

En reprenant les calculs, on voit que, pour tout $u > 0$,

$$E(e^{-u\tau}) = \left(u\int_0^\infty e^{-us}k(s)ds\right)^{-1} = \left(1 + u\int_0^\infty e^{-su}(k(s) - 1)ds\right)^{-1}. \quad (1)$$

Cela donne la loi de la v.a. τ. En particulier (par développement limité)

$$\begin{aligned} E(\tau) &= \int_0^\infty (k(s) - 1)ds \\ E(\tau^2) &= 2\left(\int_0^\infty (k(s) - 1)ds\right)^2 + E\int_0^\infty s(k(s) - 1)ds. \end{aligned}$$

Si $\mu(1, \infty) = 0$ on a $k(s) - 1 = 0$ pour $s \geq 1$, donc $E(e^{-u\tau})$ est une fonction méromorphe de u, holomorphe au voisinage de 0; il en résulte que

$$E(e^{c|\tau|}) < \infty$$

pour un $c = c_\mu > 0$.

Considérons la mesure $\mu = \mu_{a,\ell}$ égale à $a\frac{dy}{y^2}$ pour $y \geq \ell$, et nulle pour $y \leq \ell$. Ainsi $\mu(y, \infty) = a\left(\frac{1}{y} - \frac{1}{\ell}\right)^+$, et

$$k_{a,\ell}(s) = \begin{cases} \left(\dfrac{\ell}{s}\right)^a e^{as/\ell}e^{-a} & \text{pour} \quad s < \ell \\ 1 & \text{pour} \quad s \geq \ell. \end{cases}$$

Remarquons que

$$k_{a,\ell}(s) = k_a\left(\frac{s}{\ell}\right), \quad (2)$$

$$k(s) = \begin{cases} s^{-a}e^{a(s-1)} & \text{pour} \quad s \leq 1 \\ 1 & \text{pour} \quad s \geq 1. \end{cases}$$

Fixons $a < 1$, considérons la mesure $\mu = \mu_{a,\infty}$ égale à $a\frac{dy}{y^2}$ sur $\mathcal{R}^+$ et les intervalles J associés. En se restreignant aux intervalles J de longueurs $\leq \ell$, on obtient une version du recouvrement associé à $\mu_{a,\ell}$. Les formules (1) et (2) montrent que les v.a.

$$\frac{1}{\ell}\tau_\ell = \frac{1}{\ell}\inf(t \geq 0, t \notin \bigcup_{|J|\leq \ell} J)$$

ont toutes la même loi, à savoir celle de la v.a. $\tau(=\tau_1)$ telle que

$$E(e^{-u\tau}) = (1 + u\int_0^1 e^{-su}(s^{-a}e^{a(s-1)} - 1)ds)^{-1}.$$

Pour la mesure $\mu_{a,\infty}$, les grands intervalles J recouvrent p.s. $\mathcal{R}$ (parce que $\int^\infty \mu(y,\infty)dy = \infty$). Les intervalles J de longueurs ≤ 1 ont pour réunion un ouvert $G(= G_1)$ dont on désigne les composantes connexes par $\mathcal{J}$; ainsi chaque $\mathcal{J}$ est une réunion de J de longueurs ≤ 1. P.s. 0 appartient à un certain $\mathcal{J} = (-\tau', \tau)$; $\tau(=\tau_1)$ est justement le temps d'arrêt associé à $\mu_{a,1}$, et τ' est une version indépendante du même temps d'arrêt. D'où la loi de $|\mathcal{J}|$, longueur de $\mathcal{J}$:

$$E(e^{-u|\mathcal{J}|}) = \left(1 + u\int_0^1 e^{-su}(s^{-a}e^{a(s-1)} - 1)ds\right)^{-2}.$$

C'est la loi commune à tous les intervalles $\mathcal{J}$, composantes connexes de G_1. La loi des longueurs des composantes connexes de $G_\ell = \bigcup_{|J|\leq \ell} J$ est simplement la loi de $\ell|\mathcal{J}|$.

Recouvrements poissonniens de $\mathcal{R}^+$ et processus de Lévy (théorèmes de [FFS1985])

Au lieu de $\mathcal{R} \times \mathcal{R}^+$, considérons $\mathcal{R}^+ \times \mathcal{R}^+$, muni de la mesure $a\frac{dxdy}{y^2}(a < 1)$, le processus de Poisson correspondant (ensemble de points (x, y) et le recouvrement par les intervalles $J = (x, x + y)$). Comme on l'a fait pour les recouvrements sur le cercle, on peut associer au recouvrement poissonnien un opérateur Q, agissant sur les mesures localement bornées sur $\mathcal{R}$. Choisissons pour mesure

$$d\sigma(x) = \frac{x^{-a}dx}{\Gamma(1-a)}$$

et considérons la mesure aléatoire $S = Q\sigma$ et sa transformée de Laplace

$$\Phi(u) = \int_0^\infty e^{-ux}dS(x).$$

Notre propos est de montrer que la loi de la fonction aléatoire $\Phi(u)$ est la même que la loi de

$$\Psi(u) = \int_0^\infty e^{-uL(t)}dt$$

où $L(x)$ est le processus de Lévy croissant d'indice $1 - a$. La conclusion sera que $dS(x)$ est une version de l'image de la mesure de Lebesgue sur $\mathcal{R}^+$ par ce processus de Lévy.

Les étapes vont être: 1) le calcul de $E(\Phi(u_1)\cdots\Phi(u_n))$ pour $u_1 \leq \cdots \leq u_n$; 2) le calcul de $E(\Psi(u_1)\cdots\Psi(u_n))$; 3) la comparaison et la conclusion.

1) Calcul de $E(\Phi(u_1)\cdots\Phi(u_n))$.

Comme nous en avons déjà l'habitude, nous ferons des calculs formels qui ont un sens lorsque, au lieu de $\lambda\otimes\mu = a\frac{dxdy}{y^2}$, on considère $\lambda\otimes\mu_\ell = a1_{y\geq\ell}(x,y)\frac{dxdy}{y^2}$, et dont la dernière étape a un sens quand on fait tendre ℓ vers 0.

On a

$$E(\Phi(u_1)\cdots\Phi(u_n)) = \int\cdots\int E(Q(x_1)\cdots Q(x_n))e^{-u_1x_1-\cdots-u_nx_n}d\sigma(x_1)\cdots d\sigma(x_n)$$

avec

$$Q(x) = \frac{1_{G^c}(x)}{P(x \notin G)}$$

et

$$P(x \notin G) = p(x) = \exp(-\lambda\otimes\mu(A_x)) = \exp(-\int_0^x \mu(y,\infty)dy).$$

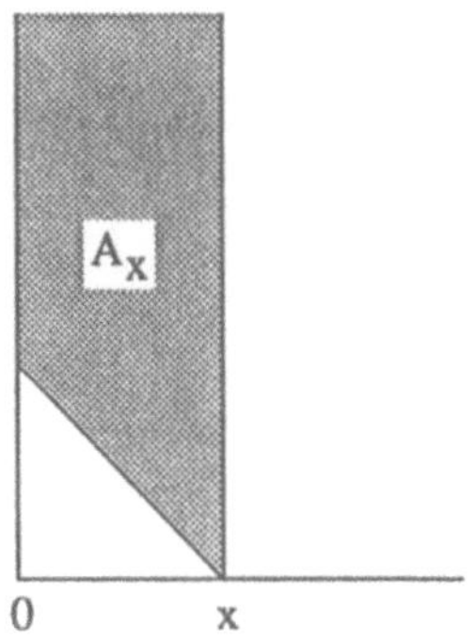

Le cas $n = 1$ est immédiat:

$$E\Phi(u) = \int_0^\infty e^{-ux}dx$$

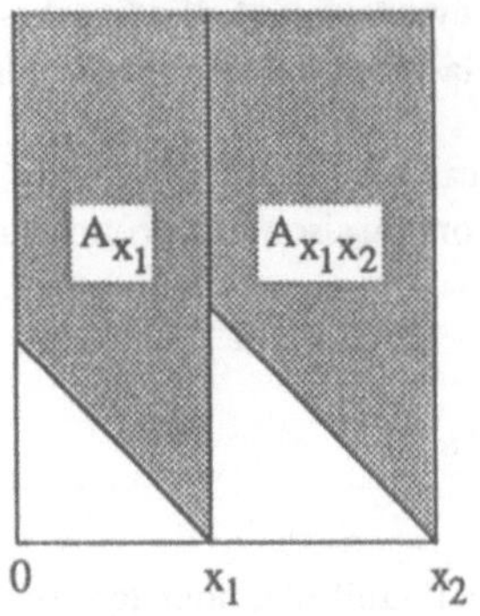

Pour $n = 2$ calculons d'abord $E(Q(x_1)Q(x_2))$ quand $x_1 \leq x_2$. On a

$$\begin{aligned} E(1_{G^c}(x_1)1_{G^c}(x_2)) &= P(x_1 \notin G)P(x_2 \notin G)|x_1 \notin G) \\ &= \exp(-\lambda \otimes \mu(A_{x_1}) - \lambda \otimes \mu(A_{x_1x_2})) \end{aligned}$$

$$\begin{aligned} E(Q(x_1)Q(x_2)) &= \exp(\lambda \otimes \mu)(A_{x_1} \cap A_{x_2}) \\ &= \exp \int_{x_2-x_1}^{x_2} \mu(y,\infty)dy. \end{aligned}$$

Après échange de x_1 et x_2 on obtient

$$E(\Phi(u_1)\Phi(u_2)) = \varphi(u_1,u_2) + \varphi(u_2,u_1)$$

$$\varphi(u_1,u_2) = \int\int_{0\leq x_1\leq x_2} e^{-u_1x_1-u_2x_2}\left(\exp\int_{x_2-x_1}^{x_2}\mu(y,\infty)dy\right)d\sigma(x_1)d\sigma(x_2).$$

Dans le cas général,

$$E(\Phi(u_1)\cdots\Phi(u_n)) = \sum_{\text{perm}} \varphi(u_{j_1},u_{j_2},\ldots u_{j_n})$$

la somme étant prise pour toutes les permutations $(j_1,\cdots j_n)$ de $(1,2,\cdots n)$, avec

$$\varphi(u_1,\cdots,u_n) = \int\cdots\int_{0\le x_1\le\cdots\le x_n} e^{-u_1x_1-\cdots-u_nx_n}\left(\exp\sum_{m=2}^{n}\int_{x_m-x_{m-1}}^{x_m}\mu(y,\infty)dy\right)$$
$$d\sigma(x_1)\cdots d\sigma(x_n)$$

Traduisons dans le cas $\mu(y,\infty) = a/y$ et $d\sigma(x) = x^{-a}dx/\Gamma(1-a)$:

$$\begin{aligned}\varphi(u_1,\cdots,u_n) &= \int\cdots\int_{0\le x_1\le\cdots\le x_n} e^{-u_1x_1-\cdots-u_nx_n}x_1^{-a}\prod_{m=2}^{n}(x_m-x_{m-1})^{-a}\frac{dx_1\cdots dx_n}{(\Gamma(1-a))^n}\\ &= u_n^{a-1}(u_n+u_{n-1})^{a-1}\cdots(u_n+\cdots+u_1)^{a-1}.\end{aligned}$$

2) Calcul de $E(\Psi(u_1)\cdots\Psi(u_n))$.

La loi de $L(\cdot)$, processus stable de Lévy d'indice $\beta = 1-a$, est donnée par

$$E(e^{-u_1L(t_1)-u_2(L(t_2)-L(t_1))-\cdots-u_n(L(t_n)-L(t_{n-1}))}) = \exp(-t_1u_1^\beta\cdots-(t_n-t_{n-1})u_n^\beta)$$

pour tout choix de $u_1\ge 0,\cdots,u_n\ge 0, 0\le t_1\le\cdots\le t_n$, et on a

$$E(\Psi(u_1)\cdots\Psi(u_n)) = E\int\cdots\int e^{-u_1L(t_1)-\cdots-u_nL(t_n)}dt_1\cdots dt_n.$$

Pour calculer le second membre on intègre d'abord sur $0\le t_1\le t_2\cdots\le t_n$, puis on permute de toutes les façons possibles $(t_1,u_1),\cdots,(t_n,u_n)$, d'où

$$E(\Psi(u_1)\cdots\Psi(u_n)) = \sum_{\text{perm}}\psi(u_{j_1},\cdots u_{j_n})$$

$$\begin{aligned}\psi(u_1,\cdots,u_n) &= E\int\cdots\int_{0\le t_1\le\cdots\le t_n} e^{-u_n(L(t_n)-L(t_{n-1}))-(u_n+u_{n-1})(L(t_{n-1})-L(t_{n-2}))\cdots}\\ &\quad dt_1\cdots dt_n\\ &= u_n^{-\beta}(u_n+u_{n-1})^{-\beta}\cdots(u_n+\cdots+u_1)^{-\beta}.\end{aligned}$$

3) On observe que

$$E(\Psi(u_1)\cdots\Psi(u_n)) = E(\Phi(u_1)\cdots\Phi(u_n))$$

ce qui est une forte présomption pour l'égalité des lois des fonctions aléatoires $\Phi(\cdot)$ et $\Psi(\cdot)$. Pour conclure, on peut considérer les transformées de Laplace des n-uples $(\Psi(u_1),\cdots\Psi(u_n))$ et $(\Phi(u_1),\cdots,\Phi(u_n))$ et montrer qu'elles sont analytiques dans un voisinage complexe de $(\mathcal{R}^+)^n$, et égales sur $(\mathcal{R}^+)^n$. Les majorations des coefficients de Taylor n'offrent aucune difficulté, à partir des formules explicites que nous avons données.

Malgré l'aspect technique des calculs, il me semble que c'est la meilleure manière d'obtenir le résultat de [FFS1985] dans le cas $d\mu = a\frac{dy}{y^2}$. La méthode s'adapte au cas général, en choisissant

$$d\sigma(x) = \gamma \exp \int_\delta^x \mu(y,\infty)dy$$

avec $\gamma = \eta(\delta)$ convenable. On remarque la grande différence entre le recouvrement à l'aide d'intervalles $(x, x+y)$ liés au processus de Poisson ponctuel d'intensité $a\frac{dxdy}{y^2}$ dans $\mathcal{R}^+ \times \mathcal{R}^+$, et le recouvrement à l'aide d'intervalles liés au processus de Poisson ponctuel d'intensité $a\frac{dxdy}{y^2}$ dans $\mathcal{R} \times \mathcal{R}^+$.

References

[F1986] Fan, Ai hua, Une condition suffisante d'existence du moment d'ordre m (entier) du chaos multiplicatif, *Ann. Sci. Math. Québec* **10** (1986), 119–120.

[F1989(1)] Fan, Ai hua, Chaos additif et chaos multiplicatif de Lévy, *C.R. Acad. Sci. Paris* **308** (1989), 151–154.

[F1989(2)] Fan, Ai hua, Sur quelques processus de naissance et de mort, *C.R. Acad. Sci. Paris* **309** (1990), 441–444.

[F1989(3)] Fan, Ai hua, Thèse, Orsay, 1989.

[FFS1985] Fitzsimmons, P.J., Fristedt, B. & Shepp, L.A., The set of real numbers left uncovered by random covering interval, *Z. Wahrsch Verw. Gebiete* **70** (1985), 175–189.

[H1979] Hawkes, J., Potential theory of Lévy processes, *Proc. London Math. Soc.* **38** (1979), 335–352.

[J1983] Janson, S., Random coverings of the circle with ares of random lenghts. *Probability and Mathematical Statistics*, Essays in honour of Carl-Gustav Essen, 1983, 62–73.

[J1986] Janson, S., Random covering in several dimensions, *Acta Math.* **156** (1986), 83–118.

[K1968] Kahane, J.-P., *Some Random Series of Functions*, 1st ed., Heath 1968.

[K1974] Kahane, J.-P., Sur le modèle de turbulence de Benoît Mandelbrot, *C.R. Acad. Sci. Paris* **278** (1974), 621–623.

[K1985] Kahane, J.-P., *Some Random Series of Functions*, 2nd ed., Cambridge Univ. Press, 1985.

[K1985(2)] Kahane, J.-P., Sur le chaos multiplicatif, *Ann. Sci. Math. Québec* **9** (1985), 105–150, voir aussi *C.R. Acad. Sci. Paris* **301** (1985), 329–332.

[K1987(1)] Kahane, J.-P., Intervalles aléatoires et décomposition des mesures, *C.R. Acad. Sci. Paris* **304** (1987), 551–554.

[K1987(2)] Kahane, J.-P., Positive martingales and random measures, *Chinese Ann. Math.* **8B1** (1987), 1–12.

[K1989] Kahane, J.-P., Random multiplications, random coverings, and multiplicative chaos, *Proceedings of the Special Year in Modern Analysis* (Ed. E. Berkson, N. Tenney Peck, J. Jerry Uhl.), London Math. Soc. Lect. Notes 137, Cambridge Univ. Press 1989, 196–255.

[KP1976] Kahane, J.-P. & Peyriere, J., Sur certaines martingales de Benoît Mandelbrot, *Adv. in Math.* **22** (1976), 131–145.

[KS1963] Kahane, J.-P. & Salem, R., *Ensembles parfaits et séries trigonométriques*, Paris Hermann, 1963.

[L1990] Lyons, R., Random walks and percolation on trees, *Ann. Probab.* **18** (1990), 931–951.

[M1972(1)] Mandelbrot, B.B., Renewal sets and random cutouts, *Z. Wahrsch. Verw. Gebiete* **22** (1972), 145–157.

[M1972(2)] Mandelbrot, B.B., Possible refinement of the log-normal hypothesis concerning the distribution of energy dissipation in intermittent turbulence, in: *Statistical Models and Turbulence.* Symposium at Univ. San Diego 1971. Lecture Notes in Physics, Springer-Verlag 1972, 333–351.

[M1974(1)] Mandelbrot, B.B., Multiplications aléatoires itérées et distributions invariantes par moyenne pondérée aléatoire, *C.R. Acad. Sci. Paris* **278** (1974), 289–292 et 355–358.

[M1974(2)] Mandelbrot, B.B., Intermittent turbulence in self-similar cascades: divergence of high moments and dimension of the carrier, *J. Fluid Mech.* **62** (1974), 331–333.

[P1974] Peyriere, J., Turbulence et dimension de Hausdorff, *C.R. Acad. Sci. Paris* **278** (1974), 567–569.

[S1972(1)] Shepp, L.A., Covering the circle with random arcs. *Israel J. Math.* **11** (1972), 328–45.

[S1972(2)] Shepp, L.A., Covering the line with random intervals, *Z. Wahrsch. Verw. Gebiete* **23** (1972), 163–170.

The Planck constant of a curve

Michel MENDÈS FRANCE
U.E.R. de Mathématiques et d'Informatique
Université de Bordeaux I
351, cours de la Libération
F-33405 Talence Cedex
France

Abstract

We discuss the dimension, entropy and confusion coefficient of rectifiable curves. These concepts have a quantum mechanic interpretation. A Heisenberg uncertainty principle applies to chaotic curves.

Introduction

I have had the opportunity to lecture on the topics of these notes many many times [16]. But each time I felt the need to modify some basic definitions in order to improve the presentation. So once again I take the liberty of performing minor changes. I hope this will not confuse the potential readers who may be familiar with my previous papers.

I must also confess that titles vary from one article to another. "Dimension and Entropy of Regular Curves", "Chaos implies Confusion" and the present paper are essentially variations of one another. The latter one is however a more expanded version.

My friend Elhanan Motzkin, a mathematician who is becoming a specialist in classic painting, once argued that changing the title of a painting can modify drastically one's appreciation. I chose the title of the present article in order to emphasize the concept of Planck's constant of a curve. It is only defined in the very last paragraphs of these notes. Driven by analogies, we shall discover Heisenberg's inequality, which in our context, indicates that on a "chaotic" curve, even though differentiable, it is not possible to locate precisely a point on the curve together with the direction of its tangent.

J. Bélair and S. Dubuc (eds.), Fractal Geometry and Analysis, 325–366.

1 The dimension of a curve

1.1 Rivers and basins

Let us imagine a hypothetical river Γ of length L and width ε. Denote the area of its bed by $\Gamma_L(\varepsilon)$ and let C_L be the length of the frontier of its basin which we will suppose to be convex.

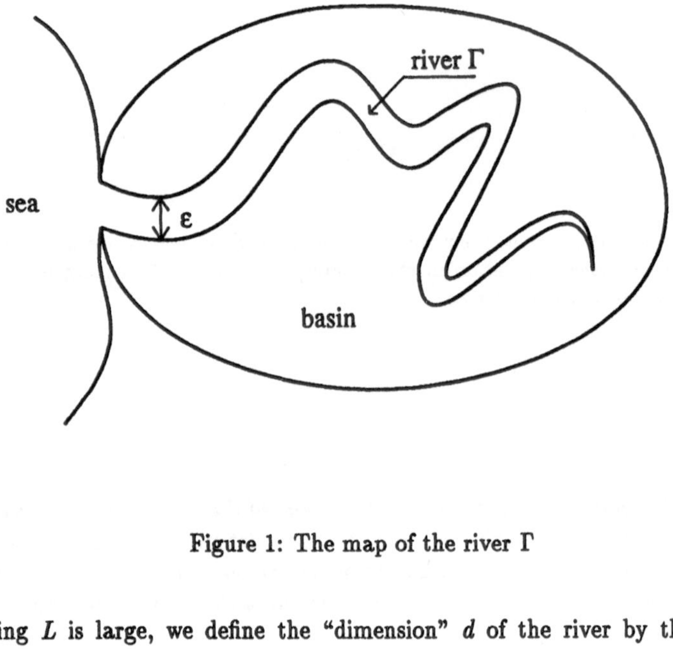

Figure 1: The map of the river Γ

Assuming L is large, we define the "dimension" d of the river by the approximate formula:

$$\Gamma_L(\varepsilon) \approx C_L^d.$$

If the river is straight, then C_L is approximately $2L$ and $\Gamma_L(\varepsilon) \approx L\varepsilon$. The dimension is then equal to 1. If, on the contrary the river meanders a great deal and resembles a swamp, then the bed of the river fills in the basin. $\Gamma_L(\varepsilon)$ is the area of the basin and scales like C_L^2. The river is two dimensional.

This rather vague discussion can actually be made more precise (and convincing) as I was told by Ricardo Rigon [20] from the Università di Venezia.

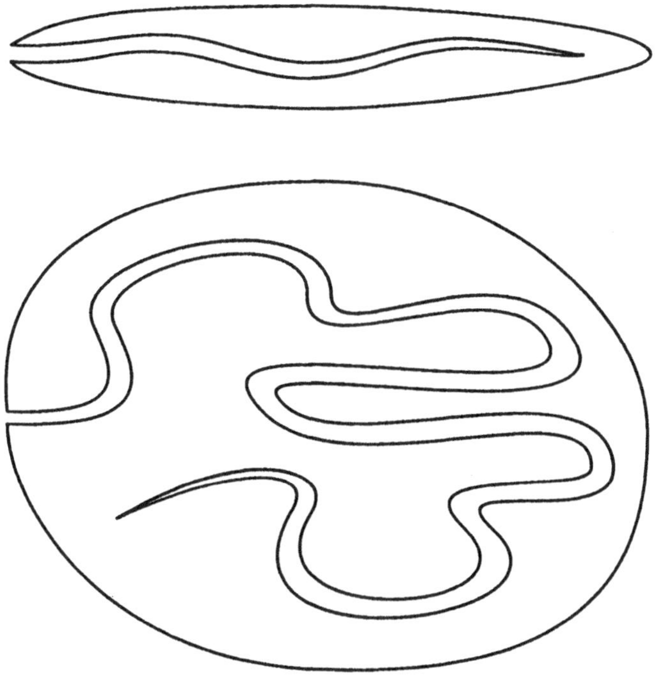

Figure 2: A straight river and a river which meanders

1.2 The dimension of a curve

The theory we shall develop concerns unbounded curves Γ in the n dimensional Euclidean space $\mathbf{R}^n$. We shall however restrict our study to the case $n = 2$ which is somehow more appealing. The general case should then cause no trouble.

The curves are assumed to be locally rectifiable (see Tricot's lectures [25]). We shall also suppose that compact sets in $\mathbf{R}^2$ contain at most a finite portion of Γ. The family of all these curves is denoted F.

Let $\varepsilon > 0$ be given. We define the ε-bed of the curve as the set of points $M \in \mathbf{R}^2$ the distance to Γ of which does not exceed ε:

$$\Gamma(\varepsilon) = \{M \in \mathbf{R}^2 | \operatorname{dist}(M, \Gamma) < \varepsilon\}.$$

(In the literature this set is also known as the ε-Minkowski sausage.) Quite obviously

$$\Gamma = \bigcap_{\varepsilon > 0} \Gamma(\varepsilon).$$

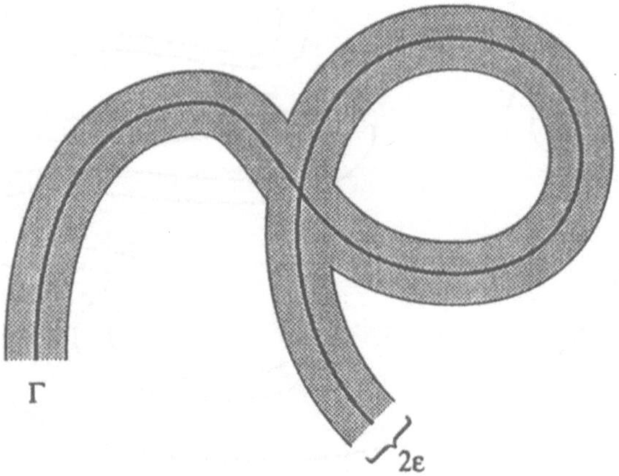

Figure 3: The bed of a curve

Let $L > 0$ be given. Γ_L represents the beginning portion of Γ of length L. $\Gamma_L(\varepsilon)$ is then the ε-bed of Γ_L:

$$\Gamma_L(\varepsilon) = \{M \in \mathbf{R}^2 | \operatorname{dist}(M, \Gamma_L) < \varepsilon\}.$$

We agree to represent the area of the set by the same symbol $\Gamma_L(\varepsilon)$. This should not create confusion.

Inspired by the first paragraph concerning rivers, we define the dimension of Γ by the formula

$$d = \dim \Gamma = \lim_{\varepsilon \searrow 0} \liminf_{L \nearrow \infty} \frac{\log \Gamma_L(\varepsilon)}{\log C_L},$$

where C_L is the length of the frontier of the convex hull of Γ_L : C_L is the "perimeter" of Γ_L. (For an algorithm to compute C_L when Γ_L is a broken line see [22].)

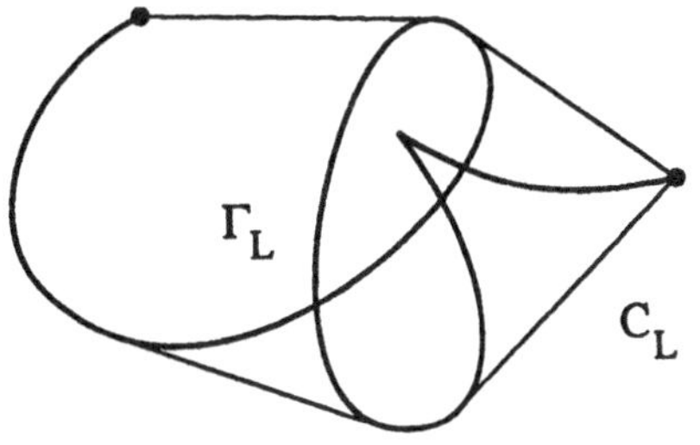

Figure 4: The curve Γ_L and the perimeter C_L

1.3 A remark due to C. Tricot

The limit

$$d(\varepsilon) = \liminf_{L \nearrow \infty} \frac{\log \Gamma_L(\varepsilon)}{\log C_L}$$

does not depend on ε.

Indeed, let $N(\varepsilon)$ be the number of squares the sides of which measure ε, which are parallel to the xy axes, and which are necessary to cover Γ_L. Then

$$\left(\frac{\varepsilon}{\sqrt{2}}\right)^2 N\left(\frac{\varepsilon}{\sqrt{2}}\right) \leq \Gamma_L(\varepsilon) \leq 9\left(\frac{\varepsilon}{\sqrt{2}}\right)^2 N\left(\frac{\varepsilon}{\sqrt{2}}\right)$$

and so

$$\Gamma_L(2\varepsilon) \leq 9\left(\frac{2\varepsilon}{\sqrt{2}}\right)^2 N\left(\frac{2\varepsilon}{\sqrt{2}}\right) \leq 9N\left(\frac{\varepsilon}{\sqrt{2}}\right) \cdot \left(\frac{2\varepsilon}{\sqrt{2}}\right)^2 \leq 9 \cdot 4\Gamma_L(\varepsilon),$$

hence,

$$\frac{1}{36}\Gamma_L(2\varepsilon) \leq \Gamma_L(\varepsilon) \leq \Gamma_L(2\varepsilon).$$

(We do not claim the factor 1/36 is optimal.) Therefore $d(\varepsilon) = d(2\varepsilon)$. Now $d(\varepsilon)$ decreases as ε tends to 0 so $d(\varepsilon)$ is constant, as claimed.

The limit $\varepsilon \to 0$ is thus unnecessary even though if only for aesthetic reasons one feels more "comfortable" in taking the limit. Mathematically, nothing would prevent us from fixing $\varepsilon = 1$ or even from letting ε tend to $+\infty$! The definition of the dimension would be equivalent, and yet it would seem to be nonsense!

$\Gamma_L(\varepsilon)$ is obviously bounded from above by the area of the convex hull of $\Gamma_L(\varepsilon)$. Hence

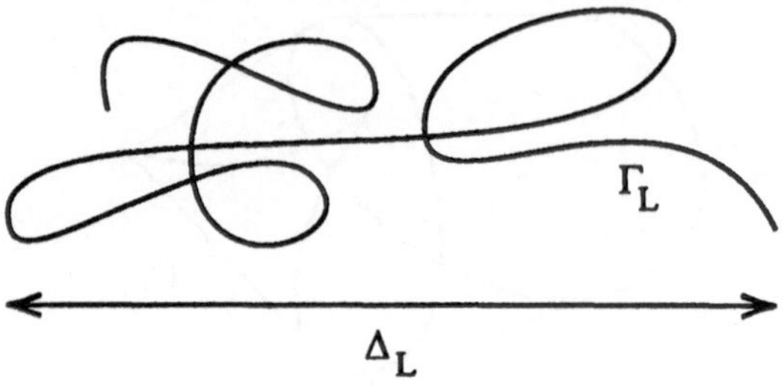

Figure 5: The diameter Δ_L

$$\Gamma_L(\varepsilon) \le \frac{\pi}{4} C_L^2 + O(C_L)$$

as L tends to infinity. Thus $d \le 2$.

On the other hand, let Δ_L be the diameter of Γ_L. Clearly $\Delta_L \le L$ and $\varepsilon\Delta_L \le \Gamma_L(\varepsilon)$, hence

$$d \ge \lim_{\varepsilon \to 0} \liminf_{L \nearrow \infty} \frac{\log \varepsilon \Delta_L}{\log C_L}.$$

But

$$C_L \le \pi \Delta_L$$

so that finally $d \ge 1$.

We have thus established the first part of the following result.

Theorem 1 *For all curves* $\Gamma \in F$,

$$1 \le \dim \Gamma \le 2.$$

Furthermore, for every $d \in [1, 2]$ *there exists a curve* $\Gamma \in F$ *such that* $\dim \Gamma = d$.

The second part of the theorem will be proved in paragraph 1.5 (see examples 1 and 2).

It is obviously difficult to compute the area $\Gamma_L(\varepsilon)$, and calculating the dimension of a curve may be a delicate matter. It is however easy to obtain an upper bound. Indeed,

$$\Gamma_L(\varepsilon) \le 2\varepsilon L + o(\varepsilon)$$

as L goes to infinity and as ε tends to zero. Then

$$\dim \Gamma \leq \liminf_{L \nearrow \infty} \frac{\log L}{\log C_L}.$$

It would be convenient to obtain a lower bound for the dimension. This we do in the following paragraph.

1.4 Resolvable curves

Let $C(M,\varepsilon)$ be the circle of center M and radius ε. If $M \in \Gamma_L(\varepsilon)$, then $C(M,\varepsilon)$ intersects Γ in at least one point. Let $N(M,\varepsilon)$ be the number of intersection points of $C(M,E)$ with Γ_L. The average number of intersection points is thus

$$\nu(L,\varepsilon) = \frac{\int\int_{\Gamma_L(\varepsilon)} N(M,\varepsilon) dM}{\int\int_{\Gamma_L(\varepsilon)} dM}$$

$$= \frac{1}{\Gamma_L(\varepsilon)} \int\int_{\Gamma_L(\varepsilon)} N(M,\varepsilon) dM.$$

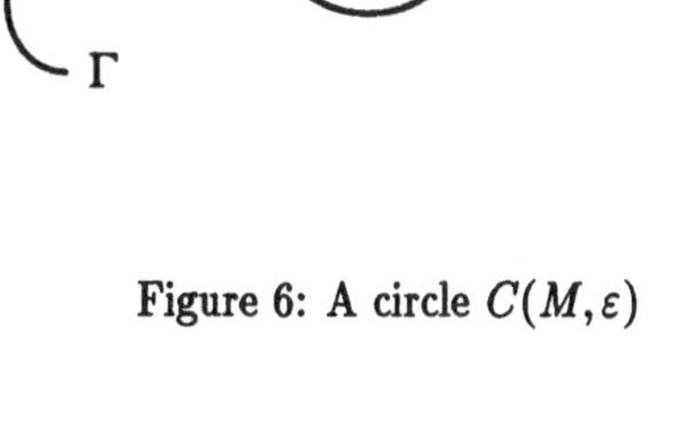

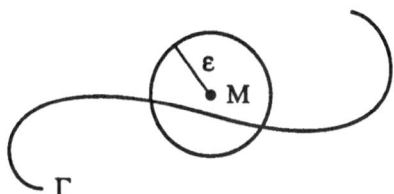

Figure 6: A circle $C(M,\varepsilon)$

We are interested in the behaviour of $\nu(L,\varepsilon)$ as L increases to infinity and ε vanishes. Define the "confusion coefficient"

$$\gamma = \gamma(\Gamma) = \limsup_{\varepsilon \searrow 0} \limsup_{L \nearrow \infty} \frac{\log \nu(L,\varepsilon)}{\log C_L}.$$

The range of γ is the closed interval $[0, +\infty]$.

Let us illustrate the meaning of γ. Suppose $\gamma > 0$. Then for some unbounded increasing sequence L_n and for some $\gamma' \in]0, \gamma]$,

$$\nu(L_n, \varepsilon) \geq C_{L_n}^{\gamma'}.$$

Therefore, however small $\varepsilon > 0$ may be, $\nu(L_n, \varepsilon)$ tends to infinity with L_n. In other words, there exist many "small" circles $C(M, \varepsilon)$ which intersect Γ in a great many points. The curve Γ is thus squeezed onto itself to the point that in the neighbourhood of infinity in the plane, it may be difficult to know on which portion of the curve one is. The branches are tightly packed. One is confused.

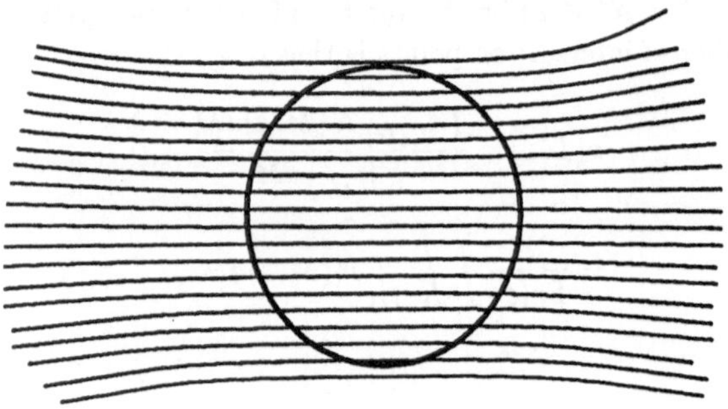

Figure 7: A circle $C(M, \varepsilon)$ which intersects Γ in many points.

We propose the following definition.

Definition If $\gamma = 0$, the curve is said to be *resolvable*. If $\gamma > 0$, the curve is *non-resolvable*. (This definition does not coincide with the one introduced in the Dekking-MF article [6] even though connections exist.)

Theorem 2 *For all curves* $\Gamma \in F$

$$\liminf_{L \nearrow \infty} \frac{\log L}{\log C_L} - \gamma \leq \dim \Gamma \leq \liminf_{L \nearrow \infty} \frac{\log L}{\log C_L}.$$

The upper bound for $\dim \Gamma$ is trivial as we have already seen at the end of paragraph 1.3. The novelty is the lower bound. The proof relies on a theorem of Poincaré's concerning geometric probability which we shall discuss in paragraph 1.7.

Corollary *If* Γ *is resolvable, then*

$$\dim \Gamma = \liminf_{L \nearrow \infty} \frac{\log L}{\log C_L}.$$

We now turn to examples to show how one actually computes dimensions, at least in some simple cases.

1.5 Examples

1.5.1 Consider the exponential spiral $\rho = \exp\theta$ (polar coordinates) which biologists call the logarithmic spiral! The length of the curve when the polar angle runs from $-\infty$ to θ is

$$L(\theta) = \int_{-\infty}^{\theta} (\rho^2 + \rho'^2)^{1/2} d\theta = \sqrt{2}\exp\theta.$$

The perimeter $C(\theta)$ is comprised between $2\pi\rho(\theta - 2\pi)$ and $2\pi\rho(\theta)$:

$$2\pi e^{-2\pi} e^{\theta} \leq C(\theta) \leq 2\pi e^{\theta}.$$

The curve is obviously resolvable so that

$$\dim \Gamma = \lim_{\theta \to +\infty} \frac{\log L(\theta)}{\log C(\theta)} = 1.$$

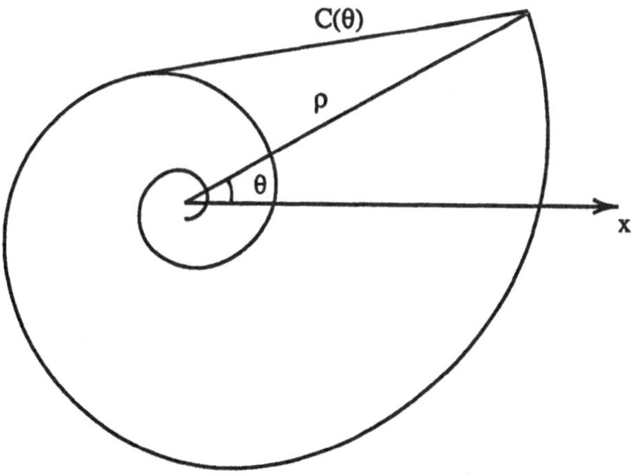

Figure 8: A spiral

1.5.2 Let $\alpha > 0$ be given and consider the spiral $\rho = \theta^\alpha$. If $\alpha \geq 1$, the curve is resolvable. A similar computation as above shows that

$$\dim \Gamma = 1 + \frac{1}{\alpha}.$$

If $0 < \alpha < 1$, then the spiral is not resolvable. In fact its confusion coefficient is $\gamma = 2(\alpha^{-1} - 1)$. Calculating the dimension from first principles is quite simple and leads to

$$\dim \Gamma = 2.$$

These examples establish that for all $d \in [1,2]$ there exists a curve (and indeed a spiral) whose dimension is d. Theorem 1 is thus completely proved.

1.5.3 Suppose Γ is an unbounded branch of an algebraic curve of degree δ, say. The circles $C(M,\varepsilon)$ intersect Γ in at most 2δ points. Therefore $\nu(L,\varepsilon) \leq 2\delta$ so that $\gamma = 0$. An unbounded algebraic curve is always resolvable.

Prove that its dimension is necessarily equal to 1.

1.6 Geometric probability

Consider a finite curve Γ_L of length L drawn on a horizontal plane, the floor for example. Toss an infinite straight line (a rod) on it and count the number of intersection points with Γ_L. If there are no intersection points we disregard the throw and start again. Let $N_1, N_2, \ldots, N_k$ be the number of intersection points of successful experiments. When k is large, we observe that

$$\frac{N_1 + N_2 + \ldots + N_k}{k} \approx \frac{2L}{C_L}.$$

This is a statistical observation (which I have never actually performed). It can be justified by probability theory. Before we do so, let us notice an important fact which will be crucial in the next chapter, namely

$$2L \geq C_L.$$

Equality holds if and only if Γ_L is a straight segment.

In the Ox, Oy plane an infinite straight line Δ is determined by its distance $\rho \geq 0$ to the origin and by its polar angle $\theta \in [0, 2\pi[$.

Points in the upper half strip $\rho \geq 0$, $0 \leq \theta < 2\pi$, in the θ, ρ plane represent straight lines in the x, y plane. (Purists with a twisted mind will notice that $(\rho, \theta) = (-\rho, \theta + \pi)$ thus identifying the family of all straight lines Δ with a Möbius manifold.) The θ, ρ plane is conveniently endowed with the Lebesgue measure $d\rho d\theta$.

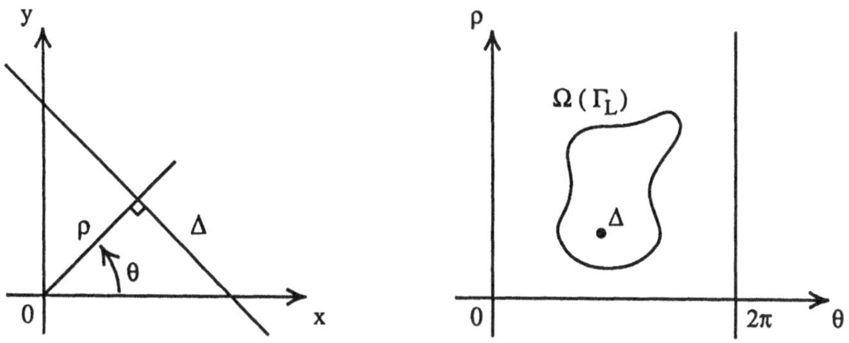

Figure 9: A straight line Δ in the x, y plane and its representation in the θ, ρ plane.

Let $\Omega(\Gamma_L)$ be the set of straight lines which actually intersect Γ_L. As Γ_L is bounded, so is $\Omega(\Gamma_L)$ in the θ, ρ plane. $\Omega(\Gamma_L)$ together with the normalized measure

$$d\rho d\theta / \text{ meas } \Omega(\Gamma_L)$$

is a probability space.

For $n \geq 1$, put

$$p_n = \text{Prob } \{\Delta \text{ intersects } \Gamma_L \text{ in exactly } n \text{ points}\}.$$

Straight lines tangent to Γ_L form a set of measure 0; it is thus irrelevant to specify whether a tangent cuts the curve in one or more points. By definition

$$\sum_{n=1}^{\infty} p_n = 1.$$

We are now able to state a beautiful result due to Steinhaus [21], [23] which justifies the statistics mentioned at the beginning of the paragraph.

Theorem 3 (Steinhaus) *The expectation of the number of intersection points of an infinite random straight line Δ with Γ_L is*

$$\sum_{n=1}^{\infty} n p_n = \frac{2L}{C_L}.$$

The proof can be found in Santaló's book [21]. It is to be compared with Buffon's needle problem and it will play a fundamental role in chapter 2.

1.7 Poincaré's theorem

At this point we should state Poincaré's theorem which is a far reaching extension of Steinhaus' result, the proof of which is also in Santaló's book [21].

Consider a planar curve $\Sigma^{\circ}_{L'}$ of length L' and all its congruent curves $\Sigma_{L'}$, obtained by translation and rotation in the plane. $\Sigma_{L'}$ is completely determined by three parameters, namely two for a translation (a,b) and one for a rotation θ. The measure on the translation rotation group is

$$d\Sigma_{L'} = dadbd\theta.$$

Let $N(\Sigma_{L'}, \Sigma_L)$ be the number of intersection points of $\Sigma_{L'}$ and Σ_L.

Theorem 4 (Poincaré)

$$\int\int\int_{\Sigma_{L'}} N(\Sigma_{L'}, \Sigma_L) d\Sigma_{L'} = 4LL'.$$

Choose $\Sigma^{\circ}_{L'}$ to be a circle of radius ε. Then Poincaré's theorem reads

$$\int\int\int N(M,\varepsilon) dM d\theta = 4L \times 2\pi\varepsilon$$

or

$$\int\int N(M,\varepsilon) dM = 4L\varepsilon.$$

We now go back to the definition of the dimension of Γ and prove Theorem 2.

Recall that $\nu(L,\varepsilon)$ is the average number of intersection points of Γ_L and circles of radius ε (see paragraph 1.4):

$$\nu(L,\varepsilon) = \frac{1}{\Gamma_L(\varepsilon)} \int\int N(M,\varepsilon) dM = \frac{4L\varepsilon}{\Gamma_L(\varepsilon)}.$$

Therefore

$$\Gamma_L(\varepsilon) = \frac{4L\varepsilon}{\nu(L,\varepsilon)}$$

and

$$\dim \Gamma = \lim_{\varepsilon \searrow 0} \liminf_{L \nearrow \infty} \frac{\log \Gamma_L(\varepsilon)}{\log C_L}$$

$$\begin{aligned}
&= \lim_{\varepsilon \searrow 0} \liminf_{L \nearrow \infty} \left(\frac{\log 4L\varepsilon}{\log C_L} - \frac{\log \nu(L,\varepsilon)}{\log C_L} \right) \\
&\geq \liminf_{L \nearrow \infty} \frac{\log L}{\log C_L} - \limsup_{\varepsilon \searrow 0} \limsup_{L \nearrow \infty} \frac{\log \nu(L,\varepsilon)}{\log C_L} \\
&\geq \liminf_{L \nearrow \infty} \frac{\log L}{\log C_L} - \gamma,
\end{aligned}$$

where γ is the confusion coefficient. This establishes Theorem 2.

2 Entropy of curves

2.1 Entropy of finite curves

We consider once again a finite curve Γ_L. Let $p_n (n \geq 1)$ be as before the probability for an infinite straight line to cut Γ_L in exactly n points. By definition the entropy of Γ_L is

$$h(\Gamma_L) = \sum_{n=1}^{\infty} p_n \log \frac{1}{p_n}$$

where as usual we agree that $0\infty = 0$. Clearly $h(\Gamma_L) \geq 0$.

If Γ_L is a straight segment, then $p_1 = 1$ and $p_n = 0$ for all $n \geq 2$. Hence $h(\Gamma_L) = 0$. If Γ_L is a closed convex curve, then $p_2 = 1$ and $p_n = 0$ for all $n \neq 2$. The entropy vanishes again. In all other cases $h(\Gamma_L) > 0$. Actually this is not true as long as we do not define properly what a curve is... and indeed we did not! If a curve is a continuous map of some interval $I \subset \mathbf{R}$ into $\mathbf{R}^2$, then it may well happen that the graph is multiple as for example $x = \cos t$, $y = \sin t$, $t \in (0, 4\pi)$ which represents the unit circle described twice. Then every straight line which intersects the curve cuts it in 4 points so that $p_4 = 1$ and $p_n = 0$ for $n \neq 4$.

From now on we shall stick to the following definition of a curve: a curve is the image of some continuous map $I \to \mathbf{R}^2$. This fine distinction was not necessary in section 1 where intuition was sufficient. We still need local rectifiability. Then $h(\Gamma_L) = 0$ if and only if Γ_L is either a straight segment or a closed convex curve.

Low entropy curves should be thought of as simple whereas high entropy curves should be thought of as complicated.

How high can entropy get? Can the series

$$h(\Gamma_L) = \sum_{n=1}^{\infty} p_n \log \frac{1}{p_n}$$

diverge? Our following result answers these questions.

Theorem 5

$$h(\Gamma_L) \le \log \frac{2L}{C_L} + \frac{\beta}{\exp \beta - 1},$$

where

$$\beta = \log \frac{2L}{2L - C_L}.$$

Remark Remember that $2L \ge C_L$ so that $\beta > 0$. Hence

$$\beta(\exp \beta - 1)^{-1} \in [0, 1].$$

The term $\log(2L/C_L)$ in the inequality is the main term.

We now prove the theorem. We maximize the function

$$h(\mathbf{p}) = h(p_1, p_2, \ldots) = \sum_{n=1}^{\infty} p_n \log \frac{1}{p_n}$$

with two constraints

$$\sum_{n=1}^{\infty} p_n = 1 \text{ and } \sum_{n=1}^{\infty} n p_n = \frac{2L}{C_L}.$$

The Lagrange multiplier technique introduces the auxiliary function

$$U = h - \alpha(\sum_{n=1}^{\infty} p_n - 1) - \beta(\sum_{n=1}^{\infty} n p_n - \frac{2L}{C_L}).$$

The conditions

$$\frac{\partial U}{\partial p_n} = 0, n = 1, 2, 3, \ldots$$

lead to

$$\log \frac{1}{p_n} - 1 - \alpha - \beta n = 0, n = 1, 2, 3, \ldots$$

$$p_n = \mathring{p}_n := e^{-1-\alpha} e^{-\beta n} = a e^{-\beta n}, n = 1, 2, 3, \ldots.$$

The two constants a and β are then determined by the two constraint conditions, hence

$$\begin{aligned} a &= e^{\beta} - 1 \\ \beta &= \log \frac{2L}{2L - C_L}. \end{aligned}$$

Finally

$$h(\mathring{\mathbf{p}}) = h(\mathring{p}_1, \mathring{p}_2, \ldots) = \log \frac{2L}{C_L} + \frac{\beta}{\exp \beta - 1}.$$

This proves that $h(\mathbf{p})$ is stationary at the point $\overset{\circ}{\mathbf{p}} = (\overset{\circ}{p}_1, \overset{\circ}{p}_2, \overset{\circ}{p}_3, \ldots)$. We now show that $h(\overset{\circ}{\mathbf{p}})$ is maximum.

Taylor's formula reads

$$h(\mathbf{p}) - h(\overset{\circ}{\mathbf{p}}) = \sum_{n=1}^{\infty} (p_n - \overset{\circ}{p}_n) \frac{\partial h}{\partial p_n}(\overset{\circ}{\mathbf{p}}) + \frac{1}{2} \sum_{n,m} (p_n - \overset{\circ}{p}_n)(p_m - \overset{\circ}{p}_m) \frac{\partial^2 h}{\partial p_n \partial p_m}(\overset{\circ}{\mathbf{p}}) + \ldots$$

The first sum vanishes and the second reduces to

$$-\frac{1}{2} \sum_{n=1}^{\infty} (p_n - \overset{\circ}{p}_n)^2 \frac{1}{\overset{\circ}{p}_n}$$

which is strictly negative when $\mathbf{p} \neq \overset{\circ}{\mathbf{p}}$. $h(\overset{\circ}{\mathbf{p}})$ is thus a local maximum. But the extremum is unique; the maximum is therefore global. Q. E. D.

In my first articles I defined the entropy of a finite curve as $\log(2L/C_L)$. This quantity appears here as an upper bound for $h(\Gamma_L)$. We shall denote it by $\hat{h}(\Gamma_L)$. (Ignore the term $\beta(\exp \beta - 1)^{-1}$.) $\hat{h}(\Gamma_L)$ plays the role of a "maximal entropy". $h(\Gamma_L)$ could be compared with the metric entropy in the theory of dynamical systems whereas $\hat{h}(\Gamma_L)$ would be the topological entropy. It is much coarser. In the following paragraph we shall see its significance and in chapter 3 we shall discover its relationship to Number Theory and exponential sums.

2.2 A thermodynamic interpretation

Imagine a gas at temperature T (the absolute temperature) in equilibrium in a box. It is made up of particles which can only take specific energies $E_1, E_2, E_3, \ldots$ (quantum statistics). Let p_n be the probability for a particle to have energy E_n. Then the average energy of the system is

$$E = \sum_{n=1}^{\infty} p_n E_n.$$

The average E is determined by the temperature T and is called the internal energy.

Equilibrium means maximal chaos in the repartition of the particles among the energy levels. We are thus led to maximizing the entropy

$$\sum_{n=1}^{\infty} p_n \log \frac{1}{p_n}$$

with the two constraints

$$\sum_{n=1}^{\infty} p_n = 1, \qquad \sum_{n=1}^{\infty} p_n E_n = E.$$

This implies

$$p_n = a\exp(-\beta E_n).$$

The coefficient β is then shown to be the inverse of the absolute temperature $\beta = 1/T$. (In Physics textbooks the formula is usually $\beta = 1/kT$, where k is Boltzmann's constant. Here we have chosen $k = 1$ which has the sole effect of changing the scale of temperature.)

Now, in our discussion of the entropy of curves, the coefficient β which appears in Theorem 5 is obtained in a very similar way. This leads us to define the *temperature* T *of the curve* Γ_L:

$$T = \frac{1}{\beta} = (\log \frac{2L}{2L - C_L})^{-1}.$$

Notice that $\beta \geq 0$ so that indeed, as it should, $T \geq 0$.

It is now tempting to push further the analogy with physics. We identify the length L of the curve with it's "volume" V. Then $1/C_L$ plays the role of the "pressure" P. Indeed, as the pressure P increases, C_L decreases: putting a curve under pressure squeezes it onto itself.

Then T, P and V are linked by the above formula which now reads

$$T = (\log \frac{2V}{2V - P^{-1}})^{-1}$$

or more conveniently,

$$PV = \frac{1}{2}(1 - \exp(-\frac{1}{T}))^{-1}.$$

This is the "equation of state" of a curve.

Suppose the temperature T vanishes. Then

$$PV = \frac{1}{2}$$

which expresses that $L/C_L = 1/2$, and which means that the curve Γ_L is a straight segment. At $T = 0$ curves freeze to straight segments. Furthermore their entropy, as we know, is 0. This is an instance of a very general law discovered by Nernst according to which at 0 temperature the entropy of a system vanishes.

We now suppose that T increases to infinity. Then

$$1 - \exp(-\frac{1}{T}) \sim \frac{1}{T}$$

and

$$PV \sim \frac{1}{2}T.$$

We recognize Boyle-Mariotte's law for perfect gases. Thus at high temperature, curves behave as a perfect gas, that is to say, as a gas in which the particles do not interact with each other. The chaos is total. In this respect, high temperature curves should be considered as random.

2.3 Miscellany

2.3.1 There is a beautiful drawing of Pablo Picasso dating from 1920 representing a horse and its trainer (Cheval et son dresseur jongleur, Musée Picasso). I counted the number of intersection points with 20 horizontal straight lines and 23 vertical straight lines. I found respectively 189 and 182 intersection points so that the average number of intersection points with one random line is

$$\overline{N} = \frac{189+182}{20+23} \simeq 8.6.$$

Recall that $\overline{N} = 2L/C_L$. The temperature of Picasso's horse and trainer is thus

$$T = (\log \frac{\overline{N}}{\overline{N}-1})^{-1} = (\log \frac{8.6}{7.6})^{-1} \simeq 8^\circ \text{ Kelvin}.$$

No doubt that Jackson Pollock's, Mathieu's and Cy Twombly's graffitis are much warmer.

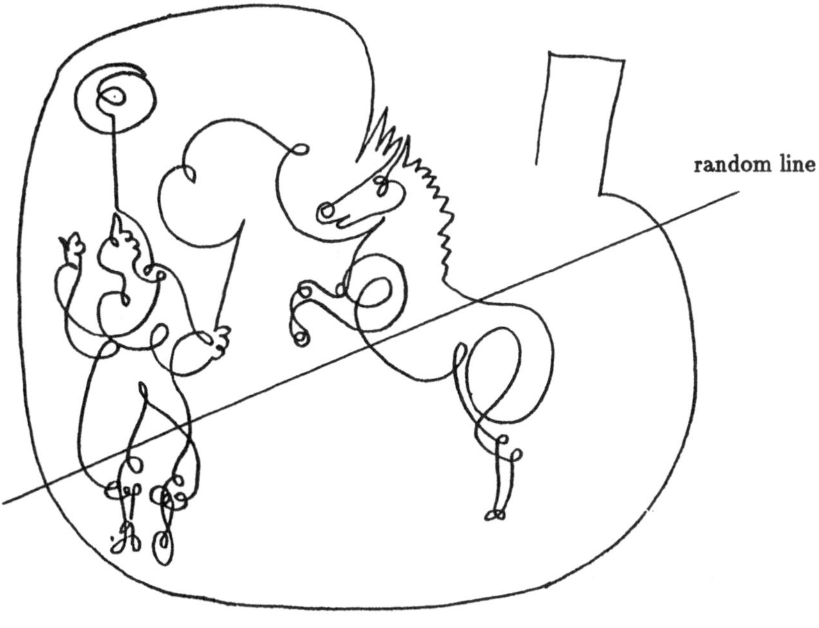

Figure 10: Picasso's drawing

2.3.2 It is possible to localize the concept of temperature. Consider a curve of finite length and three points M, A, M' on it. The temperature of the arc $\overset{\frown}{MM'}$ is known. Suppose M and M' tend to A following the curve. The upper and lower temperature at A are defined as

$$T^*(A) = \limsup_{\substack{M \to A \\ M' \to A}} T(\overset{\frown}{MM'})$$

and

$$T_*(A) = \liminf_{\substack{M \to A \\ M' \to A}} T(\overset{\frown}{MM'}).$$

In [9], Y. Dupain, T. Kamae and I prove that on a rectifiable curve, almost all points have temperature 0. There exist however rectifiable curves for which the set of nonzero temperature points is uncountable and dense on the curve. We also compute the temperature at the center of spirals and we show for example that the temperature is infinite at the center of the spiral $\rho = 1/\theta^2$. (This is reminescent of the fact that the center of a galaxy is very hot...)

2.3.3 I would like to mention an idea of René Thom ([24], p. 50) which could be relevant to our discussion.

Consider a gas in equilibrium at temperature T and let N be the number of particles. The phase space has $6N$ dimensions (each particle is determined by its location in space – 3 coordinates – and by its velocity – 3 coordinates). A point in the phase space represents the state of the system. At temperature T, the gas has a given energy E so that the points representing the gas at equilibrium lie on the hypersurface of energy E. Thom observes that the temperature, or rather its inverse $\beta = T^{-1}$ is the average on the hypersurface of the mean curvature.

In my approach to temperature I have not used the concept of curvature. Yet the ratio $2L/C_L$ does play the role of an average curvature of the curve. It would be interesting to push forward the relationship, if any.

2.3.4 Paragraph 2.2 was called "A thermodynamic interpretation". As the reader may have realized, I did not really fulfill the expectations since there is no dynamics involved in my presentation. This is certainly a pity. It would be desirable to introduce a time parameter and to consider moving curves. In this respect see the three articles of M. Gage [10], or Grayson [11], or the recent survey of B. White [29].

2.4 Entropy of infinite curves

Let Γ be a curve with infinite length. Let Γ_L be the finite portion of length L which has the same origin as Γ. The relative entropy of Γ_L is

$$h(\Gamma_L)/\log L.$$

The (relative) entropy of Γ is by definition

$$H(\Gamma) = \liminf_{L\nearrow\infty} \frac{h(\Gamma_L)}{\log L}.$$

Theorem 5 implies

$$0 \leq H(\Gamma) \leq 1.$$

It is easy to see that for every $H \in [0,1]$ there exist curves Γ for which $H(\Gamma) = H$.

We now give some examples.

2.4.1 Suppose Γ is an unbounded branch of an algebraic curve of degree d. We know that

$$h(\Gamma_L) \leq \log \frac{2L}{C_L} + 1$$

and that

$$\frac{2L}{C_L} = \sum_{n=1}^{d} np_n \leq d.$$

The above inequality comes from the fact that a straight line cannot cut Γ in more than d points. Therefore

$$h(\Gamma_L) \leq \log d + 1$$

and

$$H(\Gamma) = 0.$$

An infinite algebraic curve is "deterministic".

2.4.2 The exponential spiral $\rho = \exp\theta$ has 0 entropy.

2.4.3 Let $\alpha > 0$ be given. The spiral $\rho = \theta^\alpha$ has entropy

$$H(\Gamma) = \frac{1}{1+\alpha}.$$

2.4.4 The spiral

$$\rho = \exp(\log\log\theta)^2$$

has entropy 1.

2.4.5 The spiral $\rho = \log\theta$ has entropy 0. The spires are so closely packed together that it behaves as if it were a circle. (The discussion towards the beginning of paragraph 2.1 should be kept in mind: the logarithmic spiral is more like a circle described infinitely many times.)

2.4.6 Loosely speaking, a curve which goes rapidly to infinity will have low entropy. A curve which meanders and "hesitates" to go to infinity may have entropy $H = 1$ even though it may also have low entropy as we have seen above.

At the end of paragraph 2.1 we introduced the maximal entropy of a finite curve

$$\hat{h}(\Gamma_L) = \log\frac{2L}{C_L}.$$

In the next chapter we shall need the maximal entropy of an infinite curve

$$\hat{H}(\Gamma) = \liminf_{L\nearrow\infty}\frac{\hat{h}(\Gamma_L)}{\log L} = \liminf_{L\nearrow\infty}\frac{\log 2L - \log C_L}{\log L}.$$

This quantity is obviously easier to compute than $H(\Gamma)$. Moreover

$$0 \leq H(\Gamma) \leq \hat{H}(\Gamma) \leq 1.$$

2.5 Remarks on spirals

2.5.1 The exponential spiral $\rho = \exp\theta$ is deterministic whereas $\rho = \exp(\log\log\theta)^2$ is as "chaotic" as can be since its entropy is 1. How come the first spiral suggests order and the second disorder? The answer could be as follows.

The exponential spiral is linked with biological growth and biology is a struggle for order. Therefore the exponential spiral is an ordered spiral ($H = 0$). On the other hand, cyclones and whirlpools appear in turbulent phenomena. Could it be true that spirals of the second kind describe tornadoes?

2.5.2 In T.A. Cook's book *The Curves of Life* [2] there is a curious observation which I believe deserves to be reproduced here.

"Is the logarithmic spiral [the exponential spiral, according to my definition $\rho = \exp\theta$] the manifestation of the law which is at work in the increase of organic bodies? If so, it may be significant that Newton showed in his Principia, that if attraction had generally varied as the inverse cube instead of as the inverse square of the distance, then heavenly bodies would not have revolved in ellipses but would have rushed off into space in logarithmic spirals [exponential spirals]. Professor Goodsir therefore asked, if the law of the square is the law of attraction, and the law of the cube (that is, of the cell) the law of production?"

3 Equidistribution (mod1) and exponential sums

3.1 Equidistribution (mod1)

Let $\theta = (\theta_1, \theta_2, \theta_3, \ldots)$ be an infinite sequence of real numbers. Let I be a subinterval of $[0,1]$ and let $F(I,N)$ represent the number of terms θ_n, $0 < n \leq N$, such that $\theta_n \in I(\text{mod} 1)$. The sequence θ is said to be equidistributed (mod1) if for all subintervals I

$$\lim_{N\to\infty} \frac{1}{N} F(I,N) = \text{meas}(I).$$

In other terms, the frequency of the θ_n that fall in I coincides with the measure of I.

Examples of such sequences are (λn) where λ is an arbitrary irrational number, (λn^2) and more generally $(P(n))$ where P is a real polynomial which has at least one irrational coefficient other than the constant term. Here are other examples: (n^λ), where $\lambda > 0$ is not an integer; $(n^\lambda (\log n)^\mu)$, where either $\lambda > 0$ is not an integer and $\mu \in \mathbf{R}$, or $\lambda > 0$ is arbitrary and $\mu \neq 0$, or $\lambda = 0$ and $\mu > 1$. If λ is irrational and if p_n denotes the nth prime number, then again the sequence (λp_n) is equidistributed (mod1), and more generally, so is $(P(p_n))$, where P is a polynomial as above, ...etc ... .

For none of these examples, apart maybe for the first one, is it easy to show equidistribution.

Nonequidistributed sequences are easy to construct, for example any sequence which is nondense (mod1) (such as (λn) where $\lambda \in \mathbf{Q}$, or $(en!)$). A slowly varying sequence like $(\log n)$ is not equidistributed (mod1), ...etc ... .

It is not known whether sequences as simple as $((3/2)^n)$ are equidistributed (mod1) or not, even though it is known that for almost all $\lambda > 1$, the sequence (λ^n) is equidistributed (mod1). Actually no explicit λ is known for which (λ^n) is equidistributed (mod1). The interested reader may find many results and references on these topics in the books of Kuipers and Niederreiter [13] or Rauzy [19].

It is H. Weyl who originated the theory in a famous article in 1916 [28] by showing that the sequence $\theta = (\theta_n)$ is equidistributed (mod1) if and only if for all nonzero integers q

$$\lim_{N\to\infty} \frac{1}{N} \sum_{n=0}^{N-1} \exp 2i\pi q\theta_n = 0.$$

This condition is quite intuitive since it expresses the fact that the N points $\exp 2i\pi q\theta_n$, $n = 0, 1, \ldots, N-1$, on the unit circle have their center of gravity close to the origin.

The object of this chapter is to provide another geometric interpretation of Weyl's result which, as we shall see, involves curves and entropy. For the sake of simplicity, we chose $q = 1$ in this criterio.

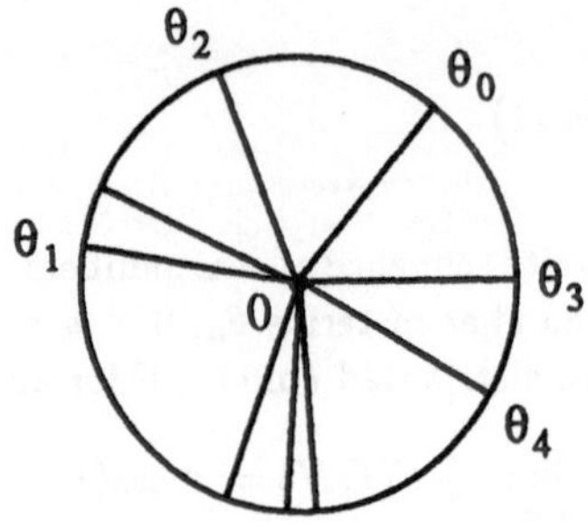

Figure 11: The unit circle is identified with $\mathbf{R}/\mathbf{Z}$ so that $\exp 2i\pi\theta_n$ coincides with θ_n.

3.2 Exponential sums

In the complex plane consider the points $z_0 = 0, z_1, z_2, \ldots, z_N, \ldots$ where

$$z_N = \sum_{n=0}^{N-1} \exp 2i\pi\theta_n.$$

Let $\Gamma = \Gamma^\theta$ be the infinite polygonal curve whose vertices are $z_0, z_1, z_2, \ldots$ in that order. Symbolically, Γ is represented by the infinite diverging sum

$$\Gamma = \sum_{n=0}^{\infty} \exp 2i\pi\theta_n.$$

We are interested in computing its maximal entropy $\hat{H}(\Gamma)$.

Theorem 6 *If, as N goes to infinity,*

$$\sum_{n=0}^{N-1} \exp 2i\pi\theta_n = O(N^\alpha)$$

for some α, then

$$1 - \alpha \leq \hat{H}(\Gamma).$$

If there exists a constant $c > 0$ such that for infinitely many N

$$\left|\sum_{n=0}^{N-1} \exp 2i\pi\theta_n\right| > cN^\beta$$

for some β, then

$$\hat{H}(\Gamma) \leq 1 - \beta.$$

One should compare the above result with Weyl's. If (θ_n) is well equidistributed (mod 1), then the center of gravity of the points $\exp 2i\pi\theta_n$, $n = 0, 1, \ldots, N-1$, is very close to the origin. More precisely, if the exponential sum is $O(N^\alpha)$ for some $\alpha < 1$, then not only can we evaluate the distance of the center of gravity to the origin, but according to our theorem, we know that $\hat{H}(\Gamma) \geq 1 - \alpha$. The smaller the α the better the sequence is equidistributed (mod1), and the higher is the entropy. In other terms, the better a sequence is equidistributed (mod1) the more it mimics randomness.

We now prove Theorem 6. Let

$$\Gamma_N = \sum_{n=0}^{N-1} \exp 2i\pi\theta_n$$

be the polygonal line of length N. The diameter of Γ_N is

$$\begin{aligned} \Delta_N &= \max_{0 \leq m < n < N} \left|\sum_{k=m}^{n} \exp 2i\pi\theta_k\right| \\ &= \max_{0 \leq m < n < N} [O(n^\alpha) + O(m^\alpha)] \\ &= O(N^\alpha). \end{aligned}$$

The perimeter C_N of Γ_N is of the same order of magnitude as Δ_N :

$$\Delta_N \leq C_N \leq \pi\Delta_N,$$

hence

$$\begin{aligned} \hat{H}(\Gamma) &= \liminf_{N \nearrow \infty} \frac{\log 2N/C_N}{\log N} \\ &\geq 1 - \limsup_{N \nearrow \infty} \frac{\log N^\alpha}{\log N} = 1 - \alpha. \end{aligned}$$

This establishes the first part of the theorem.

We now assume that for infinitely many N

$$\left|\sum_{k=0}^{N-1} \exp 2i\pi\theta_k\right| > cN^\beta,$$

so for these N,

$$C_N > cN^\beta.$$

Then, as before

$$\hat{H}(\Gamma) = 1 - \limsup_{N \nearrow \infty} \frac{\log C_N}{\log N}.$$

For infinitely many N

$$\frac{\log C_N}{\log N} > \frac{\log c + \beta \log N}{\log N}$$

so that

$$\limsup_{N \nearrow \infty} \frac{\log C_N}{\log N} \geq \beta.$$

Therefore

$$\hat{H}(\Gamma) \leq 1 - \beta. \qquad \text{Q.E.D.}$$

3.3 Examples

Computing or even estimating an exponential sum is rarely easy. Many books and articles are devoted to studying such sums. The pioneering work of H. Weyl [28], of Hardy and Littlewood [12] and of I.M. Vinogradov [27] opened the path to an important method in Number Theory which we cannot discuss here. We shall simply collect classical results and give some proofs when they are not too fastidious. Most of our examples come from an article which I wrote with M. Dekking [6].

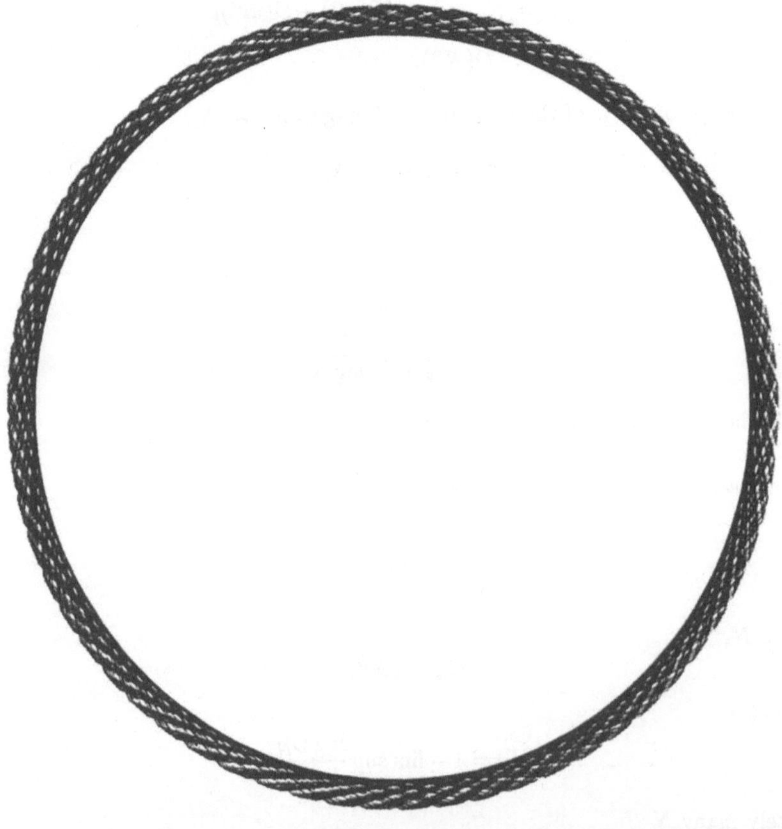

Figure 12: 400 first sides of the polygon $\sum_{n=0}^{\infty} \exp 2i\pi\lambda n$ where $\lambda = \sqrt{17}$.

3.3.1 The sequence (λn), $\lambda \in \mathbf{R} \setminus \mathbf{Z}$. Consider the sum

$$S(N) = \sum_{n=0}^{N-1} \exp 2i\pi\lambda n = \frac{1 - \exp 2i\pi\lambda N}{1 - \exp 2i\pi\lambda}.$$

Then

$$|S(N)| \leq \frac{1}{|\sin \pi\lambda|}.$$

The sum being bounded, it follows that the entropy $\hat{H}(\Gamma) = 1$ ($\alpha = 0$ in Theorem 6).

The polygon

$$\Gamma = \sum_{n=0}^{\infty} \exp 2i\pi\lambda n$$

is easy to describe. It is a star-shaped regular polygon dense in the annulus centered at $(1 - \exp 2i\pi\lambda)^{-1}$ with radii

$$\frac{1}{2|\sin \pi\lambda|} \quad \text{and} \quad \frac{1}{2}\left|\frac{\cos \pi\lambda}{\sin \pi\lambda}\right|.$$

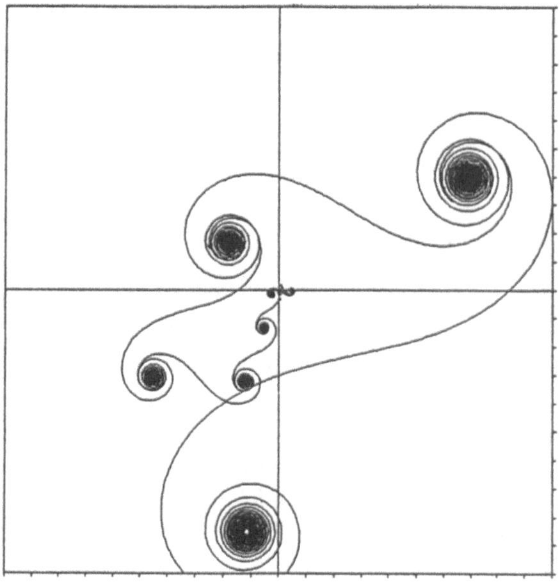

Figure 13: 4000 first sides of the polygon $\sum_{n=0}^{\infty} \exp 2i\pi n \log n$.

3.3.2 The sequence $(n \log n)$.

It is known that

$$\sum_{n=0}^{N-1} \exp 2i\pi n \log n = O(N^{1/2}).$$

The estimate is sharp (see for example volume 1, p. 199, of Zygmund's treatise on trigonometric series [30]). Therefore

$$\hat{H}(\Gamma) = 1/2.$$

In Figure 13 we have drawn the 4000 first sides of Γ. The general aspect of the curve is well understood if one considers that for large n, $\log n$ varies very slowly. Then $(n \log n)$ behaves locally like the sequence $(n\lambda)$. When $\log n \pmod 1$ is close to zero, the curve mimics a straight line (inflexion points). When on the contrary $\log n$ (mod1) is close to 1/2, the curve oscillates back and forth (center of the spirals). The spiral effect is due to a slowly moving annulus. All these remarks can be made precise.

3.3.3 The sequence $(n^{3/2})$

The polygon

$$\Gamma = \sum_{n=0}^{\infty} \exp 2i\pi n^{3/2}$$

is surprisingly regular as shown on Figure 14. It was discovered by F. Dress and its entropy was computed by J.M. Deshouillers who showed that $\hat{H} = 1/4$. For details we refer to his article [8].

Further examples are discussed below in paragraph 3.4.

3.4 The sequence (λn^2)

The curves

$$\Gamma = \sum_{n=0}^{\infty} \exp 2i\pi\lambda n^2$$

or curlicues as M. Berry calls them, are very exciting to study in that they may be described in detail and have physical interpretations; see for example M.V. Berry and J. Goldberg [1] or E.A. Coutsias and N.D. Kazarinoff [4]. We follow Berry and Goldberg's analysis which should be compared with Callot and Diener [3], Deshouillers [8] and van der Poorten [26]. See also Lehmer [14] and Loxton [15].

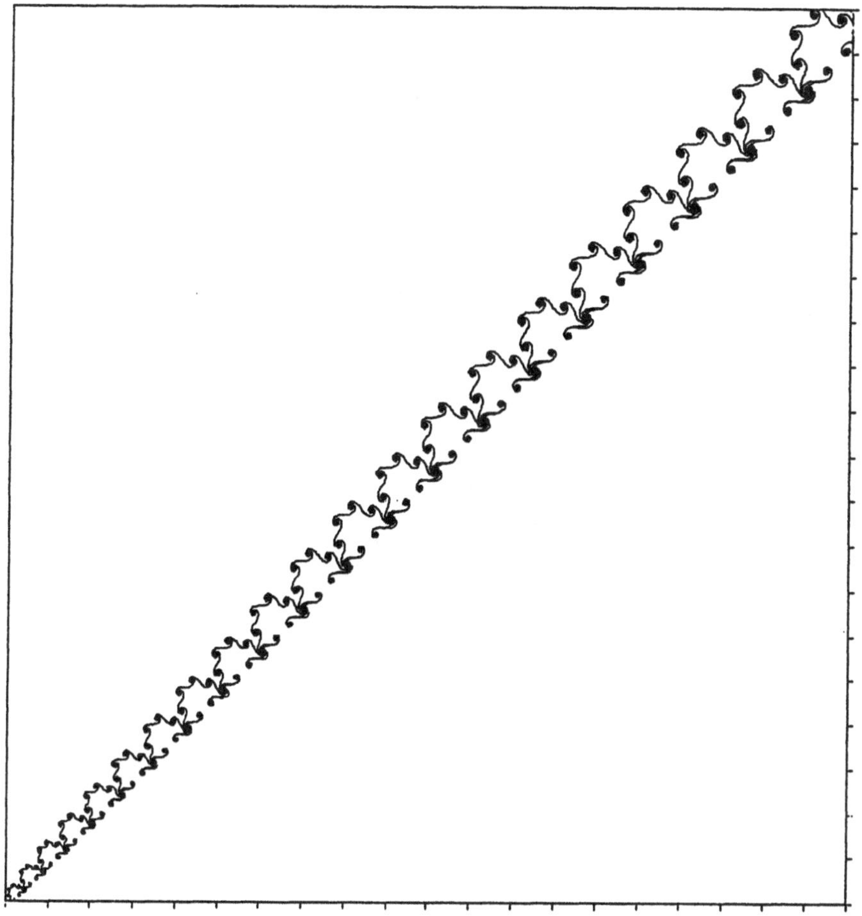

Figure 14: 20000 sides of the polygon $\sum_{n=0}^{\infty} \exp 2i\pi n^{3/2}$

Consider the finite sum

$$\Gamma(N,\lambda)=\sum_{n=0}^{N-1}\exp 2i\pi\lambda n^2.$$

Apply the Poisson summation formula

$$\sum_{n=-\infty}^{+\infty} f(n)=\sum_{n=-\infty}^{+\infty}\hat{f}(n),$$

where

$$\hat{f}(n)=\int_{-\infty}^{+\infty} f(x)\exp(-2i\pi nx)dx,$$

to the function

$$f(x)=\begin{cases}\exp 2i\pi\lambda x^2, \text{ if } \alpha<x<N-1+\alpha,\\ 0, \text{ if not.}\end{cases}$$

α is a constant, $0<\alpha<1$. Then

$$\Gamma(N,\lambda)=\sum_{n=-\infty}^{+\infty}\int_{\alpha}^{\alpha+N-1}\exp 2i\pi(\lambda x^2-nx)dx.$$

The integrals which appear on the right hand side are all negligible unless x is close to a stationary point x_n, solution of the equation

$$\frac{d}{dx}(\lambda x^2-nx)=0,$$

hence

$$x_n=\frac{n}{2\lambda}.$$

But $\alpha<x_n<\alpha+N-1$ so that

$$0\le n<[2N\lambda].$$

The sum $\Gamma(N,\lambda)$ reduces to

$$\begin{aligned}\Gamma(N,\lambda) &\approx \sum_{n=0}^{[2N\lambda]-1}\int_{\alpha}^{\alpha+N-1}\exp 2i\pi(\lambda x^2-nx)dx\\ &\approx \sum_{n=0}^{[2N\lambda]-1}\int_{-\infty}^{+\infty}\exp 2i\pi(\lambda x^2-nx)dx\\ &= \frac{\exp i\frac{\pi}{4}\mathrm{sgn}\lambda}{\sqrt{2|\lambda|}}\sum_{n=0}^{[2N\lambda]-1}\exp -2i\pi\frac{n^2}{4\lambda}.\end{aligned}$$

Finally

$$\Gamma(N,\lambda)\approx\frac{\exp i\frac{\pi}{4}\mathrm{sgn}\lambda}{\sqrt{2|\lambda|}}\Gamma([2N\lambda],-\frac{1}{4\lambda}).$$

The formula is slightly more symmetrical if one puts $\tau = 2\lambda$. Then

$$\Gamma(N, \frac{1}{2}\tau) \approx \frac{\exp i\frac{\pi}{4}\mathrm{sgn}\tau}{\sqrt{|\tau|}}\Gamma([N\tau], -\frac{1}{2\tau}).$$

The technique we have just sketched is called the method of the stationary phase.

We now observe that with no loss of generality, we can assume $0 \leq \tau \leq 1$. Indeed,

$$\Gamma(N, \frac{1}{2}\tau + 1) = \Gamma(N, \frac{1}{2}\tau)$$

$$\Gamma(N, -\frac{1}{2}\tau) = \overline{\Gamma}(N, \frac{1}{2}\tau)$$

(the bar on Γ is the complex conjugate). The two cases $\tau = 0$ and $\tau = 1$ are trivial to deal with, so we may suppose $0 < \tau < 1$. Then $[N\tau] < N$ so that the sum $\Gamma(N, \frac{1}{2}\tau)$ can be calculated in terms of a reduced sum $\Gamma([N\tau], \frac{1}{2\tau})$. This renormalization principle repeated over and over again enables one to estimate $\Gamma(N, \frac{1}{2}\tau)$; see details in Berry, Goldberg [1].

The important point I want to emphasize here is that this renormalization principle maps τ onto $-1/\tau$ which shows that the estimate will depend on the modified continued fraction expansion of τ

$$\tau = \cfrac{1}{a_1 - \cfrac{1}{a_2 - \cfrac{1}{a_3 \ddots}}}$$

This, in turn, is linked to the regular continued fraction expansion of τ and finally to the regular continued fraction expansion of $\lambda = \frac{1}{2}\tau$. To conclude, a good knowledge of the continued fraction expansion of λ provides a precise estimation of the sum $\Gamma(N, \lambda)$.

Here are some examples which have been worked out. If the partial quotients of the continued fraction expansion of λ are bounded, then

$$\Gamma(N, \lambda) = O(N^{1/2}) \text{ (sharp)}$$

and hence $\hat{H} = 1/2$. This holds for quadratic irrational λ's since the sequence of partial quotients is ultimately periodic. Figure 15 shows the first 4000 sides of the curve

$$\Gamma = \sum_{n=0}^{\infty} \exp 2i\pi\sqrt{2}n^2.$$

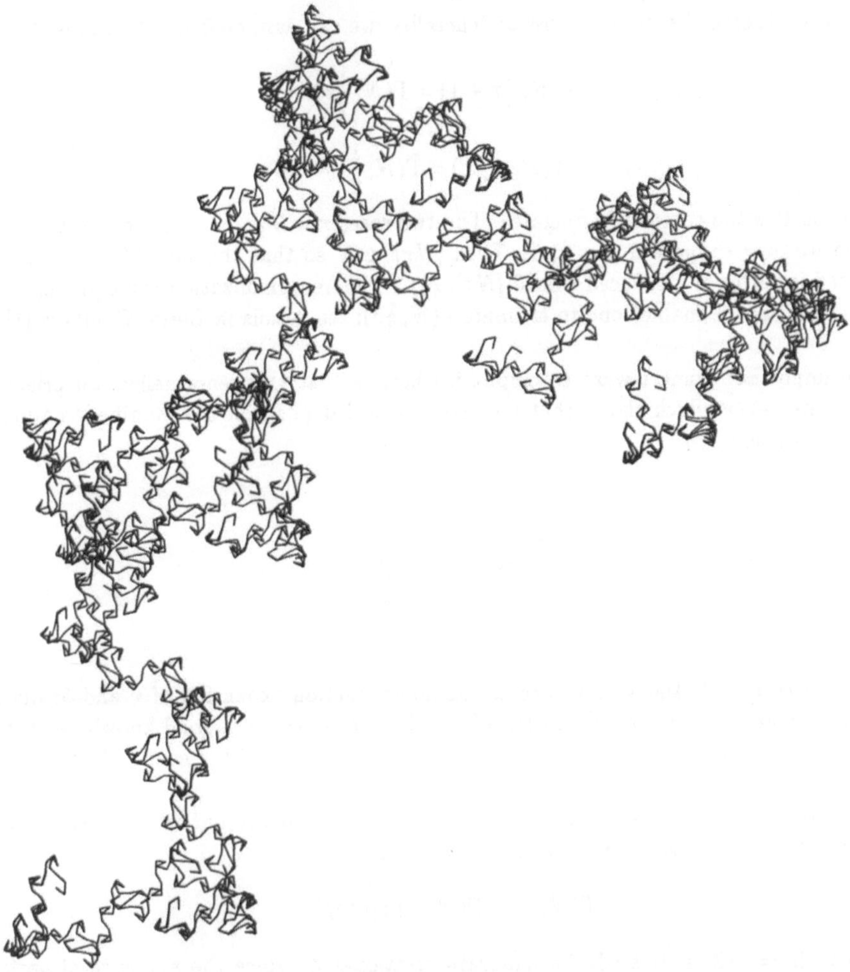

Figure 15: 4000 sides of the polygon $\sum_{n=0}^{\infty} \exp 2i\pi\sqrt{2}n^2$

We now choose $\lambda = \pi$ and consider the infinite polygon

$$\Gamma = \sum_{n=0}^{\infty} \exp 2i\pi^2 n^2.$$

As the structure of the continued fraction expansion of π is unknown we cannot compute the entropy of the curve. However curious patterns appear on the curve Γ. Let us analyze them (see Figure 16).

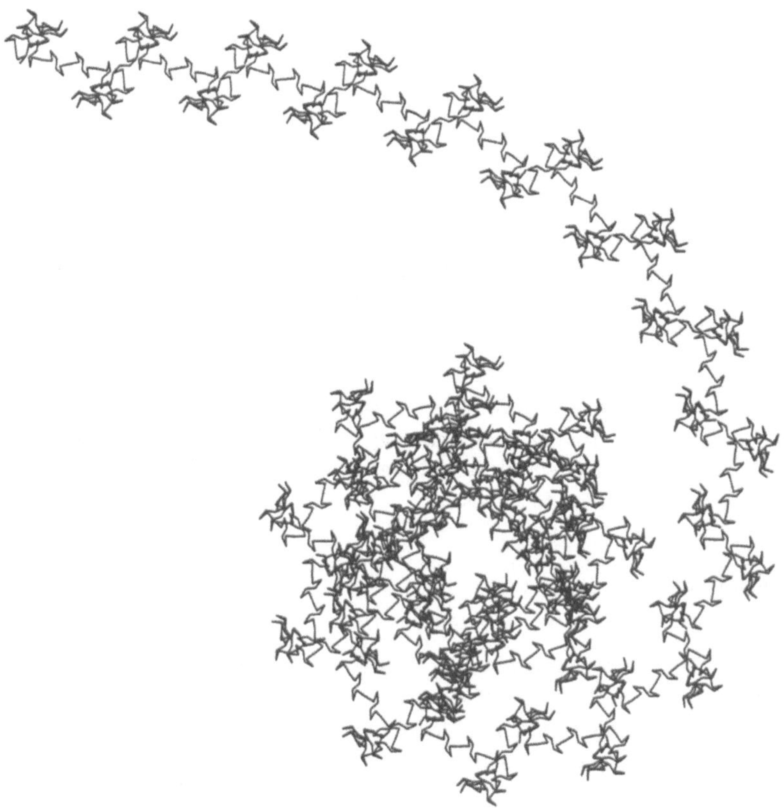

Figure 16: 4000 sides of the curlicue $\sum_{n=0}^{\infty} \exp 2i\pi^2 n^2$.

A close inspection shows a sort of periodicity and with a bit of patience one will discover that its length is 113. This obviously comes from the rational approximation of π:

$$\pi \approx \frac{335}{113} = [3, 7, 15, 1].$$

And, indeed, the curve

$$\sum_{n=0}^{\infty} \exp 2i\pi \frac{355}{113} n^2$$

is clearly periodic of period 113. The spiraling effect which perturbes the period on the curlicue Γ is due to the fact that the continued fraction expansion of π is more complicated

$$\pi = [3, 7, 15, 1, 292, 1, 1, 1, 2, 1, 3, 1, 1, 4, 2, 1, 1, 2, 2, 2, 2, 1, \ldots].$$

Cutting the expansion after the second term

$$[3, 7] = \frac{22}{7}$$

We are reminded of another well known approximation of π, and again, a close inspection of the curve Γ shows the period 7.

If one draws the curve Γ up to 20000, say, one would see the first spiral unwinding then winding up again into a new spiral very much the same as the first one, and so on. Infinitely many spirals would appear which themselves would lie on a bigger spiral and so forth. We are reminded of Richardson's famous verses

"Big whorls have little whorls
which feed on their velocity;
And little whorls have lesser whorls,
And so on to viscosity."

which I would like to read as

"Little whorls have big whorls
which feed on irrationality;
And big whorls have larger whorls
And so on to infinity."

The number of sides from one "small" spiral to its neighbour is 16551. Now $2 \times 16551 = 33102$ which happens to be the denominator of

$$[3, 7, 15, 1, 292] = \frac{103993}{33102}.$$

This shows that the period is actually formed by two spirals.

Finally, to end this paragraph I would like to say a few words on the curve

$$\sum_{n=0}^{\infty} \exp 2i\pi n^2 \log n.$$

It can be shown that

$$\sum_{n=0}^{N-1} \exp 2i\pi n^2 \log n = O(N^{5/6})$$

so that the entropy $\hat{H}(\Gamma)$ is not less than $1/6$.

This result is crude. By looking at the curve (Figure 17) one feels that indeed the entropy should be larger, probably close to 1.

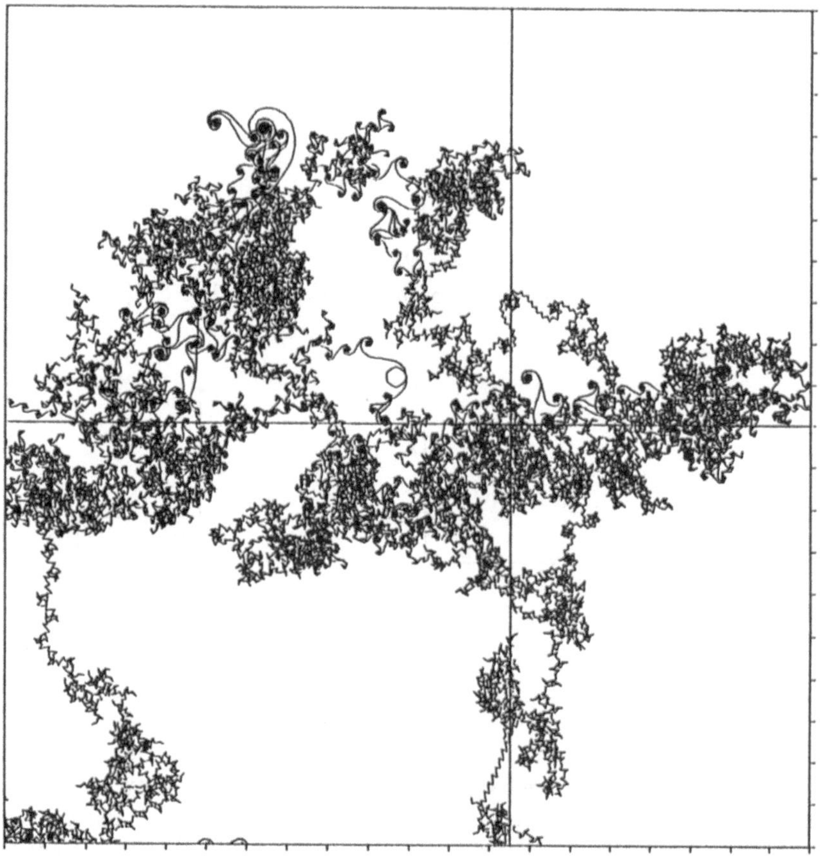

Figure 17: The curve $\Gamma = \sum_{n=0}^{\infty} \exp 2i\pi n^2 \log n$.

The very erratic behaviour of the curve can be explained as before when analyzing the sequence $(n \log n)$. The function $\log n$ is practically constant on long ranges of n. The polygon

$$\Gamma = \sum_{n=0}^{\infty} \exp 2i\pi n^2 \log n$$

is thus the union of large portions of all the curves

$$\Gamma(\lambda) = \sum_{n=0}^{\infty} \exp 2i\pi\lambda n^2, \ \lambda \in [0,1[.$$

We have seen that these curves $\Gamma(\lambda)$ depend heavily on the arithmetic structure of λ (its continued fraction expansion), so that $\Gamma(\lambda)$ may be very different from $\Gamma(\lambda')$ even though λ and λ' may be very close together. In some regions

$$\Gamma = \sum_{n=0}^{\infty} \exp 2i\pi n^2 \log n$$

may seem regular and periodic. In other regions it may seem quite chaotic.

3.5 Dragon curves

The dragon curve is generated as follows. We start off with the unit segment which we rotate to $-\pi/2$ around its endpoint A_1 to obtain the broken line OA_1A_2. Then OA_1A_2

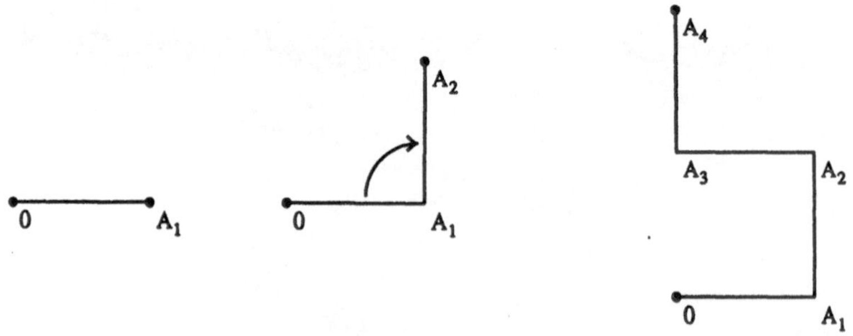

Figure 18:

is rotated around A_2 to $-\pi/2$, and we continue this process to infinity. At each step we rotate the whole figure around its endpoint A_{2^n} by $-\pi/2$. One thus obtains a complicated curve (see Figure 19) which Ch. Davis and D. Knuth [5] named the Dragon curve. They showed the surprising property according to which the curve is self-avoiding. They also studied other dragon curves obtained by rotating the figures by angles $\pm\pi/2$ arbitrarily at each step. All these curves share the self-avoiding property.

Let x_n, y_n be the Cartesian coordinates of the nth vertex of any one of these dragon curves. It can be shown that $x_n = O(\sqrt{n})$ and $y_n = O(\sqrt{n})$ which implies that $\hat{H} = 1/2$. These curves are at the midpoint between order and chaos. It is easy to prove that a self-avoiding curve on the lattice $\mathbf{Z}^2$ has entropy $\hat{H}$ at most 1/2 so that the dragon curves are as chaotic as can be. For more details, we refer to my joint paper with Tenenbaum [17] and to Folds!, a series of three articles written by Dekking, van der Poorten and myself [7].

Figure 19: The Dragon curve (after 10 steps)

4 Dimension and Entropy

4.1 Relationship between dimension and entropy

We now come back to the topics discussed in the two first chapters.

Theorem 7 *For all curves $\Gamma \in F$, the entropy $H(\Gamma)$, the dimension $\dim(\Gamma)$ and the confusion coefficient $\gamma = \gamma(\Gamma)$ are related by the inequality*

$$H(\Gamma) \leq 1 - \frac{1}{\dim(\Gamma) + \gamma}.$$

Proof: According to Theorem 2,

$$\liminf_{L \nearrow \infty} \frac{\log L}{\log C_L} \leq \gamma + \dim(\Gamma).$$

On the other hand,

$$\begin{aligned} H(\Gamma) \leq \hat{H}(\Gamma) &= \liminf_{L \nearrow \infty} \frac{\log 2L/C_L}{\log L} \\ &= 1 - \limsup_{L \nearrow \infty} \frac{\log C_L}{\log L} \\ &= 1 - [\liminf_{L \nearrow \infty} \frac{\log L}{\log C_L}]^{-1} \\ &\leq 1 - \frac{1}{\gamma + \dim(\Gamma)}. \end{aligned}$$

Q. E. D.

Remark We know that

$$\liminf_{L \nearrow \infty} \frac{\log L}{\log C_L} \geq \dim(\Gamma)$$

so that one actually has the double inequality for $\hat{H}(\Gamma)$

$$1 - \frac{1}{\dim(\Gamma)} \leq \hat{H}(\Gamma) \leq 1 - \frac{1}{\gamma + \dim(\Gamma)}.$$

Corollary *If the entropy of a curve Γ is strictly larger than 1/2, then the curve is non-resolvable.*

Proof: By hypothesis

$$\frac{1}{2} < H(\Gamma) \leq 1 - \frac{1}{\gamma + \dim(\Gamma)},$$

hence

$$\gamma + \dim(\Gamma) > 2$$

and

$$\gamma > 2 - \dim(\Gamma) \geq 0$$

so that $\gamma > 0$. The curve is thus non-resolvable. Q.E.D.

To assert $H(\Gamma) > 1/2$ is to say that the curve is somewhat chaotic ($1/2$ is the midpoint between order and chaos). On the other hand, $\gamma > 0$ indicates confusion. Our corollary thus expresses that chaos implies confusion.

The corollary could be thought of as a weak statement concerning percolation. As soon as the entropy gets larger than $1/2$, the curve becomes non-resolvable. If the curve was to be a real electric wire with nonzero width, electricity would flow in some regions as if on a surface.

The corollary can also be seen as an extension of the pigeon-hole principle which, in its crude form, tells us that if $n+1$ objects are distributed in n boxes, then one box at least contains 2 objects. It also tells us that an infinite set of points on a finite interval must have at least one cluster point. Our corollary teaches us that an infinite curve which is somehow too long must be crumpled.

We now give a last interpretation of Theorem 7.

4.2 The Planck constant of a curve

Theorem 7 reads

$$(d-1) + \gamma \geq \frac{H}{1-H},$$

where d and H are the dimension and entropy of Γ. (If $H = 1$, the ratio $H/(1-H)$ stands for $+\infty$.) A curve has topological dimension 1 so that $\delta = d - 1$ measures the excess of dimension of the curve.

Assume for simplicity that Γ has a tangent at every point. Let $\vec{V} = \vec{V}(M)$ be the unit tangent vector at $M \in \Gamma$ and let L be the length of the arc $\overset{\frown}{OM}$.

Consider the average

$$\frac{1}{L}\int_0^L \vec{V} dL = \frac{\vec{OM}}{\| \overset{\frown}{OM} \|},$$

and suppose the dimension d is strictly larger than $1, \delta > 0$. Then

$$1 < d \leq \liminf_{L \nearrow \infty} \frac{\log \| \overset{\frown}{OM} \|}{\log \|\vec{OM}\|}$$

so that for large L

$$\| \widehat{OM} \| \geq \|\vec{OM}\|^{d'}$$

for some d', $1 < d' < d$. Hence

$$\left| \frac{1}{L} \int_0^L \vec{V} dL \right| \leq \frac{\| \widehat{OM} \|}{\| \widehat{OM} \|}^{1/d'} = \| \widehat{OM} \|^{-1+1/d'}.$$

The exponent is strictly negative so that the average tends to 0 as L increases to infinity. Therefore, when $\delta > 0$ the curve has no "drift"; there is no preferred direction. The curve tends to infinity in an unpredictable way. Let us then denote δ by Δv, the uncertainty attached to the fluctuating direction of the tangent vector $\vec{V}$.

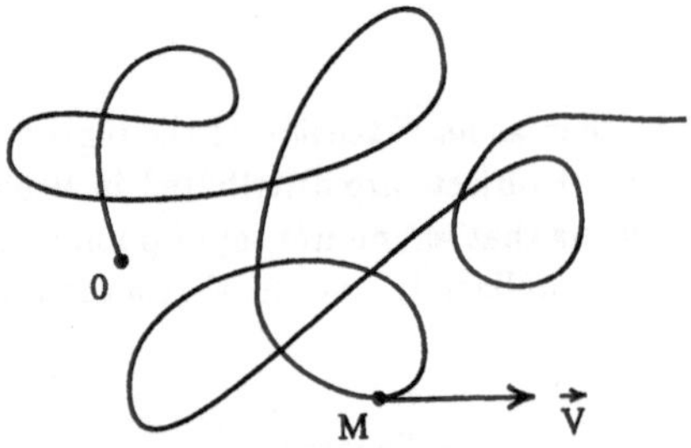

Figure 20: Curve and tangent. If $\dim(\Gamma) > 1$, the direction of the tangent averages to 0.

The confusion coefficient γ which we now write down as Δx is a measure of the uncertainty in the location of a point on the curve. Indeed, when $\gamma = \Delta x > 0$ the branches of Γ are so tightly packed together that it is not possible to decide which branch of the curve contains a given point.

Finally, let us call $H/(1 - H) = \hbar$ the Planck constant of the curve. Theorem 7 then reads

$$\Delta x + \Delta v \geq \hbar$$

which is very reminiscent of Heisenberg's inequality.

If the Planck constant $\hbar$ is nonzero (i.e., $H > 0$), then it is impossible to define with infinite precision both the location of a point on Γ and its tangent. The higher the entropy, the greater the uncertainty.

The sign + in the above inequality should not worry the reader. If $\hbar > 0$, then Δx and Δv cannot both be small. Suppose $0 < \hbar < \infty$ and suppose $\Delta x < \hbar$ and $\Delta v < \hbar$. Define

$$\Delta' x = \frac{\Delta x}{\hbar - \Delta x}, \qquad \Delta' v = \frac{\Delta v}{\hbar - \Delta v}$$

so that $\Delta' x$ and $\Delta' v$ are just some normalizations of Δx and Δv, respectively. The inequality

$$\Delta x + \Delta v \geq \hbar$$

then becomes

$$\Delta' x \cdot \Delta' v \geq 1.$$

In short,

$$\hbar > 0 \Longrightarrow \begin{cases} \text{either } \Delta x \geq \hbar \text{ or } \Delta v \geq \hbar \\ \text{or } \Delta' x \cdot \Delta' v \geq 1. \end{cases}$$

In this interpretation, Planck's constant is

$$\begin{cases} 0, \text{ if the entropy is } 0 \\ 1, \text{ if the entropy is } H > 0. \end{cases}$$

Is it true, in our Universe, that Planck's constant is related to the entropy of the Universe?

4.3 A final remark

In these notes, we have restricted ourselves to curves in the plane. Almost everything can be extended to curves in $\mathbf{R}^n$. For example, the entropy H is obtained by intersecting the curve by random hyperplanes. The dimension of the curve is comprised between 1 and n and the inequality

$$H \leq 1 - \frac{1}{\dim \Gamma + \gamma}$$

still holds. The corollary to Theorem 7 must then be replaced by

$$H > \frac{n-1}{n} \Longrightarrow \gamma > 0.$$

Heisenberg's uncertainty principle is established as for $n = 2$.

I would like to think that there is still much to do concerning the topics we have discussed here. The last chapter is certainly unfinished and one could probably extend the whole theory to p-dimensional manifolds embedded in an n-dimensional space ($n = \infty$ could be interesting to study). Actually, Alexander Nabutovsky [18] discusses a notion of crumpledness of a hypersurface in $\mathbf{R}^{n+1}$. Does his work bear any relationship to mine?

Acknowledgements

The reader may feel surprised that there is no mention of Benoît Mandelbrot in these notes. His objects are fractal, i.e., locally irregular. Mine, on the contrary are locally smooth. The curves I discuss are locally rectifiable. My topic could be thought of as "anti-Mandelbrotian" within "Mandelbrotmania". I was, I am and I hope to remain influenced by B. Mandelbrot.

Several of the curves which appear in chapter 3 have been drawn by Michel Pallard. Some appeared in Deshouillers [8] and others in Dekking, Mendès France [6] or in van der Poorten [26]. H. Cohen helped me with some computing. I warmly thank all the aforementioned friends for their help.

References

[1] M.V. Berry, J. Goldberg, Renormalization of curlicues, *Nonlinearity* **1** (1988), 1 - 26.

[2] T.A. Cook, *The Curves of Life*, Dover Publications, New York, 1979.

[3] L. Callot, M. Diener, Variations en spirale 1984; unpublished paper from the University of Oran.

[4] E.A. Coutsias, N.D. Kazarinoff, Disorder renormalization, theta functions and cornu spirals, *Physica* **26 D** (1987), 295-310.

[5] Ch. Davis, D.E. Knuth, Number representations and dragon curves I, *J. Recreational Math.* **3** (1970), 61-81; II **3** (1970), 133-149.

[6] F.M. Dekking, M. Mendès France, Uniform distribution modulo one: a geometrical viewpoint, *J. Reine Angew. Math.* **329** (1981), 143-153.

[7] F.M. Dekking, M. Mendès France, A.J. van der Poorten, Folds!, *Math. Intelligencer* **4** (1982), 130-138, 173-181, 190-195.

[8] J.-M. Deshouillers, Geometric aspect of Weyl sums, in: *Elementary and Analytic Theory of Numbers*, Banach Center Publications, vol. 17, PWN Polish Scientific Publishers, Warsaw 1985.

[9] Y. Dupain, T. Kamae, M. Mendès France, Can one measure the temperature of a curve ?, *Arch. Rational Mech. Anal.* **94** (1986), 155-163; **98** (1987), 395.

[10] M.E. Gage, An isoperimetric inequality with applications to curve shortening, *Duke Math. J.* **50** (1983), 1225-1229.

M.E. Gage, Curve shortening makes convex curves circular, *Invent. Math.* **76** (1984), 357-364.

M.E. Gage, On an area-preserving evolution equation for plane curves, *Contemporary Math.* **51** (1986), 51-62.

[11] M.A. Grayson, A short note on the evolution of a surface by its mean curvature, *Duke Math. J.* **58** (1989), 555-558.

[12] G.H. Hardy, J.E. Littlewood, The trigonometric series associated with the elliptic θ-functions, *Acta Math.* **37** (1914), 193-239.

[13] L. Kuipers, H. Niederreiter, *Uniform Distribution of Sequences*, Wiley Interscience, John Wiley & Sons, New York, 1974.

[14] D.H. Lehmer, Incomplete Gauss sums, *Mathematika* **23** (1976), 125-135.

[15] J.H. Loxton, The graphs of exponential sums, *Mathematika* **30** (1983), 153-163.

[16] M. Mendès France, Chaotic curves, in: *Rhythms in Biology and Other Fields of Applications*, Proc. Journ. Soc. Math. France, Luminy 1981, Lecture Notes in Biomathematics 49, Springer-Verlag, Berlin-Heidelberg-New York, 1983, 352-367.

M. Mendès France, Entropie, dimension et thermodynamique des courbes planes, in: *Séminaire de théorie des nombres*, Paris 1981-82, Séminaire Delange-Pisot-Poitou, Birkhäuser, 1983, 153-177.

M. Mendès France, Folding paper and thermodynamics, *Phys. Rep.* **103** (1984), 161-172.

M. Mendès France, Dimension et entropie des courbes régulières, in: *Fractals, dimensions non entières et applications* (sous la direction de G. Cherbit), Masson, Paris, 1987, 329-339.

M. Mendès France, Chaos implies confusion, in: *Number Theory and Dynamical Systems* (M. Dodson and J. Vickers, eds.), London Math. Soc. Lecture Notes 134, Cambridge Univ. Press, 1989, 137-152.

[17] M. Mendès France, G. Tenenbaum, Dimension des courbes planes, papiers pliés, suites de Rudin-Shapiro, *Bull. Soc. Math. France* **109** (1981), 207-215.

[18] A. Nabutovsky, Isotopies and non-recursive functions in real algebraic geometry, in: *Real Analytic and Algebraic Geometry* (M. Gabbiati and A. Tognoli, eds.), Lecture Notes in Mathematics 1420, Springer-Verlag, Berlin-Heidelberg-New York, 1990, 194-205.

[19] G. Rauzy, *Propriétés statistiques des suites arithmétiques*, Collection SUP, Presses Universitaires de France, 1976.

[20] R. Rigon, Private communication, 1989.

[21] L.A. Santaló, *Integral Geometry and Geometric Probability*, Addison-Wesley, 1976.

[22] A.A. Schäffer, C.J. van Wyk, Convex hulls of piecewise-smooth Jordan curves, *J. Algorithms* **8** (1987), 66-94.

[23] H. Steinhaus, Length, shape and area, *Colloq. Math.* **3** (1954), 1-13.

[24] R. Thom, *Stabilité structurelle et morphogénèse*, Deuxième édition, Inter Éditions, Paris, 1977.

[25] C. Tricot, Rectifiable and fractal sets, this volume.

[26] A.J. van der Poorten, R.R. Moore, On the thermodynamics of curves and other curlicues, in: *Proc. Conf. on Geometry and Physics (Canberra 1989)*, Proc. Center. Math. Anal., Australian Nat. Univ., 1989, 82-109.

[27] I.M. Vinogradov, *The Method of Trigonometrical Sums in the Theory of Numbers*, Interscience Publishers, 1954.

[28] H. Weyl, Über die Gleichverteilung von Zahlen mod. Eins, *Math. Ann.* 77 (1916), 313-352.

[29] B. White, Some recent developments in Differential geometry, *Math. Intelligencer* 11 (1989), 41-47.

[30] A. Zygmund, *Trigonometric Series*, 2nd ed., 2 volumes, Cambridge University Press, 1959.

Rectifiable and fractal sets

Claude TRICOT
Ecole Polytechnique de Montréal
Département de Mathématiques Appliquées
C.P. 6079, Succursale A
Montréal (Québec),
Canada H3C 3A7

Abstract

The main topics investigated are the following: rectifiability, length, Hausdorff and packing measures and dimensions, local densities, dimensional regularity of sets, index of proximity, set independance, Minkowski–Bouligand dimension, algorithms of computation.

1 Length and rectifiability

A curve Γ in $\mathbf{R}^n$ is the range of a continuous function $\gamma(t)$ defined on an interval $[a, b]$. We assume that γ is an injection: the curve Γ is simple. Its endpoints are $\gamma(a) = A$ and $\gamma(b) = B$.

1.1 Geometrical approach

A polygonal curve $\mathbf{P}$ is a *polygonal approximation* of Γ if

- $\mathbf{P}$ and Γ have the same endpoints
- All vertices of P belong to Γ, in the order induced by the parametrization.

Theorem 1. *Let $\epsilon_n \to 0$, and $\mathbf{P}_n$ be a sequence of polygonal approximations of Γ, such that the length of all segments of $\mathbf{P}_n$ is $\leq \epsilon_n$. Then*

$$\lim_{n\to\infty} \operatorname{dist}(\mathbf{P}_n, \Gamma) = 0$$

in the sense of the Hausdorff distance.

J. Bélair and S. Dubuc (eds.), Fractal Geometry and Analysis, 367–403.

Theorem 2. *Same assumptions. Let $L(\mathbf{P}_n)$ be the length of the polygonal curve $\mathbf{P}_n$. Then the sequence $L(\mathbf{P}_n)$ tends to $\sup_n L(\mathbf{P}_n)$ (finite or infinite), and this number does not depend on the particular choice of $\mathbf{P}_n$ (if $\mathbf{Q}_n$ verifies the same assumption, then $\sup_n L(\mathbf{P}_n) = \sup_n L(\mathbf{Q}_n)$).*

The number $\sup_n L(\mathbf{P}_n)$ is called *length* of Γ.

The proofs of Theorems 1 and 2 use the connectedness of Γ, and the total order relationship on Γ, image by γ of the natural order on $[a,b]$. Note that Theorem 1 could not be written as an equivalence. The only general result in the converse direction is the following:

Theorem 3. *Let (Γ_n) be a sequence of curves, such that* $\mathrm{dist}(\Gamma_n,\Gamma)) \to 0$. *Then*

$$L(\Gamma) \leq \liminf_n L(\Gamma_n) .$$

Hint of proof. Take a polygonal approximation $\mathbf{P}$ of Γ. Denote by $A_1 = A$, A_2, ...,$A_k = B$ the vertices of $\mathbf{P}$. Let $\rho_n = \mathrm{dist}(\Gamma_n,\Gamma)$. Choose n large enough to ensure that the balls $B_{\rho_n}(A_i)$ are disjoint. Then, using the fact that $\Gamma_n \cap B_{\rho_n}(A_i) \neq \emptyset$, check the following:

$$L(\Gamma_n) \geq L(\mathbf{P}) - 2\,k\,\rho_n .$$

1.2 Analytical approach

The curve Γ is parametrized by the function $\gamma(t) = (\gamma_1(t), \gamma_2(t), \ldots, \gamma_n(t))$. Then

$$L(\Gamma) = \sup_{\mathcal{P}} \sum_{j=1}^{m} \sqrt{\sum_{i=1}^{n} (\gamma_i(t_j) - \gamma_i(t_{j+1}))^2} ,$$

where the supremum is taken over all partitions $\mathcal{P} : a = t_0 < t_1 < \ldots < t_m = b$ of the interval $[a,b]$. The **variation** of a real-valued function $f(t)$ over $[a,b]$ is:

$$V_f = \sup_{\mathcal{P}} \sum_{j=1}^{m} |f(t_j) - f(t_{j+1})| .$$

The function f is of *bounded variation* if V_f is finite.

Theorem 1.

$$L(\Gamma) < \infty \Longleftrightarrow \textit{all functions } \gamma_i \textit{ are of bounded variation on } [a,b].$$

Proof. Hölder inequality $|y_i| \leq \sqrt{\sum y_i^2} \leq \sum |y_i|$ implies

$$V_{\gamma_i} \leq L(\Gamma) \leq \sum_{i=1}^{n} V_{\gamma_i} \,.$$

Theorem 2. *If γ is piecewise continuously differentiable, then*

$$L(\Gamma) = \int_a^b \sqrt{\sum_{i=1}^{n} \gamma_i'(t)^2}\, dt \,.$$

Proof. Uses essentially the mean value theorem for integrals.

Application. Let Γ be the graph of a continuous function $f(t)$. Parametrization: $\gamma(t) = (t, f(t))$. Then

$$L(\Gamma) = \int_a^b \sqrt{1 + f'(t)^2}\, dt \,.$$

Example. Theorem 2 is of no use in the following case:

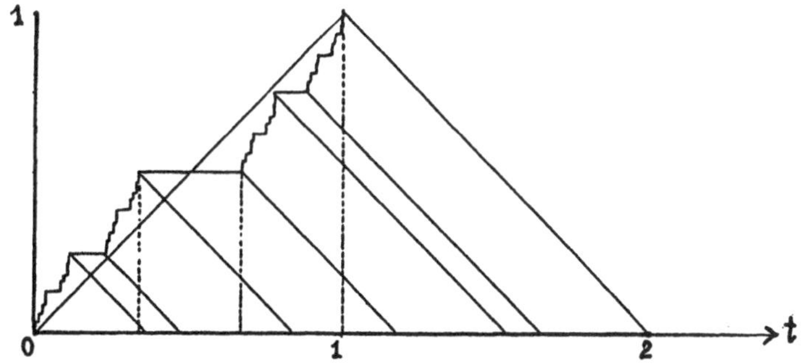

Figure 1.1:

Let C be the Cantor set in $I = [0,1]$. The devil's staircase is the graph of a continuous, increasing function $f(t)$, with no derivative on C, and such that $f'(t) = 0$ for all $t \in I - C$. The Riemann integral $\int_{I-C} \sqrt{1 + f'(t)^2}\, dt = 1$ is not equal to the length of Γ. Using polygonal approximations of Γ gives $L(\Gamma) = 2$.

In this case, the arc–length parametrization is obtained by a projection on the t–axis. The images of the points $(t, f(t))$, $t \in C$, by this projection, constitute a Cantor set (a nowhere dense, perfect set) of length 1.

1.3 Local properties

Let us call *tangent* $T(x_0)$ to the curve Γ at the point x_0 of Γ, the limit, if it exists, of the straight line passing through x_0, x, when $x \in \Gamma$, $x \to x_0$. Let us call *local cone* $C(x_0, \epsilon)$ the smallest cone of vertex x_0 containing $\Gamma \cap B_\epsilon(x_0)$, and $\theta_\epsilon(x_0)$ its angle. Finally, the *convex hull* $\mathcal{K}(x_0\frown x)$ is the smallest convex set containing the arc $x_0\frown x$ of Γ. Its volume in $\mathbf{R}^n$ is denoted $\mathcal{V}(\mathcal{K}(x_0\frown x))$.

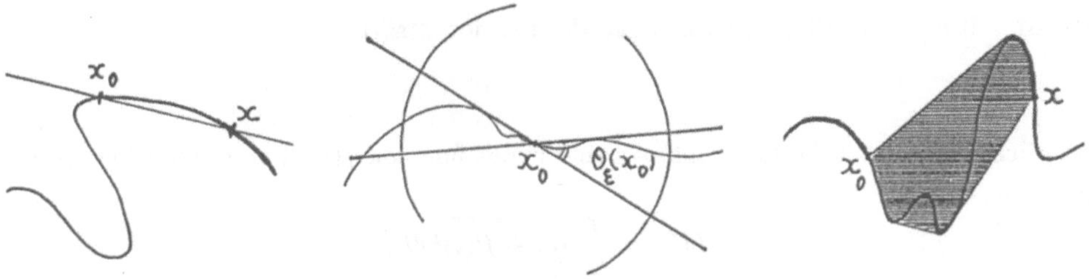

Figure 1.2:

Each of the four following properties may characterize the *local regularity* of Γ at x_0:

(P_1) $T(x_0)$ exists. as a limit.

(P_2) $\lim_{\epsilon \to 0} \theta_\epsilon(x_0) = 0$.

(P_3) $\lim_{x \to x_0} \frac{L(x_0\frown x)}{\text{dist}(x_0, x)} = 1$.

(P_4) $\lim_{x \to x_0} \frac{\mathcal{V}(\mathcal{K}(x_0\frown x))}{\text{dist}(x_0, x)^n} = 0$.

The only general relationships between these properties are

$$(P_1) \Longleftrightarrow (P_2)\,,\ (P_3) \Longrightarrow (P_4)\,.$$

When the curve Γ has a finite length, every subarc of Γ has also a finite length. A subset E of Γ has *measure zero* if, for all $\epsilon > 0$, it can be imbedded in a union of subarcs of Γ, whose total length does not exceed ϵ. A local property *true almost everywhere* on Γ is true at every $x \in \Gamma$, but for a set of measure 0.

Theorem 1. *If $L(\Gamma) < \infty$: the properties (P_1) to (P_4) are all true almost everywhere.*

1.4 Notion of perturbation

We consider a segment as a curve without perturbation. A *perturbation function* $p(\Gamma)$ may be defined as a real-valued function with the four following properties:

- $p(\Gamma) \geq \mathrm{dist}(A, B)$.
- If F is a similarity of ratio ρ: then $p(F(\Gamma)) = \rho\, p(\Gamma)$.
- $\Gamma_1 \subset \Gamma_2 \Longrightarrow p(\Gamma_1) \leq p(\Gamma_2)$.
- For some constant C,

$$p(\Gamma) \leq \mathrm{diam}\,\Gamma \left(1 + C\,\frac{\mathcal{V}(\mathcal{K}(\Gamma))}{(\mathrm{diam}\,\Gamma)^n}\right).$$

The last property implies that $p(\Gamma) = \mathrm{dist}(A, B)$ when Γ is a segment.

Example 1. $p(\Gamma) = \mathrm{diam}\,\Gamma$. This perturbation function gives the same value to the two following curves:

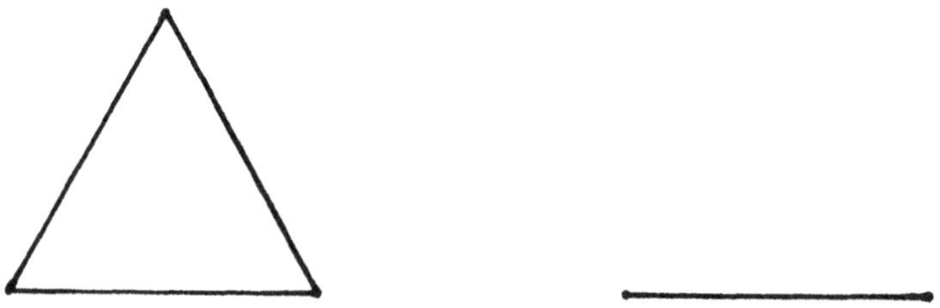

Figure 1.3:

Example 2. Let Γ be a planar curve. Then

$$p(\Gamma) = \frac{1}{2} L(\partial\mathcal{K}(\Gamma)),$$

the perimeter of the boundary of $K(\Gamma)$, is also a perturbation function.

Hint of proof. In the plane, $\mathcal{A}$ denotes the area. Construct the smallest rectangle whose one side is parallel to a diameter of Γ. Then, show the two following: $L(\partial\mathcal{K}) \leq 2(L+l)$, and, since Γ meets the four sides of the rectangle:

$$\frac{1}{2} l L \leq \mathcal{A}(\mathcal{K}(\Gamma)).$$

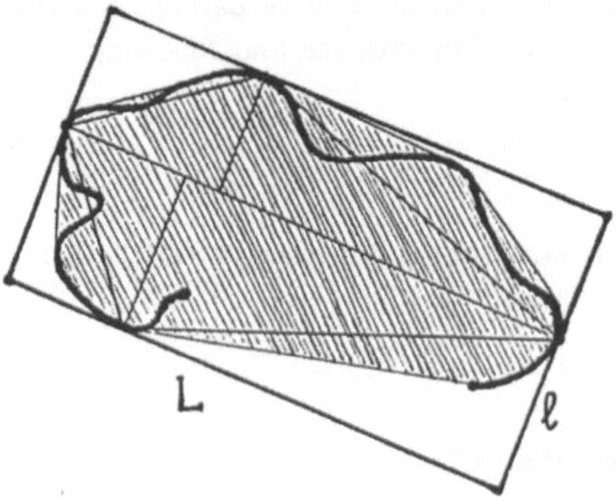

Figure 1.4:

It is interesting to recall here this result of geometric probability: considering a random line D at distance ≤ 1 from 0, then, for any convex body $\mathcal{K}$ in $B_1(0)$,

$$\frac{1}{2} L(\partial\mathcal{K}) = \pi \operatorname{Prob}(D \cap \mathcal{K} \neq \emptyset).$$

Example 3. Let Γ be a planar curve, and R be the smallest rectangle of sides parallel to the axes, circumscribing Γ. Then $p(\Gamma)$ = length of the diagonal of R, is a perturbation function.

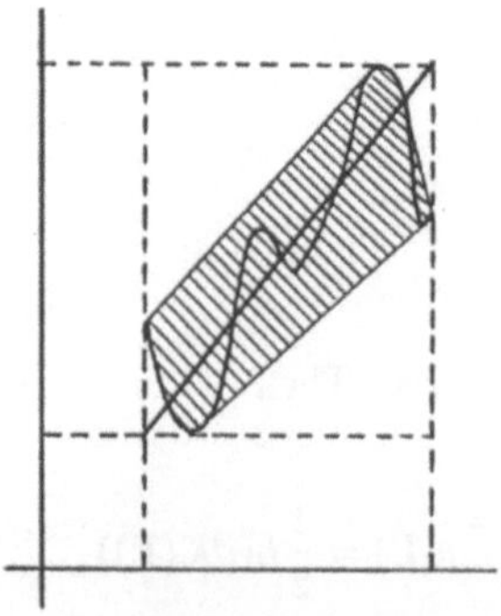

Figure 1.5:

Hint of proof. The shaded area in the figure has a perimeter ≥ 2 (length of the diagonal), so that

$$p(\Gamma) \leq \tfrac{1}{2} L(\partial \mathcal{K}(\Gamma)).$$

Theorem 1. *If $L(\Gamma) < \infty$, then $\lim_{x \to x_0} \frac{p(x_0 \frown x)}{\mathrm{dist}(x_0, x)} = 1$ almost everywhere.*

Theorem 2. *If the parametrization γ of Γ is piecewise continuously differentiable, and $\overline{p}(\tau)$ denotes the average perturbation over the curve*

$$\overline{p}(\tau) = \frac{1}{b-a} \int_a^b p(\gamma(t-\tau) \frown \gamma(t+\tau))\, dt\,,$$

then

$$L(\Gamma) = (b-a) \lim_{\tau \to 0} \frac{\overline{p}(\tau)}{2\tau}\,.$$

1.5 Hitting probability

Buffon needle. Consider a segment (the needle), of random orientation and location, of length l, on a grid made up with parallel lines at distance $\epsilon > l$ from each other. Then

$$\text{Prob(the needle hits one line)} = \frac{2\,l}{\pi\,\epsilon}\,.$$

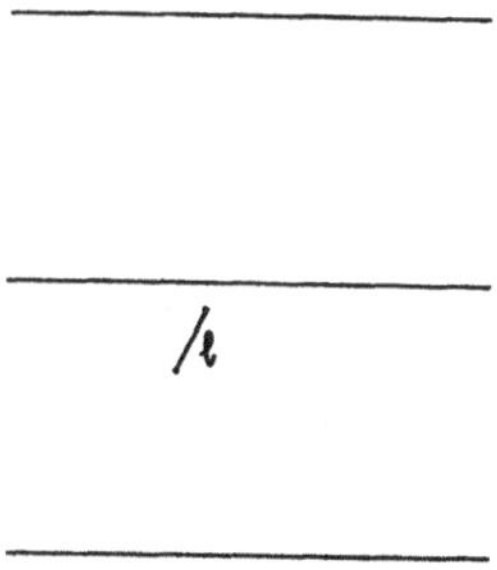

Figure 1.6:

Steinhaus length. A random line **D** intersecting $B_1(0)$ in the plane is defined by two random variables: the angle θ (uniform distribution on $[0, 2\pi]$), and the distance ρ (uniform distribution on $[0, 1]$).

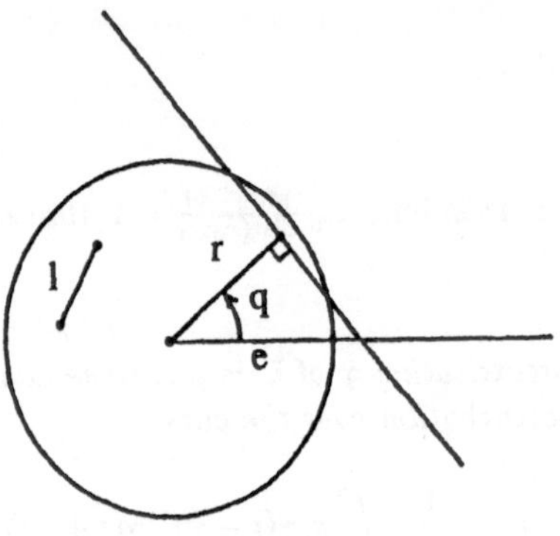

Figure 1.7:

Let Γ be a segment of length l. Using $\epsilon = 2$:

$$\mathrm{Prob}(\mathbf{D} \cap \Gamma \neq \emptyset) = \frac{l}{\pi} \,.$$

Let Γ be a polygonal curve, made up with N segments included in $B_1(0)$, of length l_1, l_2, ..., l_N. The random variable X = number of intersection points of $\mathbf{D} \cap \Gamma$ is the sum of N Bernouilli variables. The mathematical expectation $\mathcal{E}(X)$ is equal to $\sum l_i/\pi = L(\Gamma)/\pi$.

As a generalization: if Γ is any rectifiable curve in $B_1(0)$, then

$$L(\Gamma) = \pi \mathcal{E}(X) \,.$$

This can be written $\pi \sum k\, p_k$, where p_k is the probability of being hit in exactly k points.

1.6 Minkowski sausage

For every set E in $\mathbf{R}^n$, and $\epsilon > 0$, let us denote by $E(\epsilon)$ the set $\cup_{x \in E} B_\epsilon(x)$. Let C_n be the volume of the unit sphere in $\mathbf{R}^n$.

Theorem 1. *For any simple curve Γ in $\mathbf{R}^n$,*

$$L(\Gamma) = \lim_{\epsilon \to 0} \frac{\mathcal{V}(\Gamma(\epsilon))}{C_{n-1}\, \epsilon^{n-1}} .$$

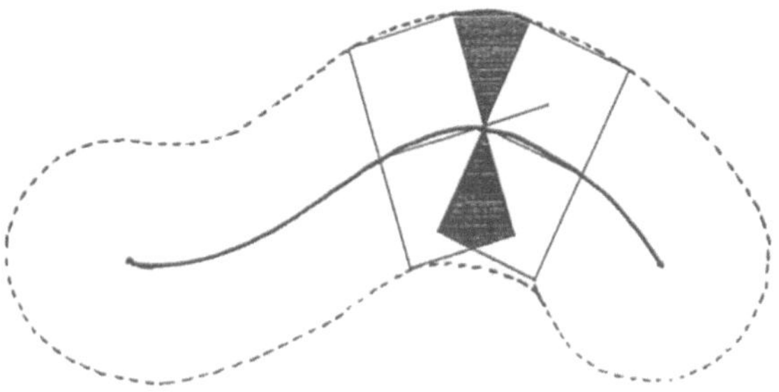

Figure 1.8:

Hint of proof. Uses the local properties (P_2) and (P_3) of Γ, to get an approximation of $\mathcal{V}(\Gamma(\epsilon))$ by cubes.

If Γ is a polygonal curve, or the boundary of a convex body in the plane, then

$$\mathcal{A}(\Gamma(\epsilon)) = 2\epsilon\, L(\Gamma) + C\,\epsilon^2 .$$

The formula

$$L(\Gamma) = \frac{1}{\epsilon_1 - \epsilon_2}\left(\epsilon_1 \frac{\mathcal{A}(\Gamma(\epsilon_2))}{2\,\epsilon_2} - \epsilon_2 \frac{\mathcal{A}(\Gamma(\epsilon_1))}{2\,\epsilon_1}\right)$$

is therefore exact for such curves.

1.7 Hausdorff measure

The Hausdorff linear measure is

$$m(\Gamma) = \lim_{\epsilon \to 0}\left[\inf\left\{\sum \operatorname{diam} U_i \ :\ \operatorname{diam} U_i \le \epsilon\,,\ \Gamma \subset \cup U_i\right\}\right] .$$

For a simple curve, it is always equal to $L(\Gamma)$. Prove it first for a segment in $\mathbf{R}$. The general case is very similar.

References.

H. Steinhaus, *Length, shape and area*, *Colloq.Math.* **3** (1954), 1–13.

L. Santaló, *Integral Geometry and Geometric Probability*, Encyclopedia of Math., Addison–Wesley (1976).

J.C. Burkhill & H. Burkhill, *A Second Course in Mathematical Analysis*, Cambridge Univ.Press (1970).

2 Hausdorff and packing measures and dimensions

2.1 Rarefaction indices

We call *rarefaction index* a set function $\mathrm{d}(E)$, defined for all bounded E in R^n, such that

- $E_1 \subset E_2 \Longrightarrow \mathrm{d}(E_1) \leq \mathrm{d}(E_2)$.
- Denoting by $\mathcal{H}$ the group of all homeomorphisms h such that, for all x,

$$\lim_{y \to x,\, z \to x} \frac{\log \mathrm{dist}(h(y), h(z))}{\log \mathrm{dist}(y, z)} = 1 ,$$

we have for any h in $\mathcal{H}$:

$$\mathrm{d} \circ h = \mathrm{d} .$$

Moreover, a rarefaction index is *stable* if

$$\mathrm{d}(E_1 \cup E_2) = \max\{\mathrm{d}(E_1), \mathrm{d}(E_2)\} ,$$

and *σ–stable* if

$$\mathrm{d}(\cup_n E_n) = \sup_n \mathrm{d}(E_n) ,$$

for any countable, bounded family (E_n).

2.2 α–Pre–Measures

A positive, monòtone, σ–subadditive set function $\mu(E)$ is a *measure*.

Theorem 1. *Any positive set function $F(E)$ (the pre–measure) gives rise to a measure, through the formula*

$$\hat{F}(E) = \inf\{\sum F(E_n), \text{ for all coverings } (E_n) \text{ of } E\}.$$

As pre–measures, we make a special use of the following:

Definition. A family (μ^α) of positive set functions, with values in $\overline{R^+}$, is a family of *α–pre–measures* if

1. $E_1 \subset E_2 \Longrightarrow \mu^\alpha(E_1) \leq \mu^\alpha(E_2)$

2. $\forall \epsilon > 0, \forall h \in \mathcal{H}$,

$$\begin{aligned} \mu^\alpha(E) &< +\infty \Longrightarrow \mu^{\alpha+\epsilon}(h(E)) = 0 \\ \mu^\alpha(E) &> 0 \Longrightarrow \mu^{\alpha-\epsilon}(h(E)) = +\infty . \end{aligned}$$

Property 2, with $h = id$, implies the existence of a Dedekind cut

$$\begin{aligned} \mathrm{d}(E) &= \inf\{\alpha : \mu^\alpha(E) = 0\} \\ &= \sup\{\alpha : \mu^\alpha(E) = +\infty\} . \end{aligned}$$

Properties 1 and 2 imply that $\mathrm{d}(E)$ is a rarefaction index, *associated to the family* (μ^α).

Moreover, the pre–measures μ^α are *subadditive* if

$$\mu^\alpha(E_1 \cup E_2) \leq \mu^\alpha(E_1) + \mu^\alpha(E_2) .$$

This implies the stability of the associated $\mathrm{d}(E)$. If they are *σ–subadditive*, that is

$$\mu^\alpha(\cup_n E_n) \leq \sum_n \mu^\alpha(E_n) ,$$

the family (μ^α) is a family of *α–measures.* The associated rarefaction index is σ–stable. For example, if the μ^α are α–pre–measures, the family $(\hat{\mu^\alpha})$ is a family of α–measures.

Finally, a general way to get a σ–stable index from any rarefaction index d is

$$\hat{d}(E) = \inf\{\sup \mathrm{d}(E_n), E \subset \cup E_n\} .$$

If d is associated to μ^α, $\hat{d}$ is asociated to $\hat{\mu^\alpha}$. Thus we have the general diagram:

$$\begin{array}{ccc} \mu^\alpha(E) & \longrightarrow & \hat{\mu^\alpha}(E) = \inf\{\sum \mu^\alpha(E_n), E \subset \cup E_n\} \\ \downarrow & & \downarrow \\ \mathrm{d}(E) & \longrightarrow & \hat{\mathrm{d}}(E) = \inf\{\sup \mathrm{d}(E_n), E \subset \cup E_n\} \end{array}$$

where the horizontal arrows indicate the operation "hat".

2.3 Hausdorff measure and dimension

Let ϕ be a continuous, increasing function such that

$$\lim_{x\to 0+}\phi(x)=0 \quad \text{and} \limsup_{x\to 0+}\phi(2x)/\phi(x)<+\infty\,.$$

The Hausdorff measure is defined as

$$m_\phi(E)=\lim_{\epsilon\to 0}\left[\inf\{\sum\phi(\operatorname{diam}U_i):\operatorname{diam}U_i\le\epsilon\,,\ E\subset\cup U_i\}\right]\,.$$

If $\phi(t)=t^\alpha$, m_ϕ is denoted m^α: it is an α–measure. The associated index (*Hausdorff dimension*) $\dim E$ is σ–stable.

Theorem 1. *Let μ be a finite measure such that $\mu(E)>0$. If*

$$\liminf_{\epsilon\to 0}\frac{\log\mu(B_\epsilon(x))}{\log\epsilon}\ \underset{(\le)}{\ge}\ \alpha \text{ for all } x\in E\ :\ \text{then } \dim E\ \underset{(\le)}{\ge}\ \alpha\,.$$

This leads to the following:

Theorem 2. *If E is analytic,*

$$\dim E=\sup_{\mu(E)>0}\left\{\inf_{x\in E}\left\{\liminf_{\epsilon\to 0}\frac{\log\mu(B_\epsilon(x))}{\log\epsilon}\right\}\right\}\,.$$

Proof. Call α_0 the right member. The inequality $\alpha<\alpha_0$ implies that there exists μ, $\mu(E)>0$, such that for all $x\in E$: $\liminf\log\mu(B_\epsilon(x))/\log\epsilon\ge\alpha$. Therefore $\dim E\ge\alpha$, so that $\dim E\ge\alpha_0$. In the other sense: use the following

Frostman Lemma. *Let F be a compact set. If $m_\phi(F)>0$, there exists a measure μ, of support F, such that for all $x\in F$, $\epsilon<1$: $\mu(B_\epsilon(x))\le\phi(\epsilon)$.*

Now, let us take $\alpha<\dim E$. If E is analytic, there exists a compact set $F\subset E$ such that $\alpha<\dim F$. The lemma implies that there exists μ, $\mu(E)>0$, such that for all $x\in F$, $\epsilon<1$: $\mu(B_\epsilon(x))\le\epsilon^\alpha$. Since F is compact, this is true also for all $x\in E$. Then $\alpha_0\ge\alpha$.

2.4 Hausdorff centred measure and dimension

Define

$$C_\phi(E)=\lim_{\epsilon\to 0}\Big[\inf\{\sum\phi(\operatorname{diam}U_i):\ \operatorname{diam}U_i\le\epsilon\,,\ E\subset\cup U_i\,,\ U_i \text{ closed balls centred in E}\}\Big]\,.$$

This set function is **not** monotone, but it is σ-subadditive.

Theorem 1. *The set function $c_\phi(E)$ defined by*

$$c_\phi(E) = \sup\{C_\phi(F) \,:\, F \subset E\}$$

is a measure.

Proof. The monotonicity of c_ϕ is trivial. If $E = \cup_k E_k$, and $F \subset E$ is such that $C_\phi(F) \geq c_\phi(E) - \epsilon$, then

$$\begin{aligned} c_\phi(E) &\leq C_\phi(F) + \epsilon && = C_\phi(\cup_k(F \cap E_k)) + \epsilon && \text{since } F \subset \cup_k E_k \\ &\leq \textstyle\sum_k C_\phi(F \cap E_k) + \epsilon && \leq \textstyle\sum_k c_\phi(E_k) + \epsilon && \text{since } F \cap E_k \subset E_k\,. \end{aligned}$$

Clearly

$$m_\phi(E) \leq C_\phi(E) \leq c_\phi(E)\,,$$

for all E. If $\phi(t) = t^\alpha$, c_ϕ is denoted c^α: it is an α–measure.

Theorem 2. *The Hausdorff dimension, and the rarefaction index associated to the α–measure c^α, are identical.*

Proof. Any covering set U, meeting E, is itself covered by a centred ball of diameter $2\,\text{diam}\,U$. This gives

$$C^\alpha(E) \leq 2^\alpha\, m^\alpha(E)\,,$$

which implies

$$c^\alpha(E) \leq 2^\alpha\, m^\alpha(E)\,.$$

2.5 Packing centred measure and dimension

Let E be bounded. Define

$$P_\phi(E) = \lim_{\epsilon \to 0}\Big[\sup\{\textstyle\sum \phi(\text{diam}\,U_i) \,:\quad \text{diam}\,U_i \leq \epsilon\,,\, U_i^\circ \cap U_j^\circ = \emptyset \text{ if } i \neq j\,,\\ U_i \text{ closed balls centred in } E\}\Big]\,.$$

This set function is monotone, subadditive, but **not** σ–subadditive. The measure $p_\phi = \hat{P}_\phi$ is called *packing measure.*

Theorem 1. *For all bounded sets E,*

$$m_\phi(E) \le C_\phi(E) \le c_\phi(E) \le p_\phi(E) \le P_\phi(E)\,.$$

Hint of proof. The only difficult inequality is

$$c_\phi(E) \le p_\phi(E)\,.$$

Use a Vitali covering argument to show that $C_\phi(E) \le P_\phi(E)$.

If $\phi(t) = t^\alpha$, denote $P_\phi = P^\alpha$ and $p_\phi = p^\alpha$.

- P^α is an α–pre–measure: the associated rarefaction index is denoted by $\Delta(E)$.

Theorem 2. *Let $\omega_\epsilon(E)$ be the number of ϵ–boxes covering E: then*

$$\Delta(E) = \limsup_{\epsilon\to 0} \frac{\log \omega_\epsilon(E)}{\log \epsilon}\,.$$

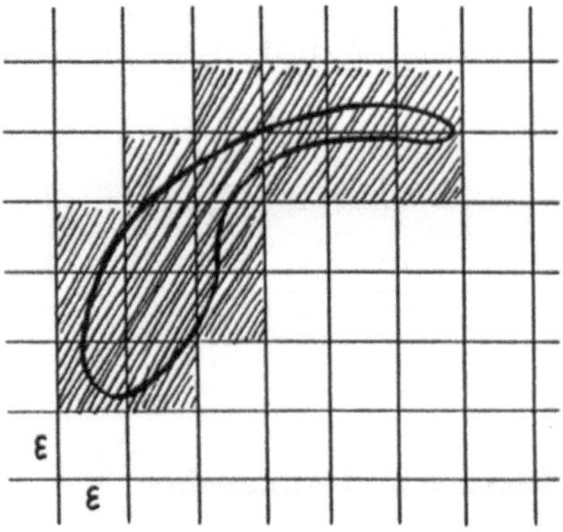

Figure 2.1:

- p^α is an α–measure: the associated rarefaction index is denoted by $\operatorname{Dim} E$. Therefore, Dim and Δ are linked by the formula

$$\operatorname{Dim} E = \inf\{\sup \Delta(E_n) \,:\, E \subset \cup E_n\}\,.$$

Theorem 1 implies directly

Theorem 3. *For all E,*

$$\dim E \leq \operatorname{Dim} E \leq \Delta(E)\,.$$

Theorem 4. *Let μ be a finite measure such that $\mu(E) > 0$. If*

$$\limsup_{\epsilon\to 0} \frac{\log \mu(B_\epsilon(x))}{\log \epsilon} \underset{(\leq)}{\geq} \alpha \text{ uniformly on } E\text{: then } \Delta(E) \underset{(\leq)}{\geq} \alpha\,.$$

By σ–stability, we deduce

Theorem 5. *Let μ be a finite measure such that $\mu(E) > 0$. If*

$$\limsup_{\epsilon\to 0} \frac{\log \mu(B_\epsilon(x))}{\log \epsilon} \underset{(\leq)}{\geq} \alpha \text{ for all } x \in E \text{ : then } \operatorname{Dim} E \underset{(\leq)}{\geq} \alpha\,.$$

This leads to

Theorem 6.

$$\operatorname{Dim} E = \inf_{\mu \text{ finite}} \left\{ \sup_{x\in E} \left\{ \limsup_{\epsilon\to 0} \frac{\log \mu(B_\epsilon(x))}{\log \epsilon} \right\} \right\}.$$

Proof. Call α_0 the right member.

$$\begin{aligned} \alpha > \alpha_0 &\implies \alpha \geq \operatorname{Dim} E : \text{ use Theorem 5} \\ \alpha > \operatorname{Dim} E &\implies \alpha \geq \alpha_0 : \text{ use the following} \end{aligned}$$

Lemma (anti–Frostman). *Let F be a bounded set. If $P_\phi(F) < \infty$, there exists a finite measure μ such that for all $x \in F$, $\epsilon \leq 1$:*

$$\mu(B_\epsilon(x)) \geq \phi(\epsilon)\,.$$

Thus, Theorem 2, Section 3, and 6, Section 5, show a fundamental symmetry between $\dim E$ and $\operatorname{Dim} E$.

References.

R.O. Davies, Subsets of finite measure in analytic sets, *Indagat.Math.* **14** (1952), 488–9.

J.P. Kahane & R. Salem, *Ensembles parfaits et séries trigonométriques*, Herman (1963).

P. Billingsley, *Ergodic Theory and Information*, J. Wiley (1965).

C.A. Rogers, *Hausdorff Measures*, Cambridge Univ. Press (1970).

C. Tricot, *Sur la classification des ensembles boréliens de mesure de Lebesgue nulle*, Thèse de doctorat, Genève (1979).

C. Tricot, Rarefaction indices, *Mathematika* **27** (1980), 46–57.

C. Tricot, Two definitions of fractional dimension, *Math. Proc. Camb. Phil. Soc.* **91** (1982), 57–74.

C. Tricot & S.J. Taylor, Packing measure, and its evaluation for a Brownian path, *Trans. Amer. Math. Soc.* **288** (1985), 679–696.

C. Tricot & X. Saint Raymond, Packing regularity of sets in n-space, *Math. Proc. Camb. Phil. Soc.* **103** (1988), 133–145.

3 Local densities and the notion of regularity

3.1 Rectifiable 1-sets

Let $\alpha = 1$, and call $m(E)$, $C(E)$, $c(E)$, $p(E)$, $P(E)$ the coresponding measures and pre–measures.

Theorem 1. *If* Γ *is a rectifiable arc:* $m(\Gamma) = \ldots = P(\Gamma) = L(\Gamma)$.

Hint of proof. Use the local property

$$\lim_{x \to x_0} \frac{L(x \frown x_0)}{\operatorname{dist}(x, x_0)} = 1$$

almost everywhere, and Vitali's Theorem.

Theorem 2. *Let* E *be a compact, arcwise connected set, such that* $m(E) < \infty$. *Then* $m(E) = p(E)$.

Proof. It is known that E can be written

$$E = (\cup E_i) \cup E^* ,$$

where the E_i are rectifiable arcs, and $m(E^*) = 0$. Since $m(\cup E_i) = p(\cup E_i)$, all we have to prove is $p(E^*) = 0$: this is done in Section 2.

3.2 Local density inequalities

Let μ be a positive, finite, borelian measure in $\mathbf{R}^n$, and for all x,

$$\underline{d}^{\phi}_{\mu}(x) = \liminf_{\epsilon\to 0} \frac{\mu(B_\epsilon(x))}{\phi(2\,\epsilon)} , \; \overline{d}^{\phi}_{\mu}(x) = \limsup_{\epsilon\to 0} \frac{\mu(B_\epsilon(x))}{\phi(2\,\epsilon)} .$$

Recall that $m_\phi(E) \leq c_\phi(E) \leq m_\phi(E)(\limsup_{t\to 0} \frac{\phi(2\,t)}{\phi(t)})$, so that

$$\begin{cases} m_\phi(E) < \infty & \Longleftrightarrow \; c_\phi(E) < \infty \\ m_\phi(E) = 0 & \Longleftrightarrow \; c_\phi(E) = 0 \, . \end{cases}$$

Theorem 1.

- *If $c_\phi(E) < \infty$: then*

$$(1) \qquad c_\phi(E) \inf_{x\in E} \overline{d}^{\phi}_{\mu}(x) \leq \mu(E)$$

$$(2) \qquad \mu(E) \leq c_\phi(E) \sup_{x\in E} \overline{d}^{\phi}_{\mu}(x) \, .$$

- *If $p_\phi(E) < \infty$: then*

$$(3) \qquad p_\phi(E) \inf_{x\in E} \underline{d}^{\phi}_{\mu}(x) \leq \mu(E)$$

$$(4) \qquad \mu(E) \leq p_\phi(E) \sup_{x\in E} \underline{d}^{\phi}_{\mu}(x) \, .$$

The proof uses Vitali covering arguments.

Notations. We will apply Theorem 1 to the restricted measures $m_{|E}$ ($m_{|E}(F) = m(E \cap F)$), $c_{|E}$, $p_{|E}$. For the Hausdorff measure, the corresponding densities are

$$\underline{d}^{\phi}_{m}(x, E) = \liminf_{\epsilon\to 0} \frac{m_\phi(E \cap B_\epsilon(x))}{\phi(2\,\epsilon)} , \; \overline{d}^{\phi}_{m}(x, E) = \limsup_{\epsilon\to 0} \frac{m_\phi(B_\epsilon(x))}{\phi(2\,\epsilon)} ,$$

and same for the measures c_ϕ and p_ϕ.

Application of Theorem 1. If E is arcwise connected, then $m(E \cap B_\epsilon(x)) \geq \epsilon$ for all $x \in E$. Hence, with $\phi(t) = t$: $\underline{d}_m(x, E) \geq 1/2$. Formula (3) with $\mu = m_{|E}$ gives: $p(E) \leq 2\,m(E)$. This is true also for any subset of E, so that $p(E^*) = 0$.

3.3 Elementary density bounds

Theorem 1.

$$(i) \quad c_\phi(E) < \infty \Longrightarrow \overline{d}_c^\phi(x,E) = 1 \; c_\phi\text{–a.s. on } E$$

$$(ii) \quad p_\phi(E) < \infty \Longrightarrow \underline{d}_c^\phi(x,E) > 0 \; c_\phi\text{–a.s. on } E$$

$$(iii) \quad p_\phi(E) < \infty \Longrightarrow \underline{d}_p^\phi(x,E) = 1 \; p_\phi\text{–a.s. on } E \, .$$

$$(iv) \quad p_\phi(E) < \infty \Longrightarrow \overline{d}_p^\phi(x,E) \geq 1 \; p_\phi\text{–a.s. on } E.$$

Proof.

(i) Take $F_k = \{x \in E \, : \, \overline{d}_c^\phi(x,E) \geq 1 + \frac{1}{k}\}$. Formula (1) with $\mu = c_{\phi|E}$, applied to the set F_k, gives: $(1+\frac{1}{k})\, c_\phi(F_k) \leq c_\phi(F_k)$. Therefore $c_\phi(F_k) = 0$: $\overline{d}_c^\phi(x,E) \leq 1 \; c_\phi$–a.s.

Then, take $G_k = \{x \in E \, : \, \overline{d}_c^\phi(x,E) \leq 1 - \frac{1}{k}\}$. Formula (2) with $\mu = c_{\phi|E}$, applied to the set G_k, gives: $c_\phi(G_k) \leq (1-\frac{1}{k})\, c_\phi(G_k)$. Therefore $c_\phi(G_k) = 0$: $\overline{d}_c^\phi(x,E) \geq 1 \; c_\phi$–a.s.

(ii) Take $H_k = \{x \in E \, : \, \underline{d}_c^\phi(x,E) \leq \frac{1}{k}\}$. Formula (4) with $\mu = c_{\phi|E}$, applied to the set H_k, gives: $c_\phi(H_k) \leq p_\phi(H_k)/k \leq p_\phi(E)/k$. Therefore $c_\phi(\cap H_k) = 0$.

(iii) Similar to (i), by replacing c_ϕ by p_ϕ, and using (3) and (4).

(iv) Follows directly from (iii).

Remark. If $\phi(t) = t^\alpha$, then

$$c^\alpha \leq 2^\alpha \, m^\alpha(E) \, ,$$

so that

$$\overline{d}_c^\alpha(x,E) \leq 2^\alpha \, \overline{d}_m^\alpha(x,E) \, .$$

Therefore, (i) implies the well-known inequality (Besicovitch):

$$m^\alpha(E) < \infty \Longrightarrow 2^{-\alpha} \leq \overline{d}_m^\alpha(x,E) \leq 1$$

m^α–a.s. on E.

3.4 Strong regularity

Theorem 1. *Let E be such that $p_\phi(E) < \infty$. The three following are equivalent:*

$$\begin{array}{ll} (i) & c_\phi(E) = p_\phi(E) \\ (ii) & \underline{d}_c^\phi(x,E) = \overline{d}_c^\phi(x,E) = 1 \quad p^\phi\text{–a.s. on } E \\ (iii) & \underline{d}_p^\phi(x,E) = \overline{d}_p^\phi(x,E) = 1 \quad p^\phi\text{–a.s. on } E\,. \end{array}$$

Theorem 2. *Let E be such that $p_\phi(E) < \infty$. The two following are equivalent:*

$$\begin{array}{ll} (i) & m_\phi(E) = p_\phi(E) \\ (ii) & \underline{d}_m^\phi(x,E) = \overline{d}_m^\phi(x,E) = 1 \quad p^\phi\text{–a.s. on } E\,. \end{array}$$

Moreover, we will see (Section 6) that, if $\phi(t) = t^\alpha$, the statements of Theorem 1 are equivalent to those of Theorem 2. This leads to the definition of strong regularity:

Definition. *E is strongly α–regular if $c^\alpha(E) = p^\alpha(E)$ ($\Longleftrightarrow m^\alpha(E) = p^\alpha(E)$).*

It is easy to construct sets which are α–regular in Besicovitch sense ($\underline{d}_m^\alpha(x,E) = 1 \quad m^\alpha$–a.s. on E), but not strongly regular. By a Cantor set procedure, we can even construct sets such that, for some α, $m^\alpha(E) = 0$ (so that $\overline{d}_m^\alpha(x,E) = 0$), and for all open set U such that $U \cap E \neq \emptyset$, $p^\alpha(U \cap E) = \infty$ (so that $\underline{d}_p^\alpha(x,E) = \infty$).

3.5 An example of a non–regular 1–set

Let E be the residual set obtained from an osculatory packing of the equilateraxl triangle of side 1, by regular hexagons. Let us estimate a few measures and densities.

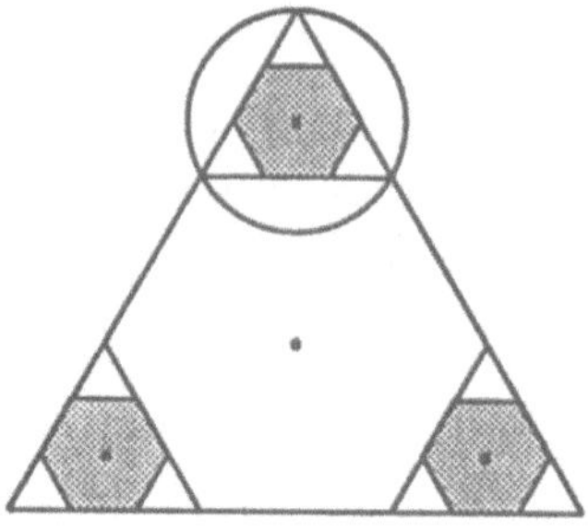

Figure 3.1:

- $m(E) = 1$:

$$\begin{array}{ll} \geq 1 & \text{by projection on one side} \\ \leq 1 & \text{with a covering of } 3^k \text{ balls of diameter } 3^{-k}\,. \end{array}$$

- If x is one of the three vertices: then

$$\inf_{\epsilon\leq 1} \frac{m(E\cap B_\epsilon(x))}{2\epsilon} = \frac{1}{4},$$

the minimum being obtained for all values $\epsilon = 2/3, 2/3^2, \ldots$.

- $\underline{d}_m(x,E) = \frac{1}{4}$ m–a.s. on E: use the self–similarity of E.

- $p(E) = 4$: deduced from the previous result, Formulas (3) and (4), with the measure $\mu = m_{|E}$.

- For all $x \in E$, $\epsilon \leq 1$,

$$\frac{m(E\cap B_\epsilon(x))}{2\epsilon} \leq \frac{4}{5}:$$

Use the geometry of E, and direct computations. The value 4/5 is certainly not the best possible. This implies $\overline{d}_m(x,E) \leq 4/5$ for all x in E.

- $C(E) \geq 5/4$: indeed, any covering $\{U_i\}$ of E by centred balls verifies

$$\sum \operatorname{diam} U_i \geq \frac{5}{4} m(E) = \frac{5}{4}.$$

- $5/4 \leq c(E) \leq 2$: since $c(E) \leq 2\,m(E)$.

Consequence 1. The measures m, c, p take distinct values on E. This shows, in particular, that c is not identical to m.

Consequence 2. $C(E)$ is not monotone. Indeed, denote by F the (countable) set of all centres of the hexagons. The set $E \cup F$ can be covered by 3^k centred balls of diameter $\frac{2}{\sqrt{3}}3^{-k}$. This gives $C(E\cup F) \leq \frac{2}{\sqrt{3}} < \frac{5}{4} \leq C(E)$.

Questions. Find the exact values of $C(E)$, $c(E)$, $\overline{d}_p(x,E)$, $\underline{d}_c(x,E)$.

3.6 The structure of strongly regular sets

Theorem 1. *If E is a set such that, for some α, $0 < c^\alpha(E) = p^\alpha(E) < \infty$, then α is an integer.*

Proof. The assumption implies that $\underline{d}^{\alpha}_{m}(x,E) = \overline{d}^{\alpha}_{m}(x,E)$ m^{α}–a.s. on E: this is possible only when α is an integer.

Theorem 2. *Let $E \subset \mathbf{R}^n$. Assume that for some integer k, $p^k(E) < \infty$. Then*

$$c^k(E) = p^k(E) \iff E \text{ is } (p^k, k)\text{–rectifiable},$$

that is $p^k(E \setminus \cup f_i(A_i)) = 0$ for some sequence (f_i) of lipschitzian maps $\mathbf{R}^k \mapsto \mathbf{R}^n$ and some sequence (A_i) of bounded subsets of $\mathbf{R}^k$.

Proof. $\mathbf{R}^n$ is (μ, k)–rectifiable $\iff \underline{d}^k_{\mu}(x) = \overline{d}^k_{\mu}(x)$ μ–a.s. on $\mathbf{R}^n$ (Federer, Mattila, Preiss). Applied to $\mu = p^k_{|E}$, this gives

$$E \text{ is } (p^k, k)\text{–rectifiable} \iff \underline{d}^k_{p}(x,E) = \overline{d}^k_{p}(x,E)\ p^k\text{–a.s. on } E\,.$$

This condition is equivalent to $c^k(E) = p^k(E)$ (Theorem 1, Section 4).

From this we deduce

Theorem 3.

$$c^{\alpha}(E) = p^{\alpha}(E) \iff m^{\alpha}(E) = p^{\alpha}(E)\,.$$

Proof. Assume $0 < c^{\alpha}(E) = p^{\alpha}(E) < \infty$. Theorem 1 implies that $\alpha = k$, an integer. Theorem 2 implies that for some sequence (E_i) of lipschitzian manifolds, $p^k(E \setminus \cup E_i) = 0$. Therefore $m^k(E \setminus \cup E_i) = 0$: E is (m^k, k)–rectifiable, so that $\underline{d}^k_{m}(x,E) = \overline{d}^k_{m}(x,E)$ m^k–a.s. on E, or c^k–a.s. on E. Since $c^k(F) = p^k(F)$ for every subset F of E, these local densities equalities are also true p^k–a.s. on E. Theorem 2, Section 4 implies now that $m^k(E) = p^k(E)$.

References.

A.S. Besicovitch, On the fundamental properties of linearly measurable plane sets of points, *Math. Ann.* **98** (1928), 422–464.

H. Federer, *Geometric Measure Theory*, Springer–Verlag (1969).

P. Mattila, Hausdorff m–regular and rectifiable sets in n–space, *Trans. Amer. Math. Soc.* **205** (1975), 263–274.

K.J. Falconer, *The Geometry of Fractal Sets*, Cambridge Univ. Press (1985).

C. Tricot & S.J. Taylor, The packing measure of rectifiable subsets in the plane, *Math. Proc. Camb. Phil. Soc.* **99** (1986), 285–296.

C. Tricot & X. Saint Raymond, Packing regularity of sets in n–space, *Math. Proc. Camb. Phil. Soc.* **103** (1988), 133–145.

D. Preiss, *Geometry of measures in* $\mathbf{R}^n$, preprint.

4 Proximity and the independence of sets

4.1 A notion of independence

Let D be a domain of volume 1 in $\mathbf{R}^n$. Let $B_\epsilon(x)$ be the random ball of radius ϵ, whose centre x follows a uniform distribution on D. Let E_1, E_2 be two subsets in D, and $E_i(\epsilon)$ be the set of all points at a distance $\geq \epsilon$ from E_i. Assume that $E_1(\epsilon)$ and $E_2(\epsilon)$ are both included in D. The event:"$B_\epsilon(x)$ *meets* E_i" has probability $\mathcal{V}(E_i(\epsilon))$.

Definition 1. E_1 and E_2 are ϵ–independent if the events "$B_\epsilon(x)$ *meets* E_1" and "$B_\epsilon(x)$ *meets* E_2" are independent, that is

$$\mathcal{V}(E_1(\epsilon) \cap E_2(\epsilon)) = \mathcal{V}(E_1(\epsilon))\, \mathcal{V}(E_2(\epsilon))\,.$$

From this, we can deduce a *weak condition of independence*:

Definition 2. E_1 and E_2 are log–independent if

$$\lim_{\epsilon \to 0} \frac{\log \mathcal{V}(E_1(\epsilon) \cap E_2(\epsilon))}{\log \mathcal{V}(E_1(\epsilon)) + \log \mathcal{V}(E_2(\epsilon))} = 1\,.$$

Classical Bouligand fractal dimensions are

$$\begin{aligned} \Delta(E) = &\ \limsup_{\epsilon \to 0} \left(n - \tfrac{\log \mathcal{V}(E(\epsilon))}{\log \epsilon}\right) \\ \delta(E) = &\ \liminf_{\epsilon \to 0} \left(n - \tfrac{\log \mathcal{V}(E(\epsilon))}{\log \epsilon}\right)\,. \end{aligned}$$

Similarly, we define the *proximity indices* of two sets (at least one of which is bounded) as

$$\begin{aligned} \Delta(E_1, E_2) = &\ \limsup_{\epsilon \to 0} \left(n - \tfrac{\log \mathcal{V}(E_1(\epsilon) \cap E_2(\epsilon))}{\log \epsilon}\right) \\ \delta(E_1, E_2) = &\ \liminf_{\epsilon \to 0} \left(n - \tfrac{\log \mathcal{V}(E_1(\epsilon) \cap E_2(\epsilon))}{\log \epsilon}\right)\,. \end{aligned}$$

Theorem 1. *If E_1, E_2 are log–independent,*

$$\delta(E_1) + \delta(E_2) - n \le \delta(E_1, E_2) \le \delta(E_1) + \Delta(E_2) - n \le \Delta(E_1, E_2) \le \Delta(E_1) + \Delta(E_2) - n\,.$$

Examples.

- In $\mathbf{R}^2$: If E_1 and E_2 are independent line segments, then $\Delta(E_1, E_2) = 0$.
- In $\mathbf{R}^3$: If E_1 is a line segment, E_2 an independent portion of a plane, then $\Delta(E_1, E_2) = 2 + 1 - 3 = 0$.
- If E_1 and E_2 are two independent portions of a plane, then $\Delta(E_1, E_2) = 2 + 2 - 3 = 1$.

4.2 Proximity indices

The following properties are true also for δ:

(i) $\Delta(E, F) = \Delta(F, E)$

(ii) $\Delta(E, \overline{F}) = \Delta(E, F)$

(iii) $E \subset F \Longrightarrow \Delta(E, F) = \Delta(E)$

(iv) $F_1 \subset F_2 \Longrightarrow \Delta(E, F_1) \le \Delta(E, F_2)$

(v) $\Delta(E \cap F) \le \Delta(E, F) \le \min(\Delta(E), \Delta(F))\,.$

Examples.

Let $f(x) = x^\alpha$, $\alpha > 1$, $0 \le x \le 1$, and Γ its graph. Then

$$\Delta(\Gamma, Ox) = \delta(\Gamma, Ox) = 2 - (1 + \frac{1}{\alpha}) = 1 - \frac{1}{\alpha}\,.$$

Let $f(x) = e^{-1/x}$, $0 \le x \le 1$, and Γ its graph. Then

$$\Delta(\Gamma, Ox) = \delta(\Gamma, Ox) = \lim \left(2 - \frac{\log(\epsilon/|\log \epsilon|)}{\log \epsilon}\right) = 1\,.$$

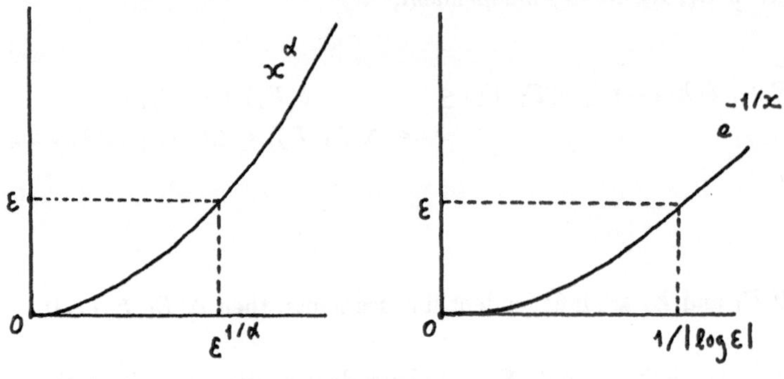

Figure 4.1:

4.3 Proximity with vertical straight lines

In $\mathbf{R}^2$, let D_x denote the line parallel to Oy, of abcissa x. For every E, we investigate the proximity indices of E and D_x.

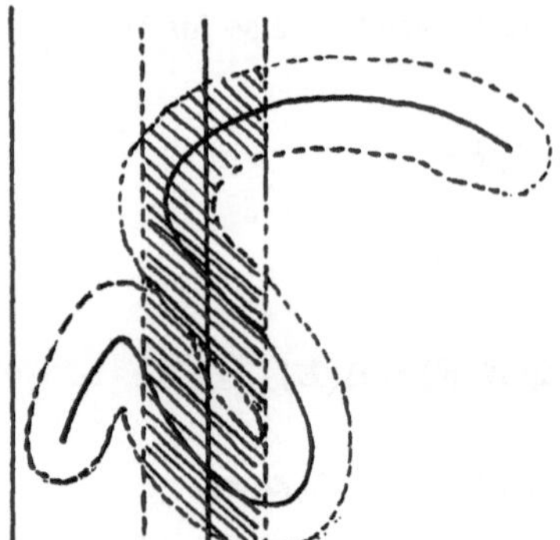

Figure 4.2:

Let $F = Proj_{Ox}E$, and

$$\overline{\beta}(E) = \sup_{x\in F} \Delta(E, D_x)\,, \quad \underline{\beta}(E) = \inf_{x\in F} \Delta(E, D_x)$$
$$\overline{\gamma}(E) = \sup_{x\in F} \delta(E, D_x)\,, \quad \underline{\gamma}(E) = \inf_{x\in F} \delta(E, D_x)\,.$$

The set functions $\underline{\beta}$, $\overline{\beta}$ $\underline{\gamma}$, $\overline{\gamma}$ are rarefaction indices. We can also consider the corresponding σ-stable indices obtained by the operation "hat":

$$\hat{\mathrm{d}}(E) = \inf\{\sup \mathrm{d}(E_n) \ : \ E \subset \cup E_n\}\,.$$

Theorem 1.

$$
\begin{array}{ll}
(1) & \dim E \le \operatorname{Dim} F + \hat{\bar{\gamma}}(E) \\
(2) & \dim E \le \dim F + \hat{\bar{\beta}}(E) \\
(3) & \operatorname{Dim} E \ge \dim F + \hat{\underline{\beta}}(E) \\
(4) & \operatorname{Dim} E \ge \operatorname{Dim} F + \hat{\underline{\gamma}}(E) \\
(5) & \operatorname{Dim} E \le \operatorname{Dim} F + \hat{\bar{\beta}}(E) .
\end{array}
$$

4.4 Application to the cartesian product

Let A, B be two sets on the line, and $E = A \times B$. Then $F = A$, $\bar{\beta}(E) = \underline{\beta}(E) = \Delta(B)$, $\bar{\gamma}(E) = \underline{\gamma}(E) = \delta(B)$.

Theorem 1.

$$\hat{\bar{\beta}}(E) = \hat{\underline{\beta}}(E) = \hat{\Delta}(B) \quad (= \operatorname{Dim} B, \textit{ the packing dimension})$$
$$\hat{\bar{\gamma}}(E) = \hat{\underline{\gamma}}(E) = \hat{\delta}(B) .$$

Recall that the rarefaction indices used in this theorem verify these general inequalities:

$$\dim E \le \hat{\delta}(E) \le \operatorname{Dim} E = \hat{\Delta}(E)$$
$$\dim E \le \delta(E) \le \Delta(E) .$$

Theorem 1 implies the following:

$$
\begin{array}{ll}
(2) \Longrightarrow & \dim A \times B \le \dim A + \operatorname{Dim} B \\
(3) \Longrightarrow & \dim A + \operatorname{Dim} B \le \operatorname{Dim} A \times B \\
(4) \Longrightarrow & \operatorname{Dim} A + \hat{\delta}(B) \le \operatorname{Dim} A \times B \\
(5) \Longrightarrow & \operatorname{Dim} A \times B \le \operatorname{Dim} A + \operatorname{Dim} B .
\end{array}
$$

Therefore

Theorem 2.

$$\dim A \times B \leq \dim A + \mathrm{Dim}\, B \leq \hat{\delta}(A) + \mathrm{Dim}\, B \leq \mathrm{Dim}\, A \times B \leq \mathrm{Dim}\, A + \mathrm{Dim}\, B\ .$$

Moreover, it is already known that

$$\dim A + \dim B \leq \dim A \times B\ .$$

Corollary. *If the set B is dimensionnally regular ($\dim B = \mathrm{Dim}\, B$), then*

$$\dim A + \dim B = \dim A \times B\ .$$

Of course, the same equality holds when A is regular, for any B.

Remark. The set function, defined for all sets in the line, by

$$\mathrm{d}(A) = \sup_{B \subset \mathbf{R}} [\dim A \times B - \dim B]$$

is a σ–stable rarefaction index, such that

$$\dim A \leq \mathrm{d}(A) \leq \mathrm{Dim}\, A\ .$$

Question: is this index d different from Dim?

4.5 Graph of lipschitzian maps

Let $f : I = [0,1] \mapsto \mathbf{R}$ be a continuous function, of graph Γ. Denote by

$$v_\epsilon(x) = \sup\{\ |f(x') - f(x'')|\ :\ |x - x'| \leq \epsilon,\ |x - x''| \leq \epsilon\}\ .$$

the ϵ-oscillation of f in x.

Theorem 1. *If $0 \leq x \leq 1$, then*

$$\begin{aligned}\Delta(\Gamma, D_x) &= 1 - \min\{\,1, \liminf \tfrac{\log v_\epsilon(x)}{\log \epsilon}\,\} \\ \delta(\Gamma, D_x) &= 1 - \min\{\,1, \limsup \tfrac{\log v_\epsilon(x)}{\log \epsilon}\,\}\ .\end{aligned}$$

Remark. If $v_\epsilon(x) \simeq \epsilon^H$ uniformly on $[0,1]$, then $\Delta(\Gamma, D_x) = \delta(\Gamma, D_x) = 1 - H$. On the other hand, we know that $\Delta(\Gamma) = \delta(\Gamma) = 2 - H$. The following theorem generalizes these relationships between the dimensions of Γ and the oscillation of f:

Notation.

$$\begin{aligned} \underline{F}(\alpha) &= \{ x \in [0,1] \, : \, \liminf \tfrac{\log v_\epsilon(x)}{\log \epsilon} \geq \alpha \} \\ \overline{F}(\alpha) &= \{ x \in [0,1] \, : \, \limsup \tfrac{\log v_\epsilon(x)}{\log \epsilon} \geq \alpha \} \, . \end{aligned}$$

All points x such that $f(x)$ is $Lip(\alpha)$ belong to $\underline{F}(\alpha)$.

Theorem 1.

$$\begin{aligned} \operatorname{Dim} \Gamma &\leq \max \{ 1 \, , 2 - \sup \{ \alpha \, : \, \operatorname{Dim}(I \setminus \underline{F}(\alpha)) = 0 \} \} \\ \dim \Gamma &\leq \max \left\{ 1 \, , 2 - \sup \{ \alpha \, : \, \operatorname{Dim}(I \setminus \overline{F}(\alpha)) = 0 \} \right\} \\ \dim \Gamma &\leq \max \{ 1 \, , 2 - \sup \{ \alpha \, : \, \dim(I \setminus \underline{F}(\alpha)) = 0 \} \} \, . \end{aligned}$$

The proof makes use of Formulas (5), (1), and (2).

Results in the other sense are more difficult to find:

Theorem 2. *If* $\liminf \frac{\log v_\epsilon(x)}{\log \epsilon} \leq \alpha$ *a.e. on* I, *for some* $\alpha \leq 1$, *then* $\operatorname{Dim} \Gamma \geq 2 - \alpha$.

No general result of such kind is known for the Hausdorff dimension. It is usually difficult to compute the exact value of $\dim \Gamma$: for example it is not known for the graph of the Weierstrass function

$$W(x) = \sum_{i=0}^{\infty} b^{-iH} \cos(b^i x) \, .$$

Remark. The inequality

$$(6) \quad \dim E \geq \dim F + \hat{\gamma}(E) \, ,$$

which would complete those of Section 3, is false in general. To show this, construct first a set B in $\mathbf{R}$ such that $\dim B = 0$ and $\hat{\delta}(B) = 1$. Let $E = I \times B$. We know that

$$\dim E \leq \dim B + \operatorname{Dim} I = 1 \, .$$

On the other hand, $\hat{\gamma}(E) = \hat{\delta}(E) = 1$, so that

$$\dim I + \hat{\gamma}(E) = 2 \, .$$

References.

J. Hawkes, Hausdorff measure, entropy, and the independence of small sets, *Proc. Lond. Math. Soc.* **28** (1974), 700–724.

C. Tricot, Two definitions of fractional dimension, *Math. Proc. Camb. Phil. Soc.* **91** (1982), 57–74.

C. Tricot, Dimensions de graphes, *C. R. Acad. Sci. Paris* **303** (1986), 609–612.

5 Bouligand dimensions

5.1 Main properties

For any bounded set E in $\mathbf{R}^n$, the Bouligand dimensions are defined as

$$\begin{aligned} \Delta(E) &= \limsup_{\epsilon\to 0}\left(n - \frac{\log \mathcal{V}(E(\epsilon))}{\log \epsilon}\right) \\ \delta(E) &= \liminf_{\epsilon\to 0}\left(n - \frac{\log \mathcal{V}(E(\epsilon))}{\log \epsilon}\right) . \end{aligned}$$

They are both rarefaction indices.

- $\Delta(E) = \Delta(\overline{E})$, $\delta(E) = \delta(\overline{E})$.

- $\Delta(E_1 \cup E_2) = \max(\Delta(E_1), \Delta(E_2))$ (Δ is stable).

- Δ is not σ–stable: for example, $\Delta(\mathcal{Q} \cap I) = 1$, $\Delta(\{\frac{1}{n}\}) = \frac{1}{2}$. This is related to the fact that the associated packing pre–measure P^α is not σ–subadditive.

- δ is even not stable.

Example. On the line, let $E = E_1 \cup E_2$, where

$$\omega(2^{n+1}, E_1) = \begin{cases} \omega(2^n, E_1) & if 4^k \leq n < 2.4^k - 1 \\ 2\,\omega(2^n, E_1) & if 2.4^k - 1 \leq n < 4^{k+1} \end{cases}$$

$$\omega(2^{n+1}, E_2) = \begin{cases} 2\,\omega(2^n, E_2) & if 4^k \leq n < 2.4^k - 1 \\ \omega(2^n, E_2) & if 2.4^k - 1 \leq n < 4^{k+1} . \end{cases}$$

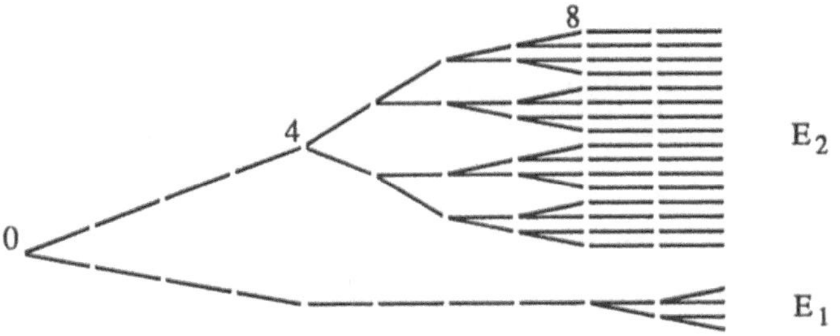

Figure 5.1:

Then

$$\delta(E_1) = \delta(E_2) = \frac{1}{3}, \; \Delta(E_1) = \Delta(E_2) = \frac{2}{3}, \; \delta(E_1 \cup E_2) = \frac{1}{2}, \; \Delta(E_1 \cup E_2) = \frac{2}{3}.$$

5.2 Other definitions

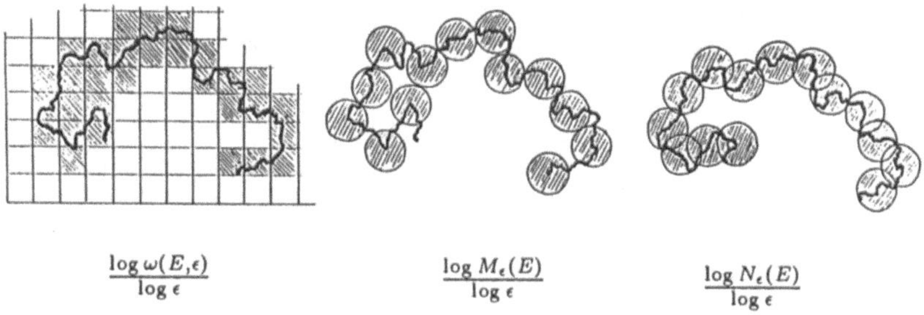

Figure 5.2:

Equivalent definitions of Δ and δ are obtained with the lim sup and lim inf of such ratios. Here, $\omega(E,\epsilon)$ is the number of disjoint boxes in an ϵ–net meeting E; $M_\epsilon(E)$ is the maximum number of disjoint balls of diameter ϵ, centred on E; $N_\epsilon(E)$ is the minimum number of balls of diameter ϵ, covering E.

Theorem 1. *In each of these definitions of Δ and δ, one can replace ϵ by a discrete sequence ϵ_k tending to 0, with the condition that*

$$\lim_{k\to\infty} \frac{\log \epsilon_{k+1}}{\log \epsilon_k} = 1 \, .$$

For any domain V, diam int V denotes the largest diameter of a disk included in V.

Theorem 2. (generalized sausages) *We consider a family* $\{ V_\epsilon(x) \ : \ x \in E, \epsilon > 0 \}$ *of domains in* $\mathbf{R}^n$, *such that every* $V_\epsilon(x)$ *has diameter* $\leq \epsilon$. *If moreover*

$$\lim_{\epsilon \to 0} \frac{\log[\inf_{x\in E} \operatorname{diam} \operatorname{int} V_\epsilon(x)]}{\log \epsilon} = 1 ,$$

then

$$\begin{aligned} \Delta(E) &= \limsup_{\epsilon\to 0} \left(n - \tfrac{\log \mathcal{V}(\cup_{x\in E} V_\epsilon(x))}{\log \epsilon}\right) \\ \delta(E) &= \liminf_{\epsilon\to 0} \left(n - \tfrac{\log \mathcal{V}(\cup_{x\in E} V_\epsilon(x))}{\log \epsilon}\right) . \end{aligned}$$

The proof makes use of the following lemma:

Lemma. *Let* $c > 1$. *Let* (B_i) *be a family of balls, and* B_i^* *be the ball of same centre, and diameter* $c \operatorname{diam} B_i$. *Then*

$$\mathcal{V}(\cup_i B_i^*) \leq c^n \, \mathcal{V}(\cup_i B_i) .$$

5.3 Parametrized curves

We come back to the notion of *perturbation* $p(\Gamma)$. If Γ is a simple curve, parametrized by the map $\gamma \ : \ [a, b] \mapsto \Gamma$,the local perturbations are

$$p(t, \tau) = p(\gamma(t - \tau)^\frown \gamma(t + \tau)) .$$

Definition. Weak sense: A curve Γ is *fractal* if $p(t,\tau)/\tau$ tends to ∞ for all t, when τ tends to 0. Strong sense: A curve Γ is *fractal* if, for some $\alpha < 1$, $\lim_{\tau\to 0} \frac{\log p(t,\tau)}{\log \tau} = \alpha$ for all t. (See Fig. 5.3.)

(a) The self–similar case :

$\Gamma = \cup_{i=1}^N \Gamma_i$, where the Γ_i have at most one point in common, and every Γ_i is similar to Γ, with a similarity ratio p_i. The curve Γ is a fractal if $\sum p_i > 1$. Let α be such that $\sum p_i^\alpha = 1$, and $\tau_i = p_i^\alpha$. The canonical parametrization of Γ attributes to every copy Γ_i the measure τ_i. If Γ' is a subarc of Γ, deduced from Γ by a similarity of ratio $p_{i_1} p_{i_2} \ldots p_{i_k} \simeq \operatorname{diam} \Gamma' \simeq p(t, \tau)$, its measure is $\tau_{i_1} \tau_{i_2} \ldots \tau_{i_k} \simeq \tau \simeq p(t, \tau)^\alpha$. Using Billingsley type results, we can interpret α as the similarity dimension $\delta(\Gamma) = \Delta(\Gamma)$. Therefore

In the self–similar case, for all t

$$\frac{\log p(t, \tau)}{\log \tau} \to \frac{1}{\Delta(\Gamma)} .$$

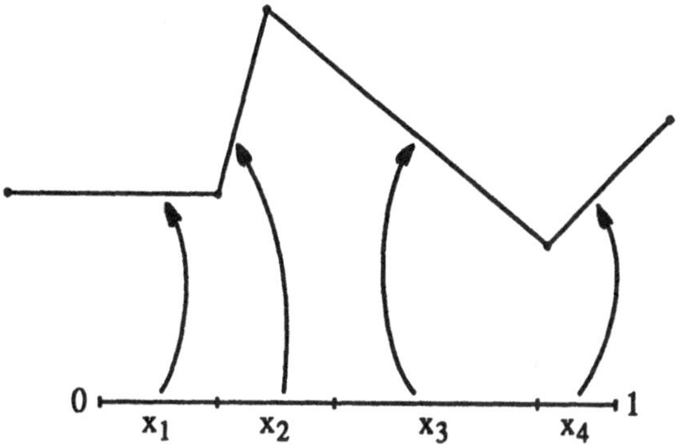

Figure 5.3:

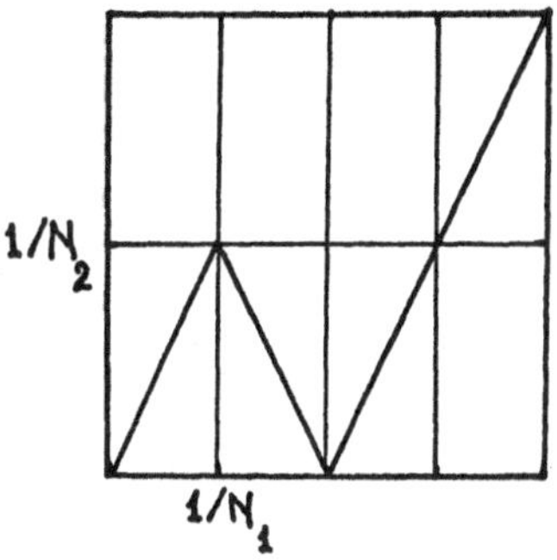

Figure 5.4:

(b) **The self–affine case :**

Let $0 < N_2 < N_1$ be two integers. Then $\Gamma = \cup_{i=1}^{N} \Gamma_i$, where the Γ_i have at most one point in common, and every Γ_i is an affine copy of Γ, shrunk horizontally by $\frac{1}{N_1}$, vertically by $\frac{1}{N_2}$. For some arrangements of Γ_i, Γ is the graph of a continuous function $f(t)$ (see Figure 5.4). The canonical parameter is the abcissa t. This parametrization induces a measure on Γ which is the projected Lebesgue measure on Ox. The local perturbation is interpreted as $v_\tau(t)$, the τ–oscillation of the function f. If $\tau = N_1^{-k}$: $v_\tau(t) \simeq N_2^{-k}$. Hence

$$\frac{\log p(t,\tau)}{\log \tau} \to \frac{\log N_2}{\log N_1} = 2 - \Delta(\Gamma)$$

for all t.

5.4 examples of continuous functions

• **Takagi function:** $f(t) = \sum_{i=0}^{\infty} 2^{-iH} g(2^i t)$, where g is the periodic, pyramidal function, and $0 < H < 1$. The graph Γ of f restricted to $[0,1]$ is self-affine, the two basic affine maps being

$$\begin{pmatrix} 1/2 & 0 \\ 1 & 1/\sqrt{2} \end{pmatrix} x \, , \quad \text{and} \quad \begin{pmatrix} -1/2 & 0 \\ 1 & 1/\sqrt{2} \end{pmatrix} x + \begin{pmatrix} 1 \\ 0 \end{pmatrix} .$$

The function f is $Lip(H)$ at every t, and

$$\delta(\Gamma) = \Delta(\Gamma) = 2 - H \, .$$

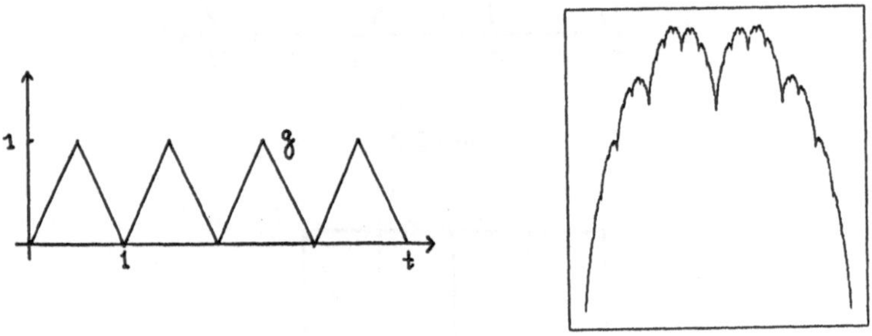

Figure 5.5:

• **Weierstrass function:** $f(t) = \sum_{i=0}^{\infty} b^{-iH} \cos(b^i t + \phi_i)$, $0 < H < 1$, $b > 1$. It is not self–affine. The function f is $Lip(H)$ at every t, and

$$\delta(\Gamma) = \Delta(\Gamma) = 2 - H \, .$$

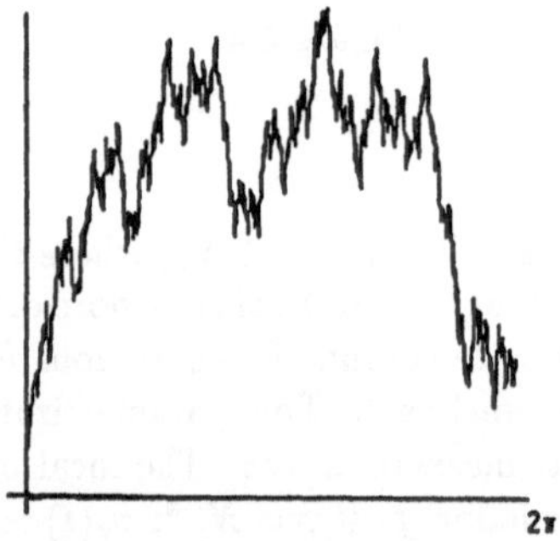

Figure 5.6:

• **Weierstrass–Mandelbrot function:** $f(t) = \sum_{i=-\infty}^{+\infty} b^{-iH}(1 - \cos(b^i t + \phi_i))$, $0 < H < 1$, $b > 1$. Same properties. It is not self–affine, but if $\phi_i = 0$ for all i, it verifies a *weak self–affinity condition:*

$$\forall \, t \, , \; f(bt) = b^H f(t) \, .$$

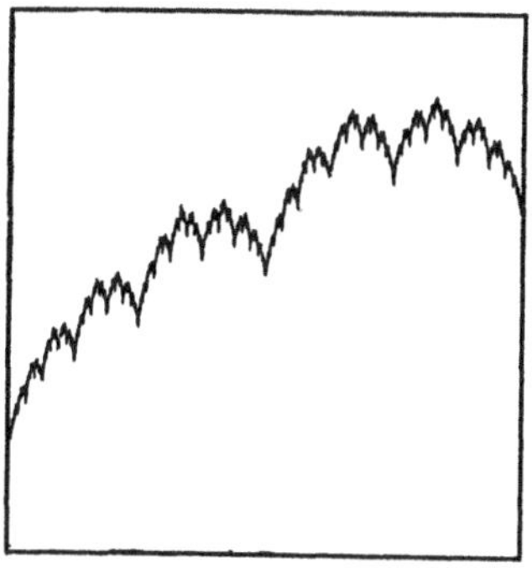

Figure 5.7:

5.5 The weak self–affinity condition

If a function $f(t)$ verifies the above condition, for some parameters H and b, then it is possible to associate to f a periodic function f^*, of period 1, defined as

$$f^*(t) = b^{-Ht} f(b^t) .$$

Conversely, every periodic function f^* of period 1 gives a weakly self–affine $f(t)$:

$$f(t) = t^H f^*(\frac{\log t}{\log b}) ,$$

corresponding to the parameters b and H. Restricted to any bounded interval, the graphs of f and f^* have similar properties: in particular, the dimensions are the same. It is easy to deduce the following:

- Not all weakly self–affine functions are fractal: in particular, if f^* is differentiable, so is f. The simplest (non–trivial) example is $f^*(t) = 1$, $f(t) = t^H$.

- As $t \to \infty$, the graph of a weakly self–affine function is expanded following t^H.

- If the spectrum of such a function f is discrete, then the power spectral density

$$G_f(s) = \lim_{T\to\infty} \frac{1}{T} \left| \int_0^T e^{-ist} f(t)\, dt \right|$$

verifies

$$G_f(bs) = b^{-2H} G_f(s) .$$

Therefore, the amplitude spectrum decreases like s^{-2H} as s tends to ∞.

- For example, the amplitude spectrum of the Weierstrass–Mandelbrot function decreases as $s^{-2H} = s^{2\Delta(\Gamma)-4}$. One should not conclude that *in general*, the rate of decreasing of the spectrum gives a direct information on the value of $\Delta(\Gamma)$. Let us choose *any* pair of parameters H, H' between 0 and 1. The function

$$f^*(t) = W(t) = \sum_0^\infty 2^{-iH} \cos(2\pi\, 2^i\, t)$$

is periodic, of period 1. The fractal dimension of its graph is $2 - H$. The corresponding

$$f(t) = t^{H'} \sum_0^\infty 2^{-iH} \cos(2\pi \frac{2^i}{\log b} \log t)$$

is weakly self-affine. Here again, $\Delta(\Gamma) = 2 - H$. But $\log G_f(s)/\log s$ tends to $-2H'$ as $s \to \infty$.

5.6 Algorithms for computing $\Delta(\Gamma)$

Let Γ be the graph of a continuous function, defined on $[a, b]$. Experience shows that Minkowski sausage method, box method, ..., are inefficient for a precise evaluation of the Bouligand fractal dimension.

1) Horizontal structural elements

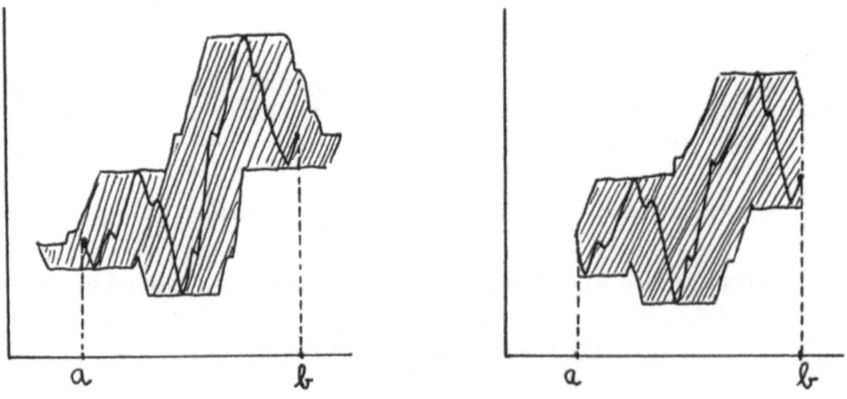

Figure 5.8:

Let $T(t,\tau)$ be the horizontal segment $[t-\tau, t+\tau] \times \{f(t)\}$, and $U(\tau) = \cup_{a \le t \le b} T(t,\tau)$ the corresponding sausage. Then

Theorem 1. *If f is not constant,*

$$\Delta(\Gamma) = \limsup_{\tau \to 0} \left(2 - \frac{\log \mathcal{A}(U(\tau))}{\log \tau} \right) .$$

The same is true by replacing $U(\tau)$ by $U^*(\tau) = U(\tau) \cap [a,b] \times \mathbf{R}$.

Theorem 2. $\mathcal{A}(U^*(\tau)) = \int_a^b v_\tau(t)\, dt$.

The function $\mathcal{A}(U^*(\tau)) = V_\tau$ is called *variation* of f, and the formula

$$\Delta(\Gamma) = \limsup_{\tau \to 0} \left(2 - \frac{\log(V(\tau))}{\log \tau} \right)$$

is the *variation method*. It is certainly the easiest and one of the most efficient of all algorithms.

2) Scanning Γ with vertical straight lines

Suppose that $\log v_\tau(t)/\log \tau$ tends to $H(t)$ uniformly on $[a,b]$. Then $\delta(\Gamma, D_t) = \Delta(\Gamma, D_t) = 1 - H(t)$. The variation V_τ can be estimated by

$$V_\tau \simeq \int_a^b \tau^{H(t)}\, dt ,$$

and the formula

$$\Delta(\Gamma) = \limsup_{\tau \to 0} \left(2 - \frac{\log(\int_a^b \tau^{H(t)}\, dt)}{\log \tau} \right)$$

is theoretically exact for all self–affine functions.

3) Scanning Γ with horizontal straight lines

Let $m = \inf_{[a,b]} f(t)$, $M = \sup_{[a,b]} f(t)$. For all $\epsilon > 0$, and $m \le y \le M$, let D_y be the horizontal line of ordinate y, and

$$\Gamma_{y,\epsilon} = \bigcup_{(t,y) \in \Gamma \cap D_y} [t - \epsilon, t + \epsilon] \times \{y\}$$

be the linear Minkowski sausage of the set $\Gamma \cap D_y$. It is a union of segments, each of length $\ge \epsilon$. We denote by $l(\Gamma_{y,\epsilon})$ the total length of $\Gamma_{y,\epsilon}$. We know that

$$\Delta(\Gamma \cap D_y) = \limsup_{\epsilon \to 0} \left(1 - \frac{\log l(\Gamma_{y,\epsilon})}{\log \epsilon} \right) .$$

This dimension is easy to evaluate. Relationships between $\Delta(\Gamma)$ and $\Delta(\Gamma \cap D_y)$ are the next:

Theorem 3.

$$\Delta(\Gamma) \geq 1 + \text{ess sup } \Delta(\Gamma \cap D_y) ,$$

where

$$\begin{aligned} \text{ess sup } z(t) = & \inf\{ a \text{ such that } l(\{ t :: z(t) \geq a \}) = 0 \} \\ = & \sup\{ a \text{ such that } l(\{ t :: z(t) < a \}) > 0 \} . \end{aligned}$$

The inverse inequality is true, if

$$\sup_{0<\epsilon\leq 1} \sup_{m\leq y\leq M} \epsilon^{\alpha-1} l(\Gamma_{y,\epsilon}) < \infty .$$

One can also try a scanning of Γ with smooth curves, in order to get more intersection points, and therefore a more representative value of $l(\Gamma_{y,\epsilon})$.

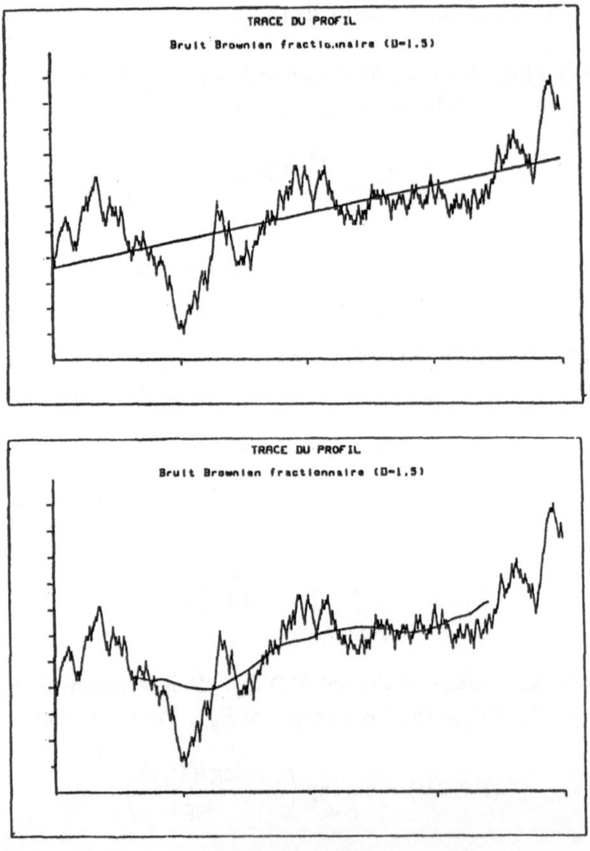

Figure 5.9:

References.

G. Bouligand, Sur la notion d'ordre de mesure d'un ensemble plan, *Bull. Sc. Math.* II **53** (1929), 185–192.

C. Tricot, *Sur la classification des ensembles boréliens de mesure de Lebesgue nulle*, Thèse de doctorat, Genève (1979).

C. Tricot, Dimension fractale et spectre, *J. Chimie Phys.* **85** (1988), 379–384.

C. Tricot, J. F. Quiniou, D. Wehbi, C. Roques–Carmes, B. Dubuc, Evaluation de la dimension fractale d'un graphe, *Rev. Phys. Appl.* **23** (1988), 111–124.

Iterated function systems: theory, applications and the inverse problem

*Edward R. Vrscay**
Department of Applied Mathematics
Faculty of Mathematics
University of Waterloo
Waterloo, Ontario,
Canada N2L 3G1

Abstract

The basic theory and properties of Iterated Function Systems are given, comparing the original approaches of Hutchinson [47] and Barnsley and Demko [8]. Some examples and applications are discussed, along with computational aspects. A generalized recurrent IFS is introduced, with suitably constructed measure space, from which the existence of an invariant measure follows. This yields a Collage Theorem for Measures on generalized RIFS. Finally, the inverse problem of fractal/measure construction is discussed. Some recent applications of Genetic Algorithms as a stochastic optimization method for (i) moment matching and (ii) Collage Theorem for measures are reported.

1 Introduction

"Iterated Function Systems" (henceforth IFS) is the name given by M. Barnsley and S. Demko [8] (and adopted by coworkers and others, for example [10-16]), originally to a system of contractive maps $\mathbf{w} = \{w_1, w_2, \ldots, w_N\}$ on a compact metric space (K, d). A unique compact set, the "attractor", $A = \cup_\sigma \; \{\lim_{n\to\infty} w_{\sigma_1} \circ w_{\sigma_2} \circ \cdots \circ w_{\sigma_n}(x)\}$, (independent of x) was shown to exist, where the union is over all half-infinite sequences $\sigma = (\sigma_1, \sigma_2, \ldots)$ where $\sigma_i \in \{1, 2, \ldots, N\}$. Associated with the maps w_i is a set of probabilities $p_i > 0$, $\sum p_i = 1$, which is then considered to define the following probabilistic dynamical system

[0]*This work has been supported by the Natural Sciences and Engineering Research Council of Canada

J. Bélair and S. Dubuc (eds.), Fractal Geometry and Analysis, 405–468.

(the "Chaos game"): Starting at a point $x_0 \in K$, define the sequence $x_{n+1} = w_{\sigma_n}(x_n)$ where the index σ_n is chosen randomly and independently from the set $\{1, 2, \ldots, N\}$ with probability $P(\sigma_n = i) = p_i > 0$. Associated with this discrete-time Markov process is a "balanced" invariant probability measure μ such that $\mu(B) = \int_K P(x, B)\, d\mu$ for all Borel subsets $B \subset \mathcal{B}(K)$, where $P(x, B)$ is the probability of transfer from $x \in K$ to the Borel subset B. Moreover, $supp(\mu) = A$.

The geometric and measure theoretic aspects of such systems of contractive maps and associated probabilities was, in fact, worked out earlier by J. Hutchinson [47], who considered the "parallel" action of $\mathbf{w}$ on K: for a set $S \subset K$, define $\mathbf{w}(S) = \cup_{i=1}^{N} w_i(S)$. The existence of a unique compact invariant set $A = \cup_{i=1}^{N} w_i(A)$ was shown using the contraction mapping principle over the complete metric space $(\mathcal{S}, h)$ where $\mathcal{S} = \{\text{compact subsets of } K\}$ and h denotes Hausdorff metric. An operator $M : X_1 \to X_1$ (X_1 = space of normalized Borel regular measures on K) was defined as follows: for $\nu \in S_1$, define $\nu' = M(\nu)$ such that $\nu'(B) = \sum p_i \nu(w_i^{-1}(B))$ for all $B \in \mathcal{B}(K)$. M was shown to be a contraction mapping on X_1, thus proving the existence of an invariant probability measure $\mu \in X_1$ satisfying $M(\mu) = \mu$. Also, $supp(\mu) = A$.

Indeed, the preceding paragraphs have outlined two major developments which can be considered as complementary in viewpoint, i.e. probabilistic *vs.* geometric-measure theoretic. These studies have provided powerful tools for the study of fractal sets [52] and their applications. The idea of the action of a system of contractive maps can actually be traced farther back to Williams [64] who looked at the fixed points of finite compositions $w_{i_1} \circ w_{i_2} \circ \cdots \circ w_{i_n}(x)$. The construction of a random walk over a Cantor-like set, essentially the "Chaos game", was also done by Karlin [48] who studied learning models. The actions of systems of contractive maps to produce fractal sets has been considered by others [4, 17, 26, 29, 44], to mention only a few of which this author is aware.

The purpose of the lectures in Montréal, summarized in these notes, was to present theoretical as well as computational aspects of IFS, along with some examples and a discussion of the inverse problem. Included below are also accounts of work which has been pursued by this author since the Montréal meeting, in collaboration with people to be acknowledged below. It is also hoped that the plots of attractors and histogram representations of invariant measures would provide the reader with some kind of picture of the action of these systems of mappings.

The plan of this paper is as follows. Section 2 is devoted to basic aspects of IFS, to show how the approaches by Hutchinson and Barnsley and coworkers are complementary. The moment recursion relations for linear IFS are also given, since they are relevant to one approach to the inverse problem. Section 3 lists some special examples, including a brief account of IFS applications to the study of fractal sets associated with complex bases, as done by W. Gilbert. I have also included two interesting aspects of IFS: (i) orthogonal polynomials over invariant measures and associated Jacobi matrices, and (ii) "missing moment" problems for nonlinear IFS (the latter was done with D. Weil, an undergraduate research assistant). Section 4 deals with recurrent IFS (RIFS), a generalization of the IFS discussed above, and is based on work with C. Cabrelli and U. Molter. It begins with a discussion of the "simple" recurrent IFS introduced in [14], with some illustrative examples and computer plots. Then, a generalized recurrent IFS over arbitrary metric spaces (K_i, d_i), $i = 1, 2, \ldots, N$, is introduced along with the construction of a suitable measure space. A generalized Markov operator on this space is defined and shown to be contractive, hence possessing a "fixed-point" invariant measure. In the case $K_i \subset \mathbf{R}$ and linear affine maps w_i, a recursion relation for moments is derived. The subject of Section 5 is the inverse problem of fractal/measure reconstruction for IFS. The Collage Theorems for Geometry (Hausdorff metric) and Measures (Hutchinson metric) are discussed, as well as a Collage Theorem for generalized RIFS, which follows from the results of Section 4. A method of matching moments of target and approximating measures used in [60] is briefly discussed, with sample calculations. This method is now justified by a result relating the Euclidean distance between moment sequences and the Hutchinson distance between corresponding measures [25]. Finally, some applications of Genetic Algorithms as a stochastic optimization method to the following problems are discussed: (i) minimization of Euclidean moment distance (continuous problem) and (ii) Collage Theorem for measures in discrete pixel space. These studies have been performed since the Montréal meeting and are still in their infancy; nevertheless, the results, particularly those of (ii), are quite encouraging.

2 Iterated function systems

In this Section, the basic features of IFS are introduced, with a comparison of the methods employed by Hutchinson [47] (Sec. 2.1) and Barnsley and Demko [8] (Sec. 2.2). For the most part, the notation follows the latter, as well as that found in [16]. Ref. [16] provides a comprehensive treatment of many aspects of IFS, aimed at a general readership, including senior undergraduates. Ref. [33] provides a detailed discussion of technical aspects of fractal sets. Falconer's recent book [34] examines the entire spectrum of research on fractal sets, from which one can see the role of IFS in perspective.

2.1 Mathematical preliminaries and introduction to IFS

In the discussion below, and indeed for most of this paper, the following notations shall be assumed: (K, d) will denote a compact metric space with metric d. In most considerations, K is either (i) a compact subset of $\mathbf{R}^n$, $n = 1,2$ and d the appropriate Euclidean metric, (ii) a bounded subset of $\mathbf{Z}^n$, $n = 1,2$ with "Manhattan metric" (this latter discrete case will be discussed in Section 2.4). The distance between a point $x \in K$ and a set $A \subset K$ will be denoted as

$$d(x, A) = \inf_{y \in A} d(x, y). \tag{2.1.1}$$

The Hausdorff metric between two sets $A, B \subset K$ is defined as

$$h(A, B) = max\,[\sup_{x \in A} d(x, B),\ \sup_{y \in B} d(y, A)]. \tag{2.1.2}$$

Let the ϵ-ball of a set $A \subset K$ be defined as $A_\epsilon = \{x \in K : d(x, A) < \epsilon\}$. Then $h(A, B) < \epsilon$ implies that $B \subset A_\epsilon$ and $A \subset B_\epsilon$. Let $\mathcal{S}$ denote the set of compact subsets of K. Then $(\mathcal{S}, h)$ is a complete metric space [33].

Now let $\mathcal{A}$ denote the σ-algebra $\mathcal{B}(K)$ of Borel subsets of K and let X be the set of all finite signed measures on $\mathcal{A}$. Let $X_1 = \{\mu \epsilon X \mid \mu(K) = 1\}$ be the subset of probability measures and define a metric on X_1 as

$$d_H(\mu, \nu) = \sup_{f \in L_1} \left[\int_K f d\mu - \int_K f d\nu\right], \qquad \mu, \nu \in X_1, \tag{2.1.3}$$

where $L_1 = \{f : K \to \mathbf{R} : \mid f(x) - f(y) \mid \leq d(x, y)\}$. Then (X_1, d_H) is a complete metric space [47]. (This metric was introduced by Hutchinson and consequently has often been referred to as the *Hutchinson metric*. The notation d_H has been adopted from Ref. [16].)

Now let $\mathbf{w} = \{w_1, w_2, \ldots, w_N\}$ denote a set of N continuous contraction maps on K, i.e. $w_i : K \to K$ and

$$d(w_i(x),\ w_i(y)) \leq s_i\, d(x, y), \qquad \forall\, x, y \in K, \quad 0 \leq s_i < 1, \quad i = 1, 2, \ldots, N. \tag{2.1.4}$$

It will be convenient to define the maximum contractivity factor of the IFS:

$$s = \max_{1 \le i \le n} (s_i) < 1 \,. \tag{2.1.5}$$

Associated with these maps is a set of non-zero probabilities, $\mathbf{p} = \{p_1, p_2, \ldots, p_N\}$, $p_j > 0$ for $j = 1, 2, \ldots, N$, and $\sum_j^N p_j = 1$, whose role will become clear below. The system $\{K, \mathbf{w}, \mathbf{p}\}$ will be referred to as a *contractive IFS* ("contractive" here replaces "hyperbolic" in [8, 16]). Now for a set $S \in K$, denote $w_i(S) = \{w_i(x) : x \in S\}$ and denote the "parallel" action of the set of maps w_i on S as

$$\mathbf{w}(S) \equiv \bigcup_{i=1}^{N} w_i(S) \,. \tag{2.1.6}$$

Also define the iteration sequence

$$\mathbf{w}^{n+1}(S) \equiv \mathbf{w}(\mathbf{w}^n(S)), \qquad n = 1, 2, \ldots \,. \tag{2.1.7}$$

(For example, if S is a singleton point, then $\mathbf{w}(S)$ is a set of at most N points, ..., $\mathbf{w}^n(S)$ is a set of at most N^n points, etc.) Two important results for contractive IFS are given below:

Theorem 2.1.1 [47]:

1. There exists a unique compact set $A \in \mathcal{S}$, the *attractor* of the IFS $\{K, \mathbf{w}\}$ (independent of $\mathbf{p}$) such that

$$A = \mathbf{w}(A) = \bigcup_{i=1}^{N} A_i = \bigcup_{i=1}^{N} w_i(A) \tag{2.1.8}$$

   and $h(\mathbf{w}^n(S), A) \to 0$ as $n \to \infty$ for all $S \in \mathcal{S}$.
2. Define the following "Markov operator" $M : X_1 \to X_1$,

$$M(\nu) = \sum_{i=1}^{N} p_i \nu \circ w_i^{-1} \tag{2.1.9}$$

   Then there exists a unique measure $\mu \in X$, termed the *invariant measure*, which obeys the fixed point condition

$$M\mu = \mu \,. \tag{2.1.10}$$

   Moreover, $supp(\mu) = A$.

Hutchinson proved 1. and 2. by the contraction mapping principle, using the fact that

1. $h(\mathbf{w}(A), \mathbf{w}(B)) \le s\, h(A, B)$,

2. $d_H(M(\mu), M(\nu)) \leq s\, d_H(\mu, \nu)$,

where s is given in Eq. (2.1.5). Note that Eq. (2.1.8) implies that A is *self-tiling*, i.e. A is a union of (distorted) copies of itself. This establishes the *geometric* nature of the IFS. Eqs. (2.1.9) and (2.1.10) characterize the IFS in terms of *measures*. The operator M is contractive and possesses a unique invariant measure μ. (See Sec. 9.6 of [16] for a graphical presentation of this latter property.) A noteworthy consequence of Eq. (2.1.9) is the following: For μ- integrable functions $f : K \to \mathbf{R}$,

$$\int_A f(x)d\mu(x) = \sum_{i=1}^{N} p_i \int_A (f \circ w_i)(x)d\,\mu(x)\,. \tag{2.1.11}$$

For the case of linear maps w_i, this invariance relation is useful in providing a recursion relation for the moments of μ (cf. Sec. 2.3 below).

Theorem 2.1.1 also provides an algorithm for producing computer generated approximations to A and μ via the iteration procedures defined in Eq. (2.1.7). For simplicity, the "seed" of this procedure will be a singleton point.

Deterministic Algorithm

1. To approximate A : Pick a "seed" $x_0 \in K$ and define $S_0 = \{x_0\}$. Then let $S_1 = \mathbf{w}(S_0) = \{w_i(x_0),\ i = 1, 2, \ldots, N\}$, $S_2 = \mathbf{w}(S_1) = \{w_i(w_j(x_0)),\ i, j = 1, 2, \ldots N\}$ etc. As $n \to \infty$, $h(S_n, A) \to 0$. For n sufficiently large, say 10-20, the set S_n of N^n points is a good approximation to A on a computer screen, where the image is represented on a discrete set of pixels. (Obviously, it is advantageous to pick a seed $x_0 \in A$, e.g. the fixed point of one of the IFS maps.) This procedure essentially involves the enumeration of an N-tree to n generations.
2. To approximate μ : Let δ_x denote a unit point mass measure at $x \in K$. Now construct the following measures $\nu_n \in X_1$, such that $supp(\nu_n) = S_n$ defined above: $\nu_1 = \delta_{x_0}$, $\nu_2 = p_1\, \delta_{w_1(x_0)} + p_2\, \delta_{w_2(x_0)} + \cdots + P_n \delta_{w_n(x_0)}$, and, in general,
$$\nu_n = \sum_{i_1, \ldots, i_n}^{N} p_{i_1}\, p_{i_2} \cdots p_{i_n}\ \delta_{w_{i_1} \circ w_{i_2} \circ \cdots \circ w_{i_n}(x_0)}\,.$$
Again, on a computer screen, the measures ν_n will provide approximations to the invariant measure.

Examples:

1. $K=[0,1]$, $w_1(x)=\frac{1}{3}x$, $w_2(x)=\frac{1}{3}x+\frac{2}{3}$, $p_1=p_2=\frac{1}{2}$: A is the ternary Cantor set on [0,1]. Note that $A_1=w_1(A)=A\cap\left[0,\frac{1}{3}\right]$, $A_2=w_2(A)=A\cap\left[\frac{2}{3},1\right]$. The invariant measure is the (uniform) Cantor-Lebesgue measure [3, p. 77].
2. $K=[0,1]$, $w_1(x)=\frac{1}{2}x$, $w_2(x)=\frac{1}{2}x+\frac{1}{2}$, $p_1=p_2=\frac{1}{2}$: $A=[0,1]$, $A_1=\left[0,\frac{1}{2}\right]$, $A_2=\left[\frac{1}{2},1\right]$, $\mu=$ uniform Lebesgue measure on [0,1].
3. $K=[0,1]\times[0,1]\in\mathbf{R}^2$, $w_1(x,y)=\left(\frac{1}{2}x,\frac{1}{2}y\right)$, $w_2(x,y)=\left(\frac{1}{2}x+\frac{1}{2},\frac{1}{2}y\right)$, $w_3(x,y)=\left(\frac{1}{2}x+\frac{1}{4},\frac{1}{2}y+\frac{1}{4}\sqrt{3}\right)$: A is the "Sierpinski gasket" shown in Fig. 1(a).
4. K as in 3, with $w_1\dots w_4$ linear affine transformations of the form in Eq. (2.3.9) below, with coefficients and probabilities listed below:

i	$a_{11}(i)$	$a_{12}(i)$	$a_{21}(i)$	$a_{22}(i)$	$b_1(i)$	$b_2(i)$	p_i
1	0.00	0.00	0.00	0.16	0.50	0.00	0.01
2	0.20	-0.26	0.23	0.22	0.40	0.05	0.07
3	-0.15	0.28	0.26	0.24	0.57	-0.12	0.07
4	0.85	0.04	-0.04	0.85	0.08	0.18	0.85

A is the *spleenwort fern* [9] attractor shown in Fig. 1(b).

Let $C(K)$ denote the Banach space of continous functions on K, and associated with the IFS $\{\mathbf{w},\mathbf{p}\}$ define the following linear operator $T:C(K)\to C(K)$,

$$Tf=\sum_{i=1}^{N}p_i(f\circ w_i)\,,\qquad f\in C(K)\,. \tag{2.1.12}$$

Now for a given $\nu\in X_1$, define the linear functional $F:C(K)\to\mathbf{R}$,

$$F(f)=\langle f,\nu\rangle\equiv\int_K f\,d\nu\,. \tag{2.1.13}$$

Then,

$$\langle Tf,\nu\rangle=\langle f,M\nu\rangle\,, \tag{2.1.14}$$

i.e. M is the adjoint operator of T. If $\nu=\mu$ the invariant measure of the IFS, then $M\mu=\mu$ and Eq. (2.1.14) is identical to Eq. (2.1.11)

Theorem 2.1.2:

$$\lim_{n\to\infty}T^nf(x)=\int f d\mu\,,\ \text{for } x\in K\,.$$

Idea of Proof: Let $\nu\in X_1$. From Eq. (2.1.14), repeated application of M gives $\langle f,M^n\nu\rangle=\langle T^nf,\nu\rangle$. From Theorem 2.1.1, $\lim_{n\to\infty}\langle f,M^n\nu\rangle=\langle f,\mu\rangle=\lim_{n\to\infty}\langle T^nf,\nu\rangle$. Now choose $\nu = \delta_x$ (point mass measure at x) and the result follows.

Note that

$$T^n f(x) = \sum_{i_1,\dots,i_n}^{N} p_{i_1} p_{i_2} \cdots p_{i_n} f(w_{i_1} \circ w_{i_2} \circ \cdots \circ w_{i_n}(x))$$

which amounts to enumerating a weighted N-tree to n generations. This property provides a means of computing integrals of functions f over IFS invariant measures, and will be discussed again later. Demko [27] has investigated the convergence of such estimates.

2.2 The IFS as a dynamical system

Barnsley and Demko [8] (henceforth BD) proved Theorem 2.1.1 by looking at the IFS as a probabilistic dynamical system. In this case, a coding, or *symbolic dynamics* approach is natural. It is instructive to outline the main ideas here as they will be relevant to later sections. As earlier, we consider a set of N continuous contraction maps $w_i : K \to K$ with associated probabilities $p_j > 0$, $\sum_j p_j = 1$.

Let $\sum_N$ denote the set of all half-infinite sequences of N symbols, i.e. $\sum_N = \{\sigma = (\sigma_1, \sigma_2, \sigma_3, \dots), \ \ \sigma_i \in \{1, 2, \dots, N\}, i = 1, 2, \dots\}$. Define the following metric on $\sum_N$:

$$d_{\sum_N} = \sum_{k=1}^{\infty} \frac{|\sigma_k - \tau_k|}{(N+1)^k}, \qquad \sigma, \tau \in \textstyle\sum_N . \tag{2.2.1}$$

Then $(\sum_N, d_{\sum_N})$ is a compact metric space [28]. Now, for an $x \in K$, and $\sigma \in \sum_N$, define the following compositions,

$$\phi(\sigma, n, x) = w_{\sigma_1} \circ w_{\sigma_2} \circ \cdots \circ w_{\sigma_n}(x), \tag{2.2.2}$$

$$\phi(\sigma, x) = \lim_{n \to \infty} w(\sigma, n, x). \tag{2.2.3}$$

Lemma 1: Let $\sigma \in \sum_N$ and D=diam $(K) = \sup_{x,y \in K} d(x, y)$. Then

$$d(\phi(\sigma, m, x_1), \phi(\sigma, n, x_2)) \leq s^{min(m,n)} D \qquad \forall m, n \in \{1, 2, \dots\}, \ \ \forall x_1, x_2 \in K. \tag{2.2.4}$$

Proof: Use contractivity property in Eq. (2.1.4).

Theorem 2.2.1[8]:

(i) For any $\sigma \in \sum_N$ the following limit

$$\phi(\sigma) = \lim_{n \to \infty} \phi(\sigma, n, x) \tag{2.2.5}$$

exists and convergence is uniform on compact subsets of K.

(ii) Moreover, $\phi(\sigma) \in A$ and is independent of $x \in K$.

(iii) The function $\phi : \sum_N \to A$ is continuous and onto.

Sketch of Proofs:

(i) As $m, n \to \infty$ in Eq. (2.2.4), $RHS \to 0$. The fact that D is independent of $x \in K$ implies uniform convergence.

(ii) $\phi(\sigma, n, x) \in \mathrm{w}^n(K)$. That the limit $\phi(\sigma) \in A$ follows from Part 1 of Theorem 2.1.1.

(iii) For $\epsilon > 0$, choose n large enough that $s^n D < \epsilon$. Let $\sigma, \tau \in \sum_N$ such that $\sigma_k = \tau_k$ $k = 1, 2, \ldots, n$. Then $d_{\sum_N}(\sigma, \tau) \leq (N+1)^{-(n+1)}$. For any $m \geq n$

$$d(\phi(\sigma, m, x), \phi(\tau, m, x)) = d(\phi(\sigma, n, x_1), \phi(\sigma, n, x_2)) \leq s^n D < \epsilon, \qquad (2.2.6)$$

where $x_1 = w_{\sigma_{n+1}} \circ \cdots \circ w_{\sigma_m}(x)$ and $x_2 = w_{\tau_{m+1}} \circ \cdots \circ w_{\tau_m}(x)$, $x_1, x_2 \in K$. Taking the limit $m \to \infty$, we have $d(\phi(\sigma), \phi(\tau)) < \epsilon$, proving continuity of ϕ.

To prove ϕ is onto, let $y \in A$. Then from Theorem 2.2.1, Part 1, there exists a sequence $\{\omega^{(n)} \in \sum_N\}$ such that $\lim_{n\to\infty} \phi(\omega^{(n)}, n, x) = y$. Compactness of $\sum_N$ implies that $\{\omega^{(n)}\}$ contains a convergent subsequence $\{\omega^{(n_k)}\}$ with limit $\omega \in \sum_N$. Then $\omega_i^{(n_k)} = \omega_i$ for $i = 1, 2, \ldots, \alpha(n_k)$ and $\alpha(n_k) \to \infty$ as $k \to \infty$. But $d(\phi(\omega, n_k, x), \phi(\omega^{(n_k)}, n_k, x)) \leq s^{\alpha(n_k)} D$. Taking limits as $k \to \infty$ implies $d(\phi(\omega), y) = 0$.

Thus, for any $y \in A$, there exists at least one sequence $\sigma \in \sum_N$ such that $y = \phi(\sigma)$. The sequence $\{\sigma_1, \sigma_2, \ldots\}$ lists the last, second-last, ... maps needed to arrive at y, thus providing an *address* or *code*:

$$\begin{aligned} y = \phi(\sigma) &= \lim_{n\to\infty} w_{\sigma_1} \circ w_{\sigma_2} \circ \cdots \circ w_{\sigma_n}(x) \\ &= w_{\sigma_1}(\lim_{n\to\infty} w_{\sigma_2} \circ \cdots \circ w_{\sigma_n}(x)) \in A_{\sigma_1} = w_{\sigma_1}(A). \end{aligned}$$

Likewise $y \in A_{\sigma_1\sigma_2} = w_{\sigma_1}(w_{\sigma_2}(A))$, etc.. Using symbolic dynamics notation, $y = \phi(\sigma) = w_{\sigma_1}(\phi(S\sigma))$, where $S : \sum_N \to \sum_N$ denotes the (Bernoulli) "shift map", $S(\sigma_1, \sigma_2, \sigma_3, \ldots) = (\sigma_2, \sigma_3, \ldots)$.

Theorem 2.2.2: Let $\{K, \mathrm{w}\}$ be a contractive IFS with attractor A and associated code space $\sum_N$. If $w_i(A) \cap w_j(A) = \emptyset$ for $\forall i, j \in \{1, 2, \ldots N\}$, $i \neq j$, then A is totally disconnected and $\phi : \sum_N \to A$ is one-to-one.

Examples:

1. $K = [0,1]$, $N = 2$, $w_1(x) = \frac{1}{3}x$, $w_2(x) = \frac{1}{3}x + \frac{2}{3}$, $A =$ ternary Cantor set on [0,1]: $y = 0$, $\sigma = (\dot{1})$; $y = 1$, $\sigma = (\dot{2})$; $y = \frac{1}{3}$, $\sigma = (1\dot{2})$. A is totally disconnected.
2. $K = [0,1]$ $N = 2$, $w_1(x) = \frac{1}{2}x$, $w_2(x) = \frac{1}{2}x + \frac{1}{2}$, $A = [0,1] : y = \frac{1}{2}$, $\sigma = \{(1\dot{2}), (2\dot{1})\}$. A is "just-touching".
3. $K = [0,1]$, $N = 2$, $w_1(x) = \frac{3}{4}x$, $w_2(x) = \frac{3}{4}x + \frac{1}{4}$, $A = [0,1] : y = \frac{1}{2}$ has an infinite number of codes. (It is a good student exercise to devise an algorithm to extract the code σ.)
4. $K = [0,1]$, $N = 10$, $w_i(x) = \frac{1}{10}x + \frac{i}{10}$, $i = 0,1,2,\ldots,9$. Then σ corresponds to the decimal expansion for $y \in [0,1]$.

We may now introduce the *Random Iteration* or "*Chaos Game*" algorithm [8, 16] which provides a convenient and fast method of generating pictures of the attractor A and invariant measure μ of a given IFS. From this dynamical and probabilistic viewpoint, the role of the probabilities p_i in determining the invariant measure μ will be seen.

Random Iteration Algorithm or "Chaos Game":

Pick an $x_0 \in K$ and define the iteration sequence

$$x_{n+1} = w_{\sigma_n}(x_n), \qquad n = 0,1,2,3,\ldots, \tag{2.2.7}$$

where the indices σ_n are chosen *randomly* and *independently* from the set $\{1,2,\ldots,N\}$ with probabilities $P(\sigma_n = i) = p_i$.

Theorem 2.2.3: Almost every orbit $\{x_n\}$ is dense on A.

Proof: Let $\phi(\tau) = y \in A$. For any $\epsilon > 0$ pick $n > 0$ so that $s^n D < \epsilon$ where $D =$ diam (K). From Eq. (2.2.6), it follows that $d(\phi(\tau, n, x_0), y) < \epsilon$. Since the sequence $\{\sigma_n\}$ in Eq. (2.2.7) is chosen randomly and independently, the finite sequence $\{\tau_n, \tau_{n-1}, \ldots, \tau_1\}$ will occur in the infinite σ sequence with probability one. (The set of sequences corresponding to periodic and preperiodic orbits is countable and has measure zero.)

The "chaos game" essentially defines a discrete-time Markov process on K defined by

$$P(x, B) = \sum_{i=1}^{N} p_i \chi_B(w_i(x)), \tag{2.2.8}$$

where $P(x, B)$ is the probability of transfer from $x \in K$ to the Borel subset $B \in \mathcal{B}(K)$, and χ_B denotes the characteristic function of $B : \chi_B(y) = 1$ if $y \in B$, and $\chi_B(y) = 0$ if $y \notin B$. BD showed that for a contractive IFS, there exists a unique probability measure μ such that

$$\mu(B) = \int_K P(x, B)\, d\mu(x)\,, \tag{2.2.9}$$

for all Borel subsets B. Basis of the BD proof: the adjoint operator $M : X_1 \to X_1$, defined in Eq. (2.1.9) (and referred to as T in [8]), is a weak* continuous mapping of a weak* compact convex set into itself. By the *Schauder* fixed-point theorem, M possesses a fixed point $\mu \in X_1$. BD referred to μ as the "**p**-balanced measure" of the IFS $\{K, \mathbf{w}, \mathbf{p}\}$ and showed that $supp(\mu) = A$.

Elton [31] has shown (for even a weaker conditon of "average contractivity" of the w_i maps) that the chaos game is *ergodic*, in the following sense:

Theorem 2.2.3: Given the contractive IFS $\{K, \mathbf{w}, \mathbf{p}\}$ with invariant measure μ, let $\{x_k\}_{k=0}^{\infty}$ denote the forward orbit of $x_0 \in K$ generated by a code $\sigma \in \sum_N$ in Eq. (2.2.7). Then for almost all code sequences $\sigma \in \sum_N$ (except for a set of measure zero)

$$\lim_{n\to\infty} \frac{1}{n+1} \sum_{k=0}^{n} f(x_k) = \int_K f(x) d\mu(x) \tag{2.2.10}$$

for all continuous and simple functions $f : K \to \mathbf{R}$.

This ergodic theorem now accounts for the fact that the chaos game can generate a picture not only of the attractor A but also of the invariant measure μ: Let f in Eq. (2.2.10) be the characteristic function χ_B for $B \subset \mathcal{B}(K)$, a Borel subset of K. Then

$$\mu(B) = \lim_{n\to\infty} \left[\frac{1}{n+1} \sum_{k=0}^{n} \chi_B(x_k) \right] . \tag{2.2.11}$$

In other words, $\mu(B)$ is the *relative visitation frequency* of B during the chaos game. For computer screen representations, B may be represented by a pixel $p(i,j)$ on the computer screen. In this way, a "histogram" approximation of the invariant measure may be obtained by counting the number of visits made to $p(i,j)$. For 2-D images, the visitation frequency may be colour coded to produce a colour representation of the measure.

In Fig. 2 are presented some histogram approximations for various IFS invariant measures on $[0,1]$, to show effects of (i) probabilities p_i and (ii) overlapping. The histograms were constructed using the random iteration algorithm. The interval $[0,1]$ was divided into 500 subintervals or "pixels". (The histograms are not normalized here.) The first two plots are associated with the maps of Example 2 above: $w_1(x) = \frac{1}{2}x$, $w_2 = \frac{1}{2}x + \frac{1}{2}$. Fig. 2(a): $p_1 = p_2 = \frac{1}{2}$, so that μ = uniform Lebesgue measure (this is a good check of the random number generator used for the "chaos game"). Fig. 2(b): $p_1 = \frac{2}{5}$, $p_2 = \frac{3}{5}$. Since p_2 is applied more often, the visitation by $\{x_n\}$ is shifted toward the point $x = 1$, but in an infinitely self-similar way over $[0,1]$. Fig. 2(c): the IFS $w_1(x) = .55x$, $w_2(x) = .55x + .45$, $p_1 = p_2 = \frac{1}{2}$. The effects of the overlapping of $w_1(A)$ and $w_2(A)$ on $[.45, .55]$ can be seen in the invariant measure.

2.3 Moment relations for linear IFS

If $\{K, \mathbf{w}, \mathbf{p}\}$ is a contractive IFS in $\mathbf{R}^n$, with attractor A and invariant measure μ, then we define the moments of μ as

$$g_{i_1 i_2 \ldots i_n} = \int_A x_1^{i_2} x_2^{i_2} \cdots x_n^{i_n} d\mu\,, \qquad g_{00\ldots 0} = \int_A d\mu = 1\,. \tag{2.3.1}$$

In most applications of the IFS method, including the inverse problem, the maps w_i used are linear, since the geometry (and dynamics) associated with such maps is quite simple. An additional advantage of linear maps is that the invariance property for integrals in Eq. (2.1.11) provides an explicit relation between moments which can be used to compute them recursively from the IFS parameters. We show this for IFS on $\mathbf{R}$ and $\mathbf{R}^2$ below.

IFS on R: Consider the following general family of linear maps on $\mathbf{R}$:

$$w_i(x) = s_i x + a_i\,, \quad |\, s_i \,| < 1\,, \quad s_i,\, a_i \in \mathbf{R}\,, \quad i = 1, 2, \ldots N, \tag{2.3.2}$$

with associated non-zero probabilities $p_i > 0$, $\sum_{i=1}^{N} p_i = 1$. From Eq. (2.1.11), setting $f(x) = x^n$, we have

$$g_n = \int x^n d\mu = \sum_{i=1}^{N} p_i \int (s_i x + a_i)^n \, d\mu \tag{2.3.3}$$

Expanding the polynomial, collecting like powers of x^n and integrating, we obtain the following well-known recursion relation

$$\left[1 - \sum_{i=1}^{N} p_i s_i^n\right] g_n = \sum_{j=1}^{n} \binom{n}{j} g_{n-j} \left(\sum_{i=1}^{N} p_i s_i^{n-j} a_i^j\right). \tag{2.3.4}$$

The coefficient of g_n in square brackets on the left cannot vanish. Thus, setting $g_0 = 1$, the moments g_n, $n \geq 1$, may be computed recursively from a knowledge of $g_0, \ldots, g_{n-1}$.

Examples:

1. $K = [0,1]$, $N = 1$, $w_1(x) = sx$, $w_2(x) = sx + (1 - s)$, $p_1 = p_2 = \frac{1}{2}$, where $0 \leq s < 1$. The first five moments are given symbolically by

$$g_0 = 1,\; g_1 = \frac{1}{2},\; g_2 = \frac{1}{2(1+s)},\; g_3 = \frac{2-s}{4(1+s)},\; g_4 = \frac{1+s^2-s^3}{2(1+s)(1+s+s^2+s^3)}\,.$$

Note some special cases:

(i) $s = \frac{1}{2} : A = [0,1]$, $\mu =$ uniform measure: $g_n = \int_0^1 x^n\,dx = \frac{1}{n+1}$.

(ii) $s = 0 : A = [0,1]$, $\mu = \frac{1}{2}(\delta_0 + \delta_1)$, where δ_a denotes point mass measure at $x = a$. $g_0 = 1$, $g_n = \frac{1}{2}$, $n = 1,2,\ldots$.

(iii) $s = \frac{1}{3} : A =$ ternary Cantor set on $[0,1]$, $\mu =$ uniform Cantor-Lebesgue measure [3, p. 77].

$$g_0 = 1,\; g_1 = \frac{1}{2},\; g_2 = \frac{3}{8},\; g_3 = \frac{5}{16},\; g_4 = \frac{87}{320},\; g_5 = \frac{31}{128},\; g_6 = \frac{10215}{46592}.$$

The derivatives of the moments with respect to the IFS parameters s_i, a_i, p_i can also be calculated recursively in closed form, by differentiating Eq. (2.3.4) implicitly. For $q = 1,2,\ldots,N$:

$$[1 - S(n,0)]\frac{\partial g_n}{\partial s_q} = \sum_{j=1}^{n}\binom{n}{j}\frac{\partial g_{n-j}}{\partial s_q}S(n,j) + \sum_{j=0}^{n}\binom{n}{j}(n-j)g_{n-j}p_q s_q^{n-j-1}a_q^j, \qquad (2.3.5a)$$

$$[1 - S(n,0)]\frac{\partial g_n}{\partial a_q} = \sum_{j=1}^{n}\binom{n}{j}\frac{\partial g_{n-j}}{\partial s_q}S(n,j) + \sum_{j=1}^{n}\binom{n}{j}j\,g_{n-j}p_q s_q^{n-j}a_q^{j-1}, \qquad (2.3.5b)$$

$$[1 - S(n,0)]\frac{\partial g_n}{\partial p_q} = \sum_{j=1}^{n}\binom{n}{j}\frac{\partial g_{n-j}}{\partial p_q}S(n,j) + \sum_{j=0}^{n}\binom{n}{j}g_{n-j}s_q^{n-j}a_q^j, \qquad (2.3.5c)$$

where

$$S(n,j) = \sum_{i=1}^{N} p_i s_i^{n-j}a_i^j.$$

This property of differentiability is a special case of a general result derived by Withers [65].

The moments g_n may thus be thought of as functions of the IFS parameters, i.e. $g_n = g_n(\mathbf{s}, \mathbf{a}, \mathbf{p})$, defined over a suitable parameter space Π. This parameter space may vary from problem to problem, depending on how we wish to scale the space $K \subset \mathbf{R}$. For example, if $K = [0,1]$, we may define Π as follows. Let $\pi \in \Pi \subset \mathbf{R}^{3N}$, where

$$\pi = (\pi_1,\ldots,\pi_{3N}) = (s_1,\ldots,s_N, a_1,\ldots,a_N, q_1,\ldots,q_n) \qquad (2.3.6)$$

is a point in the open and bounded subset $\Pi \subset \mathbf{R}^{3N}$ determined by the conditions

$$0 < s_i < 1, \qquad 0 < a_i,\; s_i + a_i < 1\,, \qquad 0 < q_i < 1\,, \qquad i = 1,2,\ldots,N. \qquad (2.3.7)$$

The first condition arises from contractivity of the $w_i(x)$; the second set insures that $w_i : [0,1] \to [0,1]$. The q_i in the third set are not exactly the probabilities: we refrain from immediately imposing the condition $\sum q_i = 1$ to keep the subspace Π open. It then follows that the moment functions $g_n : \Pi \to \mathbf{R}$ are continuous and differentiable in Π.

Special Case: Homogeneous IFS on R

The moment recursion relations in (2.3.4) simplify in form for the case of *homogeneous* IFS [32], where $s_i = s$ for $i = 1, \dots, N$ in Eq. (2.3.3), i.e. $w_i(x) = sx + a_i$. Eq. (2.3.4) then becomes

$$(1 - s^n)g_n = \sum_{j=1}^{n} \binom{n}{j} g_{n-j} s^{n-j} \sigma_j, \qquad \sigma_j = \sum_{i=1}^{N} p_i a_i^j, \tag{2.3.8}$$

and σ_j can be considered as the jth moment of a discrete measure composed of point masses at a_j with weights p_i, $i = 1, 2, \dots N$.

This case has been examined in the context of the inverse problem [1, 20, 32, 42]: given the moments g_n, determine the map parameters s and a_i, $i = 1, 2, \dots, N$. For a fixed $|s| < 1$, and known moments $g_0, g_1, \dots, g_n$, one can determine $\sigma_0, \sigma_1, \dots, \sigma_N$ from the linear system in Eq. (2.3.8). In principle, the a_i can then be found from the "moments" σ_n by classical methods.

IFS on $\mathbf{R}^2$: Now consider general IFS as defined by the affine transformations

$$w_i \begin{pmatrix} x \\ y \end{pmatrix} = \begin{bmatrix} a_{11}(i) & a_{12}(i) \\ a_{21}(i) & a_{22}(i) \end{bmatrix} \begin{pmatrix} x \\ y \end{pmatrix} + \begin{pmatrix} b_1(i) \\ b_2(i) \end{pmatrix}, \quad i = 1, 2, \dots N \tag{2.3.9}$$

with associated probabilities p_i. Eq. (2.3.9) will be written in the compact form

$$w_i(\mathbf{x}) = \mathbf{A}_i x + \mathbf{b}_i , \tag{2.3.10}$$

The matrices $\mathbf{A}_i$ must be contractive. The parameter space, Π, will now be a suitably defined open subset of $\mathbf{R}^{7N}$.

Setting $f(\mathbf{x}) = x^m y^n$ in Eq. (2.1.11), we have the following relation for the moments of the attractor:

$$\begin{aligned} g_{mn} &= \int_A x^m y^n d\mu \\ &= \sum_{i=1}^{N} p_i \int_A [a_{11}(i)x + a_{12}(i)y + b_1(i)]^m \, [a_{21}(i)x + a_{22}(i)y + b_2(i)]^n \, d\mu . \end{aligned} \tag{2.3.11}$$

Expanding the polynomials and integrating yields a complicated recursion relation for the g_{mn}:

$$g_{mn} = \sum_{k=1}^{N} \left[\sum_{i_1=0}^{m} \binom{m}{i_1} a_{11}(k)^{i_1} \sum_{i_2=0}^{m-i_1} \binom{m-i_1}{i_2} a_{12}(k)^{i_2} b_1(k)^{m-i_1-i_2} \right] \times \tag{2.3.12}$$

$$\left[\sum_{j_1=0}^{n}\binom{n}{j_1}a_{21}(k)^{j_1}\sum_{j_2=0}^{n-j_1}\binom{n-j_1}{j_2}a_{22}(k)^{j_2}b_2(k)^{n-j_1-j_2}g_{i_1+j_1,i_2+j_2}\right].$$

(Note: we draw the reader's attention to a typographical error in this relation as it appeared in Eq. (3.10) in Ref. [60]. Only one moment should appear on the RHS. The error has been corrected above.) Because of the cross terms occuring in the products of Eq. (2.3.12), we can not solve for each g_{mn} explicitly, but must proceed as follows: From $g_{00} = 1$, write down the two equations of (2.3.12) corresponding to $m = 0$, $n = 1$ and $m = 1$, $n = 0$, then solve simultaneously for the unknowns g_{01} and g_{10}. The procedure is then continued: solve a system of $M + 1$ linear inhomogeneous equations in the unknowns $\{g_{0M}, g_{1,M-1}, \dots, g_{M0}\}$ from a knowledge of the previously computed g_{mn}, $m + n < M$.

The derivatives of the moments with respect to the IFS parameters $a_{ij}(k)$, $b_i(k)$ and p_i, may also be computed in a recursive, albeit complicated manner.

2.4 IFS in discrete "pixel" space:

In this section, we consider IFS on a discrete space $\overline{K}$ as would be relevant, for example, to the study of images represented by pixels on a computer video terminal. The "bar" symbol will be used to distinguish terms in discrete case from their continuous counterparts. Consider the general space $\overline{K} \subset \mathbf{Z}^n$, i.e. $\overline{K} \doteq \{(i_1, i_2, \dots, i_n) \mid 1 \le i_k \le n_k, k = 1, 2, \dots n\}$, with (Manhattan) metric

$$\overline{d}(i,j) = \sum_{k=1}^{n} \mid i_k - j_k \mid . \tag{2.4.1}$$

$(\overline{K}, \overline{d})$ is a compact metric space. the (discrete) Hausdorff distance $\overline{h}(A, B)$ between two sets $A, B \in \overline{K}$ can then be defined as in Eq. (2.1.2): the ϵ-ball, $\epsilon = 1, 2, \dots$ of a set A is obtained by surrounding all "pixels" of A with ϵ layers of neighbouring pixels. We also define the space $\overline{X}_1$,

$$\overline{X}_1 = \{\overline{\mu} : \mathbf{Z}^n \rightarrow \mathbf{R} \mid \sum_{i_1, \dots, i_n}^{n_1, \dots, n_n} \overline{\mu}(i_1, \dots, i_n) = 1\}. \tag{2.4.2}$$

For $n = 2$, the value $\overline{\mu}(i_1, i_2)$ may represent, for example, a grey scale for the pixel $p(i_1, i_2)$ in a black-and-white image on the screen. Or, it could be the relative visitation of pixel $p(i_1, i_2)$ during a "chaos game" generation of an IFS attractor. A metric $\overline{d}_H(\mu, \nu)$ on $\overline{X}_1$ analogous to the continuous case in Eq. (2.1.3) can defined: the integration becomes a summation over all indices, and the Lip-1 property translates to $\mid f(\mathbf{i}) - f(\mathbf{j}) \mid \le 1$, where $\mathbf{i}$ and $\mathbf{j}$ represent neighbouring sites, i.e. $\overline{d}(\mathbf{i}, \mathbf{j}) = 1$.

The IFS formulation in discrete space is obviously practical, since images (including targets for the inverse problem) are usually presented in discretized form. It is more important, however, to realize that *only* on discrete spaces can distances between sets or measures be calculated practically. The accuracy can be improved by refining the grid, i.e. by increasing the number of pixels. For example, R. Shonkwiler [57] has devised fast and simple algorithms to compute Hausdorff distances in $\mathbf{Z}^n$.

Recently, a fast and extremely simple algorithm to compute the distance between measures $\overline{d}_H(\overline{\mu},\overline{\nu})$ in one dimension has been found [23]. It requires only one scan of the pixel array. Let $\overline{\mu}(i)$ and $\overline{\nu}(i)$, denote the measures on pixels $p(i)$, $i = 1, 2, \ldots N$, and S_k denote the partial sums

$$S_k = \sum_{i=1}^{k} (\overline{\mu}(i) - \overline{\nu}(i)) \, . \tag{2.4.3a}$$

Then,

$$d_H(\overline{\mu},\overline{\nu}) = \sum_{k=1}^{N-1} \mid S_k \mid \, . \tag{2.4.3b}$$

At present, the two-dimensional problem appears to be much harder, due to the extra conditions introduced by the Lipshitz property in 2D.

2.4.1 Formulation of Discrete IFS on [0,1]

In the discussions below, consider the following discretization of $K = [0,1]$: Divide $[0,1]$ into $\overline{n}$ subintervals $I_k = [(k-1)/\overline{n}, k/\overline{n})$, $k = 1, 2, \ldots, \overline{n}-1$, and $I_{\overline{n}} = [(\overline{n}-1)/\overline{n}, 1]$. (This construction permits the assignment of a point $x \in [0,1]$ to a unique interval I_k.) Then $\overline{K} = \{1, 2, \ldots, \overline{n}\}$ with metric $\overline{d}(i,j) = \mid i - j \mid$. Define $\overline{X}_1 = \{(\overline{\mu}(1), \ldots, \overline{\mu}(\overline{n})) : \overline{\mu}(i) \geq 0, \sum \overline{\mu}(i) = 1\}$. For a given $\nu \in X_1$, define its discrete counterpart $\overline{\nu} \in \overline{X}_1$, such that $\overline{\nu}(k) = \nu(I_k)$, $k = 1, 2, \ldots, \overline{n}$. There are two possibilities:

(i) *Discretization of Attractors/Invariant Measures:* Run either the *Deterministic* or *Random* algorithm on $K = [0,1]$ to generate or an approximation of either the attractor A or the invariant measure μ of the IFS. A point x will belong to the interval I_k for which $(k-1)/\overline{n} \leq x < k/\overline{n}$, *i.e.* $n\overline{x} < k \leq \overline{n}x + 1$ or $k = int(n\overline{x} + 1)$. The discrete measure $\overline{\mu}$ corresponding to the invariant measure μ will be approximated by the visitation frequency in Eq. (2.2.11).

(ii) *Define a "discrete" IFS over* $\overline{K}$: Let $\overline{x}_k = (k - \frac{1}{2})/\overline{n}$ denote the midpoint of I_k. For a contractive IFS $\{[0,1], w_i, p_i, i = 1, 2, \ldots N\}$, define an associated "discrete" IFS $\{\overline{K}, \overline{w}_i, p_i\}$, with maps $\overline{w}_i : \overline{K} \to \overline{K}$, whose action is defined as $\overline{w}_i(j) = min\{int(\overline{n}w_i(\overline{x}_j) + 1), \overline{n}\}$. (In other words, treat the subinterval I_j (and measure $\overline{\nu}(j)$) as if it were centered at $\overline{x}_j$.)
An associated "discrete" Markov operator, $\overline{M} : \overline{X}_1 \to \overline{X}_1$, can also be constructed: First, define the array $k(i,j) = \overline{w}_i(j)$, $i = 1, \ldots N, j = 1, \ldots \overline{n}$. For a given $\overline{\mu} \in \overline{X}_1$, compute $\overline{\nu} = \overline{M}(\overline{\mu})$ using the following algorithm written

in pseudocode, initializing $\overline{\nu}(i) = 0, \;\; i = 1, \ldots, \overline{n}$:

```
for i from 1 to N do
for j from 1 to n̄ do
ν̄(k(i,j)) = ν̄(k(i,j)) + p_i μ(j)
od;
od;
```

This particular discretization method is very convenient for image representations. In one dimension, the construction of invariant measures by repeated application of the Markov operator $\overline{M}$ is faster and more reliable than the "chaos game".

3 Some particular examples, applications and problems

The second lecture in this series was devoted to some examples and applications, as well as results of some preliminary studies. In order to keep the length of this section reasonable, I have decided to outline, with references, some examples in Section 3.1. The relation between IFS and tiles associated with complex bases is briefly discussed in Section 3.2. Finally, Sections 3.3 and 3.4 outline some preliminary studies of orthogonal polynomials over IFS invariant measures and "missing moment" problems, respectively.

3.1 Special examples

1. The one-complex parameter family of IFS $\{K_1, T_+, T_-\}$ on $K \subset \mathbf{C}$ compact, given by [10]

$$T_+(z) = sz + 1, \qquad T_-(z) = sz - 1, \qquad s \in \mathbf{C}, \ |s| < 1. \tag{3.1.1}$$

The attractor is given by

$$A(s) = \pm 1 \pm s \pm s^2 \pm \cdots \qquad \text{for all sequences of} \quad +, -. \tag{3.1.2}$$

In the *parameter space* $s \in \mathbf{C}$, there is an associated "Mandelbrot set",

$$M = \{s \in \mathbf{C} : |s| < 1, \qquad A(s) \quad \text{is disconnected}\}. \tag{3.1.3}$$

The boundary of M, ∂M must lie in the annular region $\frac{1}{2} \le s \le \frac{1}{\sqrt{2}}$. Let $A_\pm = T_\pm(A)$. Then if $A_+ \cap A_- = \emptyset$ then A is totally disconnected. If $A_+ \cap A_- \neq \emptyset$ then A is connected. Using this fact, a computer approximation to M was made. The boundary ∂M exhibits complicated, fractal-like structure. Hardin [43] devised an alternate method which keeps track of appropriate overlaps of circles obtained by applying sequences of $T_\pm$ to a circle C_R, with center $z = 0$ and radius R which encloses $A(s)$. See also [16, §8.1, 8.2].

2. The following variation of the IFS in (3.1.1), studied by Hardin [43, 15]:

$$H_+(z) = sz + 1, \qquad H_-(z) = s^* z - 1. \tag{3.1.4}$$

The boundary of the associated Mandelbrot set, defined as in (3.1.3), changes dramatically in nature: it is piecewise smooth .

3. A generalization of the IFS in 1. and 2. from *similitudes* (dilatation + rotation + translation) to affine maps (shear added) in $\mathbf{R}^2$ [59]:

$$S_\pm : \begin{bmatrix} x \\ y \end{bmatrix} \to \begin{bmatrix} a_{11} & a_{12} \\ a_{21} & a_{22} \end{bmatrix} \begin{bmatrix} x \\ y \end{bmatrix} \pm \begin{bmatrix} 1 \\ 0 \end{bmatrix}, \tag{3.1.5}$$

where $\mathbf{A}$ is a contractive matrix. This represents a 4 real-parameter family of IFS. Computer approximations to various "slices" of M are presented in [59].

The algorithm used follows that employed by Hardin, with the exception that it checks for overlaps of *ellipses*.

4. *Julia Sets of Rational Maps as Attractors of IFS.*
   Let $\overline{\mathbf{C}} = \mathbf{C} \cup \{\infty\}$ denote the Riemann sphere with suitably defined spherical metric and $R : \overline{\mathbf{C}} \to \overline{\mathbf{C}}$ be a rational function, $R(z) = P(z)/Q(z)$, where $P(z)$ and $Q(z)$ are polynomials with complex coefficients and no common factors, and $d = \deg(R) \equiv \max(\deg(P), \deg(Q)) \geq 2$. Define the sequence of iterates $\{R^n(z)\}$ as $R^0(z) = z$, $R^{n+1}(z) = R(R^n(z))$, and denote the inverses of $R(z)$ as $R_i^{-1}(z)$, where $i = 1, 2, \ldots d$ enumerates all branches. The *Julia set* of $R(z)$, denoted $J(R)$, can be defined as the closure of the set of all *repulsive* k-cycles of $R(z)$, $k = 1, 2, 3, \ldots$. (For a comprehensive review of complex analytic dynamics and Julia-Fatou theory, see Blanchard [22]. Devaney [28] provides an excellent introduction to the subject, and the dynamical systems theory behind it.)

   Examples:

   (i) $R(z) = z^2 : J(R) = C = \{z : \mid z \mid = 1\}$. C is the boundary between the *basins of attraction* $W(0)$ and $W(\infty)$ under forward iteration of $R(z)$, and may be regarded as a repeller set.

   (ii) $R(z) = z^2 - 2 : J(R) = [-2, 2] \subset \mathbf{R}$.

   The Julia set $J(R)$ may then be thought of as an attractor for the IFS $\{\overline{\mathbf{C}}, R_i^{-1}(z), i = 1, 2, \ldots d\}$. For the one-complex parameter family of complex maps $R_s(z) = z^2 - s$, let $w_1(z) = R_s^{-1}(z) = \sqrt{z+s}$, $w_2(z) = R_s^{-1}(z) = -\sqrt{z+s}$, where $\sqrt{z}$ denotes the principal square root of z. Then

$$J(R_s) = \sqrt{s \pm \sqrt{s \pm \sqrt{s \pm \cdots}}} \qquad \text{for all sequences of } +, -. \tag{3.1.6}$$

   (Technically, these IFS are not contractive. However, in the case $p_i = d^{-1}$, $i = 1, 2, \ldots, d$, the IFS invariant measure μ is identical to the balanced "electrostatic measure" invariant with respect to $R(z)$, and supported on $J(R)$ [13].) See also [16, §7.2].

5. *The Escape-Time Algorithm for IFS* [16, §7.1] which is motivated by the well-known procedure to picture Julia sets and the "equipotential lines" which surround them. The repelling nature of Julia sets is the basis for the beautiful pictures which can be found, for example, in Ref. [54, 55]. These references also contain discussions of the Julia set algorithm and underlying theory.

   The attractor of an IFS $\{K, w_i, i = 1, 2, \ldots N\}$ can be considered a *repeller set* under the action of the inverse maps $w_i^{-1}(z)$. An Escape-Time Algorithm can then be defined as follows: Let C_R denote a circle (square, polygon) of radius R sufficiently large to define an escape region. For any $x \in K$, define

$$\mathbf{w}^{-1}(x) = \bigcup_{i=1}^{N} w_i^{-1}(x), \qquad \mathbf{w}^{-(n+1)}(x) = \mathbf{w}^{-1}(\mathbf{w}^{-n}(x)), \quad n = 1, 2, \ldots .$$

Define the Escape Time $n(x)$ of x as follows: $n(x)$ is the minimum value of $k \in \mathbf{Z}^+$ for which all (N^k) points in $\mathbf{w}^{-k}(x)$ lie outside C_R. (This amounts to travelling down each path of an N-tree, stopping when a particular sequence $w_{i_1} \circ w_{i_2} \circ \cdots \circ w_{i_j}$ sends x outside C_R.) An Escape-Time plot for the Sierpinski gasket of Figure 1(a) is given in Figure 3. The circle C_R had radius 2 and was centered at $\left(\frac{1}{2}, \frac{\sqrt{3}}{4}\right)$. Points plotted (black, white) have escape times $n(x)$ which are (even, odd).
This algorithm could be advantageous when a magnification of a subset $B \subset A$ is desired. Suppose the "Chaos game" is used. As the area of the region in $\mathbf{R}^2$ enclosing B decreases, then so does the number of visits to B, implying that a longer time is required to produce a suitable "picture" of B. It should be mentioned that there are various other possible definitions for the escape time $n(x)$. Also, the particular nature of the IFS maps could be exploited to choose only particular maps $w_1^{-1}(x)$ depending on the position of x. For more discussion, see [56].

3.2 IFS attractor sets derived from complex bases

The IFS in Example 1 above, along with generalizations, will be seen to be relevant to the study of sets in the complex plane which are spanned by complex bases. The treatment below follows the work of Gilbert who studied the boundaries of tiling sets obtained from complex bases, including their fractal dimensions [36], and later expressed these tiling sets as attractors of appropriate IFS [37, 38]. He has also related the long division algorithm for complex bases to the Escape-Time algorithm for IFS [37].

A complex number z is said to be written in (complex) base $s \in \mathbf{C}$ if it may be expressed in the form

$$z = \sum_{j=-\infty}^{q} r_j s^j , \tag{3.2.1}$$

where q is a positive integer, and the *digits* r_j belong to an appropriate *digit set* $\mathcal{D}_s$.

This representation of z will be written as $(r_q r_{q-1} \cdots r_0 \cdot r_{-1} r_{-2} \cdots)_s$. The number $(r_q r_{q-1} \cdots r_0)_s$ to the left of the *radix point* constitutes the *integer part* of the expression. Katai and Szabo [49] showed that for the case $s = -n + i$, $n > 0$, every $z \in \mathbf{C}$ possesses a (not necessarily unique) expansion of the form in (3.2.1), with the digit set $\mathcal{D}_{-n+i} = \{0, 1, 2, \cdots, n^2\}$.

The study of complex base expansions has a geometric aspect which can be naturally expressed in terms of IFS. For a given base $s = -n + i$, consider the expansion in (3.2.1) for zero integer part, that is, the set of points given by

$$A(s) = \sum_{j=-\infty}^{-1} r_j s^j = \sum_{k=1}^{\infty} a_k s^{-k} \qquad (a_k = r_{-k} \in \mathcal{D}_{-n+i}) \tag{3.2.2}$$

for all possible sequences $a = (a_1, a_2, a_3 \ldots)$, $a \in \{0, 1, 2, \ldots, n^2\}$, i.e. $a \in \sum_N$ where $N = n^2 + 1$ ($\sum_N$ defined in Section 2.2). Then $A(s)$ is the attractor for the contractive IFS $\{w_1, w_2, \ldots, w_N\}$, where

$$w_k(z) = s^{-1}z + (k-1)s^{-1}, \qquad k = 1, 2, \ldots, N = n^2 + 1. \tag{3.2.3}$$

Addition of a given integer part $(r_q r_{q-1} \cdots r_0)_s$, to $A(s)$ in Eq. (2) merely translates $A(s)$ in the complex plane by a Gaussian integer $k_1 + k_1 i = \sum_{j=0}^{q} r_j(-n+i)^j$; $k_1, k_2 \in \mathbf{Z}$. The Katai-Szabo result implies that the set of all translations of $A(s)$ covers or *tiles* the complex plane $\mathbf{C}$. The set $A(s)$ is referred to as the *base tile*. Attractors $A(s)$ corresponding to the base tiles $-1+i, -2+i, -3+i$ are shown in Fig. 4(a-c). (Note that the base tile for $s = -1 + i$ is given a space filling twin dragon region [16, p. 312], [51, §4.1], [52, p. 67].) These sets can be compared to the tiles originally constructed by Gilbert [36] by a sequence of successive geometric approximations.

It is then natural to ask which other number fields, with appropriate digit sets, can provide valid bases for the complex numbers. If the digit set is reduced to be a set of natural numbers, then the set of possible bases is reduced. The result of Katai and Szabo is, in fact, stronger: Let β be a Gaussian integer, $\beta = k_1 + k_2 i$, and define its norm, Norm $(\beta) = k_1^2 + k_2^2 = N$, and let $\mathcal{D}_\beta = \{0, 1, 2, \ldots, N-1\}$. Then β is a valid base for $\mathbf{C}$ using $\mathcal{D}_\beta$ as digit set iff $\beta = -n + i$ for some positive integer n. A theorem for quadratic number fields using natural numbers as digits also exists. The reader is referred to [38], where such questions are discussed and other interesting examples are given.

In general, if $\beta \in \mathbf{C}$ is a valid base with the digit set $\mathcal{D}_\beta = \{d_1, d_2, \ldots d_N\}$, then, by analogy to Eq. (3.2.2), the *base tile* is given by the attractor $A(\beta)$ of the IFS defined by the maps

$$w_i(z) = \beta^{-1}z + d_i\beta^{-1}, \qquad i = 1, 2, \ldots, N.$$

The set of all translations of $A(\beta)$ formed by the integer parts will tile the complex plane. The points lying in those intersections for which $w_i(A) \cap w_j(A) \neq \emptyset, i \neq j$ do not possess unique expansions in the base β. It may also be possible that different translations of $A(\beta)$ may cover the same point.

We conclude this section with an intriguing example, given in [38]. The number $2 + i$ is a valid base with the digit set $\{0, 1, \pm i, -2 - 3i\}$. The corresponding base tile $A(\beta)$ is shown in Fig. 4(e). It is not connected, but has unit area and tiles the plane by translation along the Gaussian integers. (The reader is invited to test this latter statement with a sufficient number of copies of $A(\beta)$ on transparent plastic sheets.) This tile may be compared with the tile for the *same* base, $2+i$, but using the digit set $\{0, \pm 1, \pm i\}$, shown in Fig. 4(d).

3.3 Orthogonal polynomials relative to an IFS invariant measure

The recursive computation of the moments g_n for the invariant measure μ of a linear IFS permits a practical construction of orthogonal polynomials relative to μ. We outline the aspects of this computation below. A motivation for this study is work done on orthogonal polynomials over invariant measures supported on Julia sets of polynomial maps [5, 6, 7]. An open question – the asymptotic behaviour of recurrence coefficients – is mentioned at the end. The notation employed below follows (at least closely) that found in Wall's classic book on continued fractions [63]. The reader is also referred to the classic treatise by Szego [58].

3.3.1 Practical computation and examples

Consider a linear contractive IFS on $K \subset \mathbf{R}$, with maps $w_i(x) = s_i x + a_i$ and non-zero probabilities $p_i > 0$. Let μ denote the invariant measure of this IFS, $supp(\mu) = A$, with moments $g_n = \int x^n d\mu$. For ease of notation, we shall write $\langle f \rangle = \int f(x)\, d\mu(x)$. Define the following sequence of monic polynomials $P_n(x)$ of degree n, $n = 0, 1, 2, \ldots$ such that $P_0(x) = 1$, and

$$P_n(x) = x^n + c_{n,n-1}x^{n-1} + \cdots + c_{n1}x + c_{n0}\,, \qquad n = 1, 2, \ldots\,, \tag{3.3.1}$$

satisfy the following orthogonality relations over μ:

$$\langle P_m P_n \rangle = \int P_m(x) P_n(x)\, d\mu = N_n \delta_{mn}\,, \tag{3.3.2}$$

where $\delta_{mn} = 0$ if $m \neq n$, $\delta_{nn} = 1$. These relations are equivalent to the conditions $\langle x^m P_n \rangle = 0$, $m = 0, 1, 2, \ldots, n-1$. For a given $n > 0$, these latter conditions yield a system of n linear inhomogeneous equations for the unknowns c_{ni}, $i = 0, 1, \ldots, n-1$. A necessary condition for the existence of a unique solution to this system is that the following determinant be non-zero:

$$D_n = \begin{vmatrix} g_0 & g_1 \cdots & g_{n-1} \\ g_1 & g_2 \cdots & g_{n-2} \\ \vdots & & \\ g_{n-1} & g_n \cdots & g_{2n-2} \end{vmatrix} \neq 0\,. \tag{3.3.3}$$

A compact determinantal formula then exists for the polynomials $P_n(x)$:

$$P_n(x) = \frac{1}{D_n} \begin{vmatrix} g_0 & g_1 & \cdots & g_n \\ g_1 & g_2 & \cdots & g_{n+1} \\ \vdots & & & \vdots \\ g_{n-1} & g_n & \cdots & g_{2n-1} \\ 1 & x & \cdots & x^n \end{vmatrix} \tag{3.3.4}$$

where we also set $D_0 = 1$ so that this expression is valid for $n = 0$. In principle, this relation could be used to compute the $P_n(x)$, although the computations become tedious and unstable (in decimal arithmetic) as n becomes large.

The following three-term recurrence relation may be shown to hold:

$$P_n(x) = (b_n + x)\, P_{n-1}(x) - a_{n-1}\, P_{n-2}(x)\,, \qquad n = 2, 3, \ldots, \tag{3.3.5}$$

where

$$a_{n-1} = \frac{\langle x P_{n-1} P_{n-2} \rangle}{\langle P_{n-2} P_{n-2} \rangle}, \qquad b_n = -\frac{\langle x P_{n-1} P_{n-1} \rangle}{\langle P_{n-1} P_{n-1} \rangle}. \tag{3.3.6}$$

This relation, along with the initial values $a_0 = 1$, $P_0(x) = 1$, $P_1(x) = x - g_1$, defines a recursive procedure to compute the $P_n(x)$. In such schemes, roundoff error accumulates quickly, so computations should be performed in rational arithmetic if possible, or very high precision decimal arthmetic.

Examples: We return to the 2-map IFS examples on [0,1] given in Section 2.3: $w_1(x) = sx$, $w_2(x) = sx + (1 - s)$, $p_1 = p_2 = \frac{1}{2}$, $0 \le s < 1$.

	$s = \frac{1}{2}$	$s = \frac{1}{3}$	$s = \frac{1}{4}$
$P_0(x)$	1	1	1
$P_1(x)$	$x - \frac{1}{2}$	$x - \frac{1}{2}$	$x - \frac{1}{2}$
$P_2(x)$	$x^2 - x + \frac{1}{6}$	$x^2 - x + \frac{1}{8}$	$x^2 - x + \frac{1}{10}$
$P_3(x)$	$x^3 - \frac{3}{2}x^2 + \frac{3}{5}x - \frac{1}{20}$	$x^3 - \frac{3}{2}x^2 + \frac{23}{40}x - \frac{3}{80}$	$x^3 - \frac{3}{2}x^2 + \frac{48}{85}x - \frac{11}{340}$

Another method of computation which can be quite convenient is based on the intimate connection between orthogonal polynomials and continued fractions. Consider the following infinite *J-fraction* [63, Chap. IX]

$$J(z) = \frac{a_0}{b_1 + z} \; - \; \frac{a_1}{b_2 + z} \; - \; \cdots . \tag{3.3.7}$$

When we set $a_n = 0$, the resulting truncation of $J(z)$, the nth *convergent* or *approximant* to $J(z)$, is a rational function of z, which obeys the following recursive structure:

$$w_n(z) = \frac{A_n(z)}{B_n(z)} = \frac{(b_n + z)A_{n-1}(z) + a_{n-1}A_{n-2}(z)}{(b_n + z)B_{n-1}(z) + a_{n-1}B_{n-2}(z)}. \tag{3.3.8}$$

The $A_n(z)$ and $B_n(z)$ are polynomials which have the starting values $A_{-1}(z) = 1$, $A_0(z) = 0$, $B_{-1}(z) = 0$, $B_0(z) = 1$. Thus the denominators $B_n(z)$ are the orthogonal polynomials $P_n(x)$. It can be shown [63, Chap. XI] that $J(z)$ in Eq. (3.3.7) is the J-fraction expansion for the following power series

$$S(z) = \sum_{k=0}^{\infty} g_k z^{-(k+1)}, \tag{3.3.9}$$

whose coefficients g_k are the moments of μ ($S(z)$ is the power series expansion of the *moment generating function* $G(z) \equiv \int (z-t)^{-1} d\mu$ which is analytic on $\mathbf{C} - supp(\mu) = \mathbf{C} - A$.) The recurrence coefficients a_i and b_i may then be calculated via algorithms which compute J-fraction expansions corresponding to power series. (The *normality* condition on the series $S(z)$, $D_n \neq 0$ for $n = 1, 2, \ldots$ in (3.3.3), is necessary for the existence of a J-fraction expansion.)

In the special case where $g_{2n+1} = 0$, $n = 1, 2, \ldots$ (due, for example, to symmetry of μ with respect to inversion about $x = 0$) we also have $b_n = 0 : P_n(x)$ is (even, odd) if n is (even, odd). After a slight modification, the J-fraction may be written as

$$C(z) = \frac{a_0/z}{1} - \frac{a_1/z^2}{1} - \frac{a_2/z^2}{1} - \cdots, \tag{3.3.10}$$

which is an S-fraction [63, Chap. VI]. The quotient-difference (QD) algorithm, described beautifully and completely by Henrici [45] may be used to compute the a_i from the g_k. (Again, rational arithmetic or high precision is necessary.)

3.3.2 Orthonormal polynomials and associated Jacobi matrices

An interesting and open problem involves the elucidation of the asymptotic behaviour of the recurrence coefficients a_n, b_n in Eq. (3.3.5). In order to see the connection with symmetric Jacobi matrices, we define the corresponding sequence of orthonormal polynomials

$$\overline{P}_n(x) = P_n(x)/N_n, \tag{3.3.11}$$

where $N_n = \langle P_n P_n \rangle^{1/2}$ so that $\langle P_m P_n \rangle = \delta_{mn}$. Now substitute (3.3.11) into Eq. (3.3.5) to give a new three-term recurrence relation for the $\overline{P}_n(x)$:

$$\overline{a}_n \overline{P}_n(x) + \overline{b}_n \overline{P}_{n-1}(x) + \overline{c}_{n-1} \overline{P}_{n-2}(x) = x \overline{P}_{n-1}(x), \tag{3.3.12}$$

where $\overline{a}_n = N_n/N_{n-1}$, $\overline{c}_n = a_n N_{n-1}/N_n$. Note that multiplying (3.3.12) by the appropriate $\overline{P}_k(x)$ and integrating over μ gives (i) $\overline{a}_n = \langle x\overline{P}_{n-1}\overline{P}_n \rangle$, (ii) $\overline{b}_n = \langle x\overline{P}_{n-1}\overline{P}_{n-1} \rangle$, (iii) $\overline{c}_{n-1} = \langle x\overline{P}_{n-1}\overline{P}_{n-2} \rangle = \overline{a}_{n-1}$. Thus Eq. (3.3.12) assumes the more symmetric form

$$\overline{a}_{n-1} \overline{P}_{n-2}(x) + \overline{b}_n \overline{P}_{n-1}(x) + \overline{a}_n \overline{P}_n(x) = x\overline{P}_{n-1}(x). \tag{3.3.13}$$

Moreover, note that since $\overline{c}_n = \overline{a}_n$, then $\overline{a}_n \overline{c}_n = (\overline{a}_n)^2 = a_n$; also $\overline{b}_n = -b_n$, thus providing the connections between the respective coefficients in Eqs. (3.3.5) and (3.3.13).

If we define the infinite-dimensional vector $\psi^T = (\overline{P}_0, \overline{P}_1, \overline{P}_2, \ldots)$, then Eq. (3.3.13) may be written as the operator equation

$$J\psi = x\psi, \tag{3.3.14}$$

where J has the symmetric tridiagonal representation (Jacobi matrix)

$$J = \begin{bmatrix} \overline{b}_0 & \overline{a}_1 & 0 & \cdots \\ \overline{a}_1 & \overline{b}_1 & \overline{a}_2 & \cdots \\ 0 & \overline{a}_2 & \overline{b}_2 & \cdots \\ \vdots & \vdots & \vdots & \end{bmatrix} = \begin{bmatrix} -b_0 & \sqrt{a_1} & 0 & \cdots \\ \sqrt{a_1} & -b_1 & \sqrt{a_2} & \cdots \\ 0 & \sqrt{a_2} & -b_2 & \cdots \\ \vdots & \vdots & \vdots & \end{bmatrix}. \tag{3.3.15}$$

Following the theory outlined in Akhiezer [2, Sect. 4], since the moment problem on [0,1] is determinate, then the Jacobi matrix in (3.3.15) has a unique self-adjoint extension $\overline{J}$ with domain in ℓ_2, the Hilbert space of square-summable sequences. Moreover, the spectrum of $\overline{J}$ is $A = supp(\mu)$, and its spectral density is μ.

Some cases where μ is a balanced invariant measure supported on the Julia set $J(R)$ of a polynomial map $R : \mathbf{C} \to \mathbf{C}$ have been studied [6, 21]. For example, when $R(z) = z^3 - \lambda z$ with $\lambda \geq 3$, $J(R)$ is a Cantor-like set on the real line. The coefficients $b_k = 0$ and the $\overline{a}_k$ satisfy an "infinite memory" recursion property which implies a kind of limit periodicity. The corresponding problem for the case of linear IFS attractors, for example, the asymptotic behaviour of the $\overline{a}_n$ and $\overline{b}_n$ for orthogonal polynomials on the ternary Cantor set, is not known.

Before the results of some computations are presented, it is worthwhile to point out a physical motivation for studing the spectral properties of Jacobi matrices such as in (3.3.14-15). The eigenvalue equation in (3.3.14) can be pictured as arising from a discretized version of the quantum mechanical Schrödinger equation on a one-dimensional semi-infinite lattice of atoms. The b_i represent a potential energy of interaction between an electron and its "home" site, while the off-diagonal elements $\overline{a}_i$ represent "nearest neighbour" interactions. Corresponding to an energy eigenvalue x (if it exists) is an ℓ_2-wavefunction $(P_1(x), P_2(x), \ldots)^T$. It is of interest to elucidate the nature of the spectrum of the operator J for various properties of the potential, i.e. periodicity, almost-periodicity, randomness of the $\overline{b}_i, \overline{a}_i$.

Examples: Two-map IFS on $[-1,1]$ with $w_1(z) = s(z+1)-1$, $w_2(z) = s(z-1)+1$, $p_1 = p_2 = \frac{1}{2}$. By symmetry, all odd moments vanish, and $b_k = 0$. The recurrence coefficients a_0 to a_4 of Eq. (3.3.5) are given on the following page for $s = \frac{1}{2}, \frac{1}{3}, \frac{1}{4}$:

	$s=\frac{1}{2}$	$s=\frac{1}{3}$	$s=\frac{1}{4}$
a_0	1	1	1
a_1	$\frac{1}{3}$	$\frac{1}{2}$	$\frac{3}{5}$
a_2	$\frac{4}{15}$	$\frac{1}{5}$	$\frac{12}{85}$
a_3	$\frac{9}{35}$	$\frac{333}{910}$	$\frac{3753}{7735}$
a_4	$\frac{16}{63}$	$\frac{15286}{138047}$	$\frac{686256}{16253965}$

For the case $s=\frac{1}{2}$, the $P_n(x)$ are the (monic) Legendre polynomials on [-1,1] (μ = uniform Lebesgue measure), and $a_n = n^2[(2n-1)(2n+1)]^{-1}$. Note that $a_n \to \frac{1}{4}$ as $n \to \infty$. The coefficients a_k, $k = 1,\ldots,100$, for the two cases $s=\frac{1}{3}$ (ternary Cantor set on [-1,1]) and $s=\frac{1}{4}$ are plotted in Fig. 5. In Fig. 6 are plotted, for comparison, the recurrence coefficients a_k for polynomials orthogonal relative to the invariant measure μ supported on the (Cantor-like) Julia set J of the polynomial z^2-3. This sequence is known to be almost periodic. (In all the above calculations, the QD algorithm [45] was used to compute the S-fraction coefficients a_i in (3.3.10) from the series expansion in (3.3.9). The results in rational arithmetic were computed using the algebraic computation system MAPLE [67].)

3.4 Nonlinear IFS maps and "missing moment" problems

As mentioned earlier, the majority of applications of IFS consider only affine maps. The geometry associated with these maps is simple, and a recursive computation of the moments of the invariant measure is possible. When any or all of the maps constituting a contractive IFS are nonlinear, the recursive property for moments breaks down. To illustrate consider the following simple nonlinear analogue of the IFS given in Eq. (2.3.2), and scaled on $K=[0,1]$ (without loss of generality):

$$w_i(x) = s_i x^2 + a_i, \qquad |s_i| \le \frac{1}{2}, \qquad i = 1,2,\ldots N, \tag{3.5.1}$$

with associated probabilities p_i. The condition on the s_i follows from contractivity. Once again applying the invariance property (2.1.11) with $f(x) = x^n$, we obtain

$$g_n = \int x^n d\mu(x) = \sum_{i=1}^{N} p_i \int (s_i x^2 + a_i)^n \, d\mu(x). \tag{3.5.2}$$

Setting $g_0 = 1$, the first three equations corresponding to $n = 1,2,3$ above are

$$g_1 = g_2 \sum p_i s_i + \sum p_i a_i,$$

$$g_2 = g_4 \sum p_i s_i^2 + 2g_2 \sum p_i s_i a_i + \sum p_i a_i^2 \tag{3.5.3}$$

$$g_3 = g_6 \sum p_i s_i^3 + 3g_4 \sum p_i s_i^2 a_i + 3g_2 \sum p_i s_i a_i^2 + \sum p_i a_i^3 .$$

where the summations range over $i = 1, \ldots, N$. The relations are insufficient to permit a recursive computation of the moments, since each value of n introduces a new set of even moments g_{2k} for $n+1 \leq 2k \leq 2n$. Note, however, that the sequence of odd moments can be considered as independent degrees of freedom: all even moments may be written as *linear* functions of the odd ones, i.e.

$$g_{2k} = g_{2k}(g_1, g_3, \cdots, g_{[(2k-1)/2]}) , \tag{3.5.4}$$

where $[x] = int(x)$ denotes the integer part of x. The sequence of odd moments will be referred to as *missing moments.* Up to this point, they represent unknowns. In the spirit of Handy and Bessis [41], we now proceed to find approximations to these unknowns, using the following result from the Hausdorff moment problem [2]: Let g_n, $n = 0, 1, 2, \ldots$ denote an infinite sequence of real numbers. A necessary and sufficient condition that there exist a unique measure μ on [0,1], such that

$$g_n = \int_0^1 x^n d\mu ,$$

is that the g_n satisfy the following inequalities:

$$I(m,n) = \sum_{k=0}^{n} \binom{n}{k} (-1)^k g_{m+k} \geq 0 , \quad m, n = 0, 1, 2, \ldots . \tag{3.5.5}$$

The equality holds only when μ consists of point masses at $x = 0$ and/or 1. Without loss of generality we ignore this degenerate case, hence the equality sign. Trivially, for $n = 0, I(m, 0) = g_m > 0$ while the next set $I(m, 1) > 0$ implies the nonincreasing property $g_m > g_{m+1}$. As shown below, the Hausdorff conditions may now be applied, several at a time, to a finite set of moments to produce bounds on the missing moments (hence bounds to the even moments). We expect that the bounds on the g_{2k-1} improve as the number of Hausdorff conditions is increased.

At any one time, consider a finite number $M > 0$ of missing moments. Let $\mathbf{x}$ denote the M-vector of missing moments, i.e.

$$\mathbf{x} = (x_1, \ldots, x_M)^T = (g_1, \ldots, g_{2M-1})^T . \tag{3.5.6}$$

For the particular nonlinear IFS problem of Eq. (3.5.1), this vector uniquely defines the moment sequence $g_0, g_1, \cdots, g_{2M}$ (cf. Eq. (3.5.4)). However, the only Hausdorff inequalities which employ the missing moments in (3.5.6) are $I(m, n)$ for which $1 \leq m + n \leq 2M$, a total of $N_{max} = (M + 1)(2M + 1)$. (We ignore the trivial case $I(0, 0)$.) Some, or all, of these inequalities are then applied to the moments, producing a set of linear inequalities in terms of the missing moments:

$$\mathbf{Ax} > \mathbf{b} , \tag{3.5.7}$$

where $\mathbf{A}$ is an $N \times M$ matrix, and $N \leq N_{max}$ is the number of inequalities employed. The determination of upper and lower bounds to each missing moment $x_i = g_{2i-1}$ then becomes a linear programming problem:

$$\text{maximize (minimize)} \qquad S(\mathbf{x}) = x_i , \qquad \text{subject to } \mathbf{Ax} > \mathbf{b} , \tag{3.5.8}$$

These bounds are achieved on vertices (extreme points) of the convex polytope defined by the intersection of the hyperplanes defined by the equalities in Eq. (3.5.5).

Example: The nonlinear IFS of Eq. (3.5.1) with $s_i = \frac{1}{2}$, $a_1 = 0$, $a_2 = \frac{1}{2}$. A histogram approximation to the invariant measure μ is presented in Fig. 7. The first few even moments g_{2n} are given by the expressions

$$g_2 = 2g_1 - 1/2$$

$$g_4 = 6g_1 - 2$$

$$g_6 = 8g_3 - 12g_1 + 13/4$$

$$g_8 = -16g_3 + 98g_1 - 32$$

A little algebra reveals that the nonincreasing conditions $g_1 > g_2 > g_4$ imply the bounds

$$1/2 < g_1 < 3/8 .$$

The inequalities $I(m,n)$, $0 \le m \le 3$, $1 \le n \le 4$, in terms of missing moments, are shown below.

$I(m,n)$	$n=1$	$n=2$	$n=3$	$n=4$
$m=0$	$1-g_1$	$\frac{1}{2}$	$-\frac{1}{2}+3g_1-g_3$	$-4+14g_1-4g_3$
$m=1$	$\frac{1}{2}-g_1$	$1-3g_1+g_3$	$\frac{7}{2}-11g_1+3g_3$	
$m=2$	$-\frac{1}{2}+2g_2-g_3$	$-\frac{5}{2}+8g_1-2g_3$		
$m=3$	$2-6g_1+g_3$			

The symbolic manipulation language MAPLE [67] was used to compute this table. It proved to be quite useful, since the complexity of these expressions increases rapidly. For a relatively low number $M > 0$ of missing moments, the bounds on the moments can be computed in rational or real arithmetic using the linear programming package provided in MAPLE, Version 4.3. For larger values of M, where computation in MAPLE becomes very tedious, the linear programming problem was run in FORTRAN, using the IMSL subroutine ZX4LP. Upper and lower bounds to the missing moments obtained for the cases $M = 2, \ldots, 5$ are presented on the following page.

	$M=2$	$M=3$	$M=4$	$M=5$	g_{2n-1}
g_1	0.375000 0.312500	0.352778 0.340190	0.346963 0.344832	0.346046 0.345472	0.345701
g_3	0.250000 0.000000	0.137500 0.099138	0.121495 0.113084	0.117626 0.115620	0.116424
g_5		0.116667 0.000000	0.063785 0.038932	0.051893 0.046905	0.048834
g_7			0.058411 0.000000	0.032330 0.016168	0.027473
g_9				0.030487 0.000000	0.011441

The entries in the last column of this table are accurate estimates of the moments afforded by the sequence $y_k = T^k f(x)$ defined in Eq. (2.1.15), where $f(x) = x^{2n-1}$. For $k = 15$, the estimates y_k agreed to at least 1 part in 10^7 in all cases.

4 Recurrent iterated function systems

4.1 Introduction

The recurrent iterated function system (RIFS) [12,14, 18] represents an extension of the "usual" IFS method described to this point. The flexibility of RIFS permits the construction of more general sets and measures which do not have to exhibit the strict self-similarity of the IFS case. Its consequences and utility in image generation have been discussed in several papers. In all of these treatments, the focus was on a probabilistic interpretation of RIFS: indeed, from the very definition, this is the most natural viewpoint. Here a more general RIFS is considered from the perspective of invariant measures. First, a "Markov" operator is constructed in a fashion analogous to the usual IFS case. This operator is shown to be contractive on a complete space of measures, from which follows the existence of a "fixed point" invariant measure. This will provide a Collage Theorem for Measures for RIFS (Section 5). Also, for the case of linear maps in $\mathbf{R}^n$, the invariance of measure permits the recursive computation of moments over the unique attractor A of the RIFS.

The following Section will begin with a motivating example of a rather simple RIFS. In Section 4.3, we generalize this RIFS and even those introduced by Barnsley *et al.* in Ref. [14]. The appropriate space of measures and corresponding Markov operators are then defined and the theorems stated. In Section 4.4, for linear generalized RIFS, the invariance relations are used to derive recursion relations for moments over the attractor. This work was done with C. Cabrelli and U. Molter, and the proofs of all theorems appear in [24].

Before proceeding, it should also be mentioned that the idea of recurrence as well as generalizations is also found in the work of Dekking [26] and more recent works [4, 17]. Also, it is not possible here to discuss the very interesting concept of "mixing" in RIFS, which permits construction of more general fractal sets and images [12, 18]. Ref. [18] is a very readable account of this technique and its potential for image construction.

4.2 Simple recurrent iterated function systems

In this section we present a simple formulation of RIFS. Its connection with the usual IFS will become apparent.

As in the usual IFS, (K, d), denotes a compact metric space with metric d. Let there exist N contraction maps $w_i : K \to K$. We now associate with these maps a matrix of probabilities $\mathbf{P} = p_{ij}$ which is row stochastic, i.e. $\sum_j p_{ij} = 1$, $i = 1, \ldots, N$. From a *probabilistic* viewpoint, consider a random "chaos game" sequence

$$x_0 \in K, \qquad x_{n+1} = w_{\sigma_n}(x_n), \qquad n = 0, 1, 2, \ldots \; . \tag{4.2.1}$$

The fundamental difference between this process and the usual chaos game (Eq. (2.2.7)) is that the indices σ_n are not chosen independently, but rather with a probability that depends on the previous index σ_{n-1}:

$$P(\sigma_{n+1} = i) = p_{\sigma_n, i}, \qquad i = 1, 2, \ldots, N\,. \tag{4.2.2}$$

Thus, at each step in Eq. (2.1), to compute x_{n+1}, we look at the index σ_{n-1}. The σ_{n-1}th row of $\mathbf{P}$ then gives the probabilities of choosing the next map w_{σ_n} to apply to x_n. Clearly, in the case that all rows of $\mathbf{P}$ are identical and given by the vector $\mathbf{p}$, then the RIFS $\{K, \mathbf{w}, \mathbf{P}\}$ reduces to the usual IFS $\{K, \mathbf{w}, \mathbf{p}\}$. In all cases, we assume the matrix $\mathbf{P}$ to be *irreducible* [35], i.e. for any $1 \le i,\ j \le N$, there exists a sequence $i_1, i_2, \ldots, i_n$ with $i_1 = i$ and $i_n = j$ such that $p_{i_1 i_2} p_{i_2 i_3} \cdots p_{i_{n-1} i_n} > 0$. (In other words, for any i, j, if we apply map w_i in the sequence, there is a nonzero probability that we will apply map w_j in the future.) A major result for RIFS is the following [14]:

> There exists a unique stationary or invariant measure μ of the random walk in Eq. (4.2.1). If A is the support of μ, then there exist unique compact sets A_i, $i = 1, \ldots, N$, such that
>
> $$A = \bigcup_{i=1}^{N} A_i\,, \qquad A_i = \bigcup_{j : p_{ji} > 0}^{N} w_i(A_j)\,. \tag{4.2.3}$$

Note how the transition matrix $\mathbf{P}$ determines which maps w_i can act on A_j. The reader will note a fundamental difference between RIFS attractors and IFS attractors: The RIFS attractors need not exhibit the self-similarity or self-tiling properties characteristic of IFS attractors, where

$$A = \bigcup_{i=1}^{N} A_i\,, \qquad A_i = w_i(A)\,. \tag{4.2.4}$$

Barnsley *et al.* [14] showed that the random walk of Eq. (4.2.1) is not Markov on K itself but rather on the product $K \times \{1, 2, \ldots, N\}$. Again assuming that the matrix $\mathbf{P}$ is irreducible, it admits a stationary distribution $\{m_1, m_2, \ldots, m_N\}$, where the m_i are solutions of the linear equations

$$\sum_{j=1}^{N} p_{ji} m_j = m_i, \qquad i = 1, 2, \ldots, N, \quad \text{and} \quad \sum_{i=1}^{N} m_i = 1\,. \tag{4.2.5}$$

A convenient way to picture the RIFS is to imagine a stack of transparent planes K_i, $i = 1, 2, \ldots, N$ each of which is a copy of K, cf. [12, 14, 18]. Each $A_i \subset K_i$ and we "see" A by superimposing all planes. On the other hand, we can "see" A_i by plotting points obtained immediately after applying map w_i. During the iteration sequence in Eq. (4.2.1), motion from K_j to K_i under the action of w_i is permitted only if $p_{ji} > 0$. The invariant distribution on the indices 1 to N is $\{m_1, \ldots, m_N\}$. This may be interpreted as follows: the proportional amount of time spent by the random sequence in Eq. (4.2.1) on each plane K_i is precisely m_i.

Example 1: $N = 2,\ w_1(x) = \frac{1}{2}x,\ w_2(x) = \frac{1}{2}x + \frac{1}{2}$. Two cases:

$$(1):\ \mathbf{P} = \begin{bmatrix} 3/5 & 2/5 \\ 2/5 & 3/5 \end{bmatrix}, \qquad (2):\ \mathbf{P} = \begin{bmatrix} 2/4 & 3/5 \\ 3/5 & 2/5 \end{bmatrix}. \tag{4.2.6}$$

In both cases, $A = [0,1], A_1 = [0,1/2], A_2 = [1/2,1]$. Histogram approximations of the invariant measures are presented in Fig. 8. Qualitatively, in case (1), the measure is seen to "spread out" toward the ends of [0,1], but in a self-similar way throughout the interval. We can understand this from a look at the transition matrix in (1): when either map is applied, there is a greater probability that the same map will be applied again, thus pulling the point closer to its respective fixed point. In case (2), when a map is applied, there is a greater probability that the other map will then be applied. The result is to focus orbits toward the center. Some moments over these invariant measures will be calculated in Section 4.4.

Example 2: on $\mathbf{R}^2$, $N = 4$, the following four maps whose fixed points lie at the vertices of the unit square $[0,1] \times [0,1]$:

$$w_i(x,y) = (\frac{1}{2}x, \frac{1}{2}y) + b_i\,,\ i = 1,\ldots,4:\ b_1 = (0,0),\ b_2 = (\frac{1}{2},0),\ b_3 = (\frac{1}{2},\frac{1}{2}),\ b_4 = (0,\frac{1}{2}).$$

along with a 4×4 matrix $\mathbf{P}$. In the normal IFS case, i.e. $p_{ij} = p_j > 0,\ j = 1,\ldots,4$, the attractor A is the unit square $[0,1] \times [0,1]$.

(1) For $p_{11} = 0$ but all other $p_{ij} > 0$, the attractor A is shown in Fig. 9(a). Note that changing the nonzero p_{ij} will change the invariant measure supported on A. Its Hausdorff dimension is $\dim(A) = ln\left(\frac{1}{2}(3+\sqrt{21})\right)/(ln\,2)$.

(2) For $p_{11} = p_{22} = p_{33} = p_{44} = 0$, and all other $p_{ij} > 0$, the attractor A has the shape shown in Fig. 9(b) $\dim(A) = (ln\,3)/(ln\,2)$.

For both cases, the reader can deduce how the zero elements in $\mathbf{P}$ produce the "holes" in $[0,1]\times[0,1]$. A symbolic dynamics viewpoint helps here. The dimensions were calculated using Theorem 4.1 of [14].

4.3 Generalized RIFS and invariant measures

In this section, we consider the generalization of the RIFS introduced in Section 4.2. It is similar in form to that which first appeared in Ref. [14], Section 3.4; however, we are interested not only in the geometry but also the measures which are supported on each space. Also, we shall be using a "bar" notation, e.g. $\overline{K}$, to refer to spaces, measures etc. associated with our generalized RIFS. To begin, let $(K_1, d_1), \ldots, (K_N, d_N)$ denote compact metric spaces (they need not be copies of each other), and $\mathbf{P} = [p_{ij}]$

be an $N \times N$ row stochastic irreducible matrix for a Markov chain with state space $\{1, \ldots, N\}$. The fact that we consider the transition probability matrix instead of the index sets $I(i) = \{j \mid p_{ij} > 0\}, i = 1, \ldots, N$, represents a deviation from [14]. For each pair of indices (i,j), we let $w_{ij} : K_j \to K_i$ be a contractive map:

$$d_i(w_{ij}(x), w_{ij}(y)) \le s_{ij} d_j(x,y), \quad \forall \quad x, y \in K_j \quad 0 \le s_{ij} < 1. \tag{4.3.1}$$

We also define

$$s = \max_{1 \le i,j \le N} (s_{ij}) < 1. \tag{4.3.2}$$

(In fact, we don't need w_{ij} in the case that $p_{ji} = 0$.) In Ref. [14] it was shown that there exist unique compact sets A_i, $i = 1, \ldots, N$, with $A_i \subset K_i$, such that

$$A_i = \bigcup_{j: p_{ji} > 0} w_{ij}(A_j) \qquad i = 1, \ldots, N. \tag{4.3.3}$$

The set $A = (A_1, \ldots, A_N)$, called the *attractor* of the RIFS $\{(K_i, d_i), (p_{ij}), (w_{ij}), 1 \le i, j \le N\}$, is the fixed point of an operator $\mathbf{W}$ (cf. [14], Section 3.4) that reflects the dynamics of the w_{ij}. We shall show below that by choosing a suitable combination of measures over the spaces K_i, the action of the maps w_{ij} between these different spaces defines an invariant measure which is supported over the attractor A. This will represent a generalization of the case $K_i = K$, $\forall\ i$ and $w_{ij} = w_i$, $\forall\ j$ presented in Section 3.4 of [14].

The Markov process (or "chaos game") can be thought of as "living" in the space $\overline{K} = \bigcup_{i=1}^{N}(\{i\} \times K_i)$. (Note that the "bar" notation to be used in this section is *not* to be associated with the "bar" notation of Section 2.4 which referred to discrete pixel spaces and measures.) Starting with the element $\overline{z}_0 = (i_0, x_0) \in \overline{K}$, $1 \le i_0 \le N$, $x_0 \in K_{i_0}$, choose i_1 with the distribution given by the i_0th row of $\mathbf{P}$. Then define $x_1 = w_{i_1 i_0}(x_0)$ to give $\overline{z}_1 = (I_1, x_1)$, etc.. Note that $\overline{z}_n = (i_n, x_n)$ implies $x_n \in K_{I_n}$. $\{\overline{z}_n\}$ is a Markov process on $\overline{K}$, where the transition probability function is given by

$$p((s,x), \overline{B}) = \sum_{j=1}^{N} p_{sj} I_{\overline{B}}(j, w_{js}(x)), \tag{4.3.4}$$

which represents the probability to transfer from (s,x) to a Borel set $\overline{B} \subset \overline{K}$ in one step of the process.

Now let $\{m_1, \ldots, m_N\}$ be the stationary initial distribution of the Markov chain associated with the p_{ij}, as given by the solutions of Eq. (4.2.5). For an arbitrary metric space (K, d), define $M(K)$ as the set of Borel regular measures on K, and

$$\mathcal{M}_i = \{\mu \in M(K_i) \mid \mu(K_i) = m_i\} \tag{4.3.5}$$

and $\mathcal{M} = \mathcal{M}_1 \times \mathcal{M}_2 \times \cdots \times \mathcal{M}_N$. We define the distance between two measures $\overline{\mu}$, $\overline{\nu} \in \mathcal{M}$ as

$$\overline{d}_H(\overline{\mu}, \overline{\nu}) = \sum_{i=1}^{N} d_H^{(i)}(\mu_i, \nu_i), \tag{4.3.6}$$

where $d_H^{(i)}$ denotes the Hutchinson metric [(cf. Eq. (2.1.3))] between measures in $\mathcal{M}_i$. It is straightforward to show that $(\mathcal{M}, \overline{d}_H)$ is a complete metric space.

We now define an appropriate "Markov" operator $T : \mathcal{M} \to \mathcal{M}$ as (compare to Eq. (2.1.9))

$$T(\overline{\nu}) = T(\nu_1, \ldots, \nu_N) = (\sum_{j=1}^{N} p_{j1}\nu_j \circ w_{1j}^{-1}, \ldots, \sum_{j=1}^{N} p_{jN}\nu_j \circ w_{Nj}^{-1}). \tag{4.3.7}$$

Note that T is well defined:

$$(T\overline{\nu})_k(K_k) = \sum_{j=1}^{N} p_{jk}\nu_j(w_{kj}^{-1}(K_k)) = \sum_{j=1}^{N} p_{jk}m_j = m_k, \quad k = 1, \ldots, N. \tag{4.3.8}$$

Its construction and the proof of its contractivity are quite analogous to the original treatment by Hutchinson [47].

Proposition[24]: $T : \mathcal{M} \to \mathcal{M}$ is a contraction map in the metric $\overline{d}_H$ with constant s.

Now let $\overline{\mu}$ denote the fixed point of the Markov operator T in $\mathcal{M}$. We call $\overline{\mu}$ the *invariant measure* of the RIFS defined at the beginning of this section. The property $T\overline{\mu} = \overline{\mu}$ thus implies

$$\mu_i = \sum_{j=1}^{N} p_{ji}\mu_j \circ w_{ij}^{-1}, \qquad i = 1, \ldots, N. \tag{4.3.9}$$

Proposition [24]: Let $B = (B_1, \ldots, B_N)$, where $B_i = supp(\mu_i) \subset K_i$, $i = 1, \ldots, N$. Then

(1) $B_i = \bigcup_{j:p_{ji}>0} w_{ij}(B_j)$, $\quad i = 1, \ldots, N$, and
(2) $B = A$, the attractor of the RIFS defined in Eq. (4.3.3).

From these results, a collage theorem for invariant measures on recurrent IFS now follows, in complete analogy to that for normal IFS [9, 16]. This will be discussed in Section 5.

4.4 Moments of invariant measures of RIFS

We now consider the special case where the compact metric spaces (K_i, d_i) are subsets $K_i \subset \mathbf{R}^n$, with usual Euclidean metric. For a given RIFS with attractor A and invariant measure $\overline{\mu}$, we define the power moments of $\overline{\mu}$ by the integrals

$$g_{i_1 i_2 \dots i_n} = \int_A x_1^{i_1} \cdots x_n^{i_n} d\overline{\mu} \qquad g_{00\dots 0} = \int_A d\overline{\mu} = 1 \,. \tag{4.4.1}$$

Just as for the usual IFS, when the maps w_{ij} are linear, then the invariance property for integrals, Eq. (4.3.9) permits a recursive computation of the moments. We illustrate this property for the one dimensional case, i.e., $K_i \subset \mathbf{R}$. The maps w_{ij} will assume the following general form

$$w_{ij}(x) = s_{ij}x + a_{ij} \qquad |\, s_{ij} \,| < 1\,, \quad i = 1, \dots, N\,. \tag{4.4.2}$$

(Note that the maps w_{ij}, $j = 1, \dots, N$ are not necessarily identical, as in the usual IFS case.) We first define the power moments over A as

$$g_n = \int_A x^n d\overline{\mu} = \sum_{j=1}^{N} \int_{A_j} x^n d\mu_j = \sum_{j=1}^{N} g_n^{(j)} \,. \tag{4.4.3}$$

Using the Markov operator of Eq. (4.3.7), we have

$$g_n^{(i)} = \int_{A_i} x^n d\,\mu_i = \sum_{j=1}^{N} p_{ji} \int_{A_j} (s_{ij}x + a_{ij})^n d\,\mu_j \,. \tag{4.4.4}$$

Expanding the polynomials, and integrating, we obtain the relations

$$g_n^{(i)} = \sum_{j=1}^{N} p_{ji} \sum_{k=0}^{n} \binom{n}{j} s_{ij}^k a_{ij}^{n-k} g_k^{(j)} \,, \tag{4.4.5}$$

which for $n \geq 1$, may be rewritten as the following systems of linear equations in the moments $g_n^{(i)}$, $i = 1, \dots, N$:

$$\sum_{j=1}^{N} (p_{ji} s_{ij}^n - \delta_{ij}) g_n^{(j)} = - \sum_{k=0}^{n-1} \binom{n}{k} \left(\sum_{j=1}^{N} s_{ij}^k a_{ij}^{n-k} p_{ji} g_k^{(j)} \right) \qquad i = 1, \dots, N, \quad n \geq 1\,. \tag{4.4.6}$$

When $n = 0$, Eq. (4.4.5) yields precisely the linear equations of (4.2.5). We thus set $g_0^{(i)} = m_i$, so that $g_0 = 1$. This is in agreement with the definition of the measure space $\mathcal{M}$ constructed from the measures in Eq. (4.3.5). With this normalization, the higher moments $g_n^{(i)}$, $i = 1, 2, \dots, N$ are calculated for $n = 1, 2, \dots$ recursively. The special case

$$w_{ij}(x) = w_i(x) = s_i x + a_i\,, \quad j = 1, \dots, N\,, \tag{4.4.7}$$

corresponds to the simple RIFS of Section 4.2.

It follows that derivatives of moments with respect to the RIFS parameters could also be expressed in closed form as solutions of simultaneous linear equations. This method could be extended, in principle, to RIFS on $\mathbf{R}^2$. The relations would be rather complicated as are their counterparts for the usual IFS in two dimensions cf. Eq. (2.3.12).

4.5 Some simple moment calculations

From Eqs. (4.4.6) and (4.4.7), the first five moments over the attractor $[0,1]$ for the simple RIFS of Example 1 in Section 4.1 have been computed. The results are

$$(1): \quad g_1 = \frac{1}{2},\ g_2 = \frac{19}{54},\ g_3 = \frac{5}{18},\ g_4 = \frac{2449}{10530},\ g_5 = \frac{425}{2106}. \tag{4.4.8}$$

$$(2): \quad g_1 = \frac{1}{2},\ g_2 = \frac{7}{22},\ g_3 = \frac{5}{22},\ g_4 = \frac{783}{4510},\ g_5 = \frac{125}{902}.$$

In fact, let us extend our analysis of this simple two-map RIFS. Suppose that the transition probability matrix $\mathbf{P}$ for this system is given by the general form

$$\mathbf{P}(\epsilon) = \begin{bmatrix} \frac{1}{2}+\epsilon & \frac{1}{2}-\epsilon \\ \frac{1}{2}-\epsilon & \frac{1}{2}+\epsilon \end{bmatrix}. \tag{4.4.9}$$

The two cases given in Eq. (4.8) correspond to (1) $\epsilon = 1/10$ and (2) $\epsilon = -1/10$, respectively. Note that we have chosen to expand $\mathbf{P}$ about the "unperturbed" matrix

$$\mathbf{P}^{(0)} = \begin{bmatrix} \frac{1}{2} & \frac{1}{2} \\ \frac{1}{2} & \frac{1}{2} \end{bmatrix}. \tag{4.4.10}$$

In this case, for $\epsilon = 0$, the RIFS reduces to the IFS $\{w_1, w_2, p_1 = p_2 = \frac{1}{2}\}$ with $A = [0,1]$ and $\mu =$ uniform Lebesgue measure. Hence, the moments are

$$g_n = \frac{1}{n+1} \qquad n = 0,1,2,\ldots. \tag{4.4.11}$$

Two other special cases are easy to determine:

1. $\epsilon = \frac{1}{2}$, $\mathbf{P} = \mathbf{I}$ (not irreducible). Invariant measure $\mu = \frac{1}{2}\delta_0 + \frac{1}{2}\delta_1$ (δ_x denotes unit mass of measure at x). Hence $g_n = \frac{1}{2}$, $n > 0$.
2. $\epsilon = -\frac{1}{2}$ (two-cycle at (1/3,2/3)). $\mu = \frac{1}{2}\delta_{1/3} + \frac{1}{2}\delta_{2/3}$, with moments

$$g_n = \frac{1}{2}\left[\left(\frac{1}{3}\right)^n + \left(\frac{2}{3}\right)^n\right].$$

Using the algebraic computation language MAPLE [67], the first five moments for $\mathbf{P}(\epsilon), |\epsilon| \le \frac{1}{2}$, in Eq. (4.4.9) have been computed:

$$g_1 = \frac{1}{2}, \quad g_2 = \frac{1}{6}\frac{2-\epsilon}{1-\epsilon}, \quad g_3 = \frac{1}{4}\frac{1}{1-\epsilon},$$

$$g_4 = \frac{1}{30}\frac{(\epsilon-8)(\epsilon-3)}{(\epsilon-4)(\epsilon-1)}, \quad g_5 = \frac{1}{12}\frac{5\epsilon+8}{(\epsilon-4)(\epsilon-1)}. \tag{4.4.12}$$

5 The inverse problem of fractal/measure construction and applications

5.1 Introduction

It is convenient to classify the *inverse problems* of IFS in the following way:

1. *Geometric approximation:* Given a *target set* $S \subset K$, find an IFS $\{K, \mathbf{w}, \mathbf{p}\}$ (essentially independent of $\mathbf{p}$) whose attractor A approximates S to a prescribed accuracy, in an appropriate metric, say, the Hausdorff metric $h(A, S)$ (in continuous or discrete geometry).
2. *Approximation in measure:* Minimize the Hutchinson distance $d_H(\mu, \nu)$, where $supp(\mu) = A$ and $supp(\nu) = S$ (again, in continuous or discrete geometry).

As will be shown in the next section, the respective Collage Theorems for the above problems state mathematically what is expected geometrically: if a set (measure) can be tiled with copies of itself to an arbitrary accuracy, then it is close in metric to the attractor (invariant measure) of the IFS which produces the tiling. There also exist appropriate Collage Theorems for Geometry [14] and for Measures [24] the case of recurrent IFS.

An alternative route to the approximation of *measures* involves the matching of moments $\{g_n\}$ of an approximating IFS invariant measure μ to a given sequence of moments $\{G_n\}$ of a target measure ν. Indeed, such a moment matching was already attempted for approximating two moments by Barnsley and Demko [8, Section 3.3]. In Ref. [60], gradient methods were used to search in a parameter space of affine maps and their associated probabilities for minima of the Euclidean distance between a finite number of moments. Recently, it has been proved [25] that the Euclidean moment distance provides an upper bound to the Hutchinson distance $d_H(\mu, \nu)$ between the two measures. These ideas will be elaborated upon in Section 5.3.

The following examples illustrate that the solution to an inverse problem need not be unique. (The solution need not even be countable.)

Examples:

1. Target set $S = [0, 1]$, target measure $\nu =$ uniform Lebesgue measure on $[0, 1]$. One class of solutions: For any $N \geq 2$, define the N-map IFS, $w_i(x) = N^{-1}x + (i-1)N^{-1}$, $i = 1, 2, \ldots, N$, $p_1 = p_2 = \ldots = p_N = N^{-1}$.
2. Target set $S = [0, 1] \times [0, 1] \subset \mathbf{R}^2$, target measure $\nu =$ uniform measure. For $N = 4$, pick a point $P \in int(S)$ with coordinates (a, b) (i.e. $0 < a, b < 1$). Construct lines normal to the x and y axes which pass through P, thus partitioning S into four rectangles S_1, S_2, S_3, S_4 which have, respectively the

points $(0,0)$, $(1,0)$, $(1,1)$ and $(0,1)$ as one of their vertices. Let $w_i(x,y)$ be linear transformations (not unique!) of the form in Eq. (2.3.5), for which $w_i : S \to S_i$ $i = 1,2,3,4$, and let p_i = area (S_i).

From the practical viewpoint of image processing the representation of images in terms of IFS parameters could represent a huge degree of *data compression*: compare, for example, the storage (uncompressed) of an image on a 1000×1000 pixel array, with grey-levels or colour scales, to the possibility of representing this image by a few hundred IFS parameters. This has been discussed by M. Barnsley and coworkers [11, 16] who have achieved apparently dramatic success in IFS image compression. Unfortunately, any details of their compression scheme are proprietary and not available to the general public.

Apart from potential application to image processing, the IFS method may be promising in itself as a tool for approximating measures. Bessis and Demko [20] have applied a powerful technique of polynomial sampling to approximate a spectral measure relevant to solid state physics, the integration over which yields a thermodynamic quantity.

The plan for the remainder of this Section is as follows. In Section 5.2 are presented the basic Collage Theorems for IFS Attractors and Measures, as found in [16]. Also included is a Collage Theorem for invariant measures of the generalized recurrent IFS introduced in Section 4. Section 5.3 deals with moment methods for the inverse problem, and includes a result obtained recently: an upper bound for the Hutchinson distance between two probability measures on $[0,1]$ in terms of the Euclidean distance between their respective power moments. This bound justifies a moment matching gradient method used in Ref. [60]. Some numerical results for this method are given in Section 5.3.1. Section 5.4, considers Genetic Algorithms (GA) as a stochastic method to search for global minima in the inverse problem. The GA will be applied to moment matching as well as to the Collage Theorem for Measures in one dimension.

5.2 Collage theorems for attractors and measures

The Collage Theorems for geometry and measures are similar and are proved almost identically. At the risk of replication, all theorems will be cited, and the short proof given only in the first case.

Geometric Collage Theorem [9]: Definitions as in Section 1.1. Let $S \in \mathcal{S}$ and suppose that there exist a set of N contraction maps $w_i : K \to K$ so that

$$h(S, \mathbf{w}(S)) = h(S, \bigcup_{i=1}^{N} w_i(S)) < \epsilon. \tag{5.2.1}$$

Then

$$h(S, A) < \frac{\epsilon}{1-s}, \tag{5.2.2}$$

where $A = \mathbf{w}(A)$ is the unique attractor of the IFS $\{K, \mathbf{w}\}$.

Proof: For any $n \in \{1, 2, 3, \ldots\}$,

$$\begin{aligned} h(S, \mathbf{w}^n(S)) &\leq h(S, \mathbf{w}(S)) + h(\mathbf{w}(S), \mathbf{w}^2(S)) + \cdots + h(\mathbf{w}^{n-1}(S), \mathbf{w}^n(S)) \\ &\leq [1 + s + \cdots + s^{n-1}]h(S, \mathbf{w}(S)). \end{aligned}$$

In the limit $n \to \infty$, $h(S, \mathbf{w}^n(S)) \to h(S, A)$ and the desired result follows, since $s < 1$.

The collage theorem implies that we can forget about the attractor A (i.e. we don't have to plot it) and just concentrate on finding IFS maps $\mathbf{w}$ so that $h(S, \mathbf{w}(S))$ is minimized. Naturally, one can examine the target set S for obvious self-similarities, then determine the corresponding transformations w_i and ensure that a sufficient number of maps are employed. The use of affine maps greatly simplifies the procedure, since the geometry associated with such maps is simple: the IFS parameters become unknowns in a linear system of equations [11]. The associated probabilities, p_i may then be varied to alter the measure or shading on the attractor A. This procedure is referred to as *rendering*. These basic ideas, described in [11, 16] have been effectively developed by M. Barnsley and coworkers and illustrated with simple examples. However, few, if any, details concerning the practical algorithms have appeared, especially regarding the IFS representation of more complicated images (e.g., the "Andean girl" or the "Black Forest" colour plates in [16]). Ther are many questions which remain to be answered. For example, given a rather complicated image, perhaps composed of several smaller images, is there need for a human observer to segment these "subimages" which are then to be approximated independently? And how is the minimization of $h(S, \mathbf{w}(S))$ itself performed, regardless of S? Of course, regarding the former, it would be most desirable to eliminate the need of human intervention (e.g. graduate students, each equipped with a computer mouse; see "Not Just a Pretty Face", in Scientific American, March 1990, p. 77).

Collage Theorem for Measures: quite similar to above. Let $\nu \in X_1$ a target probability measure. Now suppose that there exists an IFS $\{K, \mathbf{w}, \mathbf{p}\}$ such that

$$d_H(\nu, M(\nu)) < \epsilon, \tag{5.2.3}$$

where M is the Markov operator defined in Eq. (2.1.9) and d_H the metric defined in Eq. (2.1.3). Then

$$d_H(\mu, \nu) < \frac{\epsilon}{1-s}, \tag{5.2.4}$$

where $\mu = M(\mu)$ is the invariant measure for the IFS $\{K, \mathbf{w}, \mathbf{p}\}$.

The idea of self-tiling expressed earlier for the geometric Collage Theorem now extends to measures. Some simple examples are discussed in [16, Section 9.6].

Collage Theorem for (Generalized) Recurrent IFS Measures [24]: Definitions as in Section 4. Let $\overline{\nu} \in \mathcal{M}$ be a measure over the metric spaces (K_i, d_i), $i = 1, \ldots, N$, as defined in Eq. (4.3.5), and suppose that there exists a RIFS $\{(K_i, d_i), \mathbf{P}, \mathbf{w},\ i = 1, \ldots, N\}$ with contractivity factor s, so that

$$\bar{d}_H(\bar{\nu}, T(\bar{\nu})) < \epsilon \tag{5.2.5}$$

Then

$$\bar{d}_H(\bar{\mu}, \bar{\nu}) < \frac{\epsilon}{(1-s)}, \tag{5.2.6}$$

where $\bar{\mu}$ is the invariant measure of the RIFS.

5.3 Moment methods for the inverse problem

The role of moments in inverse IFS problems has been explored in a number of studies [1, 8, 20, 32, 42, 60]. There is a common thread in all of these approaches: to match a prescribed finite number of moments of an IFS invariant measure μ to those of a target measure ν. The underlying principle of these procedures rests in the theory of moments over a finite interval $I \subset \mathbf{R}$ (or bounded subset of $\mathbf{R}^2$), the so-called Hausdorff problem [2]: given an infinite sequence of real numbers $g_n > 0$, which satisfy a set of positivity conditions, there exists a unique measure μ with support $S \subset I$ such that the $g_n = \int x^n d\mu$. A practical benefit of moment methods over the Collage Theorems lies in the fact that there is no need to "plot" sets or measures: assuming that the moments of the target measure are known, the moments of the invariant measure μ of an IFS can be computed directly from the IFS parameters, at least for *linear* maps.

Suppose that we are given a target measure ν with support $S \subset [0,1]$, and moments $G_n = \int x^n d\nu$. Diaconis and Shahshahani [29] proposed that by imposing the conditions $g_n = G_n$, $n = 1, 2, \ldots n_{max}$, one could solve for the IFS parameters directly. They realized, however, the problems associated with the nonlinearity of these equations in the IFS parameters. This author has also investigated this approach, using a Newton-Kantorovitch algorithm to solve the nonlinear equations. The scheme is extremely unstable, hence useless from a practical viewpoint. Moreover, its complexity increases dramatically in the two-dimensional case.

A more fruitful method of matching moments in the IFS method has been to minimize a "distance" between the target moments and the moments $g_n = g_n(\mathbf{s}, \mathbf{a}, \mathbf{p})$ of an IFS invariant measure [60]. For a fixed number, N, of IFS maps, and, M, the number of moments G_i to be employed, the objective function to be minimized was a partial sum of the squared Euclidean distance in "moment space"; for example on $\mathbf{R}$:

$$D_M^N(\mathbf{s}, \mathbf{a}, \mathbf{p}) = \sum_{i=1}^{N} (g_i(\mathbf{s}, \mathbf{a}, \mathbf{p}) - G_i)^2 . \tag{5.3.1}$$

The inverse problem thus becomes a problem of minimizing D over the appropriate parameter space $\Pi \subset \mathbf{R}^{3N}$. The original motivation to use this particular distance function lay in the fact that since the g_i are differentiable with respect to the IFS parameters $\pi_j,\ j = 1, \ldots, 3N$, then the elements of the vector **grad** D_M^N could be obtained in closed form. In this way, gradient methods for optimization [39] could be employed.

The basis of moment method matching lies in the construction of a set of approximating measures which converge weakly to a target measure. For the Hausdorff problem on $[0,1]$, we have the following theorem, based on Weierstrass approximation by polynomials:

Theorem 5.2.1 [25]: Let $K = [0,1], \nu \in X_1$ (target measure) and $\nu^{(n)} \in X_1, n = 1,2,3,\ldots$ such that

$$\int_0^1 x^k d\mu^{(n)} - \int_0^1 x^k d\nu \to 0 \text{ as } n \to \infty,\ k = 1,2,3,\ldots,. \tag{5.3.2}$$

Then $\mu^{(n)} \to \nu$ in the weak* topology, that is, $\int f d\nu - \int f d\mu^{(n)} \to 0$ for all continuous functions $f\colon K \to \mathbf{R}$.

Since K is compact, convergence in moments implies convergence in Hutchinson metric, i.e. $d_H(\mu^{(n)}, \nu) \to 0$.

A bound for the Hutchinson distance between two measures $\mu, \nu \in X_1$ may also be given [25]. Let the power moments of these two measures be denoted as

$$\mu_n = \int x^n d\mu, \qquad \nu_n = \int x^n d\nu. \tag{5.3.3}$$

Now let $P_n(x),\ n = 0,1,2,\ldots$ denote the complete set of orthonormal polynomials on $[0,1]$ relative to uniform Lebesgue measure, that is, $P_0(x) = 1$, and

$$P_n = c_{nn}x^n + c_{n,n-1}x^{n-1} + \ldots + c_{n0}, \tag{5.3.4}$$

such that $\int_0^1 P_m(x)P_n(x)dx = 1$ if $m = n$ and 0 otherwise. Now define the following polynomial moments,

$$\mu'_n = \int P_n(x)d\mu, \quad \nu'_n = \int P_n(x)d\nu,\ n = 0,1,2,\ldots \tag{5.3.5}$$

Then

$$d_H(\mu,\nu) \le \frac{1}{\sqrt{3}}\left[\sum_{k=1}^{\infty}(\mu_n - \nu_n)^2\right]^{\frac{1}{2}}. \tag{5.3.6}$$

The μ_n and μ'_n are related as follows

$$\mu'_n = \sum_{k=0}^{n} c_{nk}\mu_n. \tag{5.3.7}$$

5.3.1 Gradient method of moment matching: some numerical results

As stated earlier, the derivatives $dg_i/d\pi_j$, hence the elements of the vector $\mathbf{grad}\, D_M^N$ in parameter space, may be obtained in closed form. In Ref. [60], Davidon's gradient projection methods for optimization [39] which minimizes subject to constraints, was used. Here, the constraints are those defining the parameter space Π, cf. Eq. (2.3.7), as well as the constraint for probabilities, $q_i = p_i$, $\sum_{i=1}^{N} p_i = 1$. The function D_M^N is extremely multimodal, however, and such gradient-type methods converge only to local minima.

In Table 1 are presented some numerical results of the gradient methods applied to the problem of the ternary Cantor set in $[0,1]$ with uniform invariant measure, $M = 20$ moments being used in Eq. (5.3.1). As in Ref. [60], the Harwell library subroutine VE01AD was employed for the gradient projection. The starting point was a two-map IFS with $A = [0,1]$ and $\nu =$ uniform measure. This problem appeared in Ref. [60], but we now include values of the discrete Hausdorff and Hutchinson distances between approximating and target sets/measures. The discrete representation was performed on a grid of 1000 pixels per unit interval. The entries are somewhat different than in Ref. [60] and show better convergence, in fact. One reason may be due to the fact that the constraints on the sum of the p_i were tightened: here, $-\epsilon < 1 - \sum p_i < \epsilon$ where $\epsilon = 10^{-10}$ (as opposed to $\epsilon = 10^{-3}$ in [60].) The reader should also note that the function $D^{\frac{1}{2}}$ is being minimized here. Note that the discrete Hausdorff and Hutchinson distances, $\overline{h}(A,S)$ and $\overline{d}_H(\mu,\nu)$, respectively, do *not* necessarily decrease monotonically with moment distance. However, they eventually tend quickly to zero. Locally, the gradient method converges geometrically, as expected.

The example studied above is rather artificial and does not reveal the problems associated with multimodal objective functions. The problem of "interfering" local minima was seen in [60] for the two-dimensional case, where the target set was the "spleenwort fern" of Fig. 1(b). The initial and final (after 62 iterations) IFS attractors are shown in Figure 10(a), where the target set is shown for comparison in the background. (The initial IFS represents a slight perturbation of the target IFS.) Only $M = 10$ moments, g_{ij} $1 \leq i + j \leq 4$ were used. The search stops at a local minimum which is rather close to the fern in moment space, $D^{\frac{1}{2}} = 1.5 \times 10^{-4}$. However the error in the image is noticeable, particularly in the stem. In Figure 10(b) are shown initial and final approximations to the fern, where the initial attractor is a "tree-like" set. The minimization forces the approximating sets to "curl" in the right direction, but there are problems with the stems. (In fact, the connectedness of stems could be considered as very sensitive to perturbations, quite analogous to the structural instability of non-transversal intersections of stable/unstable manifolds in dynamical systems.)

Just to conclude with a more striking example of the limitations of the gradient method, Fig. 10(c) shows the final approximating set to the fern, when the initial IFS approximation was the unit square $[0,1] \times [0,1]$ with uniform measure (4 maps with fixed points at corners of square with contraction factor $\frac{1}{2}$ and $p_i = \frac{1}{4}$). Note that $D^{\frac{1}{2}} = 3 \times 10^{-4}$ for 10 moments. This example shatters any hopes of using gradient methods for moment matching as a global optimization method. Nevertheless, the method could still prove to be important in "fine-tuning" a set of satisfactory IFS parameters obtained from a more global search procedure.

5.4 Global (Monte-Carlo) search using genetic algorithms

As stated earlier, the "moment distance" function D_M^N in Eq. (5.3.1) can be extremely multimodal in the large space of IFS parameters. Local methods, e.g. gradients, will generally be insufficient to determine the global minima sought in the inverse problem. Such "bumpiness" would also plague the objective functions $\overline{h}(S, w(S))$ or $\overline{d}_H(\overline{\nu}, M(\overline{\nu}))$ used in a Collage method for sets or measures on discrete spaces. As such, it seems that one must turn to random search (e.g. Monte Carlo) methods. Simulated Annealing (SA) [50] is one such stochastic technique which has been employed in many optimization problems. The method is based on a statistical mechanics picture of the annealing process which is used (for example, in steel manufacturing) to produce crystalline solids of low energy configurations. Here, the system of "particles" is a system of points in an appropriate parameter space, say $\Pi \subset \mathbf{R}^n$. Each particle x_i has an energy $E(x_i) = F(x_i)$ where $F : \Pi \to \mathbf{R}$ is the objective function whose global minimum is sought.

Mantica and Sloan [53] have modified the SA approach to include *memory* which then optimizes the iteration by a form of learning. It should also be mentioned that a random search method such as Simulated Annealing *may* account, at least in part, for the apparent success in IFS image compression achieved by M. Barnsley and coworkers at Iterated Systems, Inc., but few, if indeed any, details of the computational aspects of their procedures are publically known.

Another group of stochastic methods of optimization are Genetic Algorithms (GA) [40]. GA have proved to be powerful, especially when the objective functions are highly "bumpy" and/or nonsmooth. The strategy of GA originated in the work of J. Holland [46] and is based on the current picture of evolutionary selection procedures in genetics. The algorithms consider an ensemble or population of points in appropriate parameter space, which are candidates to minimize a given objective function. The stochasticity of Monte Carlo and Simulated Annealing is coupled with memory and continued exploration of subspaces in parameter space. The three fundamental "genetic" procedures employed by this method are: (i) *parenting (crossover)*, (ii) *mutation*, and *selection.*

In what follows is but a brief outline of a particular GA scheme as applied to the inverse problem, and is in no way intended to be exhaustive. For a very readable and complete account of GA and its applications, I refer the reader to the book by Goldberg [40], from which most of the strategy was taken. At this point, I would also like to thank Professor R. Shonkwiler of Georgia Tech, for introducing me to the idea of GA. He has already been working extensively on applications of GA to several aspects of the inverse problems. The work reported below has been performed independently, and in no way has explored the full potential of GA.

5.4.1 GA strategy applied to IFS problems

In the discussion below, the goal is to find the global minimum of a nonnegative function $F : \mathbf{R}^n \to \mathbf{R}$, i.e. $F(\mathbf{x}) = F(x_1, \ldots, x_n) \geq 0$, (in fact, to find $\overline{\mathbf{x}}$ such that $F(\overline{\mathbf{x}}) = 0$). In the applications discussed below, this *objective function* will be one of the following:

(i) the moment distance function D_M^N in Eq. (5.3.1),
(ii) the collage distance between measures $\overline{d}_H(\overline{\nu}, M(\overline{\nu}))$ in discrete pixel space.

The GA considers an ensemble or "population" of $NPOP$ points in $\mathbf{R}^n$:

$$\mathbf{x}^{(i)} = (x_i^{(i)}, x_2^{(i)}, \ldots, x_n^{(i)}) \in \mathbf{R}^n, \quad i = 1, 2, \ldots NPOP. \tag{5.4.1}$$

(In general, $NPOP$ could be allowed to change in time but in this application, it remains constant.) The GA works optimally on binary sequences. Thus, we consider each $x_k^{(i)}$ in its binary representation to $ND > 0$ significant bits. The result is a population member whose binary sequence has $NB = n \times ND$ bits:

$$\mathbf{y}^{(i)} = (y_1^{(i)}, \ldots, y_{ND}^{(i)}, y_{ND+1}^{(i)}, \ldots, y_{(n-1)ND}^{(i)}, \ldots, y_{n*ND}^{(i)}) \in \{0,1\}^{NB}. \tag{5.4.2}$$

Here is the essential point. *The Genetic Algorithm considers each member* $y^{(i)}$ *of the population as a "chromosome" with NB "genes".* The "fitness" of these chromosomes is determined by how well they minimize the objective function F. The population is then allowed to "evolve" under selection and parenting: only members with high fitness serve as parents which produce offspring by the genetic "crossover rule" to populate the next generation. The strategy which has been employed here (and which roughly follows that outlined in Goldberg [40]) is as follows:

1. Initialize the population: generate $NPOP$ random sequences of $NB = n \times ND$ binary digits, which make up the $\mathbf{y}^{(i)}$ vectors. This is generation $NG = 0$.
2. "Decode" the $\mathbf{y}^{(i)}$ to give the numerical values $\mathbf{x}^{(i)}$ and evaluate $s^{(i)} = F(\mathbf{x}^{(i)})$. The "fitness" $f^{(i)}$ assigned to $\mathbf{x}^{(i)}$ is a Boltzmann-type function:

$$f^{(i)} = \exp(-\alpha s^{(i)}) \,/\, \sum_{k=1}^{NPOP} \exp(-\alpha s^{(k)}), \quad i = 1, 2, \ldots NPOP, \tag{5.4.3a}$$

where α, a scaling constant, may be unity, but in some cases, is chosen to be the reciprocal of the average value of the $s^{(k)}$, i.e.

$$\alpha = 1/\Delta, \quad \Delta = \frac{1}{NPOP} \sum_{k=1}^{NPOP} s^{(k)}. \tag{5.4.3b}$$

3. Perform "selection" of parents and "mating" to produce a new generation by repeating the following step $NW/2$ times: Pick two "parents" at random, say $\mathbf{x}^{(k_1)}$ and $\mathbf{x}^{(k_2)}$, from the population. The probability of selecting $\mathbf{x}^{(i)}$ as parents is $f^{(i)}$. With a probability p_c, the mating of these two parents will produce two offspring according to the following "crossover rule" applied to their "genes": pick an integer ic randomly from the set $\{1,2,\ldots,NB\}$, with uniform distribution. Now produce two new genes by crossing the gene sequences of both parents after the icth chromosome:

$$\mathbf{y}^{(l_1)} = (y_1^{(k_1)},\ldots,y_{ic}^{(k_1)},y_{ic}^{(k_2)},\ldots,y_{NB}^{(k_2)}), \tag{5.4.4a}$$

$$\mathbf{y}^{(l_2)} = (y_1^{(k_2)},\ldots,y_{ic}^{(k_2)},y_{ic+1}^{(k_1)},\ldots,y_{NB}^{(k_1)}). \tag{5.4.4b}$$

With probability $1-p_c$, the parents are simply copied into the offspring with no crossover.

4. Rewrite the population, with a (small) probability p_m of "mutating" each gene (i.e. replace 0 with 1 and *vice versa*) during the rewriting. This simulates errors in transcription, external "radiation", etc., to permit further random exploration. Reset NG to $NG+1$ and return to Step 2 if $NG \le NGEN$, the maximum number of iterations of the GA allowed.

The effectiveness of the GA lies in the three essential genetic features mentioned earlier. The crossover rule allows exploration of hyperplances in (binary) parameter space, keeping in memory at least some information during the search by preserving at least substrings of genes. As the population evolves, the "strong" chromosomes become dominant, passing their charcteristics on to the next generation. The population members eventually converge to variations on a single genetic trait. Needless to say, the algorithm is sensitive to the parameters listed above. Overdominance by genes which appear to be good can steer the process prematurely and halt the searching for good minima. If the mutation probability p_m is too large, then valuable sequences can be lost without adequate contribution to the population. On the other hand, if it is too small, premature convergence again can result. Also, if ND, the length of the binary string for each "gene", is too long, the algorithm can waste much time exploring rather insignificant bits. For a complete discussion, see Goldberg [40]. Finally, the importance of a good random number generator can not be overstressed.

Application to Inverse Problem on $[0,1]$: In most examples given below, the target set S, or target measure ν, will be conveniently chosen to be the attractor or invariant measure of an IFS on $K=[0,1]$. For the moment matching method, the target measures ν will have *reference* moments G_n, $n=1,2,\ldots,M$. Now suppose that we wish to consider an N-map IFS approximation. The appropriate parameter space is $\pi = (\mathbf{s},\mathbf{a},\mathbf{p}) \in \Pi \subset \mathbf{R}^{3N}$, i.e. $n=3N$ in Eq. (5.4.1). The Genetic Algorithm then considers a population of $NPOP$ such IFS *maps*, where the ith member is given by

$$\mathbf{x}^{(i)} = (s_1^{(i)},\ldots,s_N^{(i)},a_1^{(i)},\ldots,a_N^{(i)},q_1^{(i)},\ldots,q_N^{(i)}), \quad i=1,2,\ldots,NPOP. \tag{5.4.5}$$

Note that the final N elements in (5.4.5) are given by $0 \leq q_k^{(i)} \leq 1$, and hence are not true probabilities. The corresponding probabilities for the IFS maps in this *ith* population member $x^{(i)}$ are "decoded" as

$$p_k^{(i)} = q_k^{(i)} / \sum_{j=1}^{N} q_j^{(i)}, \qquad k = 1, 2, \ldots, N. \tag{5.4.6}$$

In summary, each member of the population, $\mathbf{x}^{(i)}$, $i = 1, 2, \ldots NPOP$, represents an IFS: a set of maps $w_i(x)$ with associated probabilities p_i, $i = 1, 2, \ldots N$. This member $\mathbf{x}^{(i)}$ will have a finite binary representation $\mathbf{y}^{(i)}$ of $NB = 3N \times ND$ bits, which will be used in the Genetic Algorithm outlined above.

5.4.2 Moment matching with genetic algorithms

The limitation of gradient methods to converge only to local minima provided the original motivation to explore GA as a potential global optimization method. However, when applied to the moment distance function D_M^N in Eq. (3.5.1), the GA method does not minimize as well as the gradient method. It is quite sensitive to value GA parameter values, especially ND, the number of bits per IFS parameter. The optimal value is around $ND = 8$. For higher values, much exploration is wasted on rather insignificant bits and minimization is quite poor. The value $ND = 8$, however, does not allow much refinement on the IFS parameters, and minima of $D^{\frac{1}{2}} \sim 10^{-2} - 10^{-3}$ are generally found for $M = 20$ moments (compare to $D^{\frac{1}{2}} \sim 10^{-4}$ for the gradient method). Also keeping in mind that a low D_M^N value does *not* necessarily imply closeness in Hausdorff or Hutchinson metric, it appears that GA is not well suited for this problem.

Example: Target set: Cantor set on $[0, 1]$ with uniform invariant measure. The results of a GA calculation using 20 moments is shown in Table 2, along with the GA parameters. The final column lists the discrete Hutchinson distance $\overline{d}_H(\overline{\mu}, \overline{\nu})$ between approximating and target measures (1000 pixels). The table lists only those generations for which there was an improvement in the minimization of D_M^N by the "best" member of $\mathcal{D}$ the population. Note that even though there is continued improvement, the "best" member may not always yield a measure which is close in Hutchinson distance to the target measure (or example, Step no. 23).

5.4.3 Collage theorem for discrete measures using genetic algorithms

An approach for the approximation of *discrete* target measures on $[0, 1]$, based on the Collage Theorem and using GA, will now be outlined. In the examples which follow, the target measures themselves will be invariant measures of IFS. As stated in Section 2.4.1, the direct use of the *discrete* Markov operator $\overline{M} : \overline{X}_1 \to \overline{X}_1$ is found to be faster and more accurate for the construction of invariant measures than the "ergodic chaos game" approach. In the computations shown below, an array of 1000 pixels on $[0,1]$ was used (with extra pixels on each side to accommodate IFS maps which did not map onto $[0,1]$). To compute the invariant measure $\overline{\mu}$ of a discrete IFS $\{\overline{K},\overline{w},p\}$, start with $\overline{\nu}^{(0)} =$ uniform measure (i.e. $\overline{\nu}(i) = 0.001$, $i = 1,2,\ldots,1000$) and construct the sequence $\overline{\nu}^{(k+1)} = \overline{M}\overline{\nu}^{(k)}$, $k = 0,1,2,\ldots$, using the algorithm given in Section 2.4.1. Convergence of the $\overline{\nu}^{(k)}$ to the invariant measure $\overline{\mu}$ in the appropriate discrete metric $\overline{d}_H$ was usually achieved to 1 part in 10^6 in 15-20 iterations.

Now denote the *target* measure as $\overline{\nu}$. Using the Collage Theorem for measures, it suffices to search for a discrete IFS $\{\overline{K},\overline{w},p\}$ which minimizes the distance between measures $\overline{\nu}$ and $\overline{\nu}' = \overline{M}(\overline{\nu})$, i.e.

$$\overline{d}_H(\overline{\nu},\overline{\nu}'), \qquad \overline{\nu}' = \overline{M}(\overline{\nu}) = \sum_{i=1}^{N} p_i \overline{\nu} \circ \overline{w}_i^{-1}. \tag{5.4.7}$$

In Table 3 are presented the numerical results for a GA search for the ternary Cantor set with uniform measure. The table lists only those generations for which there was an improvement in the minimization of $\overline{d}_H$ in Eq. (5.4.7). Also listed are the average values Δ for the distances $\overline{d}_H$ over the entire population, cf. Eq. (5.4.3b).

5.4.4 Comparison of various approaches to a model problem

In this concluding section are presented the results of both continuous (moment) and discrete (Collage) approaches to a more complicated IFS problem. The target measure ν is the invariant measure on $[0,1]$ of the following four-maps IFS:

$$w_1(x) = 0.2x + 0.0, \quad p_1 = 0.2; \quad w_2(x) = 0.7x + 0.3, \quad p_2 = 0.4;$$

$$w_3(x) = 0.2x + 0.3, \quad p_3 = 0.1; \quad w_4(x) = 0.5x + 0.4, \quad p_4 = 0.2. \tag{5.4.8}$$

Three "continuous" methods, each using 20 target moments were applied: (I) the gradient method of Section 5.3.1, starting with a four map IFS with uniform Lebesgue measure on $[0,1]$ as invariant measure ; (II) Genetic Algorithm method of minimizing moment distance, cf. Section 5.4.2; (III) the gradient method, using the result of (II) as a starting point, in an attempt to improve it by "fine-tuning". A final method (IV), applied Genetic Algorithms to the discrete Collage method. Here, the interval $[0,1]$ was discretized to a space of 1000 pixels. The results of these four methods are given on the following page:

	Method	$D^{\frac{1}{2}}$	$\overline{d}_H(\overline{\mu},\overline{\nu})$
I	Gradient	2.1×10^{-5}	16.7
II	GA	6.3×10^{-3}	15.8
III	GA + Gradient	4.0×10^{-6}	10.7
IV	Discrete	-	6.2

In I-III, the discrete mesure distances $\overline{d}_H(\overline{\mu},\overline{\nu})$ were computed over the lattice of 1000 pixels on $[0,1]$. The discretization essentially follows method (i) of Section 2.4.1. The GA method in (II) affords a poorer matching of moments than the gradient method of (I); nevertheless, the Hutchinson distance $\overline{d}_H$ was, in fact, improved slightly. The result of method (III) is seen to be an improvement over (I), which could suggest that the GA method may detect reasonable "valleys" through which the gradient method may find better mimima. Method (IV) employed the discrete IFS formulation (ii) of Section 2.4.1. This result is quite encouraging and suggests that Genetic Algorithms may be better suited for discrete problems. Histogram approximations to the target measure ν and the approximating measures yielded by (I)-(IV) are plotted in Figure 11.

Finally, we mention that a similar type of comparative study was performed [62] for the reconstruction of the spectral measure of an FCC crystal lattice from a knowledge of its moments. Bessis and Demko [20] (BD) originally applied a method of polynomial sampling with homogeneous IFS to perform moment matching. Using 10 moments, they could compute the zero-point energy integral over the spectral measure to an accuracy which was at least two orders of magnitude better than results obtained from Padé approximants. A gradient method applied to the BD IFS breaks the homogeneity of the IFS and improves the estimate of the integral. A gradient method starting with a four-map IFS $w_i(x) = \frac{1}{4}x + \frac{1}{4}(i-1)$, $i = 1 \ldots 4$ does not achieve the accuracy of BD. Numerical results are presented in [62].

5.5 Concluding remarks on the inverse problem

Several methods of measure reconstruction using IFS have been examined: moment matching for continuous space, Collage Theorem for discrete space. The use of Genetic Algorithms for moment matching appears to be limited. The results obtained by applying GA to the discrete collage method are quite encouraging. A natural extension of this work to two-dimensional problems is currently in progress. At the time of writing, Professor R. Shonkwiler of Georgia Tech has informed me of early success for inverse problems on the line, using GA with parallel computation. The method demonstrates extremely fast convergence, and is quite promising. Another path being explored is the inverse problem using recurrent IFS, for which a Collage Theorem for measures now exists (Section 5.2, Eq. (5.2.5)).

Acknowledgements

It is a sincere pleasure to thank Professors C. Cabrelli and U. Molter, who are currently visiting Waterloo from the University of Buenos Aires, Argentina, for the many stimulating and fruitful conversations, ongoing collaboration, and helpful reading of this manuscript. I also wish to thank Professor R. Shonkwiler of Georgia Tech and Dr. G. Mantica (currently at CEN Saclay, France) for very helpful correspondence, and Professors B. Forte and W. Gilbert (Waterloo) for ongoing discussions and encouragement.

The typesetting of this manuscript was performed by Mrs. A. Puncher. Her conscientious efforts, in light of the short notice given, are greatly appreciated. The support of this research by a grant from the Natural Sciences and Engineering Research Council of Canada is gratefully acknowledged.

References

[1] S. Abenda and G. Turchetti, Inverse problem of fractal sets on the real line via the moment method, *Nuovo Cimento* **104B**, 213-227 (1989).

[2] N.I. Akhiezer, *The Classical Moment Problem*, Hafner, NY (1965).

[3] R.B. Ash, *Measure, Integration and Functional Analysis*, Academic Press, NY (1972).

[4] C. Bandt, Self-similar sets. I. Topological Markov chains and mixed self-similar sets, *Math. Nachr.* **142**, 107-123 (1989).

[5] M.F. Barnsley, J.S. Geronimo and A.N. Harrington, Orthogonal polynomials associated with invariant measures on Julia sets, *Bull. AMS* **7**, 381-384 (1982).

[6] M.F. Barnsley, J.S. Geronimo and A.N. Harrington, Infinite-dimensional Jacobi matrices associated with Julia sets, *Proc. AMS* **88**, 625-630 (1983).

[7] M.F. Barnsley and A.N. Harrington, Moments of balanced measures on Julia sets, *Trans. AMS* **284**, 271-280 (1984).

[8] M.F. Barnsley and S. Demko, Iterated function systems and the global construction of fractals, *Proc. Roy. Soc. London* **A399**, 243-275 (1985).

[9] M.F. Barnsley, V. Ervin, D. Hardin and J. Lancaster, Solution of an inverse problem for fractals and other sets, *Proc. Nat. Acad. Sci. U.S.A.* **83**, 1975-1977 (1985).

[10] M.F. Barnsley and A.N. Harrington, A Mandelbrot set for pairs of linear maps, *Physica* **15D**, 421-432 (1985).

[11] M.F. Barnsley and A.D. Sloan, A better way to compress images, *BYTE Magazine*, January issue, 215-223 (1988).

[12] M.F. Barnsley, M.A. Berger and H.M. Soner, Mixing Markov chains and their images, *Prob. Eng. Inf. Sci.* **2**, 387-414 (1988).

[13] M.F. Barnsley, S.G. Demko, J. Elton and J.S. Geronimo, Invariant measures for Markov processes arising from iterated function systems with place-dependent probabilities, *Ann. Inst. H. Poincaré* **24**, 367-394 (1988).

[14] M.F. Barnsley, J.H. Elton and D.P. Hardin, Recurrent iterated function systems, Constr. Approx. **B5**, 3-31 (1989)

[15] M.F. Barnsley and D.P. Hardin, A Mandelbrot set whose boundary is piecewise smooth, *Trans. AMS* **315**, 641-659 (1989).

[16] M.F. Barnsley, *Fractals Everywhere*, Academic Press, NY (1988).

[17] T. Bedford, Dimension and dynamics for fractal recurrent sets, *J. Lond. Math. Soc. (2)* **33**, 89-100 (1986).

[18] M.A. Berger, Images generated by orbits of 2-D Markov chains, in CHANCE, New Directions for Statistics and Computing, Vol. 2, No. 2, 18-28 (1989).

[19] M.A. Berger and H.M. Soner, Random walks generated by affine mappings, *J. Theoret. Probab.* **1**, 239-254 (1988).

[20] D. Bessis and S. Demko, Stable recovery of fractal measures by polynomial sampling, CEN-Saclay preprint, PhT 89-150 (1989).

[21] D. Bessis, M.L. Mehta and P. Moussa, Orthogonal polynomials on a family of Cantor sets and the problem of iterations of quadratic mappings, *Lett. Math. Phys.* **6**, 123-140 (1982).

[22] P. Blanchard, Complex analytic dynamics on the Riemann sphere, *Bull. AMS* **11**, 88-144 (1984).

[23] J. Brandt, C. Cabrelli and U. Molter, An algorithm for the computation of the Hutchinson distance, preprint submitted to *Inform. Process. Lett.* (1990).

[24] C. Cabrelli, U. Molter and E.R. Vrscay, Recurrent iterated function systems: invariant measures, a collage theorem and moment relations, preprint submitted to FRACTAL 90 Conference, Lisbon, June (1990).

[25] C. Cabrelli, U. Molter and E.R. Vrscay, in preparation.

[26] F.M. Dekking, Recurrent sets, *Adv. in Math.* **B44**, 78-104 (1982).

[27] S. Demko, Euler-Maclaurin-type expansions for some fractal measures, preprint submitted to FRACTAL 90 Conference, Lisbon, June (1990).

[28] R. Devaney, *An Introduction to Chaotic Dynamical Systems*, Addison Wesley (1986).

[29] P. Diaconis and M. Shahshahani, Products of random matrices and computer image generation, in *Random Matrices and Their Applications*, Vol. 50, *Contemp. Math.*, AMS, Providence, RI (1986).

[30] S. Dubuc and A. Elqortobi, Approximations of fractal sets, *J. Comput. Appl. Math.* **29**, 79-89 (1990).

[31] J. Elton, An ergodic theorem for iterated maps, *Ergodic Theory Dynamical Systems* **7**, 481-488 (1987).

[32] J. Elton and Z. Yan, Approximation of measures by Markov processes and homogeneous affine iterated function systems, *Constr. Approx.* **5**, 69-87 (1989).

[33] K.J. Falconer, *The Geometry of Fractal Sets*, Cambridge University Press (1985).

[34] K.J. Falconer, *Fractal Geometry, Mathematical Foundations and Applications*, Wiley (1990).

[35] W. Feller, *An Introduction to Probability Theory and Its Applications*, Vol. 1, Wiley (1957).

[36] W.J. Gilbert, The fractal dimension of sets derived from complex bases, *Canad. Math. Bull.* **29**, 495-500 (1986).

[37] W.J. Gilbert, The division algorithm in complex bases, preprint (1989).

[38] W.J. Gilbert, Gaussian integers as bases for exotic number systems, to appear in *The Mathematical Heritage of C.F. Gauss*, G.M. Rassias ed., World Scientific (1990).

[39] P.E. Gill, W. Murray and M.H. Wright, *Practical Optimization*, Academic Press, NY (1981).

[40] D. Goldberg, *Genetic Algorithms in Search, Optimization and Machine Learning*, Addison-Wesley (1989).

[41] C.R. Handy and D. Bessis, Rapidly convergent lower bounds for the Schrodinger equation ground state energy, *Phys. Rev. Lett.* **55**, 931 (1985).

[42] C. Handy and G. Mantica, Inverse problems in fractal construction: moment method solution, *Physica D* **43**, 17-36 (1990).

[43] D.P. Hardin, *Hyperbolic Iterated Function Systems and Applications*, Ph.D. Thesis, Georgia Institute of Technology (1985).

[44] M. Hata, On the structure of self-similar sets, *Japan J. Appl. Math.* **2**, 381-414 (1985).

[45] P. Henrici, *Applied and Computational Complex Analysis*, Wiley. Vol. 1, Sec. 7.6 (1974), Vol 2, Chap. 12. (1977).

[46] J. Holland, *Adaptation in Natural and Artificial Systems*, Univ. Mich. Press (1975).

[47] J. Hutchinson, Fractals and self-similarity, *Indiana Univ. Math. J.* **30**, 713-747 (1981).

[48] S. Karlin, Some random walks arising in learning models, I, *Pacific J. Math.* **3**, 725-756 (1953).

[49] I. Katai and J. Szabo, Canonical number systems for complex integers, *Acta Sci. Math. (Szeged)* **37**, 255-260 (1975).

[50] S. Kirkpatrick, C.D. Gelatt and M.P. Vecchi, Optimization by simulated annealing, *Science* **220**, 671 (1983).

[51] D.E. Knuth, *The Art of Computer Programming, Vol. 2, Seminumerical Algorithms*, Addison-Wesley (1981).

[52] B. Mandelbrot, *The Fractal Geometry of Nature*, Freeman (1983).

[53] G. Mantica and A. Sloan, Chaotic optimization and the construction of fractals: solution of an inverse problem, *Complex Systems* **3**, 37-62 (1989).

[54] H.O. Peitgen, Fantastic deterministic fractals, in *The Science of Fractal Images*, H.O. Peitgen and D. Saupe, Editors, Springer Verlag (1988).

[55] H.O. Peitgen and P.H. Richter, *The Beauty of Fractals, Images of Complex Dynamical Systems*, Springer Verlag (1986).

[56] P. Prusinkiewicz and G. Sandness, Koch curves as attractors and repellers, *IEEE Comp. Graphics and Appl.*, 26-40 (Nov. 1988).

[57] R. Shonkwiler, An image algorithm for computing the Hausdorff distance efficiently in linear time, *Inform. Process. Lett.* **30**, 87-89 (1988).

[58] G. Szego, *Orthogonal Polynomials*, AMS (1939).

[59] E.R. Vrscay, Mandelbrot sets for pairs of affine transformations in the plane, *J. Phys.* **A19**, 1985-2001 (1986).

[60] E.R. Vrscay and C.J. Roehrig, Iterated function systems and the inverse problem of fractal construction using moments, in *Computers and Mathematics*, E. Kaltofen and S.M. Watt ed., Springer Verlag (1989), 250-259.

[61] E.R. Vrscay and D. Weil, "Missing moment" and perturbative methods for polynomial iterated function systems, to appear in *Physica D* (1991).

[62] E.R. Vrscay, Moment and collage methods for the inverse problem of fractal construction with iterated function systems, preprint submitted to FRACTAL 90 Conference, Lisbon, June (1990).

[63] H. Wall, *Analytic Theory of Continued Fractions*, Van Nostrand (1948).

[64] R.F. Williams, Composition of contractions, *Bol. Soc. Brasil. Mat.* **2**, 55-59 (1971).

[65] W.D. Withers, Differentiability with respect to parameters of average values in probabilistic contracting dynamical systems, preprint.

[66] W.D. Withers, Newton's method for fractal approximation, *Constr. Approx.* **5**, 151-170 (1989).

[67] B.W. Char, K.O. Geddes, G.H. Gonnet, M.B. Monagan, S.M. Watt, *MAPLE Reference Manual*, 5th Edition, WATCOM Pub. (1988).

Tables

Step	s_1	s_2	a_1	a_2	p_1	p_2	$D^{1/2}$	$\bar{h}$	$\bar{d}_H$
1	0.50000	0.50000	0.00000	0.50000	0.50000	0.50000	2.957D-01	166	83.381
5	0.50582	0.48135	0.01638	0.53763	0.49374	0.50626	6.887D-02	166	83.120
10	0.52743	0.36070	-0.10807	0.64214	0.48690	0.51310	1.794D-02	196	111.853
15	0.47511	0.36568	-0.07069	0.63741	0.48946	0.51054	8.335D-03	135	34.914
20	0.33679	0.33011	-0.00060	0.66969	0.50204	0.49796	5.033D-04	3	3.252
25	0.33837	0.33213	-0.00153	0.66789	0.50122	0.49878	6.984D-05	4	2.310
30	0.33279	0.33342	0.00020	0.66658	0.49991	0.50009	1.111D-05	1	0.238
35	0.33334	0.33334	-0.00001	0.66666	0.49999	0.50001	2.046D-06	0	0.007
40	0.33335	0.33333	0.00000	0.66667	0.50001	0.49999	3.908D-07	0	0.005
50	0.33333	0.33333	0.00000	0.66667	0.50000	0.50000	3.845D-08	0	0.000
60	0.33333	0.33333	0.00000	0.66667	0.50000	0.50000	1.151D-09	0	0.000
70	0.33333	0.33333	0.00000	0.66667	0.50000	0.50000	3.800D-11	0	0.000

Table 1. Moment matching using Gradient method. M=20 moments used. Target: Cantor set on [0,1], uniform measure.

Step	$D^{1/2}$	Δ	s_1	s_2	a_1	a_2	p_1	p_2	$\bar{d}_H$
1	0.21647	163.80682	0.15625	0.32031	0.79688	0.26953	0.37226	0.62774	176.812
2	0.08327	0.57058	0.51563	0.04688	0.47266	0.03516	0.69343	0.30657	73.815
7	0.07173	0.40079	0.05469	0.48047	0.92578	0.12500	0.27778	0.72222	113.517
9	0.05533	0.36332	0.44922	0.30078	0.03125	0.69922	0.54208	0.45792	66.541
14	0.02234	0.28511	0.23828	0.30078	0.04688	0.69531	0.51656	0.48344	31.751
17	0.01741	0.18378	0.38672	0.30078	0.00000	0.69531	0.51961	0.48039	29.752
18	0.01302	0.20609	0.23828	0.30078	0.04688	0.69531	0.51333	0.48667	29.948
19	0.01300	0.18398	0.29297	0.30078	0.03125	0.69531	0.51656	0.48344	24.631
22	0.01244	0.13098	0.22266	0.30078	0.04688	0.69531	0.50649	0.49351	28.452
23	0.01149	0.11079	0.01563	0.30078	0.17188	0.69531	0.51333	0.48667	71.912
24	0.01137	0.10150	0.26563	0.30078	0.04688	0.69531	0.51333	0.48667	27.623
27	0.00921	0.09629	0.22266	0.30078	0.05469	0.69531	0.50980	0.49020	30.770
31	0.00811	0.10892	0.26953	0.30078	0.05469	0.69922	0.52619	0.47381	36.004

Table 2. Genetic Algorithm applied to moment matching. Target: Cantor set, uniform measure. M=20 moments used. GA parameters: $NPOP$=150,ND=8,p_c=0.6,p_m=0.007.

Step	$\bar{d}_H(\nu, M(\nu))$	Δ	s_1	s_2	a_1	a_2	p_1	p_2
1	117.87541	679.58802	0.77344	0.04297	0.38867	0.20605	0.62876	0.37124
2	71.48226	467.88086	0.29199	0.27441	0.75391	0.05957	0.39668	0.60332
4	60.53484	265.35926	0.29199	0.27441	0.75391	0.03125	0.53039	0.46961
5	60.34892	205.41018	0.29199	0.27441	0.75391	0.03125	0.53008	0.46992
9	57.75408	191.34608	0.35645	0.34082	0.69141	0.05957	0.42390	0.57610
10	56.06380	182.55103	0.34863	0.33691	0.69141	0.05957	0.42390	0.57610
11	55.39377	175.90981	0.34863	0.33691	0.70703	0.05957	0.46251	0.53749
13	49.86771	157.54014	0.33789	0.27051	0.58789	0.01758	0.51316	0.48684
14	49.75656	142.82110	0.34863	0.33691	0.69141	0.05762	0.45455	0.54545
15	40.87947	151.67805	0.35645	0.33301	0.58008	0.00195	0.49505	0.50495
17	39.81594	120.99704	0.35645	0.33301	0.58008	0.00195	0.49716	0.50284
22	25.82601	122.39637	0.35156	0.33496	0.70508	0.00195	0.49716	0.50284
23	18.28054	122.02409	0.35645	0.34082	0.62695	0.00195	0.49716	0.50284
40	13.62865	89.93239	0.34082	0.33301	0.68750	0.00098	0.50136	0.49864
42	10.38938	83.48315	0.31738	0.33691	0.68945	0.00098	0.50238	0.49762
65	10.00238	57.55462	0.32520	0.33203	0.67480	0.01660	0.50000	0.50000
67	8.42923	33.91037	0.31348	0.32715	0.68848	0.00098	0.49734	0.50266
68	8.19991	31.82397	0.31250	0.32715	0.68848	0.00098	0.49734	0.50266
70	7.95417	40.21205	0.31348	0.32715	0.68750	0.00098	0.49734	0.50266
75	5.56505	50.11270	0.30957	0.33301	0.68750	0.00098	0.49934	0.50066

Table 3. Collage Theorem for discrete measures using Genetic Algorithm. Target: Cantor set, uniform measure. GA parameters: *NPOP*=100, *ND*=10, p_c=0.6, p_m=0.007.

Figures

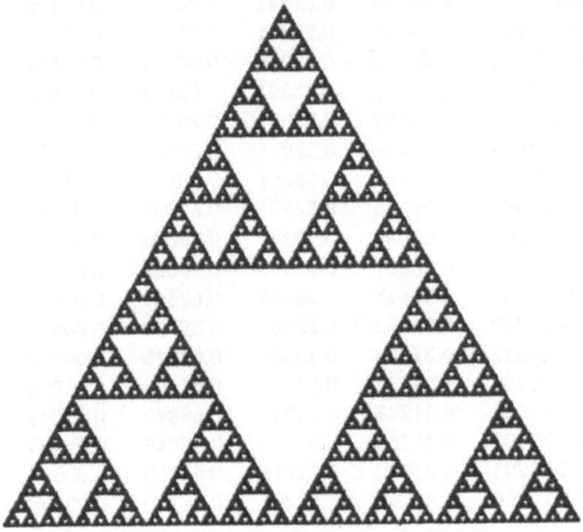

Figure 1(a). Sierpinski gasket

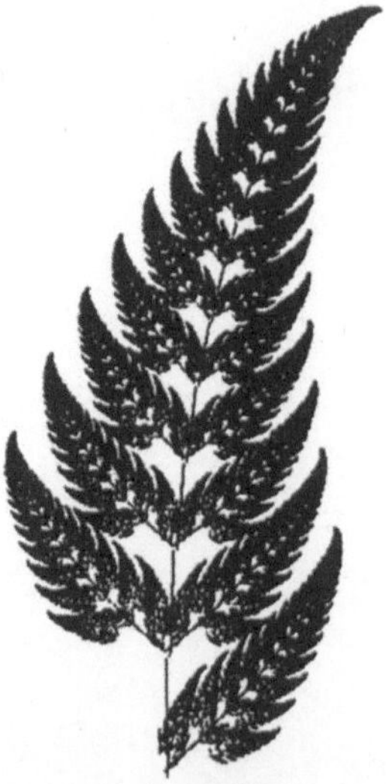

Figure 1(b). Spleenwort fern

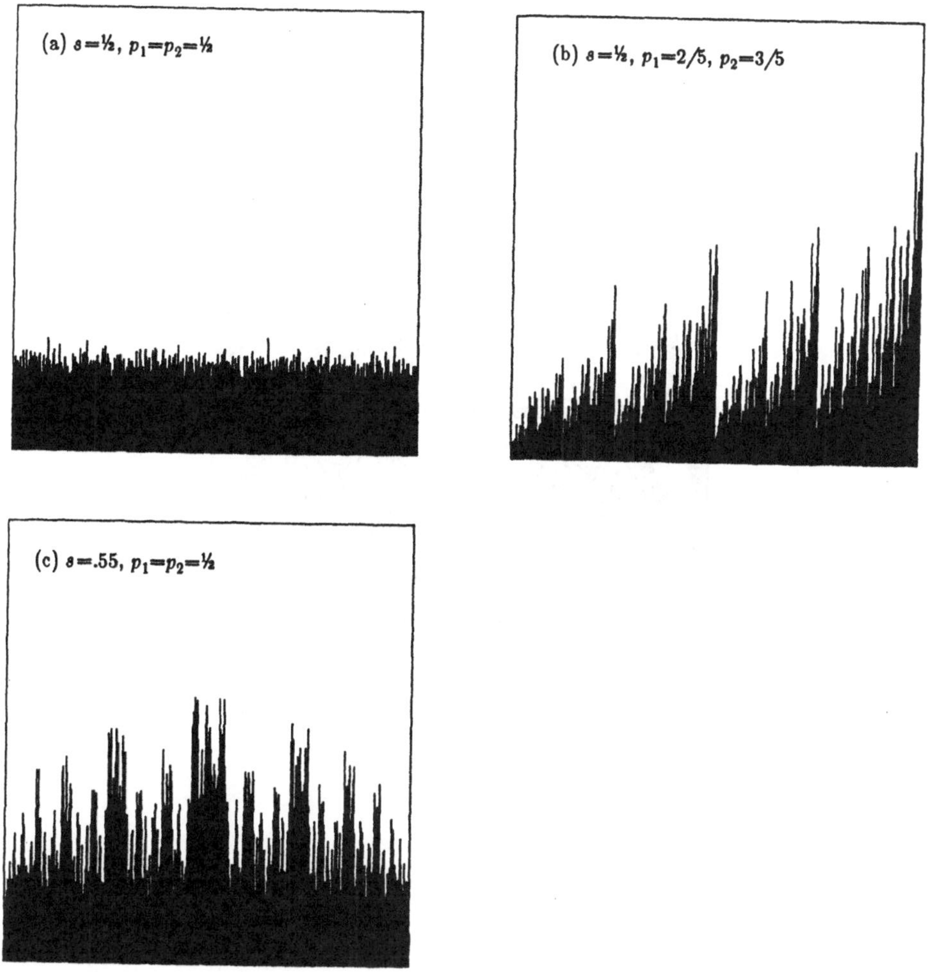

Figure 2. Histogram approximations to invariant measures on $[0,1]$ for IFS of form $w_1(x)=sx$, $w_2(x)=sx+(1-s)$, with indicated probabilities, as given in Examples of Section 2.2.

Figure 3. Escape-time algorithm applied to Sierpinski gasket of Figure 1(a).

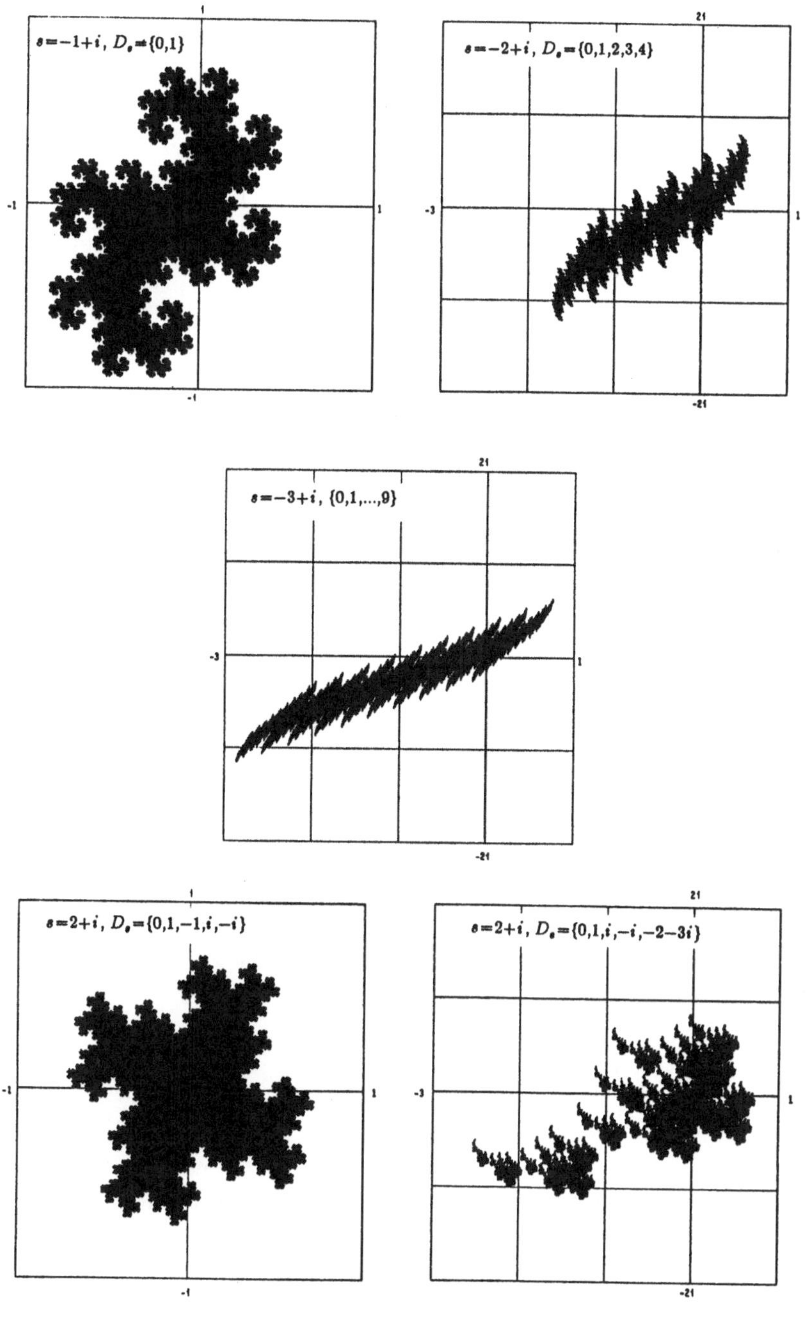

Figure 4. Base tiles for complex numbers $s \in \mathbb{C}$, with digit sets D_s as indicated.

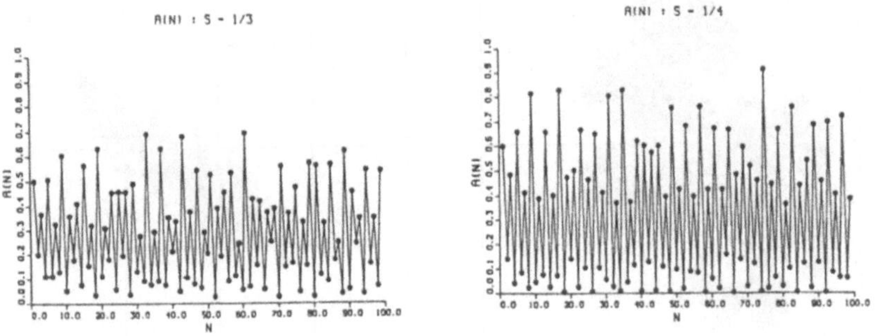

Figure 5. Plot of coefficients a_n, $n=0,1,...,100$ in recursion relation (3.4.5) for orthogonal polynomials over invariant measure of IFS $\{[-1,1], w_1(x)=s(x-1)+1, w_2(x)=s(x+1)-1\}$: (a) $s=1/3$ (ternary Cantor set on $[-1,1]$), (b) $s=1/4$.

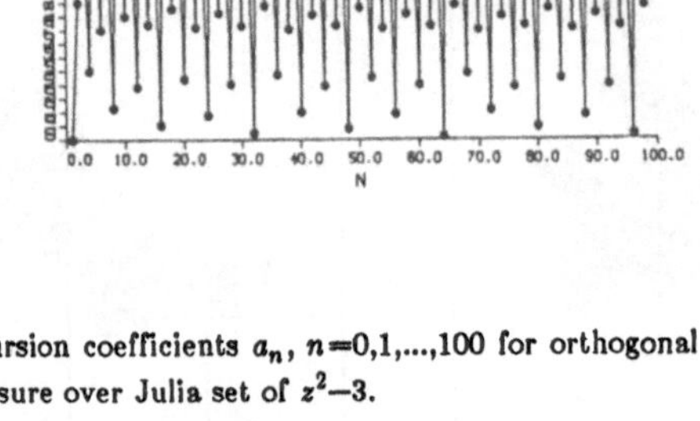

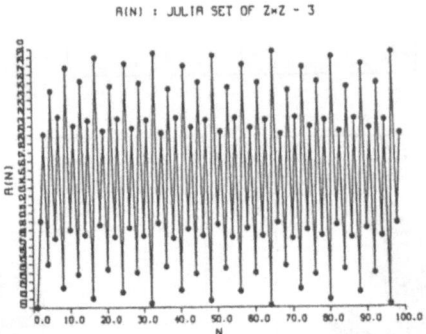

Figure 6. Recursion coefficients a_n, $n=0,1,...,100$ for orthogonal polynomials relative to invariant measure over Julia set of z^2-3.

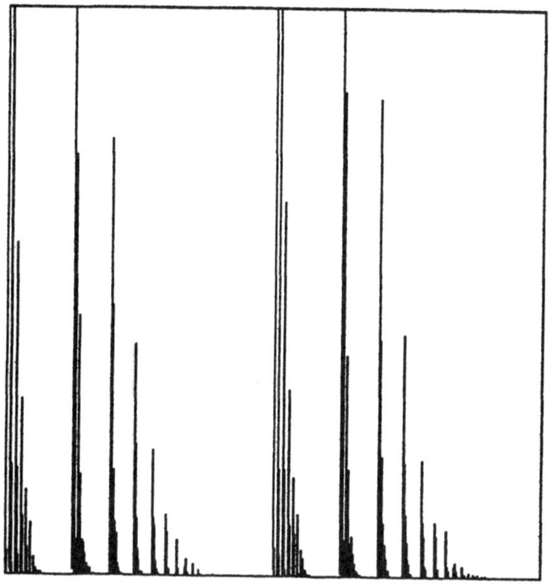

Figure 7. Histogram approximation to invariant measure on [0,1] for nonlinear IFS $w_1(x)=\frac{1}{2}x^2$, $w_2(x)=\frac{1}{2}x^2+\frac{1}{2}$.

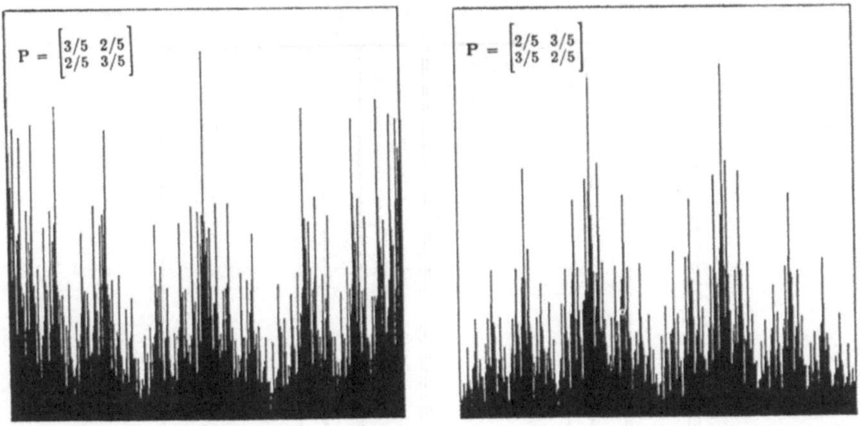

Figure 8. Histogram approximations of invariant measures on [0,1] for recurrent IFS of Example 1, Section 4.2, with transition probability matrices **P** indicated.

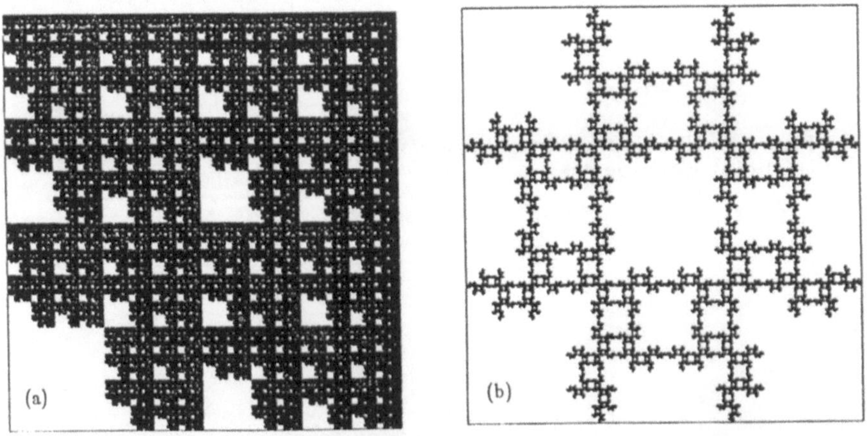

Figure 9. Attractors on [0,1]×[0,1] for recurrent IFS of Example 2, Section 4.2.

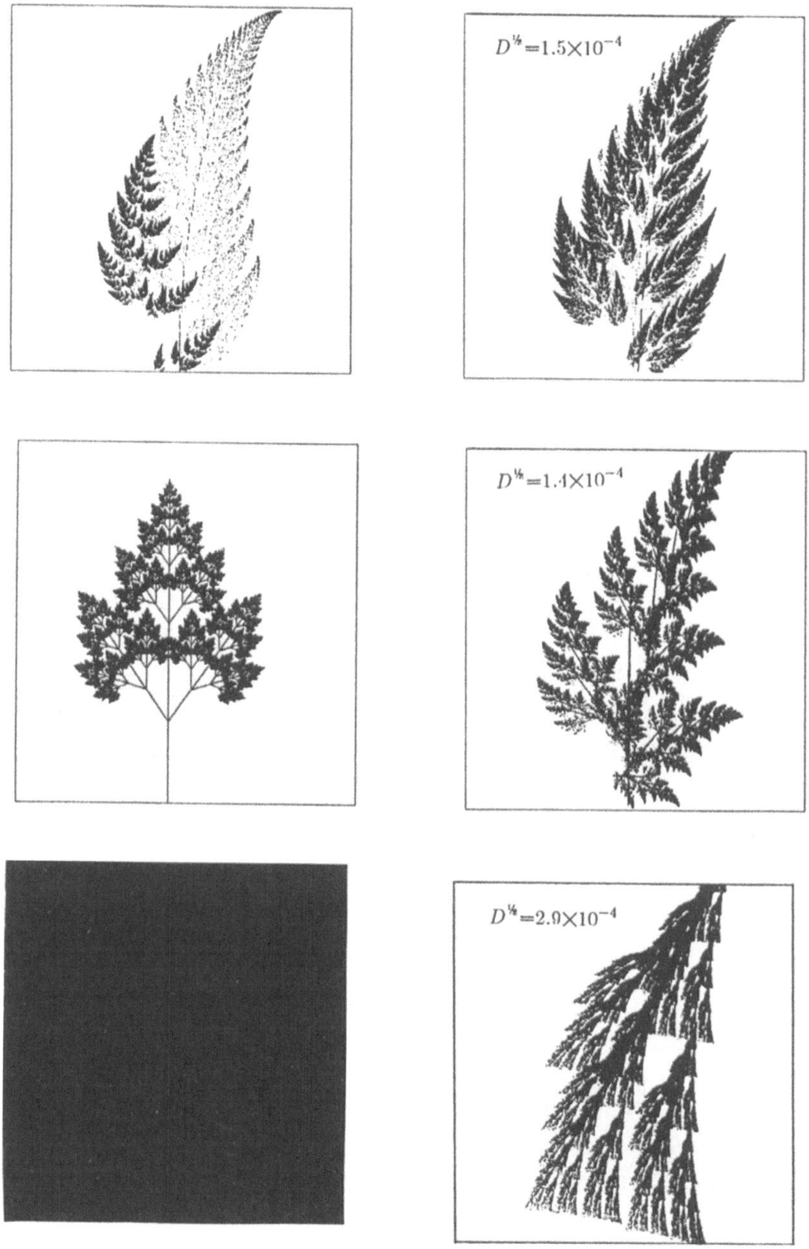

Figure 10. Initial sets and resulting approximations to *Spleenwort fern* target measure obtained by gradient method of moment matching in $\mathbf{R}^2$. M=20 moments used: g_{ij}, $1 \leq i+j \leq 4$. Initial sets: (a) perturbation of fern, (b) a "generic" tree, (c) unit square with uniform measure. Moment distance $D^{1/2}$ for final approximations indicated.

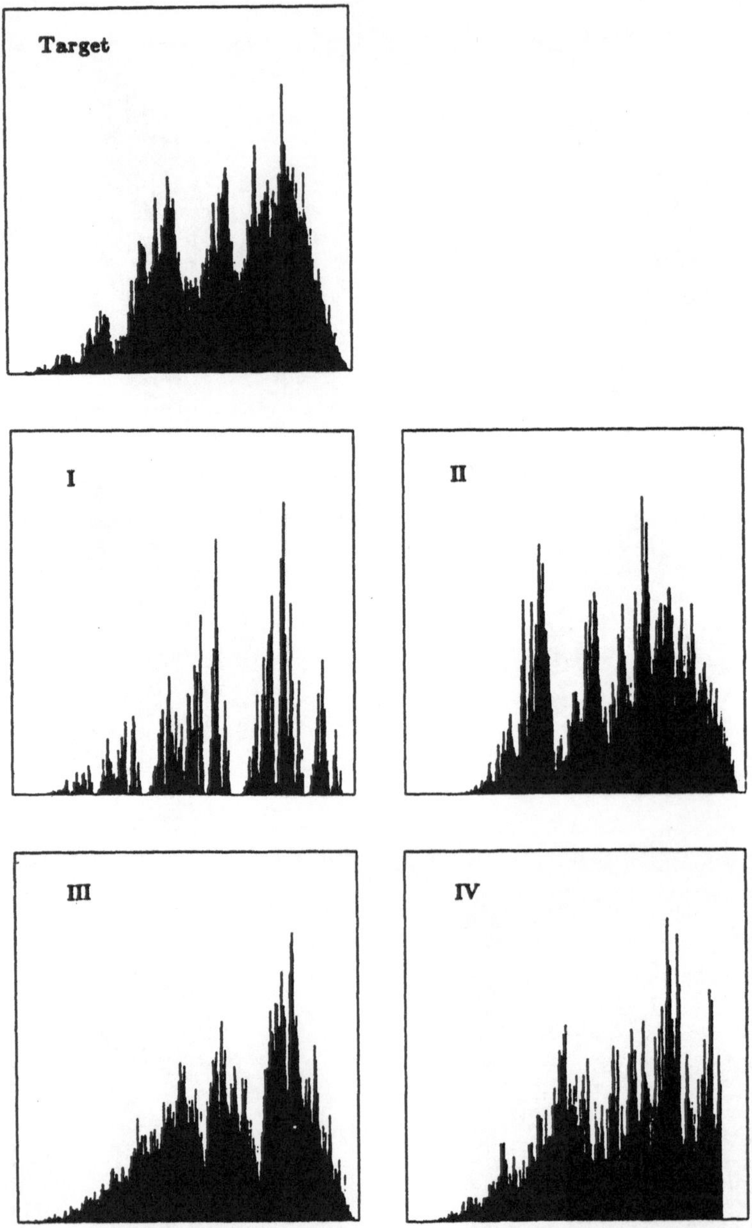

Figure 11. Histogram approximations of invariant measures on [0,1] for Target IFS in Eq. (5.4.8), and approximations afforded by methods I-IV listed in Section 5.4.4.

Index